Institute for Nonlinear Science

Leon Glass
Peter Hunter
Andrew McCulloch
Editors

Theory of Heart

Biomechanics, Biophysics, and
Nonlinear Dynamics of Cardiac Function

With 183 Illustrations

Springer-Verlag
New York Berlin Heidelberg London
Paris Tokyo Hong Kong Barcelona

Leon Glass
Department of Physiology
McGill University
Montreal, Quebec H3G 1Y6
Canada

Andrew McCulloch
Department of Applied Mechanics and
Engineering Science—Bioengineering
University of California, San Diego
La Jolla, CA 92093 USA

Peter Hunter
Department of Engineering Science
University of Auckland
Auckland, New Zealand

Library of Congress Cataloging in Publication Data
Theory of heart: biomechanics, biophysics, and nonlinear dynamics
  of cardiac function / Leon Glass, Peter Hunter, Andrew McCulloch
    p. cm
  Based on conference held under the auspices of the Institute
for Nonlinear Science at the University of California, San Diego during
July 10–18, 1989
  Includes bibliographical references.
  ISBN 0-387-97483-0 -- ISBN 3-540-97483-0
  1. Heart--Muscle--Congresses. 2. Heart--Contraction--Mathematical
  models--Congresses. 3. Human mechanics--Congresses. I. Glass,
Leon, 1943-   . II. Hunter, Peter, 1948-   III. McCulloch,
Andrew. 1961- IV. Institute for Nonlinear Science.
  [DNLM: 1. Heart--physiology--congresses. 2. Models,
Cardiovascular--congresses. WG 202 T396 1989,
  QP113.2.T48   1991
  612..1'71--dc20
  DNLM/DLC
  for Library of Congress   90-10457

Printed on acid-free paper.

Camera-ready copy provided by the author using LaTeX.
Printed and bound by R.R. Donnelley & Sons, Harrisonburg, VA.
Printed in the United States of America.

9 8 7 6 5 4 3 2 1

ISBN 0-387-97483-0 Springer-Verlag New York Berlin Heidelberg
ISBN 3-540-97483-0 Springer-Verlag Berlin Heidelberg New York

To the memory of B. van der Pol and
J. van der Mark—pioneers in the application of engineering
and nonlinear dynamics to the heart

Preface

The purpose of the heart is to pump blood. To do so continuously for 70 years or more at a rate varying from 5 to 35 liters/min requires a very sophisticated organ. Electrical excitation originating in specialized regions of the cardiac muscle spreads over the myocardium to activate muscular contraction; all of this is under feedback control regulating the cardiac output. Because of its importance to human health, the heart is most frequently studied by those trained in medical science or allied basic science disciplines. The theoretical mathematical training of these workers is often minimal, and it is therefore not surprising that most research in cardiac physiology has a nonmathematical flavor. However, in recent years there has been a growth of interest in studying the heart from a perspective of the physical sciences. Such work has been directed at either the mechanical or electrical aspects of cardiac function, but rarely have the two been combined.

Because of the recent advances in the development of theory of the heart, and the difficulties that young workers have in entering this interdisciplinary field, a conference was held under the auspices of the Institute for Nonlinear Science at the University of California, San Diego during July 10–18, 1989. The object of the conference was to bring together scientists and clinicians with young graduate students and postdoctoral fellows, all of whom had an interest in the theory of the heart. Invited speakers were asked to prepare written documents that concentrated on providing didactic and background information in their areas, as well as giving some indication of more recent advances. The present volume is the final product of this effort.

The first nine chapters of this book are concerned with the mechanics of the heart wall. Although there are many empirical models describing cardiac mechanical performance as a function of various inputs such as ventricular pressure and heart rate, these "lumped parameter" models do not provide any insight into the underlying relationships between structure and function in the heart muscle. Nor are they able to describe regional variations in mechanical function. Such heterogeneities are important because cardiac disorders such as coronary heart disease are frequently characterized by localized abnormalities in the wall. To address problems of this nature, the theoretician in mechanics turns to a "continuum" approach, in which regional mechanics are described by continuous distributions of stress and strain in the heart wall. To determine these distributions, a mathematical model must describe the material properties of the heart muscle, the geometry and structure of the wall, and the boundary conditions. It must then use this knowledge together with the conservation laws of mass, energy, and momentum to yield a predictive tool with clinical applications.

The greatest problem in cardiac mechanics theory is obtaining a mathematical description for the material properties of the heart wall (myocardium). For a complex, nonlinear, anisotropic material such as myocardium, obtaining an adequate functional form for this "constitutive law" and then evaluating its parameters is an extremely difficult problem. However, much can be learned about the mechanical properties of tissues by studying their microscopic structure. In Chapter 1, Smaill and Hunter describe the structural arrangement of the cardiac muscle cells in the ventricular wall and the complex connective tissue matrix. Then, in Chapter 2, Horowitz presents a microstructural model for the mechanical properties of the myocardium based on this type of information. Although this approach promises to provide a link between the microstructure of the wall and the function of the organ as a whole, much of the detailed information on the properties of the individual tissue components remains incomplete. In Chapter 3, taking a more phenomenological approach (nevertheless motivated by structural knowledge), Humphrey and colleagues outline a formalism for identifying the constitutive law for the resting heart muscle based on an extensive program of mechanical testing.

Whereas the previous chapters concentrated on properties of the resting heart tissue, Chapter 4 deals with the mechanics of the myocardium as a muscle capable of active contraction and relaxation. Here, Nielsen and Hunter describe new experimental and analytical methods for identifying the time-varying properties of active cardiac muscle.

With a suitable description of the material properties, a model can be used to predict the stresses and strains in the heart wall, but experimental measurements are needed to validate the predictions. To date, the experimental measurement of force or stress in the intact myocardium has been restricted by technical difficulties. However, the measurement of regional shape changes or strains has been more successful. In Chapter 5, McCulloch and Omens describe the regional mechanics of the intact heart during passive filling, and they compare the predictions of a simple model of the left ventricle with experimental measurements of wall strains. If a model is to represent accurately the complex geometry and architecture of the heart wall, however, computational techniques are required. The use of finite element methods in cardiac mechanics modeling is reviewed by Guccione and McCulloch in Chapter 6.

The complex three-dimensional structure and time-varying mechanical properties of the intact ventricular wall give rise to significant heterogeneities of mechanical function in the normal heart. This can pose substantial difficulties to the cardiologist trying to diagnose regional disorders from radiographic images. In Chapter 7, Waldman uses the theoretical framework of kinematics as the basis for measuring three-dimensional strains throughout the wall of the beating heart—both normal and diseased. These methods rely on the use of radiopaque implants in experimental animals, but natural landmarks on the heart offer the prospect of measuring regional mechanics in patients using conventional clinical imaging techniques. In Chapter 8, Young describes an approach for analyzing ventricular strains using coronary arteriograms obtained by cardiac catheterization. The functional consequences of the regional heterogeneities in mechanics that occur in heart disease are discussed by Lew in Chapter 9.

The understanding of the electrical properties of the heart has developed independently of the study of the mechanical properties. Current research ranges from studies of the dynamics of single ion channels in the heart to studies of the mathematics of complex arrhythmias.

The electrical activity of the heart is caused by the opening and closing of ionic channels in the cardiac membrane. Theoretical studies are often directed at developing nonlinear ordinary differential equations describing the various ionic currents underlying cardiac activity. An introduction to this approach is provided by Guevara in Chapter 10. In Chapter 11, Clark and colleagues address the formulation of ionic models for pacemakers in cardiac tissue,

An alternative approach to the formulation of detailed ionic models is to develop low-dimensional finite difference equations reflecting the response of cardiac tissue to various inputs, such has periodic stimulation. Glass and Shrier outline the basics of this approach in Chapter 12. The application of this strategy to study atrioventricular heart block is discussed from a purely theoretical standpoint by Guevara in Chapter 13. In Chapter 14, Jalife and Delmar discuss the ionic basis of heart block combining experimental data with finite difference equations to analyze in vitro dynamics. One-dimensional finite difference equations also figure prominently in analysis of another arrhythmia, parasystole, which is discussed by Bélair and coworkers in Chapter 15.

The preceding chapters do not deal explicitly with the spread of excitation in the heart. Studying this problem raises several novel problems from theoretical, experimental, and clinical perspectives. Indeed, since reentrant arrhythmias that are believed to involve self-maintaining excitation such as ventricular tachycardia and ventricular fibrillation are often life-threatening or fatal, understanding the theory of these arrhythmias is of potential clinical significance. An introduction to the physiology of the spread of excitation in the heart is given by Kootsey in Chapter 16, and in Chapter 17, Keener provides a mathematician's perspective. Since numerical integration of realistic ionic models of the spread of cardiac excitation is extremely time-consuming, alternative simplified numerical approaches are being developed. In Chapter 18, Saxberg and Cohen discuss the use of cellular automata models to study cardiac excitation and the onset of ventricular fibrillation. Theoretical insight into the onset of ventricular fibrillation is provided by Winfree in Chapter 19, who discusses the computation of the threshold electrical current needed to induce ventricular fibrillation. A related problem concerning the currents needed for ventricular defibrillation is treated by Ideker and coworkers in Chapter 20.

One of the striking observations to emerge from the conference is that research involving mechanical and electrical properties of the heart is largely disjoint and there is little overlap, even though in the functioning of the heart both the electrical and mechanical aspects of cardiac function are tightly coupled and essential. In Chapter 21, Lab and Holden begin to address the interrelationships betweeen mechanical and electrical activity.

Aside from the many questions relating to basic science, one eventual goal of the research is to make a contribution to the practice of medicine by finding novel methods to diagnose and treat cardiac disease. A number of clinicians were present at the conference and it is clear that interdisciplinary collaboration between clinicians and basic scientists is needed. In Chapter 22, Goldberger and Rigney describe striking dynamic aspects of cardiac arrhythmias. Finally, in Chapter 23 Kovács offers a clinical perspective on the theory of heart, indicating areas in which he expects future advances.

Although we have tried to include what we believe is the most important work dealing with theory of heart, we recognize that volumes such as this must reflect the biases of the editors and the authors. We apologize for any important omissions in the study of the theory of heart. In particular, we have been unable to include anything dealing with hydrodynamic aspects of cardiac function. Important progress in these areas is being made by Charles Peskin and others, and readers interested in these topics will have to look elsewhere.

Bringing a project such as this to completion requires the assistance of many unsung heroes. First among these is Henry Abarbanel, director of the Institute for Nonlinear Science at the University of California, San Diego. Henry had the original idea to have a conference on the theory of heart, and he provided the administrative and financial support needed to bring this about. Thanks also to Peggy Orr who handled all of the administrative details of the conference and Robert Siverson, Ernie Lee, and Sang Im, who helped prepare the manuscripts for publication. Both the Department of Energy and Medtronic Corporation provided financial support. Mr. Earl Bakken of Medtronic Corporation has been particularly supportive of this and other projects of a similar nature and gets a special thanks.

Leon Glass, Montréal, Québec, Canada
Peter Hunter, Auckland, New Zealand
Andrew McCulloch, La Jolla, California, USA
June 1990

Contents

1

Structure and Function of the Diastolic Heart: Material Properties of Passive Myocardium

Bruce Smaill[1]
Peter Hunter[2]

ABSTRACT A considerable body of indirect evidence indicates that the characteristics and extent of the extracellular connective tissue matrix in the heart are important determinants of ventricular function. An appropriate constitutive law for passive ventricular myocardium should therefore incorporate the most important features of its microstructure. In the first part of this chapter we outline the current understanding of cardiac microstructure. The organization and classification of the connective tissue hierarchy are reviewed and the contributions of the different collagen types constituting the extracellular tissue matrix are considered. We present recent morphological findings that indicate that ventricular myocardium is a layered composite rather than a uniformly branching continuum, and that the extent of coupling between adjacent layers of cells varies through the wall of the ventricle. In the second part of the chapter we describe the results of biaxial mechanical tests on thin sections of ventricular myocardium. For specimens from the midwall and subepicardium of the left ventricle, stress-extension relations in the fiber direction were nonlinear, with maximum fiber extensions between 15 and 25%. In the cross-fiber direction large extensions were measured in midwall specimens, together with rate dependence and hysteresis. This was not observed in subepicardial specimens. In both cases, however, cross-fiber stress-extension relations were highly nonlinear. The variation in biaxial mechanical behavior in midwall and subepicardial specimens is seen to reflect differences in microstructure at these two sites. Finally, we demonstrate that although the observed mechanical behavior can be accurately reproduced by a simple phenomenological constitutive law, it is difficult to identify unique material parameters with such a constitutive formulation.

---

[1] Department of Physiology, School of Medicine, University of Auckland, New Zealand

[2] Department of Engineering Science, School of Engineering, University of Auckland, New Zealand

1.1   Introduction

In order to use the methods of continuum mechanics to analyze stress and deformation in the diastolic heart it is necessary to formulate appropriate constitutive relations for passive myocardium. Any such constitutive formulation must faithfully reproduce the macroscopic response of the myocardium to the possible loads that may be imposed on it in vivo. However, the formulation should also be compact, representing observed mechanical behavior with as few parameters as possible.

To date, most of the material laws formulated for soft biological tissues have been empirical, using functional relations that best fit the stress-strain behavior observed in experimental studies. There are, however, a number of problems associated with this phenomenological approach. For instance, it is often difficult to identify unique material parameters for the highly nonlinear functions that have been used to represent the mechanical properties of soft tissues. In addition, and this problem is particularly acute for ventricular myocardium, it is not always possible to obtain the experimental data necessary to characterize the material behavior of soft biological tissues. Because the right and left ventricles are thick-walled shells with complex geometry and substantial transmural variation of muscle-fiber orientation, the systematic mechanical tests required for direct constitutive law identification cannot be performed on intact ventricular myocardium. Some inferences about the mechanical properties of ventricular myocardium may be drawn from in vitro studies of isolated papillary muscles and trabeculae in which one-dimensional mechanical testing systems have been used. However, there are significant structural differences between papillary muscle and ventricular myocardium. Moreover, the mechanical properties of passive myocardium are not isotropic and therefore cannot be fully characterized by one-dimensional mechanical tests. Biaxial testing procedures have been used to study the material properties of myocardial tissue excised from the walls of the ventricles. Specimens are cut so that the muscle-fiber axis lies in the plane of the section and the fiber orientation is relatively uniform through the thickness of the sample, but the extent to which these results reflect the material properties of intact myocardium remains uncertain. In order to make these tests it is necessary to disrupt the structural integrity of the myocardium, and the specimen may be further damaged by contracture induced by ischemia or by calcium released from cells injured during the cutting process. Finally, it is not possible to reproduce either the compressive loading or the shear loading that occurs in vivo using these testing methods.

Attempts have also been made to formulate constitutive laws for soft tissues that are based on the observed structure of these materials and that incorporate knowledge of the mechanical properties of their main constituent elements. This microstructure-based approach has a number of attractions. Because they incorporate a priori information, such constitutive formulations should, in principle, require fewer fitted parameters to represent material behavior. Moreover, microstructure-based constitutive laws provide a framework for representing material inhomogeneity due to regional variation of structure, and for modeling mechanical properties that cannot be characterized readily using conventional testing procedures. The formulation of a realistic microstructure-based constitutive law for passive ventricular myocardium, however, requires detailed information about the organization of cardiac microstructure and a clear understanding of the way in which the rearrangement of this structure is reflected in macroscopic mechanical responses.

In this chapter, we (1) critically review the current understanding of cardiac microstructure, (2) consider the extent to which the observed mechanical behavior of passive myocardium can be related to this underlying structure, and (3) briefly illustrate the advantages and disadvantages of using a phenomenological constitutive law to represent the material properties of passive ventricular myocardium.

1.2   The Microstructure of the Heart

The cardiac muscle cell or myocyte is the main structural component of myocardium occupying around 70% of ventricular wall volume under normal circumstances. Cardiac myocytes resemble ellipsoid cylinders with a major-axis dimension of 10 to 20 μm and a length of 80 to 100 μm. Myocytes insert end to end and each is connected with several others to form a three-dimensional network of cells. The interface between adjacent cells is referred to as the intercalated disc and the structure and properties of this region are of considerable importance [13,18]. Cardiac myocytes have an array of fingerlike projections at each end. The interdigitation of these processes, which have a high concentration of gap junctions, together with the specialized properties of adhesion in the region ensures the physical and electrical connectivity of cardiac cell networks [13].

The complex organization of cardiac muscle fibers is mirrored by the dense vascular network of the coronary microcirculation. One or more capillaries run along the boundary of each myocyte parallel to the long axis of the cell, while other blood vessels lie in the cleavage planes between bundles of cells. Coronary blood vessels and cardiac myocytes are embedded in a complex extracellular matrix that consists of collagen, elastin, glycosaminoglycans, and glycoproteins.

1.2.1   THE CARDIAC CONNECTIVE TISSUE HIERARCHY

The extracellular collagen matrix of the heart was first observed using light microscopy by Holmgren [16] and Benninghoff [2]. Caulfield and Borg [8] exploited the resolution and depth of field of the scanning electron microscope

(SEM) to reveal the basic organization of this connective tissue network. They described the following three classes of connective tissue organization: (1) interconnections between myocytes, (2) connections between myocytes and capillaries, and (3) a collagen weave surrounding groups of myocytes. Caulfield and Borg [8] noted that myocytes were connected to adjacent myocytes by numerous bundles of collagen 120 to 150 nm in diameter. These collagen cords were quite straight.[3] They were seen to have a relatively uniform circumferential distribution and to be inserted into the basement membrane of the myocyte close to the Z-line of the sarcomere. Radial collagen cords of similar dimension and appearance also extended between myocytes and capillaries. Finally, Caulfield and Borg [8] also observed a meshwork of collagen bundles that surrounded groups of three or more myocytes. This external weave was complex and it was noted that enclosed myocytes were connected to the encompassing network by numerous short, straight, collagen cords. However, Caulfield and Borg [8] observed that adjacent complexes were only loosely coupled by sparse and relatively long collagen fibers.

Robinson and his coworkers [34] went on further to categorize the components of this extracellular connective tissue matrix in an elegant series of studies using light microscopy, together with transmission and scanning electron microscopy. They classified the hierarchy of cardiac connective tissue organization as "endomysium," "perimysium," and "epimysium," using terminology more commonly associated with skeletal muscle.

Endomysium. The cardiac endomysium was seen to incorporate the system of radial collagen cords described by Caulfield and Borg [8], together with a pericellular network of fibers that encompass the myocyte and a lattice of collagen fibrils and microthreads. Robinson and colleagues [34] observed that the radial cords consist of helically wound collagen fibrils that divide close to the cell surface and ramify as part of the pericellular network. In addition to this fine pericellular weave of connective tissue, distinct pericellular cuffs were also seen using light microscopy [34,36]. It was argued that components of the radial collagen cords also insert into the cell membrane and are tethered to the contractile apparatus at the Z-line [36].

Perimysium. The term perimysium was used to describe the extensive meshwork of connective tissue that surrounds groups of cells and the connections between contiguous cell bundles. Robinson and colleagues [34] observed large coiled bundles of collagen fibers oriented parallel to the long axis of the myocytes in left ventricular myocardium and papillary muscle. These were associated with groups of cells and were defined as "coiled per-

---

[3] Caulfield and Borg [8] coined the term "strut" to describe these collagen cords. This usage is widespread, but inappropriate. In engineering mechanics, a strut is a structural element that bears compressive load, whereas the radial collagen cords probably support tensile load only.

imysial fibers." A more detailed study of the organization of the coiled perimysial fibers in rat papillary muscle was recently reported [38]. A branched
network of coiled perimysial fibers 1 to 10 μm in diameter was seen to diverge from the muscle-tendon interface and a regular array of radial cords
interconnected these fibers with adjacent myocytes and with the endocardial membrane surrounding the papillary muscle.

Epimysium. The epimysium was defined as "the sheath of connective
tissue that surrounds entire muscles, for example, papillary muscle and
trabeculae" [35]. The extension of this definition to the ventricular wall is
less clear, but presumably epicardium and endocardium should be classified
as epimysium. It was observed that the epicardium and much of the endocardium consists of a thin layer of endothelial cells overlaying a network of
randomly oriented collagen fibrils and elastin fibers.

1.2.2   TRANSMURAL VARIATION OF CONNECTIVE TISSUE ORGANIZATION IN VENTRICULAR MYOCARDIUM

The most detailed descriptions of connective tissue architecture in the heart
have been obtained for papillary muscle [34,38]. Although differences between the extracellular collagen matrix in papillary muscle and ventricular
myocardium have been reported [8,34], there has been no systematic attempt to characterize their distinct features. Moreover, there have been
no comparative studies of connective tissue architecture at different sites
within the ventricular wall as far as we are aware.

We have begun to study the transmural variation of connective tissue
organization in ventricular myocardium at the University of Auckland. Dog
hearts are arrested in diastole, rapidly excised, and fixed in an unloaded
state. Full-thickness wedges are removed from the lateral free wall of the
left ventricle and subdivided into five transmural samples of approximately
equal thickness. The samples are prepared for SEM using conventional
techniques and examined using an ISI DS 130 scanning electron microscope.
Some preliminary observations are presented.

It has been stated that the epicardium is a biaxial network of collagen
cords overlaid by a thin layer of endothelial cells. The epicardium is tightly
coupled to the adjacent myocytes of the subepicardium by an extensive
system of thick coiled fibers that insert into the epicardium and penetrate
obliquely into the ventricular wall (Figure 1.1). Up to 1 mm below the
epicardial surface, the subepicardium is characterized by closely packed
bundles of cells and a relatively uniform connective tissue organization.

Further into the ventricular wall, the architecture of the myocardium is
distinctly different. Groups of myocytes are organized in layers that are
three or four cells deep (25–40 μm thick) and surrounded by an extensive
meshwork of perimysial connective tissue. These layers are of considerable
extent and appear to run transmurally from endocardium to epicardium.
Adjacent layers are interconnected by a network of long convoluted col-

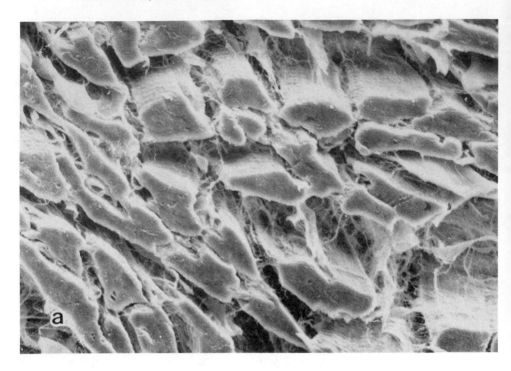

FIGURE 1.1. Electron micrographs of cardiac microstructure. (A) Transverse midwall section from the left ventricle showing the radial collagen cords that interconnect myocytes and the pericellular network of connective tissue surrounding individual myocytes, ×1050;

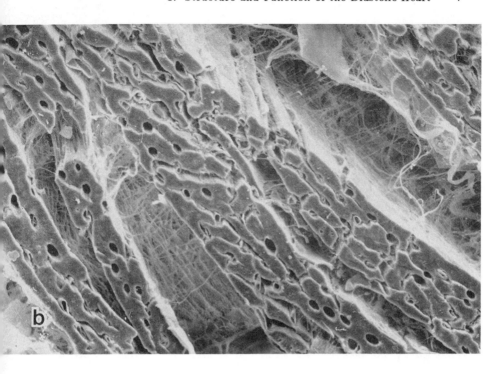

FIGURE 1.1. (B) Transverse midwall section from the left ventricle showing the meshwork of connective tissue that surrounds groups of myocytes and the network of collagen fibers that interconnect adjacent bundles of cells, ×510;

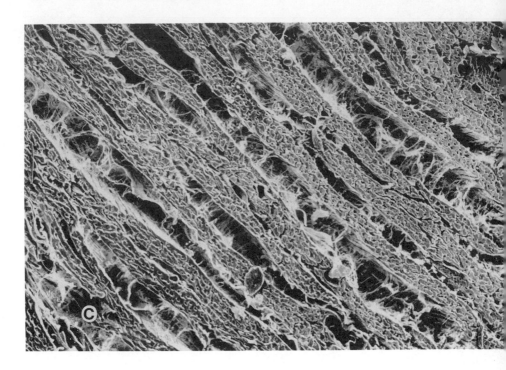

FIGURE 1.1. (C) Transverse midwall section showing the appearance of cell layers in the midwall of the left ventricle. Note the extent of the planes of cleavage between adjacent layers, ×145;

FIGURE 1.1. (D) Longitudinal freeze fracture of left ventricular subepicardial specimen showing tight coupling between myocytes and thick coiled collagen fibers oriented obliquely to the epicardial surface, ×145.

lagen fibers interspersed with straplike collagen bundles (see Figure 1.1). Preparation of samples for SEM expands the spaces between layers of cells, and gaps of up to 40 μm have been observed without apparent damage to the perimysial network. This indicates that the network of collagen fibers interconnecting layers of cells is able to undergo substantial deformation. Although this organization is consistent with the observation of bundles of cells in the left ventricular myocardium made by Caulfield and Borg [8], neither they nor Robinson and colleagues [34] described the discrete layers of cells shown here. More recently, however, Abrahams and colleagues [1] reported a comparable organization in the normal heart of the long-tailed macaque.

The subendocardium appears to be characterized by large spaces between bundles of cells and the connective tissue network between adjacent bundles is relatively sparse. Moreover, the coupling of the endocardium to myocytes in the subendocardial layer is demonstrably less tight than is the case for epicardium.

On the basis of these observations it appears to us that the conventional view of myocardium as a three-dimensional continuum of uniformly branching myocytes [41] is incorrect. The ventricular myocardium is a composite material bounded by the membranes of the epicardium and endocardium. Whereas common components of the connective tissue hierarchy are observed throughout the myocardium, it appears that the extent of coupling between groups of myocytes varies at different sites across the ventricular wall.

1.2.3   COLLAGEN TYPES IN MYOCARDIUM

Collagen types I, III, IV, and V have been identified in myocardium. Whereas types IV and V collagen are components of the basement membrane of cardiac cells, types I and III collagen are the main constituents of the extracellular connective tissue matrix, contributing around 75 to 80% and 15 to 20% of total collagen content, respectively [26,44]. The proportion of type I collagen to the total of type I and type III collagen is least in the fetal heart and increases with age [26]. However, the fraction of type III collagen may increase with myocardial hypertrophy [26,44].

The mechanical and structural properties of type I and type III collagen are quite different. Type I collagen forms striated fibrils 20 to 100 nm in diameter that may aggregate to compose larger collagen fibers. Type I collagen is the main component of tendon and skin. Mechanical tests carried out on tendon reveal that type I collagen has a high tensile strength and stiffness. On the other hand, type III collagen is the principal component of the fine collagen fibrils that compose the highly deformable reticular networks characteristic of loose connective tissue and the media of blood vessels.

Studies with monoclonal antibodies [3,4,5,36,36,38] indicate that type I and type III collagen are present in both the fine pericellular network of the endomysium and the radial cords that connect myocytes with adjacent myocytes and capillaries. It was demonstrated [36] that type I and type III collagen aggregate in the radial collagen cords to form a copolymer. The intensity of antibody staining for type III collagen is stronger for the pericellular network than for the radial collagen cords. On the basis of these results it could be argued that the fine pericellular networks of the cardiac connective tissue hierarchy are predominantly composed of type III collagen. Although it has not been established directly, a number of lines of indirect evidence support this view. In bovine skeletal muscle, Light and Champion [24] found that the proportion of type I collagen to total type I and type III collagen is 84% in epimysium, 72% in perimysium, and 38% in endomysium. In nonhuman primates the proportion of type III collagen to total collagen increases in the evolutionary phase of the response to chronic ventricular pressure overload [44]. This is coincident with the proliferation of fine fibrils in the connective tissue weaves of the endomysium and the perimysium.

1.2.4   Cardiac Mechanical Function in Relation to Connective Tissue Organization

There is a considerable body of indirect experimental evidence that suggests that the characteristics and extent of the extracellular connective tissue matrix are important determinants of diastolic and systolic ventricular function:

1. Borg and colleagues [6] observed that the perimysial network was qualitatively less extensive in hamsters than in age-matched rats, with fewer lateral collagen cords between adjacent bundles of cells. It was suggested that this may account for the substantially reduced passive left ventricular stiffness for hamster.

2. Extensive disruption of the extracellular collagen matrix has been reported for hearts exposed to prolonged ischemia [9,39] and in hearts "stunned" by repeated periods of transient ischemia [47]. Zhao and colleagues [47] argued that uncoupling of the connective tissue network may contribute to the mechanical dysfunction associated with ischemic injury and play an important role in the impaired systolic and diastolic function seen in the "stunned" heart.

3. Increased loading of the ventricles leads to wall thickening and an associated increase in the volume fraction of collagen. It has been proposed that the changes in connective tissue organization that accompany myocardial hypertrophy in chronic ventricular pressure overload cause increased passive stiffness and may contribute to impaired systolic mechanical function. The changes in the connective tissue matrix

that occur with hypertension have been studied in rat [21,26,27], non-human primate [1,44], and man [7]. In all cases, the perimysial weave became more extensive, with a progressive increase in the number and thickness of collagen fibers in the meshwork surrounding groups of myocytes, and in the network between cell bundles. Caulfield [7] also reported an increase in the diameter of radial collagen cords interconnecting myocytes in man, but this has not been observed in animal models of hypertension. It has been argued that the remodeling of the connective tissue matrix is a necessary adaptation to distribute and sustain the increased contractile force generated by the hypertrophied myocytes. The enzyme lysyl oxidase, which is responsible for collagen and elastin cross-linking, may be inhibited by dietary copper deficiency [5] or by treatment with the lathyritic agent β-amino-proprionitrile (BAPN) [4]. Inhibition of lysyl oxidase disrupts the adaptive reorganization of the connective tissue matrix that normally accompanies hypertrophy. There is a reduction in the extent of the endomysial and perimysial networks with distinct separation of individual myocytes. This distortion of structure was associated with focal necrosis, ventricular aneurysm, and eventual rupture.

Robinson and his coworkers [34–38] proposed specific mechanical roles for the cardiac connective tissue matrix. The collagen network was seen as a "strain-locking" system, which resists the extension of myocytes beyond sarcomere lengths of 2.25 μm while allowing relatively free extension up to this length. Thus, when myocardium is stretched, the biaxially arranged connective tissue networks of the epimysium, perimysium, and endomysium are rearranged to align more closely with the muscle-fiber direction (cargo net hypothesis). In addition, coiled perimysial fibers in ventricular myocardium and papillary muscle, which are obliquely or longitudinally aligned with respect to the long axis of muscle fibers, become less convoluted. Robinson and colleagues [34,38] noted that the end-points of both these changes in connective tissue configuration are correlated with a striking increase in the axial stiffness of papillary muscles when extension exceeds 15% of slack length.

Robinson and colleagues [37] suggested that the extracellular connective tissue matrix may also store energy during systole and contribute to the elastic recoil of the ventricles during rapid filling. One of their lines of argument involves a straightforward corollary of the cargo net hypothesis. Since the volume occupied by the contractile lattice is relatively constant, the cross-sectional area of the myocyte increases when shortening occurs. It was argued that this dimensional change will lead to a realignment of the pericellular network of the endomysium with collagen fibrils now preferentially oriented transverse to the muscle-fiber direction. (A similar reorganization was also seen to apply for the perimysial network surrounding groups of muscle cells.) There was further speculation that the systolic deforma-

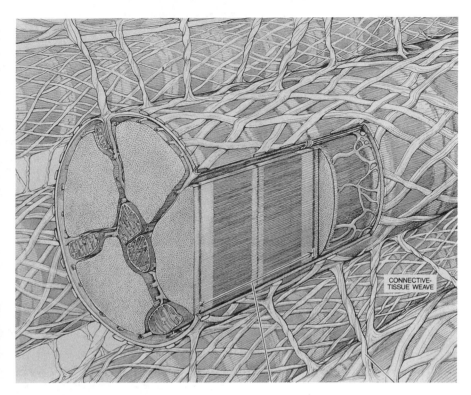

FIGURE 1.2. Schematic representation of cardiac microstructure incorporating the collagen network surrounding myocytes and the radial collagen cords that interconnect myocytes. (Reproduced from Robinson et al. [37] with permission from *Scientific American*.)

tion of the ventricle could impose tensile load on the collagen cords that tether adjacent myocytes [37] and compress the coiled fibers and tendons of the perimysium [38]. It was postulated that the potential energy storage associated with any of these mechanisms could oppose further shortening and tend to restore the myocardium to its resting state.

The development of an appropriate conceptual model is a necessary first step toward understanding how cardiac microstructure influences the material properties of passive myocardium. One such model, outlined by Robinson and colleagues [37] and illustrated in Figure 1.2, is the basis of the material law proposed by Horowitz and associates [17] (see Chapter 2) and was employed in a recent analysis of ventricular mechanics by Ohayon and Chadwick [29]. The myocardium is represented as an assembly of parallel myocytes cross-linked by a uniformly distributed array of radial collagen cords.

This model incorporates endomysial structure only and represents the ventricular myocardium as a continuum. In fact, myocardium is a composite material in which bundles of cells appear to be organized in discrete layers through most of the ventricular wall. It is most probable that each of the components of the connective tissue hierarchy contributes to the mechanical properties of passive myocardium. Caulfield and Borg [8] argued that the short radial collagen cords of the endomysium ensure that adjacent myocytes remain in registration providing a mechanism for uniform distribution of strain at the cellular level. They saw the perimysial network as organizing individual myocytes into a functional unit and providing a framework for distributing the contractile force within the ventricular wall. Finally, it was suggested that the loose collagen connections between groups of myocytes accommodate the large shearing deformations that are necessary for normal diastolic and systolic ventricular function. There is a need, therefore, to develop a comprehensive description of the microstructure of ventricular myocardium that incorporates systematic information about the organization of the various components of the connective tissue hierarchy through the ventricular wall.

1.3   Mechanical Properties of Myocardium

1.3.1   UNIAXIAL TESTING

Stress-extension relations in the muscle-fiber direction have been determined in passive papillary muscle and ventricular trabeculae carnae using a variety of experimental procedures [31–33]. This work established that cardiac muscle is a pseudo-elastic material with nonlinear, time-dependent mechanical properties [14]. However, the accuracy of the mechanical properties reported in these studies is questionable, because the nonlinear distribution of strain in the samples tested introduced artifacts that were not accounted for adequately [30]. The most reliable uniaxial stress-extension data have been obtained by measuring length changes in isolated papillary muscle and trabeculae carnae at the center of the specimens, where strain is most uniformly distributed. Using these techniques, sarcomere length-force relations for passive trabeculae carnae and papillary muscle were shown to be highly nonlinear—very compliant for sarcomere lengths shorter than 2.1 μm, but essentially inextensible at a sarcomere length of 2.3 μm [22,42]. It is evident that this "strain-locking" behavior reflects the connective tissue architecture previously described.

1.3.2   BIAXIAL TESTING

The three-dimensional material properties of passive ventricular myocardium cannot be fully characterized by one-dimensional tests carried out on papillary muscle or trabeculae carnae. However, it is possible to obtain more complete information using biaxial testing procedures. This approach was used by Lanir and Fung [23] to measure the dynamic mechanical properties of rabbit skin. In-plane forces were applied to square samples via threads attached at up to 17 points along each edge and dimensional changes were monitored in a central region of the sample. Employing very similar techniques, Demer and Yin [10] and Yin and colleagues [46] studied the mechanical properties of thin sections of noncontracting myocardium under cyclic biaxial and uniaxial loading (see Chapter 3 by Humphrey et al.).

A two-dimensional testing apparatus that offers some improvements over the system used by Demer and Yin [10] and Yin and colleagues [46] has recently been developed by our group [25]. A simple tethering arrangement is employed with four attachment points on each side of the sample. The advantages of this design are that specimens can be mounted quickly and that relatively small samples may be tested. Individual point forces can be adjusted independently to ensure that stresses at the center of the sample are uniform. Loading is imposed with electromagnetic vibrators and the applied forces are controlled using a flexible software control system. This equipment has been used to study the biaxial mechanical properties of passive canine myocardium [40]. The hearts of six dogs arrested in diastole were rapidly excised, perfused with a cardioplegic solution containing 1 mM EGTA, and cooled to 10°C. Thin myocardial sections were sliced from the right and left ventricular free walls at various transmural depths using a dermatome. Specimens 25-mm square cut from the slices were oriented so that one pair of sides was parallel to the myocardial fiber direction. Four short lengths of fine suture material were passed through the specimen so that their cut ends marked the corners of a 1.5 to 2.0-mm central square. The specimen was then mounted in the biaxial testing apparatus. Testing was carried out at approximately 10°C in an oxygenated Krebs-Henseleit solution to which 1 mM EGTA had been added. Samples were subjected to 10 cycles of equibiaxial or uniaxial force loading at three different loading rates.

In Figure 1.3 we present results for a specimen taken from the middle of the left ventricular wall subjected to 10 cycles of equibiaxial force loading with a cycle period of 30 sec. The mechanical behavior at cycle periods of 2 sec and 10 sec was qualitatively similar. Stress-extension relations in the fiber direction were consistent throughout the test and exhibited little hysteresis. In the cross-fiber direction, however, reproducible stress-extension relations were obtained only after three load cycles. Moreover, the steady-state behavior was rate-dependent and exhibited some hysteresis. After testing, elastic recovery was incomplete in the cross-fiber direction, indicating that irreversible reorganization of structure may have occurred. For all results presented subsequently, stress-extension data relate to the tenth (and last) load cycle, whereas extension ratios are referred to marker

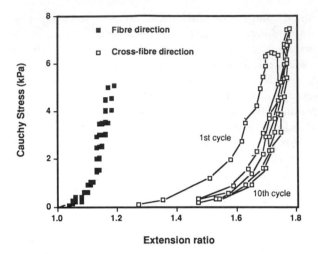

FIGURE 1.3. Stress-extension relations for left ventricular midwall specimen during 1st, 9th and 10th cycles of equibiaxial loading. Cycle period 30 sec and specimen thickness 1.83 mm. The solid lines indicate the order of loading in the cross-fiber direction. For clarity these are omitted in the fiber direction.

dimensions obtained with the sample in an unloaded state prior to the application of any cyclic mechanical loading.

The response of midwall specimens to equibiaxial loading was consistent (Figure 1.4). Maximum fiber extensions ranged between 15 and 25% and all specimens were highly compliant in the cross-fiber direction, although "strain-locking" was seen at extensions of 60 to 80%. Observation of the surface of a typical sample at high magnification suggests that cross-fiber strains are associated with the elastic deformation of muscle fiber bundles up to 140 μm across, together with a partly reversible rearrangement of these fiber bundles. In Figure 1.5, comparable results are given for a set of midmyocardial specimens under uniaxial load. Stress-extension data for the fiber direction again exhibit little hysteresis or rate-dependence and maximum fiber extensions were between 15 and 25%.

The results for a sample of subepicardial left ventricular myocardium, with epicardium intact, are given in Figure 1.6. For the fiber direction, the stress-extension behavior is similar to that observed for midmyocardial specimens. However, this is not the case for the cross-fiber direction. Under equibiaxial loading, the maximum cross-fiber extension was around 18% and this limit was reached at a relatively low stress, so that all further extension took place in the fiber direction. In addition, the cross-fiber stress-extension relation showed no sign of hysteresis or creep. To determine the extent to which the epicardial membrane constrains the mechanical behavior of composite subepicardial/epicardial specimens, the epicardium

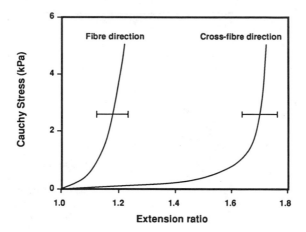

FIGURE 1.4. Mean stress-extension relations for left ventricular midwall specimens undergoing equibiaxial loading. Cycle period 10 sec, $n = 4$, and mean specimen thickness 2.06 mm. The bars indicate ± 1 standard deviation (σ_{n-1}).

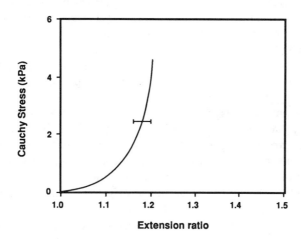

FIGURE 1.5. Mean stress-extension relations for left ventricular midwall specimens undergoing uniaxial loading in muscle fiber direction. Cycle period 10 sec, $n = 3$, and mean specimen thickness 2.17 mm. The bar indicates ± 1 standard deviation (σ_{n-1}).

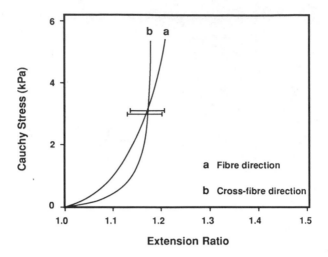

FIGURE 1.6. Mean stress-extension relations for left ventricular subepicardial specimens undergoing equibiaxial loading. Cycle period 10 sec, $n = 4$, and mean specimen thickness 1.76 mm. The bars indicate ± 1 standard deviation (σ_{n-1}).

was removed from the wall of the left ventricle by blunt dissection in two cases and tested under equibiaxial loading. The typical results shown in Figure 1.7 demonstrate highly nonlinear biaxial mechanical behavior, with "strain-locking" observed at extensions of 35 to 50%.

These findings reveal a clear anisotropy associated with the fiber direction. Moreover, the nature of this anisotropy apparently depends on the transmural site from which the sample was removed. In all cases, passive stress-extension relations in the fiber direction were similar to those obtained in papillary muscle and trabeculae carnae by Julian and Sollins [22] and ter Keurs and colleagues [42]. Extensions were limited to approximately 20%, consistent with the view that the connective tissue networks of the endomysium and perimysium prevent the extension of sarcomeres in ventricular myocardium beyond their optimum length. Midmyocardial specimens exhibited limited capacity to carry load in the cross-fiber direction and substantial extensions occurred before "strain-locking" was observed. The fact that stress-extension relations for uniaxial loading closely resemble those obtained under equibiaxial load indicates that there was little mechanical coupling between the fiber and cross-fiber directions in our midmyocardial specimens. We believe that cross-fiber extensions reflect the deformation of individual bundles or layers of muscle fibers, together with the rearrangement of these layers. Thus, the substantial cross-fiber extensions in midmyocardial specimens may be related to the loose coupling between the adjacent layers of myocytes in this region. On the other hand, the reduced cross-fiber extensions for composite subepicardial/epicardial specimens are

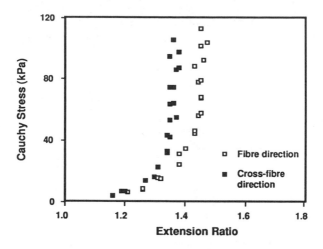

FIGURE 1.7. Stress-extension relations for excised left ventricular epicardium undergoing equibiaxial loading. Cycle period 10 sec and mean specimen thickness 0.04 mm.

consistent with a tighter linkage between subepicardial muscle fiber bundles. The extent to which the biaxial stress-extension relations obtained in the latter case are influenced by the mechanical properties of the epicardium is uncertain. Our results for excised epicardium, which are similar to those reported by Humphrey and Yin [19], indicate that subepicardial cross-fiber extension would not be constrained by the epicardium. However, there is evidence to suggest that the epicardium is under residual biaxial extension when the adjacent myocardium is in an apparently unloaded state. If this is correct then the epicardium may play an important role in determining the material properties of passive ventricular myocardium.

Given the similarity of the studies, it is important to establish the sources of the difference between our results and those reported by Demer and Yin [10] and Yin and colleagues [46]. A possible explanation is the artifact potentially introduced by cutting injury. When thin sections are cut from myocardium, damage to cell membranes causes contracture, supercontracture, and further cell damage [28]. We prevented the onset of contracture in our study with the calcium buffer EGTA and by testing at around 10°C. No steps were taken to inhibit contracture in the studies of Demer and Yin [10] and Yin and colleagues [46], which may explain the high fiber and cross-fiber stiffness reported by them. Also, in our study extension ratios were referred to the unloaded state prior to the application of any cyclic loading. The midmyocardial specimens exhibited irreversible creep in the cross-fiber direction during testing and, thus, maximum extensions in this direction would be considerably less had they been referred to the unloaded dimensions measured after testing.

The observations above highlight problems associated with the use of biaxial testing to characterize the material properties of myocardium. Ventricular myocardium is a three-dimensional composite and its structure is disrupted when thin sections of tissue are removed for two-dimensional mechanical testing. However, there are few alternatives. Two-dimensional sheets of myocytes can be cultured and tested mechanically, but the relationship between this preparation and ventricular myocardium remains unclear. To develop a more complete understanding of the three-dimensional mechanical properties of passive ventricular myocardium, it will be necessary to characterize biaxial behavior at representative sites within the myocardium and to relate these observations to the local microstructure.

1.3.3 CONSTITUTIVE LAW MODELING

It is necessary to formulate constitutive laws that provide a compact representation of the material properties of passive myocardium and can be used for mechanics modeling. In the chapters that follow, constitutive formulations are outlined that incorporate some knowledge of the microstructure of myocardium. However, before moving to this work it is instructive to consider the advantages and disadvantages of wholly phenomenological formulations in more detail.

The strain energy function below was advanced by Fung [15] as an appropriate general constitutive law for soft tissues:

$$W = B_{mnkl}E_{mn}E_{kl} + 0.5C(e^Q - 1) \quad \text{(summation implied over } m,n,k,l\text{)}$$

$$(1.1)$$

where $Q = A_{mnkl}E_{mn}E_{kl}$ and E_{mn} are components of the Green's strain tensor. Versions of this function have been used to model the material properties of skin [43], pericardium [45], and passive myocardium [46]. Although excellent fits to biaxial stress-strain data have been obtained, the fitted material parameters are highly variable and it is difficult to relate the individual parameters to specific aspects of material behavior. Moreover, using a single exponential function such as Equation (1.1), it is not possible to represent accurately biaxial stress-strain data in which the extent of nonlinearity along each axis differs markedly.

We have used a modified strain energy function based on a proposal by Fung [14, p. 251] to represent the observed material behavior of passive ventricular myocardium.

$$W = C_1(e^{Q_1} - 1) + C_4(e^{Q_2} - 1) \tag{1.2}$$

where

$$Q_1 = C_2E_{11}^2 + C_3(E_{12}^2 + E_{13}^2) \tag{1.3}$$

and

$$Q_2 = C_5(E_{22} + E_{33})^2 + C_6(E_{22}E_{33} - E_{23}^2). \tag{1.4}$$

For this transversely isotropic law the strain components are referred to a coordinate system in which the direction 1 is aligned with the fiber axis. Thus, the stiffness in the fiber direction and in the transverse plane are governed by C_2 and C_5, respectively, the shearing properties in the (1,2) and (1,3) coordinate planes are determined by C_3, while C_6 governs the shearing behavior in the (2,3) plane.

The stress σ_{ij} is given by

$$\sigma_{ij} = \frac{1}{2}\left(\frac{\partial W}{\partial E_{ij}} + \frac{\partial W}{\partial E_{ji}}\right) - p\delta_{ij} \tag{1.5}$$

where p is the hydrostatic pressure. For a material under biaxial traction in the (1,2) plane, $\sigma_{33} = 0$ and, thus,

$$\sigma_{11} = 2C_1C_2E_{11}e^{Q_1} - C_4[(2C_5 + C_6)E_{22} + 2C_5E_{33}]e^{Q_2}$$
$$\sigma_{22} = C_4C_6(E_{33} - E_{22})e^{Q_2}. \tag{1.6}$$

$E_{11} = 0.5(\lambda_1^2 - 1)$ and $E_{22} = 0.5(\lambda_2^2 - 1)$, where λ_1 and λ_2 are the extension ratios in the 1 and 2 directions, respectively. Assuming incompressibility, $E_{33} = 0.5[(\lambda_1\lambda_2)^{-2} - 1]$.

It should be possible, therefore, to fit the material parameters $C_1, C_2, C_4, C_5,$ and C_6 using the biaxial test data presented. The parameter C_3 is omitted because it cannot be identified in this case. In practice, it is difficult to fit Equation (1.6) by minimizing the differences between observed and estimated stresses because there is significant error in the estimated strain and, owing to the nonlinearity of the stress-strain relations, small fluctuations in strain can produce enormous variations in predicted stress. For this reason, the material parameters for passive ventricular myocardium were fitted by minimizing the errors between observed and predicted extensions. Least-squares optimization was used and Equations (1.6) were inverted using a modified Newton-Raphson method.

The constitutive formulation accurately reproduces the biaxial data obtained for specimens from the midwall of the left ventricle and a typical result is shown in Figure 1.8. The best-fit material parameters for each midwall specimen are presented in Table 1.1, which reveals considerable variability. Despite this, the constitutive law predicts uniaxial stress-extension relations in the fiber direction with reasonable accuracy. In Figure 1.9, experimental data are overlaid on the uniaxial stress-extension relation for the fiber direction predicted from the averaged material parameters in Table 1.1. The biaxial mechanical behavior of the subepicardial composite specimens is also faithfully represented (Figure 1.10). The best-fit material parameters for the subepicardium are given in Table 1.2 and are marginally more robust than those for the midwall. Note, however, that the parameters C_5 and C_6 are often negative, which is physically unacceptable.

This study demonstrates that it is possible to represent a complex repertoire of biaxial mechanical behavior using a strain energy function with

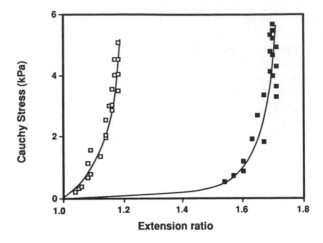

FIGURE 1.8. Best-fit stress-extension relations for left ventricular midwall specimen undergoing equibiaxial loading. Cycle period 10 sec and mean specimen thickness 1.83 mm. The symbols represent experimental data, while the solid lines are predicted by the constitutive equation, Equation (1.2), with material parameters $C_1 = 0.0175$, $C_2 = 48.02$, $C_3 = 0.0$, $C_4 = 0.00099$, $C_5 = 0.0413$, $C_6 = -12.98$.

TABLE 1.1. Best-fit material parameters for left ventricular midwall specimens.

	C_1	C_2	C_3	C_4	C_5	C_6
	0.0532	24.13	0.00	0.0078	2.404	−5.06
	0.0175	48.02	0.00	0.99×10^{-3}	0.041	−12.98
	0.0607	159.3	0.00	0.0976	6.492	−43.95
	0.0803	23.53	0.00	0.4587	1.429	−1.09
mean	0.0529	178.73	0.00	0.1413	2.5916	−15.77
σ_{n-1}	0.0262	64.17	0.00	0.2162	2.7750	19.43

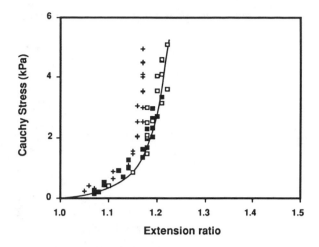

FIGURE 1.9. Predicted stress-extension relations for left ventricular midwall specimens undergoing uniaxial loading in muscle-fiber direction. The solid lines are given by Equation (1.2), using the mean, best-fit material parameters obtained from equibiaxial test data (see Table 1.1). The symbols represent data from individual uniaxial tests on different left ventricular midwall specimens. Cycle period 10 sec and mean specimen thickness 2.17 mm.

TABLE 1.2. Best-fit material parameters for left ventricular subepicardial specimens.

	C_1	C_2	C_3	C_4	C_5	C_6
	0.1404	8.42	0.00	0.0101	44.98	−69.52
	0.3963	12.50	0.00	0.0621	−26.47	−30.90
	0.2062	25.58	0.00	0.0269	−56.00	−50.63
	0.0696	35.55	0.00	0.0628	−56.27	−57.19
mean	0.2030	20.49	0.00	0.0402	−23.44	−52.07
σ_{n-1}	0.1410	12.40	0.00	0.0261	47.70	16.13

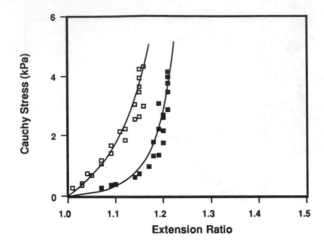

FIGURE 1.10. Best fit stress-extension relations for left ventricular subepicardial specimen undergoing equibiaxial loading. Cycle period 10 sec and mean specimen thickness 1.46 mm. The symbols represent experimental data, while the solid lines are predicted by Equation (1.2), with material parameters $C_1 = 0.2062$, $C_2 = 25.58$, $C_3 = 0.0$, $C_4 = 0.0269$, $C_5 = -56.27$, $C_6 = -57.19$.

only five fitted parameters. This empirical constitutive formulation has the advantage that it is straightforward to evaluate. Moreover, because the material parameters can be associated with specific aspects of mechanical behavior, the effects of altered material properties may be modeled in a relatively direct fashion. In fact, the microstructural studies referred to earlier suggest that an orthotropic, rather than a transversely isotropic, constitutive model is appropriate for intact myocardium. However, to estimate the parameters of an orthotropic constitutive law requires a more comprehensive investigation than the biaxial tests reported here.

The major problem with the constitutive formulation is the lack of uniqueness of the material parameters. This cannot be interpreted as real material variability because the biaxial test data are quite reproducible. It is due to the use of exponential functions to represent the highly nonlinear stress-strain behavior in passive myocardium. In the strain limit, similar stresses are predicted by different combinations of scaling and exponent parameters. The problem is exacerbated further by basing the strain energy law on strain invariants that are themselves fourth-degree functions of the extension ratio.

1.4   Concluding Remarks

It appears impractical to adopt a completely empirical approach to formulating a constitutive law for passive ventricular myocardium. The systematic tests required for full characterization of material properties cannot be performed on intact myocardium and it is necessary to disrupt the myocardium to obtain the isolated specimens suitable for controlled mechanical testing. At the least, some knowledge of the microstructure of ventricular myocardium is required to interpret and integrate the results of such tests.

In order to understand how cardiac microstructure influences the macroscopic material behavior of passive myocardium, it is necessary to obtain detailed information about connective tissue architecture throughout the wall of left and right ventricles. Unfortunately, SEM, which has revealed so much about the organization of cardiac connective tissue, does not lend itself easily to quantitative morphology. As a consequence, classification of the components of the connective tissue hierarchy has been largely descriptive. Moreover, while very complete descriptions of connective tissue organization have been presented for papillary muscle [38], comparable studies have not been reported for ventricular myocardium. There is an urgent need to address each of these points.

Questions can be raised about the usefulness of *in vitro* mechanical testing in general and biaxial testing in particular. It must be acknowledged that initial studies of the biaxial mechanical properties of noncontracting myocardium [10,46] failed to account for the artifacts introduced by removing thin sections of tissue from the ventricular wall. Cutting injury to cardiac cell membranes leads to contracture, which causes further cell damage [28] and disruption of cardiac microstructure. In addition, repeated cyclic loading may produce irreversible rearrangement of connective tissue organization. Despite these problems, *in vitro* mechanical testing can yield information about the mechanical behavior of the various components of the connective tissue hierarchy. It also provides a pathway for at least qualitative comparison of biaxial mechanical properties at different sites within the ventricular wall. However, appropriate steps must be taken to protect myocardium against cutting injury [20,28]. In addition, methods should be developed to assess the extent of structural disruption associated with *in vitro* testing of isolated myocardial specimens.

We have already stated that any realistic constitutive formulation for passive myocardium must incorporate microstructural information. It is worth reiterating that an appropriate constitutive law should be compact, with unique material parameters. We have demonstrated that the last of these requirements cannot easily be met by phenomenological constitutive formulations that use highly nonlinear functions. However, it is unlikely that the first requirement could be satisfied by a constitutive law based completely on cardiac microstructure. Thus, a constitutive formulation that incorporates the most advantageous features of both phenomenologically and microstructure-based approaches appears the best option at present.

Acknowledgements: Much of the work on the extracellular collagen matrix cited in this chapter was by Dr. Tom Robinson and his collaborators. Tom kindled our interest in this field during a visit to Auckland and we were most upset to learn of his recent death. Tom is widely missed as a good friend and a respected scientist. We gratefully acknowledge the assistance of Professors John Gavin and Steve Edgar, Department of Pathology, University of Auckland Medical School, with our studies of connective tissue organization in ventricular myocardium. The biaxial testing was carried out with Tony Shacklock. Finally, we thank the National Heart Foundation of New Zealand and the Auckland Medical Research Foundation for supporting our work.

REFERENCES

[1] C. Abrahams, J.S. Janicki, and K.T. Weber. Myocardial hypertrophy in macaca fascicularis: Structural remodeling of the collagen matrix. *Lab. Invest.*, 56:676–683, 1987.

[2] A. Benninghoff. Das perimysium internum. Handbuch der mikrosk. *Anatomie von v. Mollendorf*, 6:192–196, 1930.

[3] T.K. Borg, J. Buggy, I. Sullivan, J. Laks, and L. Terracio. Morphological and chemical characteristics of the connective tissue network during normal development and hypertrophy. *J. Mol. Cell. Cardiol.*, 18(Suppl):247, 1986.

[4] T.K. Borg, R.E. Gay, and L.D. Johnson. Changes in the distribution of fibronectin and collagen during development of the neonatal rat heart. *Collagen Rel. Res.*, 2:211–218, 1982.

[5] T.K. Borg, L.M. Klevay, R.E. Gay, R. Siegel, and M.E. Bergin. Alteration of the connective tissue network of striated muscle in copper deficient rats. *J. Mol. Cell. Cardiol.*, 17:1173–1183, 1985.

[6] T.K. Borg, W.F. Ranson, F.A. Moslehy, and J.B. Caulfield. Structural basis of ventricular stiffness. *Lab. Invest.*, 44:49–54, 1981.

[7] J.B. Caulfield. Alterations in cardiac collagen with hypertrophy. *Pers. Cardiovasc. Res.*, 8:49–57, 1983.

[8] J.B. Caulfield and T.K. Borg. The collagen network of the heart. *Lab. Invest.*, 40:364–371, 1979.

[9] J.B. Caulfield, W. Ranson, J.C. Xuan, and S.B. Tao. Histologic correlates of altered ventricular strain patterns following coronary artery ligation. *J. Mol. Cell. Cardiol.*, 16(Suppl 1):11, 1984.

[10] L.L. Demer and F.C.P. Yin. Passive biaxial mechanical properties of isolated canine myocardium. *J. Physiol. (London)*, 339:615–630, 1983.

[11] M. Eghbali, M.J. Czaja, M. Zeydel, et al. Collagen chain mRNAs in isolated heart cells from young and adult rats. *J. Mol. Cell. Cardiol.*, 20:267–276, 1988.

[12] M. Eghbali, S. Seifter, T.F. Robinson, and O.O. Blumenfeld. Enzyme-antibody histochemistry. A method for detection of collagens collectively. *Histochemistry*, 87:257–262, 1987.

[13] M.S. Forbes and N. Sperelakis. Intercalated disk of of the mammalian heart: A review of structure and function. *Tissue Cell*, 17:605–648, 1985.

[14] Y.C. Fung. *Biomechanics: Mechanical Properties of Living Tissues.* Springer-Verlag, New York, 1981.

[15] Y.C. Fung. Biorheology of soft tissues. *Biorheology*, 10:139–155, 1973.

[16] E. Holmgren. Uber die trophospongien der quergestreiften muskelfasern, nebst bermerkungen uber den allgemeinen bau dieser fasern. *Arch. Mikrosk. Anat.*, 71:165–247, 1907.

[17] A. Horowitz, I. Sheinman, and Y. Lanir. Nonlinear incompressible finite element for simulating loading of cardiac tissue—part II: Three dimensional formulation for thick ventricular wall segments. *ASME J. Biomech. Eng.*, 110:62–68, 1988.

[18] R.H. Hoyt, M.L. Cohen, and J.E. Saffitz. Distribution and three-dimensional structure of intercellular junctions in canine myocardium. *Circ. Res.*, 64(3):563–574, 1989.

[19] J.D. Humphrey and F.C.P. Yin. Biaxial mechanical behavior of excised epicardium. *ASME J. Biomech. Eng.*, 110:349–351, 1988.

[20] P.J. Hunter and B.H. Smaill. The analysis of cardiac function: A continuum approach. *Prog. Biophys. Mol. Biol.*, 52:101–164, 1989.

[21] J.E. Jalil, C.W. Doering, J.S. Janicki, R. Pick, S.G. Shroff, and K.T. Weber. Fibrillar collagen and myocardial stiffness in the intact hypertrophied rat left ventricle. *Circ. Res.*, 64:1041–1050, 1989.

[22] F.J. Julian and M.R. Sollins. Sarcomere length-tension relations in living rat papillary muscle. *Circ. Res.*, 37:299–308, 1975.

[23] Y. Lanir and Y.C. Fung. Two-dimensional mechanical properties of rabbit skin. I. Experimental system. *J. Biomech.*, 7:29–34, 1974.

[24] N.D. Light and A.E. Champion. Characterization of muscle epimysium, perimysium, and endomysium collagens. *Biochem. J.*, 219:1017–1026, 1984.

[25] A.S.D. Mayne, G.W. Christie, B.H. Smaill, P.J. Hunter, and B.G. Barratt-Boyes. An assessment of the material properties of leaflets from four second generation porcine bioprostheses using biaxial testing techniques. *J. Thorac. Cardiovasc. Surg.*, 98:170–180, 1989.

[26] I. Medugorac and R. Jacob. Characterization of left ventricular collagen in the rat. *Cardiovasc. Res.*, 17:15–21, 1983.

[27] J.B. Michel, J.L. Salzmann, M.O. Nlom, P. Bruneval, D. Barres, and J.P. Camilleri. Morphometric analysis of collagen network and plasma perfused capillary bed in the myocardium of rats during evolution of cardiac hypertrophy. *Basic Res. Cardiol.*, 81:142–154, 1986.

[28] L.A. Mulieri, G. Hasenfuss, F. Ittleman, E.M. Blanchard, and N.R. Alpert. Protection of human left ventricle from cutting injury with 2,3-butanedione monoxime. *Circ. Res.*, 65:1441–1444, 1989.

[29] J. Ohayon and R.S. Chadwick. Effects of collagen microstructure on the mechanics of the left ventricle. *Biophys. J.*, 54:1077–1088, 1989.

[30] J.G. Pinto. Some mechanical considerations in the selection and testing of papillary muscles. *J. Biomech. Eng.*, 102:62–66, 1980.

[31] J.G. Pinto and Y.C. Fung. Mechanical properties of the heart muscle in the passive state. *J. Biomech.*, 6:597–616, 1973.

[32] J.G. Pinto and P.J. Patitucci. Creep in cardiac muscle. *Am. J. Physiol.*, 232:H553–H563, 1977.

[33] J.G. Pinto and P.J. Patitucci. Visco-elasticity of passive cardiac muscle. *J. Biomech. Eng.*, 102:57–61, 1980.

[34] T.F. Robinson, L. Cohen-Gould, and S.M. Factor. Skeletal framework of mammalian heart muscle: Arrangement of inter- and peri-cellular connective tissue structures. *Lab. Invest.*, 49:482–498, 1983.

[35] T.F. Robinson, L. Cohen-Gould, S.M. Factor, M. Eghbali, and O.O. Blumenfeld. Structure and function of connective tissue in cardiac muscle: Collagen types I and III in endomysial struts and pericellular fibers. *Scan. Microsc.*, 2:1005–1015, 1988.

[36] T.F. Robinson, S.M. Factor, J.M. Capasso, B.A. Wittenburg, O.O. Blumenfeld, and S. Seifter. Morphology, composition, and function of struts between cardiac myocytes of rat and hamster. *Cell Tissue Res.*, 249:247–255, 1987.

[37] T.F. Robinson, S.M. Factor, and E.H. Sonnenblick. The heart as a suction pump. *Scientif. Am.*, 254:84–91, 1986.

[38] T.F. Robinson, M.A. Geraci, E.H. Sonnenblick, and S.M. Factor. Coiled perimysial fibres in papillary muscle in rat heart: Morphology, distribution and changes in configuration. *Circ. Res.*, 63:577–592, 1988.

[39] S. Sato, M. Ashraf, R.W. Millard, H. Fujiwara, and A. Schwartz. Connective tissue changes in early ischaemia of porcine myocardium: An ultrastructure study. *J. Mol. Cell. Cardiol.*, 15:261–267, 1983.

[40] A.J. Shacklock. *Biaxial Testing of Cardiac Tissue*. Master's thesis, University of Auckland, New Zealand, 1987.

[41] D.D. Streeter. Gross morphology and fiber geometry of the heart. In Berne R.M., editor, *Handbook of Physiology*, American Physiological Society, Bethesda, MD, 1979. Section 2.

[42] H.E.D.J. ter Keurs, W.H. Rijnsburger, R. van Heuningen, and M.J. Nagelsmit. Tension development and sarcomere length in rat cardiac trabeculae: Evidence of length-dependent activation. *Circ. Res.*, 46:703–714, 1980.

[43] P. Tong and Y.C. Fung. The stress-strain relationship for skin. *J. Biomech.*, 9:649–657, 1976.

[44] K.T. Weber, J.S. Janicki, S.G. Shroff, R. Pick, R.M. Chen, and R.I. Bashley. Collagen remodeling of the pressure-overloaded, hypertrophied nonhuman primate myocardium. *Circ. Res.*, 62:757–765, 1988.

[45] F.C.P. Yin, P.H. Chew, and S.L. Zeger. An approach to quantification of biaxial tissue stress-strain data. *J. Biomech.*, 19:27–37, 1986.

[46] F.C.P. Yin, R.K. Strumpf, P.H. Chew, and S.L. Zeger. Quantification of the mechanical properities of noncontracting canine myocardium under simultaneous biaxial loading. *J. Biomech.*, 20:577–589, 1987.

[47] M. Zhao, H. Zhang, T.F. Robinson, S.M. Factor, and E.H. Sonnenblick. Profound structural alterations of the extracellular collagen matrix in postischemic dysfunctional ("stunned") but viable myocardium. *J. Am. Coll. Cardiol.*, 10:1322–1324, 1987.

2

Structural Considerations in Formulating Material Laws for the Myocardium

Arie Horowitz[1]

ABSTRACT Specification of material laws for the heart muscle, mainly for the purpose of mechanical modeling, has been pursued intensively during the last three decades. Structural studies of the heart have an even longer history, but only recently, however, have these studies revealed the intercellular connections in the myocardium and provided a clearer picture of the myofiber pattern in the ventricular walls. Gradually increasing consideration of these findings in mechanical models of the heart has improved their validity and their predictive capabilities. A recent formulation of a material law presented in this chapter is based on analysis of the arrangement of the myofibers and the network of collagen fibers that interconnect them. Application of this material law to the description of the mechanical response of thin subepicardial slabs subjected to biaxial stretch has indicated the possible structural origin of several mechanical properties of the myocardium: its overall stiffness, the exponential-like shape of its stress-strain curves, and the mechanical coupling between the fiber and the cross-fiber directions. Further insights into the mechanical behavior of the myocardium were provided by simulating loading of thick ventricular wall segments, using a finite element formulation based on the material law mentioned above. This simulation has contributed to the understanding of the relation between the myofiber direction and the local anisotropy of the wall, the comparative magnitude of its tensile, compressive, and shear stiffnesses, and the possible effect of the collagen fiber network on the blood perfusion in the myocardium. Problems involved in future extensions of this approach by implementing even more detailed newly available structural data in models of the mechanics of the heart are also discussed.

2.1 Introduction

Mechanical characterization of materials is strongly dependent on the extent to which their structure is known. This is particularly true of biological

[1]Department of Bioengineering, University of Washington, Seattle, Washington 98195

materials, which often display complex mechanical behavior and are made up of numerous constituents. The myocardium poses a high degree of difficulty in that sense because of its intricate fibrous structure. Since the heart focused many anatomical studies, its muscular architecture is relatively well known. Until this century, however, many of these studies used vague descriptive terms to explain their findings. A few recent studies were carried out more methodically, in an attempt to obtain a comprehensive picture of the fiber pattern in the ventricles [23,28,29]. While further clarifying the picture, these works were still qualitative, and did not accompany their findings with quantitative measurements.

Hort [10], and later Streeter (see [26] for a review) were the first to include systematic measurements of fiber orientations. Streeter excised plugs from different locations in the walls of canine left ventricles and measured the fiber orientations on planes parallel to the epicardial surface. His main finding, confirming that of Hort, was that fiber directions generally vary in a continuous manner from $+60°$ on the endocardium to $-60°$ on the epicardium. Although clearly a simplification, many of the models of ventricular mechanics were and still are based on this assumption.

Although Streeter's work gave a clearer quantitative description of fiber pattern, it still did not document fiber directions methodically over the whole left ventricle. Since the location of the excised plugs was not recorded in a reproducible manner, it was difficult to confirm his results by studying additional specimens. A systematic quantitative study of fiber directions in the adult mammalian heart is only now being reported [17]. This study aims to produce a comprehensive data set of myofiber directions in the whole heart. Such data have already been compiled for the adult mouse heart and is planned to be extended to primates and possibly to humans. The findings of this study have not yet been implemented in models of the mechanics of the left ventricle.

The material laws initially proposed for the myocardium were based on uniaxial tests, mostly performed with papillary muscles. The common approach in formulating these material laws was to specify a functional relation between stress and strain that would best fit the measured stress-strain curves [14,19,20]. As such, this approach has not explicitly utilized structural data of the myocardium. Moreover, in view of the known fiber pattern in the heart, generalization of the uniaxial stress-strain relation to two or three dimensions is inappropriate, due to its assumption of isotropy. This was confirmed also in the biaxial tests carried out by Demer and Yin [3] and by Yin and colleagues [33], which have clearly shown that the myocardium is anisotropic.

The biaxial tests were followed by a proposition of a set of constitutive relations that incorporated the strains in the fiber and in the cross-fiber directions, and related them to the corresponding stresses [33]. Yet, the approach employed for this purpose was the same phenomenological one previously used for the characterization of uniaxial tests. A different set of constitutive relations was subsequently proposed by Humphrey and Yin [11,12] and applied to the same biaxial tests performed by Yin and colleagues [33]. This formulation represented the structure of the myocardium by composing the strain energy function of the tissue of two parts, one representing the contribution of the myofibers and another standing for the contribution of the noncontractile constituents, which were assumed to have an isotropic arrangement. However, the postulation of each of those two parts was phenomenological, and employed strain energy function of forms that were used before for other soft tissues by Demiray [4] and by Wilson [31].

The results of the biaxial tests of Yin and colleagues were analyzed also by Horowitz and colleagues [7], who proposed a material law based on the microstructural observation of several investigators [2,24]. The derivation of this law, which followed the approach of Lanir [15], will be elaborated further in this chapter. The structural evidence that served for putting together the stress-strain relations will be presented, as well as the consequences of the application of these relations to the results of the biaxial tests. This is followed by a finite element formulation that incorporates that material law. Finally, the results of new structural studies and some tentative suggestions for further mechanical characterization of the myocardium will be discussed.

2.2   Structural Background

The studies of Streeter [26] are still the most systematic and quantitative analyses of muscle-fiber orientations published at present. His method of investigation involved excision of wall plugs at different meridional locations between the base and the apex of the left ventricle. Although this approach did not produce a comprehensive documentation of the fiber pattern over the whole ventricle, it gave a clear picture of its main characteristics: (1) fiber orientations are generally tangential to the endocardial and epicardial surfaces, (2) fiber paths follow a helical geometry, and the helices can be described approximately as geodesics lying on a family of nested toroidal surfaces, and (3) fiber angle changes smoothly in a manner that corresponds to a transition from a right-handed helix on the endocardium to a left-handed one on the epicardium (Figure 2.1). The fiber pattern suggested by Streeter has been widely implemented in numerous models of ventricular mechanics.

For a review of the histology of myocardium, see Chapter 1 by Smaill and Hunter.

FIGURE 2.1. Parallel sections through a block of canine myocardium excised from the free wall of the left ventricle. (Reproduced with permission from Streeter and Hanna [27].)

2.3   Formulation of Stress-Strain Relations

The approach presented here is structural in the sense that its underlying principle is to obtain the total strain energy of the tissue in a given unit of volume by summing the individual strain energies of its constituents. Such an approach has been formulated and applied by Lanir [15] to fibrous connective tissues. Unlike phenomenological approaches, a structural approach attempts to gain information not only from the gross mechanical behavior of macroscopic specimens of the tissue under various loading conditions, but also from structural observations at the microscopic scale. One of the main anticipated advantages of the structural approach over phenomenological ones is that analysis of the configuration of the tissue constituents in the stress-free state may virtually lead to stress-strain relations that are, to a large extent, independent of the particular boundary conditions (i.e., loads and kinematic constraints). Consequently, the choice of the general forms of the strain energy functions should be guided initially by structural findings before considering some particular loading protocols. The interpretation of the structural observations described in the previous section and in Chapter 1 into guidelines for formulating an appropriate strain energy function is given in the next subsections.

2.3.1   BASIC ASSUMPTIONS

The structural data constitute a vast amount of detail that needs to be somewhat generalized prior to its incorporation into mathematical stress-strain relations. This can be facilitated by postulating several basic assumptions concerning the morphological and mechanical properties of the myocardium:

1. The myocardium may be treated as a hyperelastic material, that is, a strain-energy function is proposed from which the stresses are derived. The myocardium is found, though, to exhibit some strain-history dependence [20]. Consequently, it would be more appropriate to term it, similarly to other soft bilogical tissues, as pseudoelastic [6]. However, a supportive argument for the neglect of viscoelastic behavior under in vivo conditions is that the typical loading cycle, namely, the duration of a single heart beat, is quite short compared to the characteristic relaxation time of the myocardium.

2. The main structural elements of the myocardium are the interconnected networks of muscle fibers and collagen fibers, and the fluid matrix that embeds them.

3. The fibers that comprise the myocardium are thin and extensible and, consequently, cannot resist compression and buckle immediately as soon as compressive forces are applied.

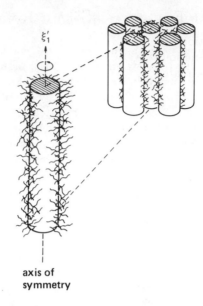

axis of
symmetry

FIGURE 2.2. Schematic representation of the arrangement of muscle fibers and their surrounding network of collagen fibers. (Reproduced with permission from Horowitz et al. [7].)

4. The fibers that comprise the myocardium may, in part, be undulated in the stress-free stress of the tissue; therefore, they do not resist tensile forces until being straightened.

5. The interstitial fluid matrix carries only hydrostatic pressure, which, in turn, is affected by length and configuration changes of the fibers. These cause pressure gradients, which may result in flow of the matrix. However, since the permeability of the myocardium is low, the fluid flow within the tissue is negligible for the duration of a cardiac cycle. Consequently, the myocardium may be assumed to be both locally and globally incompressible.

6. Numerous interconnections exist between the fibers, therefore the uniaxial strain of each fiber may be obtained from the overall strains of the tissue by tensorial transformation.

7. The distribution of collagen struts around each myocyte is axisymmetric (Figure 2.2).

These basic assumptions will be subsequently supplemented by additional structural and mechanical arguments.

2.3.2   GENERAL FORMULATION

The general derivation of the contravariant stress components, S^{ij}, from the strain energy function, $W(\gamma_{ij})$, and from the hydrostatic pressure, P, of an incompressible material that undergoes large deformations is [18]:

$$S^{ij} = (g/G)^{1/2} \frac{\partial W}{\partial \gamma_{ij}} + PG^{ij} \tag{2.1}$$

where the components of the second Piola-Kirchoff stress tensor S^{ij} and the Green's strain tensor γ_{ij} are referred to some general material coordinates, ξ_i, and where g and G are the determinants of the metric tensors before and after the deformation, respectively.

In the case of the myocardium, the total strain energy in a unit of volume is assumed to be the sum of the strain energies of the myofibers and of the collagen struts. Similar to the case of fibrous connective tissues [15], the contribution of each family of fibers to the total strain energy is determined by their volumetric fraction in the tissue and by their morphological characteristics: their spatial orientations and their waviness in the unstressed state. The large dispersion of both the orientations and the waviness of the fiber networks commonly observed in soft biological tissues calls for description of the variation of these properties through the tissue by using a stochastic approach, rather than a deterministic one.

Consequently, the variation of both the orientation and the waviness of each of the two types of fibers will be described by density distribution functions. The spatial orientation of fibers of type k ($k = 1$ for myofibers, $k = 2$ for collagen struts) may be ascribed some distribution function $R_k(\hat{n})$, where $\hat{n}$ is a unit vector tangent to the fiber. Thus, the relative portion of type k fibers confined within a spatial sector $\Delta\Omega$ is $R_k(\hat{n})\Delta\Omega$. Similarly, the waviness of fibers of type k may be ascribed another distribution function, $D_{k,\hat{n}}(x)$, so that the portion of fibers of this type aligned along $\hat{n}$ and having a straightening strain between x and $x + \Delta x$ is $D_{k,\hat{n}}\Delta x$. Generally, the distribution functions may be determined by histological studies, or, if such data are not available, they may be assumed a priori and validated later by fitting the stresses calculated on the basis of these distribution functions to stresses measured in mechanical tests.

Since the strain energy of an individual fiber depends on its uniaxial strain, it is necessary to relate this strain to the global ones. This is carried out by a tensorial transformation of the global strain components, γ_{rs}, to the uniaxial one, γ'_{11}:

$$\gamma'_{11} = \frac{\partial \xi^r}{\partial \xi'_1} \cdot \frac{\partial \xi^s}{\partial \xi'_1} \cdot \gamma_{rs} \tag{2.2}$$

where ξ'_i is a fiber coordinate system in which ξ'_1 coincides with the fiber direction. Thus, the uniaxial stress referred to the initial cross section of a straight fiber would be derived from its individual strain energy function w_k by:

$$f_k(\gamma'_{11}) = \frac{\partial w_k}{\partial \gamma'_{11}}. \tag{2.3}$$

However, the strain that actually determines the strain energy of the fiber is only the one beyond the straightening strain. This strain, termed here as "true strain," is related to the total fiber strain γ'_{11} by:

$$\gamma'_{11t} = \frac{\gamma'_{11} - x}{1 + 2x}. \tag{2.4}$$

To derive the uniaxial fiber stress corrected for waviness, we need to introduce also the waviness distribution function and to integrate from zero to the upper limit, which is the total fiber strain:

$$f^*_k(\gamma'_{11}) = \int_0^{\gamma'_{11}} D_{k,\hat{n}}(x) f_k(\gamma'_{11t}) dx. \tag{2.5}$$

To calculate the strain energy of all the fibers of type k in the tissue, we must sum over the entire range of spatial directions containing all the fibers of that type:

$$W_k = s_k \int_\Omega R_k(\hat{n}) w^*_k(\gamma'_{11}) d\Omega \tag{2.6}$$

where w^*_k is the strain energy of undulated fibers of type k [which is related to the strain energy of straight fibers in the same way as the stress in Equation (2.5)], and s_k is their volumetric fraction in the tissue. To obtain the strain energy in all the fiber types contained in the tissue per unit of undeformed volume, we simply need to sum over k:

$$W = \sum_k s_k \int_\Omega R_k(\hat{n}) w^*_k(\gamma'_{11}) d\Omega. \tag{2.7}$$

Finally, the stress components in the tissue, referred to the initial configuration, can be derived from the total strain energy by:

$$S^{ij} = (g/G)^{1/2} \sum_k s_k \int_\Omega R_k(\hat{n}) f^*_k(\gamma'_{11}) \frac{\partial \gamma'_{11}}{\partial \gamma_{ij}} d\Omega + P G^{ij}. \tag{2.8}$$

Equation (2.8) yields the stresses as a function of the global strains, the uniaxial stress-strain relation of each fiber type, the volumetric fraction of the fibers, and their waviness and orientation. Since we employ a large deformation approach, this relation should be valid in principle even when the fibers reorient and change their waviness because of global deformations of the tissue. However, this is true only as long as these alterations do not contradict the basic assumptions regarding the general morphology of the fibers, namely, the form of the orientation and waviness distribution functions.

2.3.3 AN EXAMPLE: APPLICATION TO BIAXIAL TESTS ON THIN MYOCARDIAL SLABS

The general forms obtained in the previous subsection have to be made more specific in order to characterize a particular type of tissue. This involves specification of the form of the uniaxial stress-strain relation and of the orientation and waviness distribution functions of each fiber type, including the material constants that these functions may contain [7]. This process was carried out for mechanical tests performed by Yin and colleagues [33], in which canine subepicardial slabs were subjected to various loading protocols. The slabs were square and thin ($4 \times 4 \times 0.1$–0.2 cm) and cut so that the muscle fibers were oriented in one predominant direction, parallel to two opposing edges, as judged by eye. Nevertheless, the slabs still contained some variation in fiber directions, ranging between 10 and 15°.

The results that will be analyzed here correspond to three specimens that underwent three loading protocols. The tissue slabs were stretched in each protocol at a constant ratio between the fiber and cross-fiber strains. These ratios varied between 1:3 and 3:1, including an equibiaxial test. Each protocol produced two stress-strain curves, one for the fiber and the other for the cross-fiber direction.

Fiber Orientation Distribution Functions

It is convenient to express the orientation distribution functions in a global spherical coordinate system, (r, θ, ϕ), in which the polar axis coincides with the ξ_1 axis of the cartesian coordinate system of the tissue specimen (Figure 2.3).

Muscle Fibers: Muscle fiber directions are assumed to vary by 15° through the thickness of the tissue slabs. However, the exact form of this variation is not documented. In view of Streeter's findings [26] and of the small thickness of the slabs, it seems reasonable to assume that fiber orientations are distributed symmetrically relative to the principal direction (ξ_1) and that they vary linearly. Therefore, the appropriate distribution function, that in the general case would depend on the angle α between each individual fiber and the ξ_1 direction (Figure 2.3), is in this case simply a constant:

$$R_1(\alpha) = \frac{12}{\pi}. \tag{2.9}$$

Collagen Struts: It is more convenient to describe first the distribution of the collagen struts in a local spherical coordinate system (r', θ', ϕ'), whose polar axis coincides with the axis of the myofiber to which the collagen struts are attached, rather than directly in the global spherical coordinates.

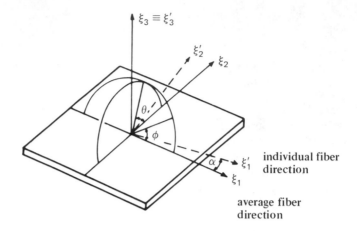

FIGURE 2.3. The alignment of the coordinate system relative to the tissue specimen. (Reproduced with permission from Horowitz et al. [7].)

The overall distribution function in the (θ', ϕ') space is given by the product of the respective distribution functions in the θ' plane and in the ϕ' plane:

$$R_2(\theta', \phi') = R_2(\theta') \cdot R_2(\phi'). \qquad (2.10)$$

The assumption of axisymmetric distribution of collagen struts around myofibers [basic assumption (7) in Section 2.3.1] implies that the appropriate distribution function in the θ' plane is uniform:

$$R_2(\theta') = \frac{1}{2\pi}. \qquad (2.11)$$

There is no clear indication, however, what form of distribution to assume in the ϕ' plane. For lack of other evidence, we have chosen to employ a normal (Gaussian) distribution. Such a distribution is widely used to describe large populations of a finite variance, which seems to apply to this case as well. We have initially assumed that most collagen struts are normal to the myofiber to which they are attached, and, therefore, that a normal distribution with a mean of $\pi/2$ should be suitable. However, this distribution yielded a poor fit to the measured stress-strain curves and was substituted by a double-peaked normal distribution, in which both the mean and the variance are free parameters to be determined by fitting the calculated stresses to the measured ones. The reasons that lead to this choice will be explained later.

Since there is no reason to give preference to one direction along any myofiber over the opposite direction, the two means of the distribution (namely, the location of its two peaks) are assumed to be symmetrically placed relative to the normal to the fiber. Thus, if one mean is $m_{\phi'}$, its counterpart would be $m_{\phi'} - \phi'$, and the distribution function is:

$$R_2(\phi') = \frac{1}{2(2\pi)^{1/2}\sigma_{\phi'}} \exp - \frac{(m_{\phi'} - \phi')^2}{2\sigma_{\phi'}^2} \qquad (2.12)$$

for $0 \leq \phi' \leq 2/\pi$, and:

$$R_2(\phi') = \frac{1}{2(2\pi)^{1/2}\sigma_{\phi'}} \exp - \frac{(\pi - m_{\phi'} - \phi')^2}{2\sigma_{\phi'}^2} \qquad (2.13)$$

for $2/\pi \leq \phi' \leq \pi$. The division by 2 is required to keep the total area between 0 and π equal to 1.

The meridional angle ϕ' in the myofiber coordinate system has to be expressed in terms of the global spherical coordinates. This first involves transformation to the myofiber Cartesian system (ξ_i'), then to the global Cartesian system (ξ_i), and finally to the global spherical system. The relation between ϕ' and θ, ϕ and α (α being the variable degree of each individual myofiber in the $\xi_1 - \xi_2$ plane) is:

$$\phi' = \cos^{-1}(\sin\alpha \, \cos\theta + \cos\alpha \, \cos\phi). \qquad (2.14)$$

Substituting ϕ' from Equation (2.14) in Equations (2.12) and (2.13) gives the final form of the orientation distribution function.

Fiber Waviness Distribution Functions

In the absence of quantitative data regarding the distributions of the waviness of both myofibers and collagen struts, we have adopted here normal distribution functions, for the same reason mentioned before. In the present formulation, the waviness of the fibers is assumed to be independent of their direction. Thus, the distribution functions are:

$$D_k(x) = \frac{1}{(2\pi)^{1/2}\sigma_k} \exp - \frac{(m_k - x)^2}{2\sigma_k^2} \qquad (2.15)$$

where the mean m_k and the variance σ_k receive different values for myofibers and for collagen struts.

Fiber Uniaxial Stress-Strain Relations

Both myofibers [5] and connective tissues exhibit exponential stress-strain relations. However, this behavior may be interpreted as a result of gradual straightening of fibers that have linear stress-strain relations when devoid of undulations [16]. Consequently, the uniaxial stress-strain relations for both fiber types are:

$$f_1(\varepsilon_{kt}) = C_k \varepsilon_{kt} \qquad (2.16)$$

where ε_{kt} is the fiber uniaxial true strain and C_k is a material constant. The value of C_1 that can be deduced from tests performed on single myofibers [5] is 1 MPa, and the value of C_2 that has been measured for collagen fibers is 1 GPa [30].

Stress Analysis of the Specimens

Owing to the manner of loading in the biaxial tests of Yin and colleagues [33], the specimens are assumed to be subjected only to normal stresses in the fiber and in the cross-fiber directions S^{11} and S^{22}, respectively (see, however, [11], for critical discussion of this assumption). Each of these components is made up of the contributions of the myofibers, collagen struts, and hydrostatic pressure:

$$S^{11} = S_1^{11} + S_2^{11} + PG^{11} \qquad (2.17)$$

and similarly for S^{22}. Since the top and bottom of the specimens were free of loads, and since the specimens are thin, it is assumed that the normal stress in the perpendicular direction, S^{33}, is constant and equal to zero (plane stress state). Owing to the planar arrangement of the myofibers, only the collagen struts and the fluid matrix contribute to S^{33}:

$$S^{33} = S_2^{33} + PG^{33} = 0 \qquad (2.18)$$

Consequently, in this case the hydrostatic pressure is not an independent variable and can be expressed in terms of the fiber contributions.

Owing to the incompressibility of the tissue, the value of the determinant of the metric tensor after the deformation is $G = 1$, and the thickness strain can be determined on the basis of the two measured in-plane strains:

$$\gamma_{33} = \frac{-2\gamma_{11}\gamma_{22} - \gamma_{11} - \gamma_{22}}{(2\gamma_{11} + 1)(2\gamma_{22} + 1)}. \qquad (2.19)$$

Using the tensorial transformation in Equation (2.2), the uniaxial strain of the myofibers can be calculated from the global strains by:

$$\varepsilon_1 = \gamma_{11} \cos^2\alpha + \gamma_{22} \sin^2\alpha \qquad (2.20)$$

and, similarly for the uniaxial strain of the collagen struts:

$$\varepsilon_2 = \gamma_{11} \cos^2\phi + \gamma_{22} \cos^2\theta \sin^2\phi + \gamma_{33} \sin^2\theta \sin^2\phi. \qquad (2.21)$$

The final forms of the stress expressions are obtained after substituting the appropriate distribution functions and uniaxial stress-strain relations. Thus, the contributions of the myofibers to each of the two normal stresses are:

$$S_1^{ii} = s_1 \int_{\frac{-\pi}{24}}^{\frac{\pi}{24}} f_1^*(\varepsilon_1) R_1(\alpha) \frac{\partial \varepsilon_1}{\partial \gamma_{ii}} d\alpha, i = 1, 2 \qquad (2.22)$$

(no summation over i) and the contributions of the collagen struts to the same stresses are:

$$S_2^{ii} = s_2 \int_{\frac{-\pi}{24}}^{\frac{\pi}{24}} \int_0^{2\pi} \int_0^\pi f_2^*(\varepsilon_2) R_2(\alpha, \theta, \phi) \frac{\partial \varepsilon_2}{\partial \gamma_{ii}} J d\phi d\theta d\alpha, i = 1, 2 \qquad (2.23)$$

(no summation over i), where J is the determinant of the transformation of θ' and ϕ' to θ and ϕ. The integration over α in Equation (2.23) is carried out in order to sum up the contributions of the collagen struts attached to all the myofibers, whose orientation varies between $\pi/24$ to $-\pi/24$ relative to ξ_1.

The stresses given by Equations (2.22) and (2.23) are the Piola-Kirchhoff mathematical components, which still have to be converted to the physical ones:

$$S_{(ij)} = S^{ij} (G_{jj}/g^{ii})^{1/2} \qquad (2.24)$$

(no summation over i and over j) in order to be compared to the stresses measured in the mechanical tests (which were also referred to the initial configuration).

The Results and their Significance

The fitting of the stress expressions to the stress-strain curves measured in the biaxial tests of Yin and colleagues [33] requires carrying out a nonlinear least-squares procedure in order to estimate the values of the parameters that appear in those expressions. Initially, there were eight parameters, three for the muscle fibers (C_1, m_1, σ_1) and five for the collagen struts $(C_2, m_2, \sigma_2, m_{\phi'}, \sigma_{\phi'})$. However, because of the large difference between the stiffnesses of the myofibers and the collagen struts (the former are smaller by about three orders of magnitude than the latter), the influence of the parameters associated with the myofibers on the convergence of the least-squares procedure was insignificant compared to that of the parameters associated with the collagen struts. We have chosen, therefore, to omit the contribution of the myofibers to the tissue stresses and to estimate only the five parameters associated with the collagen struts. The high nonlinearity of the relations between some of the parameters and the stresses (see also Horowitz and colleagues [7]) demanded to fit the stress expressions simultaneously to all the three protocols of each specimen, rather to only one at a time.

The values that were obtained for the five estimated parameters (Table 2.1) provided a very good fit to the stresses measured in the biaxial tests (Figure 2.4). Owing to the structural approach employed here, the values of the parameters may be interpreted as having physical meaning, and lead to the following deductions:

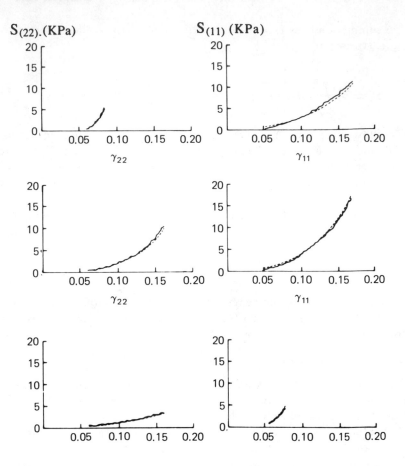

FIGURE 2.4. Comparison of the measured (continuous line) and calculated (dashed line) stresses for three loading protocols of one of the myocardial specimens (right side—fiber direction; left side—cross-fiber direction). (Reproduced with permission from Horowitz et al. [7].)

TABLE 2.1. Material constant values and mean square errors (Msq) for the three specimens. The data by which each set of values is determined consist of three pairs of stress-strain curves, measured in the fiber and cross-fiber directions in three different protocols [33] (the corresponding specimen numbers from the original paper are indicated in the table). Fifteen data points are sampled from each curve, resulting in a total of 90 points for each specimen. The three sets of material constant values are obtained by repeatedly performing nonlinear least-squares fitting of the stress expressions to each of the three data groups (see also text).

	C_2 (GPa)	M_2	σ_2	$M_{\phi'}$ (rad)	$\sigma_{\phi'}$ (rad)	Msq (KPa)
Specimen #3	1.35	0.28	0.11	0.69	0.14	0.40
Specimen #5	1.72	0.28	0.11	0.59	0.12	0.56
Specimen #2	2.89	0.26	0.08	0.78	0.28	0.28

1. The value of the uniaxial stress-strain constant C_2 averages 2 GPa, similar to the reported value [30]. The difference between this value and that of the myofibers suggests that the collagen matrix is the main load-bearing component in the myocardium. This conclusion is in agreement with histological studies, which identify the collagen matrix as a skeletal network that provides the tissue its structural integrity [2,24].

2. The mean waviness of the collagen struts m_2 is approximately 0.3, indicating that these fibers are significantly slack in the load-free state.

3. The mean inclination angles of the orientation distributions of the collagen struts $m_{\phi'}$ is between 35 and 45 degrees. This may imply that the struts that play the main load-bearing role in the biaxial tests are aligned mostly diagonally to the myofibers that they interconnect.

4. The diagonal alignment of collagen struts may provide a possible mechanism for the mechanical coupling observed in the biaxial tests between the fiber and the cross-fiber directions. This coupling, which was the reason for employing the double-peaked distribution of collagen struts in the stress expressions, is expressed in the measured stress-strain curves. Referring, for example, to one of the specimens

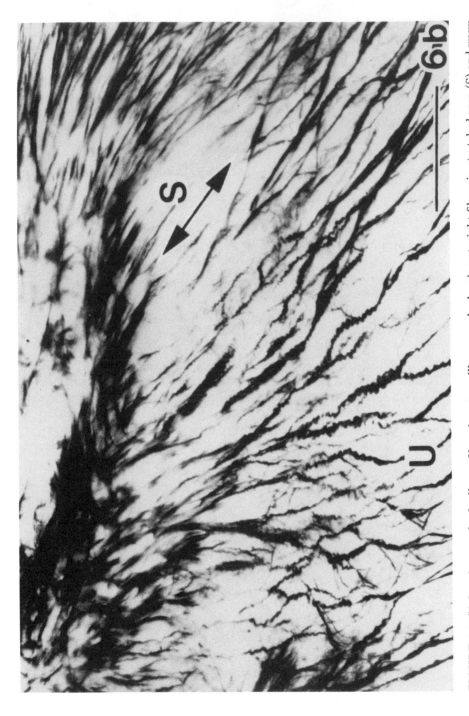

FIGURE 2.5. Light micrograph of collagen fibers in rat papillary muscle showing straight fibers in stretched area (S) and wavy in unaffected area (U). Double-headed arrow indicates direction of stretch ×80. (Reproduced with permission from Robinson

(Figure 2.4), we may notice that even when the fiber direction strain is low the stiffness in this is high. At the same time, the cross-fiber stiffness seems to be relatively low, even though the strain in this direction is large. An opposite effect appears when the fiber direction undergoes large strains and the cross-fiber strains are low, but in this case the fiber direction stiffness is higher than the cross-fiber direction stiffness in the previous case. Evidently, straining one direction stiffens the other, and the fiber direction is affected more by this than the cross-fiber direction. In the equibiaxial test, the fiber direction stiffness seems to be somewhat higher than the cross-fiber direction stiffness.

A predominantly diagonal alignment of the collagen struts with inclination angles that are mostly lower than 45° may explain this behavior: the struts are initially slack, and because the above inclination angle they contribute more to the load bearing in the fiber than in the cross-fiber direction; therefore, the resistance of the tissue to cross-fiber stretch is low, but at the same time large stretch in that direction eventually straightens a large portion of the struts and increases the stiffness in the fiber direction.

It should be emphasized, however, that the existence of normally oriented struts need not necessarily be excluded. Moreover, structural findings [25] indicate that most of the struts are perpendicular to the myofibers. Even if the particular mechanism suggested above is inexact, the coupling phenomenon itself probably does originate from reorientation and interaction between the struts and the collagen weave along the myofibers. Reorientation of collagen fibers in the myocardium has indeed been reported [25] (Figure 2.5).

2.4   Simulation of a Ventricular Wall Segment Subjected to Various Loading Conditions

The application of the material law presented in the previous chapter in models of the left ventricle, with its irregular three-dimensional shape, results in an elasticity problem that cannot be solved analytically. The common alternative approach is to solve it numerically, by the finite element method (see [32] for a review of some of the work in this area). An appropriate three-dimensional finite element has been formulated and used for simulating a ventricular wall segment under various loading cases [8]. The formulation of this element and the results of the simulations are presented below.

2.4.1   FINITE ELEMENT FORMULATION

The derivation of the element equations of motion is based on the requirement of minimum potential energy [18]. The total potential energy of the element is:

$$\Pi = U + V \tag{2.25}$$

where U is the strain energy stored in the element and V is the work performed on it by external loads.

Owing to the assumption of incompressibility, the strain energy function $W_{(e)}$ has to be augmented by the Lagrange constraint of setting the third invariant of Green's deformation tensor to 1:

$$W = W_{(e)} + P(I_3 - 1) \tag{2.26}$$

where the Lagrange multiplier P is physically interpreted as the hydrostatic pressure (HP) in the element. The total strain energy of the element is obtained by integrating the strain energy function over its volume. Thus, the total potential energy of the element is:

$$\Pi(\gamma_{ij}, P) = \int_{0_v} [W_{(e)} + P(I_3 - 1)]^0 dv + V. \tag{2.27}$$

The stationarity of the total potential energy requires that its first variation in respect to the strains and the HP has to vanish:

$$\delta\Pi = \frac{\partial\Pi}{\partial\gamma_{ij}}\delta\gamma_{ij} + \frac{\partial\Pi}{\partial P}\delta P = 0 \tag{2.28}$$

and, the strains and the HP being unrelated, each of the two variations has to vanish separately:

$$\frac{\partial\Pi}{\partial\gamma_{ij}} = 0 \tag{2.29}$$

$$\frac{\partial\Pi}{\partial P} = 0. \tag{2.30}$$

The first equality leads to the equation of motion of the element, and the second one leads to the equation of incompressibility. The derivation of the equation of motion (see, e.g., [1]), whose notation is used here, produces the following general expression:

$$\int_{0_v} {}^{t+\Delta t}_{0}S^{ij} \, \delta \, ({}^{t+\Delta t}_{0}\gamma_{ij})^0 dv = {}^{t+\Delta t}R \tag{2.31}$$

where the second Piola-Kirchoff stresses, ${}^{t+\Delta t}_{0}S^{ij}$, and the Green's strains, ${}^{t+\Delta t}_{0}\gamma_{ij}$, are referred to the initial configuration (Lagrangian approach).

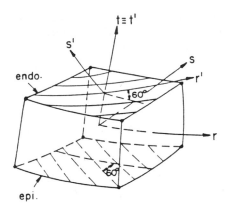

FIGURE 2.6. The global and fiber-attached coordinate systems in a finite element, which represents an endocardium-to-endocardium ventricular wall segment. (Reproduced with permission from Horowitz et al. [7].)

Because of the material and geometric nonlinearities, the equation of motion is nonlinear and is converted to a sequence of linear equations by incremental decomposition. The material law is incorporated in the formulation through the stress and the elastic matrix expressions. The reader is referred to Bathe [1] and to Horowitz and colleagues [8,9] for a more detailed presentation of this process.

The isoparametric finite element formulated here uses Lagrangian interpolation for both the nodal incremental displacements and the HP, which are the variables to be solved. The interpolation order used for the displacements is parabolic, resulting in a maximally 21 nodes finite element. The HP, however, should preferably have a lower interpolation order than the displacements [18], so that the maximal number of nodes for the HP is 8.

The elemental displacements, referred to the element Cartesian coordinates (r, s, t) (Figure 2.6), are given by:

$$u(r, s, t) = \mathbf{H}\mathbf{u} \tag{2.32}$$

where $\mathbf{H}$ is the displacement interpolation function matrix [1] and $\mathbf{u}$ is the vector of incremental nodal displacements, containing $3N$ components (N being the number of nodes for interpolation of displacements). Similarly, the incremental HP in the element is given by:

$$P(r, s, t) = \mathbf{H}_p\mathbf{P} \tag{2.33}$$

where $\mathbf{H}_p$ is the displacement interpolation function vector and $\mathbf{P}$ is the vector of incremental nodal HP, containing M components (M being the number of nodes for interpolation of pressures, so that $M < N$).

It still remains to derive the element equation of incompressibility, which is carried out by equating to zero the first variation of the total potential energy [Equation (2.27)] in respect to the HP. After expressing the HP by its vector of interpolation functions, the resulting equation is:

$$\int_{0_v} {}_0^t H_{P_k} (I_3 - 1)^0 dv = 0, (k = 1, 2, ..., M).$$ (2.34)

Equation (2.34) is also brought to incremental form by perturbing the strains [8].

The complete elemental system of equations for solving the N incremental displacements and the M increments of HP is obtained by assembling the equations of motion and the equations of incompressibility:

$$\begin{bmatrix} \mathbf{K}_u & \mathbf{K}_p \\ \mathbf{K}_p^T & 0 \end{bmatrix} \begin{bmatrix} \mathbf{u} \\ \mathbf{P} \end{bmatrix} = \begin{bmatrix} \mathbf{F} \\ 0 \end{bmatrix}$$ (2.35)

where $\mathbf{K}_u$ is the $N \times N$ matrix of coefficients of the incremental displacements in the equation of motion, $\mathbf{K}_p$ is the $N \times M$ matrix of coefficients of the incremental HP in the same equation, and $\mathbf{F}$ is the vector of nodal forces. The matrix on the left-hand side of Equation (2.35) is the elemental stiffness matrix. This matrix is symmetric, but because of the constraint of incompressibility, it is not necessarily positive definite.

The system of equations is solved by an incremental-iterative method that is based on Newton's method [1]. The iterative process in each load increment is terminated once the Euclidean norm of the displacements converges, and if the new configuration maintains incompressibility.

2.4.2 THE RESULTS AND THEIR SIGNIFICANCE

The present finite element allows the user to employ varying fiber distributions and may consequently be used to simulate passive wall segments of the left ventricle under various loading conditions [8]. Similar to the common approach in other models, the fiber direction is assumed to vary linearly between 60° on the endocardium to −60° on the epicardium (however, it is equally feasible to implement other fiber distributions in this element). The inclination angle of the muscle fibers is measured relative to one of the local axes of the element. Accordingly, that direction corresponds to the circumferential aspect of the ventricular wall, and the direction normal to it in a plane parallel to the fibers corresponds to the longitudinal aspect (Figure 2.6).

The results of the loading cases performed in the present study lead to several conclusions regarding the mechanical properties of the myocardium:

1. Uniaxial tension applied separately in the circumferential and in the longitudinal directions indicates that the former is the stiffer one.

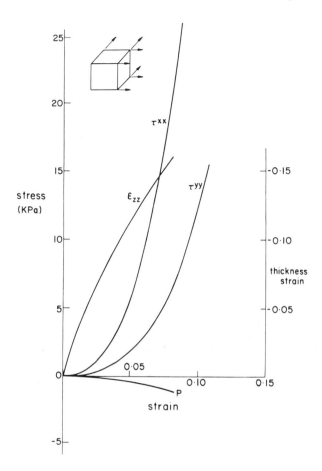

FIGURE 2.7. Stress-strain, hydrostatic pressure (P), and thickness strain curves of a myocardial segment under biaxial stretch. (Reproduced with permission from Horowitz et al. [7.])

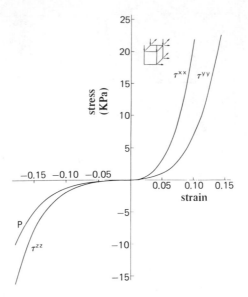

FIGURE 2.8. Stress-strain, hydrostatic pressure (P), and thickness strain curves of a myocardial segment under biaxial stretch combined with vertical compression. (Reproduced with permission from Horowitz et al. [7].)

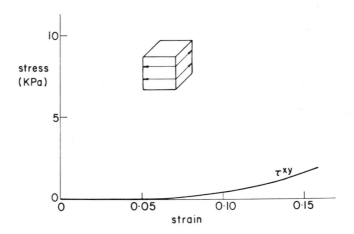

FIGURE 2.9. Stress-strain curve of a myocardial segment under simple shear. (Reproduced with permission from Horowitz et al. [7].)

However, the difference is significantly smaller than for a parallel fiber arrangement [9]. Subjecting the wall segment to equibiaxial tension stiffens both directions, compared to the uniaxial case. The increase is small, however (Figure 2.7).

2. The fiber distribution through the thickness of the wall modifies the ratio between the circumferential and the longitudinal stiffness. Expectedly, the difference between the two directions becomes larger as the fiber distribution is made more unidirectional.

3. When the wall segment is subjected to biaxial tension combined with compression in the vertical direction (which resembles the loading pattern in the ventricle) the HP becomes the predominant compressive stress component in that direction (Figure 2.8). The compressive stiffness of the myocardium turns out to be lower, although not excessively, than its tensile stiffness.

4. The shear stiffness of the myocardium is significantly lower than its tensile and compressive stiffness (Figure 2.9). A possible structural basis for this result might be the low resistance of the myofiber-collagen strut connections to angular deformation. This property might moderate the stresses that develop in the walls when the left ventricle twists during the cardiac cycle [13].

Because of its ability to account for material and geometric nonlinearities, anisotropy and incompressibility, the present finite element may serve as a tool for more accurate simulation of the mechanics of the left ventricle than previously possible. The element has not been incorporated yet in a model of the whole ventricle.

2.5   Discussion

Both structural studies and mechanical characterization of the myocardium have been progressing gradually. The acquisition of new and increasingly detailed histological data poses, however, an increasing difficulty for its representation in constitutive relations. Despite that seemingly problematic aspect, it is likely that the clarification of the concrete structural picture of the myocardium will eventually lead to more realistic and specific material laws that will hopefully capture the unique mechanical characteristics of the myocardium. Some recent structural findings and some notions concerning present and future approaches in mechanical modeling are discussed below.

2.5.1 New Structural Findings and their Possible Implications

A study that employs a quantitative and systematic methodology aimed at producing a three-dimensional reconstruction of the myofiber pattern in the mammalian heart is currently being carried out [21]. This is done by systematic sectioning of the heart, both at the macroscopic and microscopic scale, which identifies and delineates the directional organization of myocytes in the ventricular walls.

The pattern that emerges from three-dimensional reconstruction of the myofiber orientations in the mouse heart [17] seems to be more complex than the previous concepts and differs from the myofiber patterns described in other mammals (such as dog and human). Whereas previous studies (e.g., [26]) reported that fibers are arranged in a fanlike manner and that circumferential fibers are mostly found in the midwall, the new study finds a significant number of transverse and longitudinal fibers across most of the thickness of the ventricular wall. The myofiber pattern across the septum was found to be particularly intricate, with many fibers running longitudinally.

The new study has also examined the correlation between the direction of macroscopic fiber bundles exposed in previous studies (e.g., [29]) and myofiber orientations at the microscopic level in one of those bundles. Surprisingly, the myofibers were found to be arranged in widely varying directions, and not only in the direction observed at the macroscopic level [22].

These findings challenge the validity of the relatively simple fiber patterns assumed in most models of ventricular mechanics. Although a direct and exact representation of the microscopic myofiber pattern in the models is clearly unfeasible, the new systematic studies eventually may produce a general myofiber architecture that is different from the present concepts.

2.5.2 Notions for Further Advances

The structural approach presented here for the myocardium yielded good fit to the mechanical behavior of the passive myocardium in biaxial tests and provided additional insights into the possible structural source of this behavior. It also produced several predictions concerning the mechanical properties of ventricular wall segments.

Further application of the same approach implies incorporation of new histological data as it emerges from ongoing structural studies. However, this will inevitably result in an increasing mathematical and computational complexity. Clearly, a one-to-one representation of all tissue elements is not feasible.

The structural approach is also advantageous in the sense that its assumptions and predictions may be verified in principle by histological observations. Yet, some of the structural properties that were employed in the present study (e.g., mean waviness of fibers, form of orientation distribution of collagen struts around myofibers) might be practically difficult to assess. This has necessitated the use of tentative assumptions instead, such as distribution functions. At that point, the distinction between the "structural" and "phenomenological" approaches becomes only a matter of the level at which the a priori assumptions are introduced.

Between the extremes of no consideration of any structural data and of literal representation of all relevant tissue constituents, it might be preferable to attempt to develop material models that capture the unique mechanisms and properties that dominate the mechanical behavior of the material. These properties may be deduced from both structural and mechanical studies. The results of the present study indicate that these characteristic properties are:

1. Mechanical coupling between fiber and cross-fiber directions.

2. Anisotropy determined by the local distribution of fiber directions in the ventricular wall.

3. Load-dependent straightening and realignment of fiber networks.

4. Low shear stiffness.

If these properties, partially supported by structural findings, are further substantiated, mechanical models of the myocardium should account for them.

REFERENCES

[1] K.J. Bathe. *Finite Element Procedures in Engineering Analysis.* Prentice-Hall, Englewood Cliffs, NJ, 1982.

[2] J.B. Caulfield and T.K. Borg. The collagen matrix of the heart. *Lab. Invest.*, 40:364–372, 1979.

[3] L.L. Demer and F.C.P. Yin. Passive biaxial properties of isolated canine myocardium. *J. Physiol.*, 339:615–630, 1983.

[4] H. Demiray. Stresses in ventricular wall. *ASME J. Appl. Mech.*, 43: 194–197, 1976.

[5] D. Fish, J. Orenstein, and S. Bloom. Passive stiffness of isolated cardiac and skeletal myocytes in the hamster. *Circ. Res.*, 54:267–276, 1984.

[6] Y.C. Fung. *Biomechanics: Mechanical Properties of Living Tissues.* Springer-Verlag, New York, 1981.

[7] A. Horowitz, Y. Lanir, F.C.P. Yin, M. Perl, I. Sheinman, and R.K. Strumpf. Structural three-dimensional constitutive law for the passive myocardium. *ASME J. Biomech. Eng.*, 200–207, 1988.

[8] A. Horowitz, I. Sheinman, and Y. Lanir. Nonlinear incompressible finite element for simulating loading of cardiac tissue—Part II: Three dimensional formulation for thick ventricular wall segments. *ASME J. Biomech. Eng.*, 110:62–68, 1988.

[9] A. Horowitz, I. Sheinman, Y. Lanir, M. Perl, and S. Sideman. Nonlinear incompressible finite element for simulating loading of cardiac tissue—Part I: Two dimensional formulation for thin myocardial strips. *ASME J. Biomech. Eng.*, 110:57–61, 1988.

[10] W. Hort. Makroskopische und mikrometriche untersuchungen am myokard verschiedenstark gefuller linker kammern. *Virchows Archiv Path. Anat.*, 329:694–731, 1960.

[11] J.D. Humphrey and F.C.P. Yin. Biomechanical experiments on excised myocardium: Theoretical considerations. *J. Biomech.*, 22:377–383, 1989.

[12] J.D. Humphrey and F.C.P. Yin. On constitutive relations and finite deformations of passive cardiac tissue: I. A pseudostrain-energy function. *ASME J. Biomech. Eng.*, 109:298–304, 1987.

[13] N.B. Ingels, D.E. Hansen, G.T. Daughters, E.B. Stinson, E.L. Alderman, and E.C. Miller. Relation between longitudinal, circumferential, and oblique shortening and torsional deformation in the left ventricle of the transplanted human heart. *Circ. Res.*, 64:915–927, 1989.

[14] A. Kitabatake and H. Suga. Diastolic stress-strain relations of nonexcised blood perfused canine papillary muscle. *Am. J. Physiol.*, 234:416–419, 1978.

[15] Y. Lanir. Constitutive equations for fibrous connective tissue. *J. Biomech.*, 16:1–12, 1983.

[16] Y. Lanir. A structural theory for the homogenous biaxial stress-strain relationship in flat collagenous tissues. *J. Biomech.*, 12:423–436, 1979.

[17] M. McLean, M.A. Ross, and J.W. Prothero. A three-dimensional reconstruction of the myofiber pattern in the fetal and neonatal mouse-heart. *Anat. Rec.*, 224:392–406, 1989.

[18] J.T. Oden. *Finite Elements of Nonlinear Continua*. McGraw-Hill, New York, 1989.

[19] Y.C. Pao, G.K. Nagendra, R. Padiyar, and E.L. Ritman. Derivation of myocardial fiber stiffness equation based on theory of laminated composite. *ASME J. Biomech. Eng.*, 102:252–257, 1980.

[20] J.G. Pinto and Y.C. Fung. Mechanical properties of the heart muscle in the passive state. *J. Biomech.*, 6:597–616, 1973.

[21] J.S. Prothero and J.W. Prothero. Three-dimensional reconstruction from serial sections: IV. The reassembly problem. *Comp. Biomed. Res.*, 19:361–373, 1986.

[22] J.W. Prothero. Personal communication, 1989.

[23] J.S. Robb. The structure of the mammalian ventricle. *Med. Woman's J.*, 41:65–72, 1934.

[24] T.F. Robinson, L. Cohen-Gould, and S.M. Factor. Skeletal network of mammalian heart muscle. *Lab. Invest.*, 49:482–498, 1983.

[25] T.F. Robinson, S.M. Factor, J.M. Capasso, B.A. Wittenberg, O.O. Blumenfeld, and S. Seifer. Morphology, composition and function of struts between cardiac myocytes of rat and hamster. *Cell Tissue Res.*, 249:247–255, 1987.

[26] D.D. Streeter. Gross morphology and fiber geometry in the heart. In R.M. Berne, editor, *Handbook of Physiology, Vol. 1: The Heart. Section 2: The Cardiovascular System*, pages 61–112, American Physiological Society, Bethesda, MD, 1979.

[27] D.D. Streeter Jr. and W.T. Hanna. Engineering mechanics for successive states in canine left ventricular myocardium: I. Cavity and wall geometry. *Circ. Res.*, 33:639–655, 1973.

[28] C.E. Thomas. The muscular architecture of the ventricles of hog and dog hearts. *Am. J. Anat.*, 101:17–57, 1957.

[29] F. Torrent-Guasp. *The Cardiac Muscle*. Juan March Foundation, Madrid, 1973.

[30] A. Viidik. Functional properties of collagenous tissue. In D.A. Hall and D.S. Jackson, editors, *International Review of Connective Tissue Research*, volume 6, pages 127–215, Academic Press, New York/London, 1973.

[31] T.A. Wilson. Mechanics of the pressure-volume curve of the lungs. *Ann. of Biomed. Eng.*, 9:439–449, 1981.

[32] F.C.P. Yin. Ventricular wall stress. *Circ. Res.*, 49:829–842, 1981.

[33] F.C.P. Yin, R.K. Strumpf, P.H. Chew, and S.L. Zeger. Quantification of the mechanical properties of non-contracting myocardium. *J. Biomech.*, 20:577–589, 1987.

3

Toward a Stress Analysis in the Heart

Jay Humphrey[1]
Robert Strumpf[2]
Henry Halperin[2]
Frank Yin[3]

ABSTRACT In this chapter, we briefly discuss the continuum mechanics approach to determining biomechanical constitutive relations and performing stress analyses, with particular emphasis on applications to the non-contracting heart. Examples taken from our own work illustrate possible avenues toward the eventual goal of estimating mechanical stresses in the heart and using this information to understand better certain aspects of cardiac mechanics.

3.1   Introduction

A recurring theme in this book is that mechanical stresses are important determinants of various aspects of cardiac physiology and pathophysiology. Moreover, since these stresses cannot be measured directly, there is a need for a reliable technique for estimating the state of stress in both normal and pathologic hearts. One such approach is to calculate these stresses using the classical balance relations of continuum mechanics and knowledge of the anatomy, physiology, deformations, boundary (loading) conditions and material properties of the heart. The fundamental relations of continuum mechanics are well known and can be found in many monographs [23]. The gross structure, geometry, and physiology of the heart are also well documented [2,22]. Data on the complex three-dimensional finite deformations experienced by the heart remain incomplete, but significant advances have been recently reported [24]. Information on the boundary conditions rele-

---

[1] Department of Mechanical Engineering, University of Maryland, Baltimore, Maryland 21228

[2] Department of Medicine, Cardiology Division, The Johns Hopkins Medical Institutions, Baltimore, Maryland 21205

[3] Departments of Medicine and Biomedical Engineering, The Johns Hopkins Medical Institutions, Baltimore, Maryland 21205

vant to the heart is also less than complete. For example, the effects of the parietal pericardium, great vessels, and lungs on the behavior of the heart are not known with certainty. Perhaps the least known of all, however, is a rigorous description of the mechanical behavior of heart tissue. Thus, quantification of the material behavior of cardiac tissue remains essential to the continued development of cardiac mechanics and physiology.

There are two basic approaches for identifying constitutive relations. First is a microstructural approach, wherein one formulates a constitutive relation based on the material properties of the individual constituents that comprise the tissue and knowledge of the architecture of and interactions between the constituents. This approach has been advocated by Lanir and his colleagues, and its application to describe myocardial behavior is discussed in Chapter 2 by Horowitz. The second approach is the phenomenological approach, wherein one develops a constitutive relation based primarily on observations of gross material behavior. Various applications of this approach are described elsewhere in this text (Chapter 1 by Smaill and Hunter and Chapter 5 by McCulloch and Omens).

In this chapter, we describe a general methodology for developing a phenomenological constitutive relation. For illustrative purposes, we outline both the development of Hooke's law for linearly elastic materials and our own approach to finding a constitutive relation for noncontracting myocardium. The latter is based partly on experimental findings that have been reported over the years by different investigators. Once a constitutive relation is known, one can then begin to calculate stresses using the methods of continuum mechanics. Therefore, we conclude by discussing two general approaches that have been used to calculate stresses in the heart. Again, we illustrate one of these approaches by presenting some of our previous work. Because of space limitations, our discussion of each topic is brief and the reader is referred to the original papers for details.

3.2   Quantifying Material Properties

3.2.1   GENERAL APPROACH

There are four basic steps in a continuum mechanics approach to determining a mechanical constitutive relation. First is identification of the general characteristics of the material [6]. That is, one must determine whether the material is a solid, a fluid, or a mixture. If the material is a solid, for example, one must then determine whether it behaves elastically, viscoelastically, or plastically. Similarly, one must ascertain whether the material is homogeneous or heterogeneous in composition, compressible or incompressible, isotropic or anisotropic, or if its behavior is linear or nonlinear. These and many similar questions should be answered definitively, usually by performing simple experiments, before one moves on to the second step.

The second step is establishment of a theoretical framework for the purposes of developing appropriate constitutive relations. For example, the principle of material frame indifference and restrictions associated with material symmetries and kinematic constraints (e.g., incompressibility) must be respected. Theories are important because they restrict the general functional forms of the constitutive relations and thereby reduce the scope of the required experimental exploration of the material properties. Moreover, theories are essential in guiding the experimentalist in identifying appropriate experiments to perform and important quantities to measure, as well as providing a framework for interpreting the collected data.

The third step is identification of a specific functional form of the constitutive relation. This can be accomplished either experimentally or theoretically on the basis of the information learned in steps one and two. Examples of this are the Rivlin-Saunders and Neo-Hookean relations for rubberlike materials that were determined from experimental data and statistical long-chain molecular theories, respectively (see Atkin and Fox [1]). Many functional forms of phenomenological relations are also postulated in an ad hoc fashion, and their appropriateness checked by comparison of theory and data.

Fourth is the determination of the particular values of the material parameters and testing the predictive capability of the final constitutive relation. Most parameter values come from experimental data, although theoretical arguments can be used to determine some values directly, or at least to define limits on the magnitudes or the signs of the parameters. Although it is seldom done in biomechanics, one should always explore and respect theoretical restrictions on the values of the parameters. The predictive capability of a final constitutive relation can be checked theoretically by evaluating whether physically realistic behavior results. The final test of any constitutive relation, however, depends on the agreement between theory and a wide variety of experimental data.

These four steps provide one with a guide for quantifying the mechanical behavior of many materials, but because of the complexity of most materials, and in particular cardiac tissue, this process must be iterative. That is, it is often necessary to repeat certain steps in order to refine iteratively the constitutive formulation until the final results are acceptable.

3.2.2   EXAMPLE FROM LINEAR ELASTICITY

One of the most useful and well accepted constitutive relations in solid mechanics is the generalized Hooke's law for linearly elastic materials undergoing small strains and rotations. This relation describes the behavior of many metals and alloys extremely well, and has been known for well over a century. Because of its proven usefulness, general simplicity, and wide recognition, Hooke's law has also been employed in soft tissue biomechanics. Hooke's law was derived for materials very different from soft tissues, however, and therefore is not appropriate in general for characterizing the behavior of most tissues. Nonetheless, we discuss it here to illustrate how the above steps can be used to determine a constitutive relation.

Step 1: Many *preliminary experiments* have revealed that most engineering metals are essentially isotropic, compressible, homogeneous solids that exhibit a linearly elastic stress-strain behavior as long as the strains are small.

Step 2: It can be shown theoretically that the various definitions of stress and strain all reduce to single measures provided that both the strains and rotations are small. Moreover, in the absence of any kinematic constraints, the state of stress can be determined completely from the deformation, or strain. Thus, based on the observations in step 1, an appropriate constitutive relation for engineering metals can be formulated by writing the stress ($\tilde{\sigma}$) directly as a function of the strain ($\tilde{\epsilon}$).

Step 3: For example, we can postulate that stress is a polynomial function of strain, such as

$$\tilde{\sigma} = f_0 \tilde{I} + f_1 \tilde{\epsilon} + f_2 \tilde{\epsilon}^2 + \cdots . \tag{3.1}$$

To maintain the equation dimensionally correct, the f_i $(i = 1, 2, \ldots)$ are at most scalar functions of the material parameters and strains. Since the observed stress-strain behavior is linear, however, only terms up to and including linear terms in strain are needed (i.e., $f_i = 0$ for all $i = 2, 3, \ldots$). Moreover, since the behavior is isotropic, the functions f can depend on the strains only through the coordinate invariants (J_1, J_2, J_3). This latter observation actually results from theoretical constructs in Step 2.

In order to keep the stress-strain relation linear, f_0 can be a function of J_1 at most and f_1 must be a constant at most,

$$\tilde{\sigma} = (c + J_1 \lambda)\tilde{I} + 2\mu\tilde{\epsilon} \tag{3.2}$$

where c, λ, and μ are constants. The stress must be zero at zero strain, however, and thus $c = 0$; this is a simple example of how theoretical restrictions can be used to determine the value of a material parameter. The final constitutive relation is

$$\tilde{\sigma} = \lambda J_1 \tilde{I} + 2\mu\tilde{\epsilon} \quad (\text{or, } \sigma_{ij} = \lambda\epsilon_{kk}\delta_{ij} + 2\mu\epsilon_{ij}). \tag{3.3}$$

Only two constants are needed, therefore, to quantify the material behavior of most metals and alloys. The more familiar Young's modulus, E, and Poisson's ratio, ν, can be expressed in terms of these two constants (Step 2), namely,

$$E = \frac{\mu(3\lambda + 2\mu)}{\lambda + \mu}, \quad \nu = \frac{\lambda}{2(\lambda + \mu)}. \tag{3.4}$$

Note that Equation (3.3) describes a general class of materials, not just a single material, and that only very simple preliminary experiments coupled with theoretical developments were needed to arrive at this conclusion.

Step 4: The particular values of the two material parameters must be determined from well-designed experiments for each of the individual materials defined by Equation (3.3). It is easy to show (theoretically, Step 2) that a simple uniaxial tension test yields, in a central gage length, a one-dimensional state of stress and a three-dimensional state of strain, wherein the actual values of stress and strain are both well represented by experimentally measured values (i.e., the fields are homogeneous within the measurement area). Consequently, by measuring the *uniaxial stress* as applied force (P) per cross-sectional area (A) of the specimen in the gage length and two of the three extensional strains, the values of the constants can be determined directly from experimentally measurable quantities, namely,

$$E = \frac{\frac{P}{A}}{\epsilon_{\text{axial}}}, \quad \nu = -\frac{\epsilon_{\text{transverse}}}{\epsilon_{\text{axial}}}. \tag{3.5}$$

Although other experimental set-ups could be used to determine the values of the parameters, they may or may not yield as reliable and easily measured data. It is for this reason that experiments should be well thought out and not performed simply because of experimental convenience. Note, too, that if the material were incompressible, measurement and theory would both reveal that $\nu = 0.5$. This is another example of how theory can be used to determine a particular value of a material parameter and thereby reduce the number of experiments or measurements that are needed.

The predictive capability of the final constitutive relation [Equation (3.3) with the particular values of the material parameters] can and must be checked by performing additional experiments (e.g., torsion, bending, etc.). This requires that theoretical solutions (Step 2) of the experimental situations are available so that the experiments can be interpreted properly. We see, therefore, that theory and experiment must progress iteratively, and in harmony, for constitutive relations to be developed rigorously. Although Hooke's law does not strictly apply to any biological soft tissue, there are many methodological lessons to be learned from the rich history of theoretical and experimental linear elasticity, and the interested reader is encouraged to explore this field in some depth.

3.3 Characteristics of Cardiac Tissue

3.3.1 STRUCTURAL INFORMATION

Knowledge of the morphology of the heart can be very useful in accomplishing Step 1 above. Information on whether cardiac tissue might be homogeneous or isotropic can be inferred from gross and microscopic examinations of the tissue structure. For example, we know that most of the heart wall consists of myocardium, but that delimiting connective tissue membranes are present on the inner and outer surfaces (endocardium and epicardium, respectively). Moreover, we know that there exists a connective tissue "ring" about the base of the heart where the valves are located. This information alone reveals that the heart is a *composite structure*, and therefore is probably regionally heterogeneous in material properties and behavior.

Microscopic observations further reveal that myocardium consists of locally parallel muscle fibers, a complex vascular network, and a dense plexus of connective tissue. The myocardial collagen, albeit nearly randomly oriented, is organized into two major groups: collagen struts that provide myocyte-to-myocyte and myocyte-to-capillary connections, and a collagen weave that encloses the muscle fibers and runs along their axes. These data further reveal that myocardium is a *composite material*, with a seemingly preferred direction defined by muscle-fiber orientations that vary throughout the heart wall. That is, myocardium is unlikely to behave isotropically. Micrographs from various regions in the heart show a similar ultrastructure, thereby suggesting that myocardium may be locally homogeneous while regionally heterogeneous. The interested reader is referred to Streeter [22], Borg and Caulfield [3], Robinson [20], and Chapter 1 herein for detailed structural and geometric information. Our main message here is simply that qualitative and quantitative morphological information is vital to the development of a constitutive relation (Step 1).

Historically, there have been two phenomenological approaches for studying the mechanical properties of cardiac tissue. First, pressure-volume measurements of the intact heart, isolated ventricles, or atria have been used extensively to study the overall response of the heart. Second, stress-strain experiments on excised tissues have been used to identify characteristic behavior and the associated constitutive relations.

3.3.2   PRESSURE-VOLUME MEASUREMENTS

Quantifying the global mechanical behavior of whole ventricles, atria, or hearts via simultaneous measurements of cavity pressure and volume has been an extremely useful tool in cardiac physiology and mechanics for many years. In particular, this method has been championed by Sagawa and his colleagues in both basic science and clinically relevant studies of the heart. The interested reader is referred to the text by Sagawa and coworkers [21] and Chapter 5 herein for details.

Pressure-volume analyses have also been used to determine the values of the material parameters within various constitutive relations. Briefly, simplifying assumptions are usually invoked about the geometry, boundary conditions, and deformations of the heart, a constitutive relation is selected, and then the appropriate balance equations are used to calculate the associated pressures and volumes. By comparing predicted and experimental values, best-fit constitutive parameters can be calculated. Although appropriate for determining global behavior and certain features of the characteristic tissue behavior, pressure-volume studies are not as useful in studying local phenomena such as regional behavior or a constitutive relation (which are local relations between field quantities). The reader is encouraged to consider the remarks of Moriarty [18] in this regard.

3.3.3 UNIAXIAL TESTS

By far the most commonly performed experiment on isolated, excised cardiac tissue is the uniaxial test. Moreover, although some tests have been performed on uniaxial strips of myocardium, most have been performed on papillary muscles and trabeculae carnae. Uniaxial data have been extremely useful in identifying general characteristics of the tissue behavior (Step 1), including both the nonlinear quasistatic material behavior of quiescent muscle and the Frank-Starling behavior of contracting tissue. The review article by Mirsky [17] should be consulted for details on uniaxial data. Owing to the anisotropy of cardiac tissue, however, uniaxial data are not sufficient for the rigorous construction of multidimensional constitutive relations.

3.3.4 BIAXIAL TESTS: MYOCARDIUM

Since most cardiac tissue is incompressible or nearly so, biaxial tests can be useful in identifying certain three-dimensional constitutive relations. The first biaxial experiments on thin excised slabs of noncontracting myocardium were performed by Demer and Yin [5]. Biaxial specimens are usually obtained by isolating the ventricular free-walls and slicing them tangential to the epicardial surface using a commercial rotary meat slicer. Thin (1–3 mm) rectangular specimens result, wherein the muscle-fiber directions remain in the plane of the tissue. The specimens are then mounted in a biaxial testing apparatus and stretched in two in-plane orthogonal directions. The biaxial force-length (stress-strain) data are collected, with the deformation measured in a central gage length to avoid end effects. The data obtained by Demer and Yin were important because they suggested that noncontracting myocardium behaved in a nonlinear, anisotropic, nearly pseudoelastic and perhaps regionally heterogeneous fashion. These observations are intuitively consistent with both structural and uniaxial data.

Yin and colleagues [29] later reported additional data from a single region in five canine hearts. These data were qualitatively similar to the Demer data, but suggested further that the muscle was consistently stiffer in the fiber direction than the orthogonal cross-fiber direction. Again, these observations are consistent with the observed ultrastructure. For additional details, the reader is referred to Humphrey and Yin [14]. Also, biaxial data collected in the lab of Smaill and Hunter are discussed in Chapter 1, whereas Humphrey and colleagues [12] recently reported new biaxial data collected under experimental protocols different from those used previously. These latter data and the associated constitutive relation are discussed below.

3.3.5   BIAXIAL TESTS: EPICARDIUM

As mentioned, myocardium is bounded on the inner and outer surfaces by connective tissue membranes. Yet, the mechanical properties of endocardium and epicardium have not been studied in much detail and little is known of their potential role in cardiac mechanics. This is somewhat surprising since the parietal pericardium has attracted a great deal of interest recently, and is now thought to play a significant role in the mechanics and physiology of the heart in certain conditions [4].

We recently performed biaxial tests on excised sheets of visceral pericardium, or epicardium, and found that it behaves mechanically similar to parietal pericardium, but different from myocardium [10,13]. That is, both atrial and ventricular epicardium appear to be initially very compliant and isotropic, but stiffen rapidly and become anisotropic near the limit of the tissues' extensibility. These data suggest, therefore, that epicardium behaves both qualitatively and quantitatively different from noncontracting myocardium, and that the heart wall should be considered to be heterogeneous in principle. Whether or not the epicardium and endocardium play a significant mechanical role in cardiac mechanics has not been determined, however, and there is a need for continued investigation in this area.

3.3.6   TRIAXIAL TESTS

Although biaxial tests are useful for studying certain characteristics of the behavior of cardiac tissue and for identifying two-dimensional or three-dimensional constitutive relations, there are inherent disadvantages. For example, a good biaxial preparation for studying the active properties of myocardium continues to elude investigators. Three-dimensional tests can be useful, but are typically difficult to design and interpret rigorously. A recent experiment reported by Halperin and colleagues [7] does provide, however, a promising first step in that direction.

Briefly, isolated, perfused interventricular canine septa were mounted in a *biaxial* experimental apparatus and subjected to orthogonal in-plane loads, while a punch indented a small portion of the top surface of the specimen in the third orthogonal direction. Simultaneously, in-plane biaxial stresses and strains were measured along with the associated indentation force and punch penetration depth. From these data, a number of interesting empirical findings were obtained. First, the indentation force-depth relationship was linear regardless of the in-plane state of stress or strain as long as the indentation was small. Second, the slope of the indentation force-depth relationship increased with increasing in-plane stress/strain. Third, this slope, called a measure of transverse stiffness, was linearly related to the in-plane stress for both contracting and noncontracting septum. Fourth, contracting and noncontracting septal behavior could be distinguished by significantly different intercepts in the transverse-stiffness versus in-plane stress data. These findings suggest, therefore, that indentation tests may be useful in distinguishing different types of cardiac material behavior. Indeed, this observation was supported by a recent analytical study of the basic mechanics of indentations superimposed on finite in-plane deformations [9], and by new (unpublished) data on tissue having varying degrees of contractility. Finally, a promising feature of indentation tests is that since they are relatively easy to perform, they may be able to be used to study regional differences in material behavior in intact hearts.

3.4   A Myocardial Constitutive Determination

Although the wealth of experimental data has increased our basic understanding of the mechanical behavior and properties of cardiac tissue, identification of well-accepted and proven constitutive relations remains an ambitious goal. Reviews of some of the previously postulated constitutive relations can be found elsewhere [14,16,17,27]. Here, we briefly review some of our recent work on the determination of a constitutive relation for noncontracting myocardium within the general framework discussed above. See Humphrey and coworkers [11,12] for details.

Step 1: We draw on the *preliminary* experimental data that are available, and assume that excised noncontracting myocardium is nonlinearly pseudoelastic, locally homogeneous but regionally heterogeneous, incompressible, and locally transversely isotropic. We also assume that the preferred direction coincides with the muscle fiber orientations and that the tissue experiences large (finite) deformations.

Step 2: We borrow from the large body of knowledge that exists on the theory of large elastic deformations. We find that a pseudostrain-energy function (W) for a general class of incompressible, transversely isotropic materials undergoing large deformations has been shown to be a function of at most four coordinate invariant measures of the deformation and an unspecified number of material parameters. Since it is extremely difficult to determine a specific functional form of such a W, however, we defined a subclass of these materials by,

$$W = W(I, \alpha) \tag{3.6}$$

where I_1 is the first principal invariant of the Cauchy-Green deformation tensors and α is the stretch ratio in the preferred direction. The general constitutive relation is, therefore,

$$\tilde{t} = -p\tilde{I} + 2W_1\tilde{B} + \left(\frac{W_\alpha}{\alpha}\right)\tilde{F} \cdot \mathbf{N} \otimes \mathbf{N} \cdot \tilde{F}^T \tag{3.7}$$

where $\tilde{t}$ is the Cauchy stress, p is a Lagrange multiplier enforcing incompressibility, $\tilde{B}$ is the left Cauchy-Green deformation tensor, F is the deformation gradient, N is a unit vector defining the undeformed muscle-fiber direction, and

$$W_1 = \frac{\partial W}{\partial I_1} \quad \text{and} \quad W_\alpha = \frac{\partial W}{\partial \alpha}. \tag{3.8}$$

Step 3: The specific functional form of the phenomenological constitutive relation can be either chosen in an ad hoc fashion or determined from well-designed experiments. We prefer, when possible, to find the functional form of W directly from experimental data rather than making an ad hoc selection. Based on theoretical considerations, we found that convenient experiments for determining the specific functional form of a W consist of biaxial tests on thin (2–3 mm) slabs of myocardium excised from the midwall of the left ventricular free wall, tangential to the epicardial surface. Midwall specimens contain a single predominant muscle-fiber direction that does not vary much through the thickness of the slab. Moreover, the tests should be performed such that there are minimal in-plane shearing strains, and the invariants (I_1, α) are separately maintained at constant values. Finally, we placed the specimen in the device such that the muscle-fiber direction of the specimen coincided with a stretching axis. For such an experimental situation,

$$N = (1,0,0) \quad \text{and} \quad \tilde{F} = \text{diag}[\lambda_1, \lambda_2, \lambda_3] \tag{3.9}$$

where λ_i $(i = 1,2,3)$ are stretch ratios, and the in-plane normal components of the Cauchy stress are

$$t_{11} = 2W_1(\lambda_1^2 - \lambda_3^2) + \lambda_1 W_\alpha \tag{3.10}$$

$$t_{22} = 2W_1(\lambda_2^2 - \lambda_3^2) \tag{3.11}$$

with $\lambda_3 = 1/(\lambda_1 \lambda_2)$. With two equations, we can solve for W_1 and W_α, namely,

$$W_1 = \left[\frac{t_{22}}{2(\lambda_2^2 - \lambda_3^2)} \right] \tag{3.12}$$

$$W_\alpha = \left[\frac{t_{11} - t_{22}(\lambda_1^2 - \lambda_3^2)}{\lambda_2^2 - \lambda_3^2} \right] \left(\frac{1}{\lambda_1} \right). \tag{3.13}$$

That is, the partial derivatives of the requisite pseudostrain-energy function can be calculated as a function of the deformation directly from experimentally measurable quantities [i.e., right hand sides of Equations (3.12) and (3.13)]. Thus, the characteristics of the specific functional form of W can be determined directly from data, without the need to assume the form in a completely ad hoc fashion. Moreover, by performing experiments wherein each of the two invariants, I_i and α, are maintained separately at constant values during the different protocols, the functional form of W is motivated immediately.

Based on our experimental data, we found the following characteristics of the partial derivatives of W with respect to the invariants: that W increases linearly with increasing I_i but decreases nearly linearly with increasing α, and W_α increases nonlinearly with increasing α but decreases nearly linearly with increasing I_i. This suggested, therefore, that a polynomial form of W should be studied. Utilizing requirements that, for example, $W = 0$ at zero strain and $\tilde{t} = \tilde{0}$ at zero strain provided additional restrictions on the polynomial form of W. We concluded, therefore, that

$$W = C_1(\alpha - 1)^2 + C_2(\alpha - 1)^3 + C_3(I_1 - 3) + C_4(I_1 - 3)(\alpha - 1) + C_5(I_1 - 3)^2$$
$$(3.14)$$

was consistent with the qualitative behavior of the myocardial tissues studied; C_i ($i = 1, 2, 3, 4, 5$) are material parameters.

Note: The functional form of Hooke's law was based entirely on theoretical constructs, whereas the functional form of Equation (3.14) for noncontracting myocardium was developed based on theoretically motivated experimental data. Owing to the linearity associated with Hooke's law, it is a unique constitutive relation for a large class of materials. In contrast, because of the nonlinearities and complex anisotropy, myocardial constitutive relations will not be unique. Rather, one must seek relations that are sufficiently *useful*, not unique. Finally, uniaxial data yield one equation for stress and therefore are insufficient for solving for the two partial derivatives of W [see Equations (3.10)–(3.13)] and thus determining a functional form of W from data.

Step 4: Finally, the specific values of the five material parameters contained in Equation (3.14) were determined using a nonlinear regression [11]. That is, the best-fit parameters were determined by minimizing the difference between theoretically predicted and experimentally inferred biaxial stresses in a nonlinear least-squares sense. We found that the associated fit to data was quite reasonable, and that the predictive capability was better than we have obtained previously. Again, the reader is referred to the original papers for further details, and to Yin and colleagues [28] for details on nonlinear regression and nonparametric statistical analysis of the parameter values.

As a final comment, we emphasize that it is difficult to impossible to determine rigorously a specific functional form of a constitutive relation from in vivo data. Rather, well-defined and controlled in vitro experiments are necessary for this aspect of the constitutive formulation. Determination of the values of the material parameters can be accomplished using in vitro, in situ, or in vivo data, however, provided that theoretical solutions of the experimental set-ups are available. This is an important observation since the particular values of the material parameters determined from, for example, data collected from excised potassium-arrested myocardium may not be representative of any physiologic state. Alternatively, it is possible, and hoped, that the functional form is valid for both physiologic and in

vitro states. Thus, the functional form of physiologically relevant constitutive relations should be determined from in vitro data and the values
of the parameters from apppropriate in situ or in vivo data. The material
parameters determined from in vitro data remain useful, however, because
they can be used to check rigorously the constitutive relation in the in
vitro setting and can be used as initial guesses in the nonlinear parameter
estimations based on the physiologic data.

3.5 Stress Analysis

3.5.1 ANALYTICAL APPROACHES

Since Woods ([25], see Yin [27]) first employed LaPlace's relation to estimate the average stresses in the wall of the heart, analytical approaches
employing simplified geometries and constitutive relations have been extremely useful in advancing our knowledge of the mechanics of the heart.
Many of these works are reviewed in detail in Huisman and colleagues [8]
and Yin [27]. Analytical solutions promise to continue to provide insight
into many areas of cardiac mechanics and should not be forsaken. An example of our own analytical work is described below.

An Example

Humphrey and Yin [15] recently proposed a new analytical approach for
calculating transmural distributions of stress in a noncontracting left ventricle. Briefly, we assumed that (1) the equatorial region of a noncontracting
left ventricle is nearly cylindrical in shape, (2) the wall is thick, (3) the muscle fiber directions lie in constant radial surfaces but vary in orientation as
a function of radius, (4) the myocardium is locally transversely isotropic
with the preferred directions coinciding with the muscle-fiber directions, (5)
the right ventricular pressures and pericardial influence are the same order
of magnitude during diastole, and therefore can be approximated by a single nonzero external pressure, and (6) the deformations in the wall can be
measured experimentally. Clearly, this is a gross simplification of the actual
heart, and potentially important effects such as residual stresses, the separate properties of the endocardium and epicardium, and regional variations
in myocardial properties and boundary conditions were not included (primarily due to a current lack of complete data). The problem formulation
is reasonably general, however, so that these effects can be incorporated as
they become better known, and despite the many simplifications, various
aspects of myocardial mechanics can be studied.

The main feature of our approach is employment of the semi-inverse
approach of finite elasticity, wherein the deformations and known loading
conditions are specified and the stresses determined by using the constitu-

tive relation and satisfying the equilibrium equations. Waldman and colleagues [24] have shown that the deformations in the heart wall are finite, fully three-dimensional, and very complex throughout the cardiac cycle. In particular, their data suggest that even in diastole, the heart wall inflates, extends, twists, and experiences transmural shearing. A possible description of this complex deformation is

$$r = r(R), \quad \theta = \Theta + \psi Z + \omega(R), \quad z = \Lambda Z + w(R) \qquad (3.15)$$

where (R, Θ, Z) are locations of material particles in an unloaded configuration that become (r, θ, z) after deformation. Moreover, ψ is a twist per unloaded length, Λ an axial stretch ratio, and ω and w are radially dependent parts of the circumferential and axial displacements, respectively. The physical components of the deformation gradient, and therefore various strain tensors, are easily determined. For example, the components of the Green's strain are

$$E_{RR} = \frac{r'^2 + (r\omega')^2 + w'^2 - 1}{2}$$

$$E_{\Theta\Theta} = \frac{\left(\frac{r}{R}\right)^2 - 1}{2}$$

$$E_{ZZ} = \frac{(r\psi)^2 + \Lambda^2 - 1}{2}$$

$$E_{R\Theta} = \frac{(r\omega')\left(\frac{r}{R}\right)}{2}$$

$$E_{RZ} = \frac{(r\omega')(r\psi)}{2}$$

$$E_{\Theta Z} = \frac{\left(\frac{r}{R}\right)(r\psi)}{2}. \qquad (3.16)$$

Thus, if the components of the Green's strain are experimentally measurable at a single radial location in the wall, then we can calculate the various deformation parameters in Equation (3.15). That is,

$$r = R\sqrt{1 + 2E_{\Theta\Theta}}$$

$$\psi = \left(\frac{R}{r^2}\right)(2E_{\Theta Z})$$

$$\Lambda = \sqrt{1 + 2E_{ZZ} - (r\psi)^2}$$

$$\omega' = \left(\frac{R}{r^2}\right)(2E_{R\Theta})$$

$$w' = \frac{2E_{RZ} - (r\omega')(r\psi)}{\Lambda}. \qquad (3.17)$$

Since ψ and Λ do not vary with radial location, these equations yield their values at all radii. Similarly, $r' = dr/dR$ is easily evaluated at any radial location using the incompressibility constraint. Finally, using the values of ω' and w' at a single radial location from Equation (3.17), we can determine the remaining values from the circumferential and axial equilibrium equations. When integrated, these equations become nonlinear coupled algebraic equations, in general, which can be decoupled and solved using a Newton-Raphson method. Finally, the radial equilibrium equation provides information on the Lagrange multiplier in the constitutive relation Equation (3.7). Given this information, all six components of the Cauchy stress can be calculated at each radial location in the wall. Details and numerical examples are in Humphrey and Yin [15].

We emphasize, however, that because the many simplifying assumptions, one must be cautious in interpreting the physiological implications of stresses or strains using this or other current approaches. Rather, the current utility of this method lies in the ease of performing parametric studies to investigate the potential importance of various assumptions. For example, this analysis can be used to perform error analyses to suggest the necessary accuracy of experimental measurements of, for example, ventricular geometry or deformations. Finally, analytical solutions can be very useful in checking finite element analyses.

3.5.2   FINITE ELEMENT METHODS

Owing to the complexities inherent to the heart (material behavior, geometry, ultrastructure, etc.), the finite element method [30] is probably the best available tool for estimating cardiac stresses. A detailed review of many previous finite element analyses of the heart are in Yin [26], Hunter and Smaill [16] and Chapter 6 of this book. Yet, one must be cautious when using or interpreting (e.g., assigning any physiological significance) current finite element predictions because of the lack of data. That is, regardless of the computational sophistication, the results are only as good as the reference configurations, geometry, material properties, and boundary conditions that are used in the analysis. For example, (1) the true stress-free state of the heart is only now being studied [19], (2) possible regional variations in myocardial material properties are suspected, but have not been verified or quantified, (3) the material properties of the endocardium and epicardium have not been quantified, (4) the precise boundary conditions are still not known, and the importance of the two-phased nature of the myocardium is not well known (see Chapter 5). Thus, although we must continue to develop finite element programs for performing stress analyses in the heart, we must not be too zealous in overinterpreting the physiologic implications of the associated calculations until the necessary data are available.

3.6 Conclusions

Although tremendous strides have been made toward a better understanding of the basic mechanics of the heart, there remains a grave need for new theoretical frameworks, the associated experimental data, and a coupling of the mechanical and electrical aspects of the heart so that we will be able to exploit fully the power of the finite element method for calculating stresses and interpreting their physiological significance in the normal and pathologic heart.

Acknowledgements: Partial financial support of our previous and ongoing work from the American Heart Association and the National Institutes of Health (NHLBI) is gratefully acknowledged. Also, we thank Mr. John Downs and Ms. Donna McCready for their technical expertise and dedication to our laboratory investigations. Finally, our thanks to Dr. Peter Hunter and Dr. Leon Glass for organizing this important interdisciplinary workshop on the theory of the heart.

REFERENCES

[1] R.J. Atkin and N. Fox. *An Introduction to the Theory of Elasticity.* Longman Press, London, 1980.

[2] R.M. Berne and M.N. Levy. *Cardiovascular Physiology.* The C.V. Mosby Co., St. Louis, MO, 1986.

[3] T.K. Borg and J.B. Caulfield. Collagen in the heart. *Tex. Rep. Biol. Med.*, 39:321–333, 1979.

[4] P.H. Chew, F.C.P. Yin, and S.L. Zeger. Biaxial stress-strain properties of canine pericardium. *J. Mol. Cell. Cardiol.*, 18:567–578, 1986.

[5] L.L. Demer and F.C.P. Yin. Passive biaxial mechanical properties of isolated canine myocardium. *J. Physiol. (London)*, 339:615–630, 1983.

[6] A.E. Green and J.E. Adkins. *Large Elastic Deformations.* Oxford University Press, Oxford, 1970.

[7] H.R. Halperin, P.H. Chew, M.L. Weisfeldt, K. Sagawa, J.D. Humphrey, and F.C.P. Yin. Transverse stiffness: A method for estimation of myocardial wall stress. *Circ. Res.*, 61:695–703, 1987.

[8] R.M. Huisman, P. Sipkema, N. Westerhof, and G. Elzinga. Comparison of models used to calculate left ventricular wall force. *Med. Biol. Eng. Comput.*, 18:133–144, 1980.

[9] J.D. Humphrey, H.R. Halperin, and F.C.P. Yin. Small indentation superimposed on a finite equibiaxial stretch: Implications to cardiac mechanics. *ASME J. Appl. Mech.*, 1990. (submitted).

[10] J.D. Humphrey, R.K. Strumpf, and F.C.P. Yin. Biaxial mechanical behavior of excised ventricular epicardium. *Am. J. Physiol.*, 259:H101–H108, 1990.

[11] J.D. Humphrey, R.K. Strumpf, and F.C.P. Yin. Determination of a constitutive relation for passive myocardium: II. Parameter identification. *ASME J. Biomech. Eng.*, 112:340–346, 1990.

[12] J.D. Humphrey, R.K. Strumpf, and F.C.P. Yin. Determination of a constitutive relation for passive myocardium: I. A new functional form. *ASME J. Biomech. Eng.*, 112:333–339, 1990.

[13] J.D. Humphrey and F.C.P. Yin. Biaxial mechanical behavior of excised epicardium. *ASME J. Biomech. Eng.*, 110:349–351, 1988.

[14] J.D. Humphrey and F.C.P. Yin. Biaxial mechanical properties of passive myocardium. In A. Yettram, editor, *Material Properties and Stress Analysis in Biomechanics*. Manchester University Press, Manchester, England, 1989.

[15] J.D. Humphrey and F.C.P. Yin. Constitutive relations and finite deformations of passive cardiac tissue: II. Stress analysis in the left ventricle. *Circ. Res.*, 65:805–817, 1989.

[16] P.J. Hunter and B.H. Smaill. The analysis of cardiac function: A continuum approach. *Prog. Biophys. Mol. Biol.*, 52:101–164, 1988.

[17] I. Mirsky. Assessment of passive elastic stiffness of cardiac muscle: Mathematical concepts, physiologic and clinical considerations, directions of future research. *Prog. Cardiovasc. Dis.*, 18:277–308, 1976.

[18] T. Moriarty. The law of Laplace: Its limitations as a relation for diastolic pressure, volume or wall stress of the left ventricle. *Circ. Res.*, 46:321–331, 1980.

[19] J.H. Omens and Y.C. Fung. Residual stress in the left ventricle. In A.H. Erdman, editor, *1987 ASME Advances in Bioengineering*. New York, 1987.

[20] T.F. Robinson. The physiological relationship between connective tissue and contractile elements in heart muscle. *Einstein Q.*, 1:121–127, 1983.

[21] K. Sagawa, L. Maughan, H. Suga, and K. Sunagawa. *Cardiac Contraction and the Pressure-Volume Relationship*. Oxford University Press, New York, 1988.

[22] D.D. Streeter. *Handbook of Physiology*, Volume 1. American Physiological Society, Bethesda, MD, 1979. Section 2.

[23] C. Truesdell and W. Noll. The nonlinear field theories of mechanics. In S. Flugge, editor, *Handbuch der Physik*, Volume III, Springer-Verlag, Berlin, 1965.

[24] L.K. Waldman, Y.C. Fung, and J.W. Covell. Transmural myocardial deformation in the canine left ventricle. *Circ. Res.*, 57:152–163, 1985.

[25] R.H. Woods. A few applications of a physical theorem to membranes in the human body in the state of tension. *J. Anat. Physiol.*, 26:262–270, 1892.

[26] F.C.P. Yin. Applications of the finite-element method to ventricular mechanics. *CRC Crit. Rev. Biomed. Eng.*, 12:311–342, 1985.

[27] F.C.P. Yin. Ventricular wall stress. *Circ. Res.*, 49:829–842, 1981.

[28] F.C.P. Yin, P.H. Chew, and S.L. Zeger. An approach to quantification of biaxial tissue stress-strain data. *J. Biomech.*, 19:27–37, 1986.

[29] F.C.P. Yin, R.K. Strumpf, P.H. Chew, and S.L. Zeger. Quantification of the mechanical properties of noncontracting canine myocardium under simultaneous biaxial loading. *J. Biomech.*, 20:577–589, 1987.

[30] O.C. Zienkiewicz. *The Finite Element Method*. McGraw Hill, New York, 1979.

4

Identification of the Time-Varying Properties of the Heart

Poul Nielsen[1]
Ian Hunter[1]

ABSTRACT The mechanical behavior of both skeletal and cardiac muscle exhibits nonlinear time-varying properties. These nonlinearities and time variations present major difficulties to researchers attempting to identify a constitutive law for muscle to use in continuum models. Traditionally, quasi-static identification methods have been used on tetaninized skeletal muscle to determine its mechanical properties at various levels of activation. Such approaches are unsuitable for cardiac muscle where tetanic contractions are difficult or impossible to achieve. In this case one must resort to nonlinear time-varying identification methods. Here we present an identification scheme, together with the associated experimental apparatus, suitable for characterizing muscle models under time-varying conditions. Although this technique is used to identify the time-varying mechanical properties of isolated skeletal muscle throughout the time course of a single twitch, this approach is directly applicable to characterizing the constitutive properties of cardiac muscle.

4.1 Introduction

A detailed understanding of the heart requires knowledge of its functional constituents at many levels. In order to determine the relationships that exist between these constituents, and their effects on the performance of the heart as a whole, they must be placed within a common framework. Continuum modeling provides such a unifying framework allowing one to investigate the complicated interplay between mechanical, electrical, thermal, and chemical effects over a wide range of scales, from groups of cells up to the level of the entire organ [1,7,13]. Continuum models may be divided into three general components: First, the geometry and kinematics, which deal with the mathematical description of the deformation and motion of

---

[1]Biomedical Engineering, McGill University, Montréal, Québec, H3A 2B4

the body; second, the balance laws, which include conservation of linear and angular momentum, mass, and energy; and finally, the constitutive relations, which give a functional form to the stress tensor, free energy, and heat flux in terms of the deformation, temperature, and state of tissue activation. It is clear [10] that any realistic model of the heart that attempts to examine regional variations of myocardial behavior must include accurate descriptions of the large strains, irregular geometry, and the constitutive properties of cardiac tissue.

Although the theory of large deformations has been understood for some time [2], and detailed measurements of cardiac architecture have been made [11], the problem of obtaining accurate descriptions of the constitutive properties of myocardium remains. This is due, first, to the fact that cardiac muscle cannot be tetanized and, second, because there are few sites where uniform manageable portions of cardiac tissue may be extracted without causing significant tissue damage.

This chapter focuses on the problem of estimating the time-varying parameters that characterize dynamic systems such as the heart. In particular, a technique is presented for identifying systems in which the characteristics of the system are changing in the same order of time as its dynamics. We present estimates, obtained using this method, of the time-varying stiffness of skeletal muscle throughout the course of a single isometric twitch. The techniques developed to analyze skeletal muscle are directly applicable to cardiac muscle. We conclude with a brief discussion about how these techniques may be applied to estimate dynamic constitutive properties of the heart.

4.2   Theory

Models used to characterize the input-output relationships of systems may be generally classified as belonging to one of two categories, depending on whether the outputs are expressed as implicit or explicit functions of the inputs. Differential equations and convolution integrals are examples of implicit and explicit models, respectively. Implicit models are appropriate where one wishes to incorporate a priori knowledge of the system structure into the model. Conversely, where the system is considered as a "black box" an explicit representation is more appropriate. A powerful class of explicit models suitable for describing general nonlinear input-output relationships are Volterra series representations. With the Volterra series, the output, $y(t)$, of a time invariant system is related to the input, $x(t)$, by

$$y(t) = h_0 + \int h_1(\tau)x(t - \tau)d\tau$$

$$+ \iint h_2(\tau_1, \tau_2)x(t - \tau_1)x(t - \tau_2)d\tau_1 d\tau_2 + \dots . \tag{4.1}$$

With such a representation the system is completely characterized by the Volterra kernels $h_0, h_1(\tau), h_2(\tau_1, \tau_2), \ldots$

In the case of a time-varying system, where the intrinsic properties of the system are not constant but depend on some parameter α, the Volterra series must be modified to incorporate the time variation. Here, the output, $y(t)$, of a time invariant system is related to the input, $x(t)$, by

$$y(t) = h_0(\alpha) + \int h_1(\alpha, \tau) x(t - \tau) d\tau$$

$$+ \iint h_2(\alpha, \tau_1, \tau_2) x(t - \tau_1) x(t - \tau_2) d\tau_1 d\tau_2 + \ldots \qquad (4.2)$$

where the Volterra kernels, $h_0(\alpha), h_1(\alpha, \tau), h_2(\alpha, \tau_1, \tau_2), \ldots$, are functions of the time-varying parameter, α. In general, determining the kernels, h_i, is a difficult problem. Where α changes slowly with respect to h_i, a piecewise approach may be taken. In this case α is assumed to be constant over the support of the Volterra kernels. Where α changes very rapidly with respect to h_i severe difficulties exist in determining the kernels from $x(t)$ and $y(t)$. However, if changes in α occur in the same order of time as h_i then techniques exist to effectively "freeze" α. Where the time variation can be repeated many times and may be initiated experimentally, although not necessarily controlled, a particularly powerful identification method becomes applicable. This method is outlined below.

Consider the case where an experimental trial is repeated many times. We denote the time within each trial by t_1. The trial number, or the time across trials, is denoted t_2. In this case the input, $x(t_1, t_2)$, and the corresponding system output, $y(t_1, t_2)$, are functions of both t_1 and t_2. Now consider the case where the input possesses a lag symmetry defined by

$$x(t_1 - t, t_2) = x(t_1, t_2 - t). \qquad (4.3)$$

Further, if the time variation, α, can be initiated at some constant value of t_1 for all values of t_2, then at any value of t_1 we have a constant for all values of t_2. The Volterra series description of our system may now be expressed as

$$y(t_1, t_2) = h_0(\alpha) + \int h_1(\alpha, \tau) x(t_1 - \tau, t_2) d\tau$$

$$+ \iint h_2(\alpha, \tau_1, \tau_2) x(t_1 - \tau_1, t_2) x(t_1 - \tau_2, t_2) d\tau_1 d\tau_2 + \ldots \quad (4.4)$$

and with the input lag symmetry

$$y(t_1, t_2) = h_0(\alpha) + \int h_1(\alpha, \tau) x(t_1, t_2 - \tau) d\tau$$

$$+ \iint h_2(\alpha, \tau_1, \tau_2) x(t_1, t_2 - \tau_1) x(t_1, t_2 - \tau_2) d\tau_1 d\tau_2 + \ldots \quad (4.5)$$

It should be noted that if t_1 is held constant, α will be constant. Since the convolution integrals are performed over t_2 the above equation may be solved at fixed values of t_1, and hence α, using conventional nonlinear time invariant identification techniques [4,8,9].

4.3   Apparatus

In this section, we provide some details of an experimental apparatus for performing dynamic mechanical tests on isolated muscle. The experimental apparatus comprises a muscle bath, a muscle length and force control unit, a sarcomere length measuring system, and a muscle temperature control unit (Figure 4.1). All of these elements are attached to a vibration isolated optical table possessing a natural oscillation frequency of 1.0 Hz.

The muscle sits within a bath of Ringer's solution measuring 22 mm × 8 mm × 2 mm. The major faces are formed by two microscope cover glasses separated at the top by a 2-mm thick aluminum heat conductor with integral temperature sensor. The remaining two sides and bottom are open to the air. Ringer's solution is held within the bath volume by capillary action of the cover slips. With this arrangement the muscle and associated on-axis probes may be inserted directly into, and removed from, the bath without any need for fluid seals. Accurate force and end-muscle position measurements may thus be made using transducers placed outside the bath.

Stiff 0.1-mm diameter stainless steel probes are hooked into both tendinous ends of the muscle. One probe is directly connected to a full-bridge diffused-beam piezoresistive semiconductor force transducer (Entran Devices ESDB4-14-350) capable of measuring forces up to 0.1 N. The position of the other probe is monitored by an optical emitter-detector assembly. A narrow-beam subminiature light emitting diode (Hewlett Packard HEMT-6000) is attached to a rigid housing. The beam from the emitter impinges on a one-dimensional lateral effect photodiode position sensing detector (SiTek Electro Optics 1L10). This arrangement provides position measurement over a range of 10 mm, with a bandwidth up to 1.75 MHz, linearity of 0.1%, and resolution better than 1 μm. The position sensing assembly is mounted on a linear electromagnetic motor (Brüel and Kjær type 4810 minishaker) to provide control over muscle length along the muscle axis. The motor and position sensor assembly has a displacement range of 6 mm, a force rating of 10 N, and a flat frequency response to 300 Hz.

The sarcomere length measuring system monitors diffraction patterns produced by a continuous 8-mW transversely polarized helium-neon laser (Uniphase 1105P) passing through a portion of the muscle. The angle of one of the first-order diffraction peaks is measured using a removeable length-calibrated screen. A measure of sarcomere length within the region of muscle illuminated by the laser may be calculated from the position of

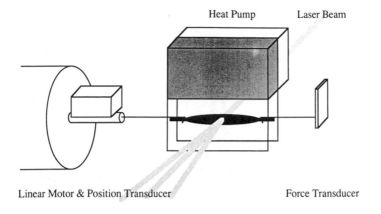

FIGURE 4.1. Muscle testing apparatus.

the first-order diffraction peak using the grating equation.

The temperature of the Ringer's solution is monitored by a miniature thermistor probe (Yellow Springs Instrument Co. YSI 44203) mounted at the top center of the muscle bath. Deviations from a preset temperature are measured through a control circuit connected to a 10-W solid state thermoelectric heat pump (Marlow Industries MI1023T-03AC). One face of the heat pump is attached to a massive aluminum heat sink maintained at ambient temperature. The other surface abuts against a small aluminum temperature conductor, one face of which forms the top of the bath and is in intimate contact with the Ringer's solution. With the above configuration bath temperature may be kept to within 0.3°C of a preset value.

All sequencing, component identification, and control, apart from maintenance of muscle temperature, are handled digitally by a laboratory computer. Muscle end position and force are monitored by the computer through A/D converters, while the linear motor is similarly controlled using D/A converters feeding into a power amplifier. Muscle stimulation is effected under computer control via small stainless steel electrodes implanted within the muscle bath. Calibration, preconditioning, and experimental procedures are established by the user through the computer via a Common Lisp environment.

4.4   Method

These experiments were designed to measure the change in mechanical dynamics of muscle throughout single isometric muscle twitches. The mechanical dynamics are largely determined both by the numbers of active cross-bridges (myosin molecule heads) as well as their dynamics, which in

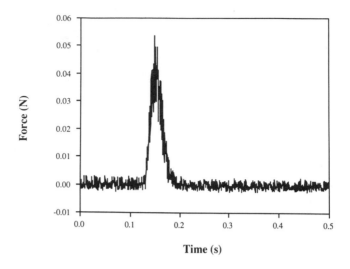

FIGURE 4.2. Muscle force versus time during a single twitch.

turn depend on the time-várying calcium ion concentration.

A frog posterior semitendinosus muscle measuring approximately 20 mm in length and 1.5 mm in diameter was mounted within the apparatus and maintained at a temperature of 20°C. One end of the muscle was subjected to an input of stochastic length perturbations generated using the technique presented by Hunter and Kearney [3]. Five hundred trials, each lasting 500 ms, were performed to build up an input ensemble with the required lag symmetry. Data acquired at 2.5 kHz from each trial corresponded to signals varying with t_1, at constant t_2. Figure 4.2 shows a typical output force response of a trial prior to, during, and following a single twitch. Notice the force fluctuations resulting from the stochastic displacement perturbations superimposed on the usual twitch response. The bath fluid surrounding the muscle was replaced with fresh oxygenated Ringer's solution after each trial in order to minimize muscle fatigue. The entire experiment of 500 trials was completed within 35 min of mounting.

4.5   Results

The length perturbation input and muscle force output data ensembles were analyzed using the method outlined above to yield instantaneous measures of dynamic muscle stiffness throughout an isometric twitch. Dynamic muscle stiffness was defined as the zero lag point on the length perturbation input/force perturbation output impulse response function. Figure 4.3 clearly shows dynamic muscle stiffness rising and then decaying during the time

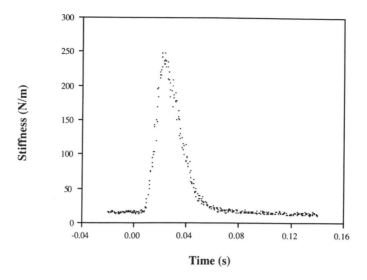

FIGURE 4.3. Dynamic muscle stiffness versus time after activation.

course of a twitch. Stiffness as a function of force is presented in Figure 4.4. Note that there is no hysteresis and that the relationship is remarkably linear. This is not an unexpected result. The dynamic muscle stiffness may be assumed to consist of both active and passive components. The passive components include stiffness contributions from all of the noncontractile elements of the muscle. These contributions are thus independent of muscle contractility, defining the zero force offset on the stiffness versus force curve. In contrast to the passive stiffness contributions, the active stiffness contributions arise from interactions between the contractile elements of the muscle. In the crossbridge model of muscle contraction calcium released from the sarcoplasmic reticulum binds to troponin which, in turn, causes the tropomyosin molecule to undergo a change in conformation, thus exposing sites on the actin chain to which crossbridges may bind. The crossbridges rotate, generating force from the hydrolysis of ATP to form ADP. The total force generated by the muscle at any instant is the sum of the forces generated by each crossbridge. Since each crossbridge acts in parallel, it contributes an increment of stiffness to the muscle. Thus, like the total force generated by a muscle, the active dynamic muscle stiffness is proportional to the total number of attached crossbridges resulting in the linear stiffness-force relationship observed in these experiments.

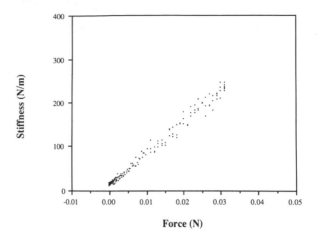

FIGURE 4.4. Dynamic muscle stiffness versus active force developed.

4.6   Discussion

The theory, apparatus, and method outlined above should be directly applicable to the estimation of instantaneous dynamic stiffness of cardiac muscle. Since the theory provides a framework in which to analyze the instantaneous properties of time-varying systems it is well suited to studying muscle where, like the myocardium, tetanic contractions are difficult or impossible to achieve. The above apparatus has been designed to perform mechanical tests on muscles in the order of 10 mm in length. The only sites within the mammalian heart where muscles of comparable size with uniform structure may be obtained are the trabeculae and papillary muscles. Unfortunately, there appear to be some structural differences between these muscles and tissue located within the ventricular wall [12]. It is possible, therefore, that the constitutive relations identified from these sites may not be representative of myocardium at other sites. Other difficulties traditionally experienced with the use of trabeculae and papillary muscles for isolated muscle experiments include nonuniform strain distributions due to end clamping, and effects of tissue damage at the cut ends. Steps may be taken to reduce these effects by monitoring strains at central regions of the samples, away from the clamps and damaged tissue. Despite these difficulties it should be possible, using the above approach, to obtain reasonable estimates of the constitutive properties of active myocardium suitable for use in continuum models of the whole heart.

At a somewhat lower level we have also been interested in determining the mechanical properties of muscle cells, independent of the extracellular connective tissue. A promising approach, which we have been pursuing, involves studying the mechanics of single isolated muscle cells using a somewhat more sophisticated experimental apparatus [5]. By combining telemicrorobotic devices with a very high resolution three-dimensional laser vision system and a dedicated parallel control supercomputer the mechanical, electrochemical, and optical properties of single active muscle cells may be investigated using adaptive identification techniques [6]. Using this technique the constitutive properties of individual muscle cells may be identified, free from the problems associated with isolated whole muscle experiments outlined above. High resolution studies such as this promise accurate characterizations of the time-varying properties of muscle.

Acknowledgements: We wish to thank the Medical Research Council of Canada and the Natural Sciences and Engineering Research Council of Canada for support of this work.

REFERENCES

[1] D.A. Bergel and P.J. Hunter. The mechanics of the heart. In D.R. Gross, N.H.C. Hwang, and D.J. Patel, editors, *Quantitative Cardiovascular Studies*, pages 151–213, University Park Press, Baltimore, 1979.

[2] A.E. Green and W. Zerna. *Theoretical Elasticity, 2nd Edition*. Oxford University Press, London, 1968.

[3] I.W. Hunter and R.E. Kearney. Generation of random sequences with jointly specified probability density and auto correlation functions. *Bio. Cybernet.*, 47:141–146, 1983.

[4] I.W. Hunter and M.J. Korenberg. The identification of nonlinear biological systems: Wiener and Hammerstein cascade models. *Bio. Cybernet.*, 55:135–144, 1986.

[5] I.W. Hunter, S. Lafontaine, P.M.F. Nielsen, and P.J. Hunter. Manipulation and dynamic mechanical testing of microscopic objects using a tele-micro-robot system. *IEEE Control Systems*, 10:3–9, 1990.

[6] P.J. Hunter and I.W. Hunter. Triaxial testing of soft tissue: Parameter estimation (preprint).

[7] P.J. Hunter and B.H. Smaill. The analysis of cardiac function: A continuum approach. *Prog. Biophys. Mol. Biol.*, 52:101–164, 1988.

[8] M.J. Korenberg. Statistical identification of volterra kernels of high order systems. *IEEE International Symposium on Circuits and Systems*, 2:570–575, 1984.

[9] M.J. Korenberg and I.W. Hunter. The identification of nonlinear biological systems: Wiener kernel approaches. *Ann. Biomed. Eng.*, 18:629–654, 1990.

[10] A.D. McCulloch. *Deformation and Stress in the Passive Heart*. PhD thesis, University of Auckland, New Zealand, 1986.

[11] P.M.F. Nielsen, I.J. LeGrice, B.H. Smaill, and P.J. Hunter. A mathematical model of the geometry and fibrous structure of the heart. *Am. J. Physiol.*, submitted.

[12] T.F. Robinson, M.A. Geraci, E.H. Sonnenblick, and S.M. Factor. Coiled perimysial fibres in papillary muscle in rat heart: Morphology, distribution and changes in configuration. *Circ. Res.*, 63:577–592, 1988.

[13] F.C.P. Yin. Ventricular wall stress. *Circ. Res.*, 49:829–842, 1981.

5

Factors Affecting the Regional Mechanics of the Diastolic Heart

Andrew D. McCulloch[1]
Jeffrey H. Omens[2]

ABSTRACT The mechanics of the diastolic heart are introduced in the context of ventricular pump function. Since systolic contraction is directly influenced by end-diastolic volume, we discuss factors that influence the diastolic pressure-volume relation. Turning to an engineering approach, the experimental investigation of regional ventricular mechanics is described, including the analysis of local stresses and strains, and the mechanical effects of coronary perfusion. The notion of residual stress is introduced and recent measurements of residual strains are given, shedding light onto the true stress-free reference state of the left ventricular myocardium. We conclude with a brief description of a recent theoretical model for predicting wall stress in the passive left ventricle, with properties such as residual stress, fiber orientation, and measured myocardial deformations incorporated into the analysis.

5.1 Introduction

The heart is a hollow muscular pump that converts metabolic energy to mechanical work in generating the pressure that drives the blood throughout the body's circulation system. The cyclic mechanical function of the heart can be appreciated by following the time-course of blood pressure and volume in the left ventricular chamber, as shown in Figure 5.1. During each cycle, the left ventricle is filled from the left atrium through the mitral valve until the deceleration of the inflowing blood reverses the pressure gradient across the valve leaflets. This causes the inlet valve to close suddenly at point A on the curves with no reflux in the normally functioning heart. At the same time, contraction of the ventricular muscle causes the pressure

---

[1] Department of Applied Mechanics and Engineering Sciences (Bioengineering), University of California, San Diego, La Jolla, California, 2093

[2] Department of Medicine (Cardiology), University of California, San Diego, La Jolla, California 92093

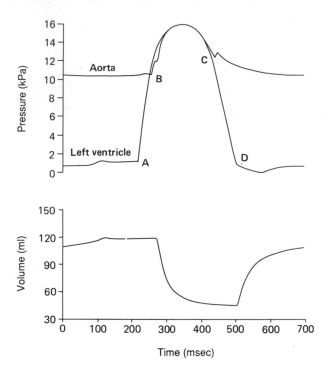

FIGURE 5.1. Left ventricular pressure, aortic pressure, and left ventricular volume during a single cardiac cycle showing the times of mitral valve closure (A), aortic valve opening (B), aortic valve closure (C), and mitral valve opening (D).

in the closed chamber to rise rapidly during the interval between points A and B (isovolumic contraction). At point B, the left ventricular pressure exceeds the aortic pressure, the aortic valve opens, and blood is ejected into the systemic circulation. The interval from B to C is the ejection phase of the cardiac cycle. The left ventricular volume tracing in Figure 5.1 shows that most of the cardiac output is ejected within the first quarter of this phase before the pressure has peaked. The aortic valve closes at point C after the ventricular pressure falls below the aortic pressure owing to the deceleration of the ejecting blood. The ventricular pressure falls during isovolumic relaxation (C to D) until the mitral valve reopens at point D owing to the pressure gradient created by the elastic recoil of the ventricular wall. The chamber refills, rapidly at first and then more slowly (diastasis), and the cycle continues. The pressure and volume curves for the right ventricle look essentially the same; however, the right ventricular and pulmonary artery pressures are much lower than the corresponding pressures on the left side of the heart. Studies of ventricular function usually focus on the left ventricle, but many of the conclusions apply equally to right ventricular function.

The various phases of the cardiac cycle are customarily separated into two major divisions: "systole" and "diastole." The end of diastole—the start of systole—is commonly defined as the time of mitral valve closure. It coincides with the R-wave peak of the QRS complex on the electrocardiogram (ECG), which corresponds to depolarization of the ventricular myocardium. However, definitions of end-systole vary among different authors. Classically, Wiggers [78] defined mechanical end-systole as the end of ejection, and this definition is usually invoked in the context of ventricular function and clinical practice. But Brutsaert and colleagues [10] have pointed out that during isovolumic relaxation and well into the ventricular filling phase there remains considerable myofilament interaction. Since the tension in the myocardium is still significantly greater than the resting tension, they proposed extending systole until the onset of diastasis. In this chapter, the distinction is important: from the point of view of muscle mechanics, the myocardium is still active for much of diastole, therefore experimental and theoretical models of the passive or diastolic arrested heart apply only to the period between the end of relaxation and the end of diastole. Thus, we will retain the traditional definition of diastole, but consider the ventricular myocardium to be "passive" only in the final slow-filling stage of diastole.

5.2   The Left Ventricular Pressure-Volume Relation

A useful alternative means of displaying the pressure and volume changes in the left ventricle is the pressure-volume loop, shown schematically in Figure 5.2, where the left ventricular pressure at each instant during the cardiac cycle is plotted against the volume at the same instant. This state-space representation of cardiac function was probably first used by Otto Frank in his famous 1898 paper on frog ventricular function [14]. However, only since the mid-1970s have the properties and value of this view been fully appreciated and investigated. The ventricular pressure-volume relationship has been explored extensively in the last decade, particularly by Sagawa, H. Suga and coworkers, who recently published a comprehensive book championing the approach and demonstrating the many useful physiological and clinical parameters that can be characterized [59]. And recently, applications of the pressure-volume relation in the clinical setting have been reviewed by Kass and Maughan [34].

The points of valve opening and closure shown in Figure 5.1 are also labeled in Figure 5.2. The isovolumic phases of the cardiac cycle can be recognized as the vertical segments of the loop. The portion from D to A represents ventricular filling, and B to C is the ejection phase. How the pressure-volume relation depends on ventricular loading conditions has been studied in a variety of preparations, but the most accurately controlled experiments have used the isolated cross-circulated canine heart. In the preparations of Sagawa and colleagues [59], the coronary circulation of a heart donor dog was perfused with arterial blood from a larger support dog. The cross-circulated heart was excised and placed over a funnel, which collected coronary venous blood from the right heart and returned it to the support dog via a venous catheter. After venting the left ventricle at the apex, an oversized water-filled balloon with a Konigsberg pressure transducer was fitted into the cavity through the left atrium. The balloon was attached to a connector that was sutured into the mitral annulus and coupled to a servopump for controlling ventricular volume. In this way, the ventricle fills and ejects through the mitral orifice against the volume-loading of the pump, which, in more recent times, is driven by a computer to simulate the dynamic loading of the circulation [65]. This preparation has allowed these workers to study the ventricular pressure-volume relation under a wide and well controlled range of input and output pressure loading conditions.

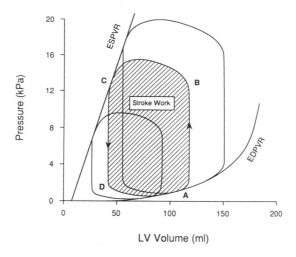

FIGURE 5.2. Schematic diagram of left ventricular pressure-volume loops, with the end-systolic pressure-volume relation (ESPVR) and the end-diastolic pressure volume relation (EDPVR) shown. The external stroke work of a single heart beat is defined by the area within the loop. The three loops show the effects of changing end-diastolic pressure (preload) and end-systolic pressure (afterload).

5.2.1   THE END-SYSTOLIC PRESSURE-VOLUME RELATION AND INDICES OF VENTRICULAR FUNCTION

If the afterload seen by the isolated left ventricle—the pressure against which it is ejecting—is increased, the volume ejected by the ventricle decreases in a predictable manner. The locus of end-ejection points (C in Figure 5.2) forms the end-systolic pressure-volume line, which is approximately linear in a wide variety of situations. Indeed, connecting points at corresponding times in the cardiac cycle results in a relatively linear relationship throughout systole with the intercept on the volume axis remaining very nearly constant. Thus, the slope of this line can be plotted as a function of time $E(t)$ through systole and reaches a peak at end-systole.

Therefore, the slope of the end-systolic pressure-volume relation (ESPVR)—denoted as E_{max}—has acquired considerable significance and has been proposed as an index of cardiac contractility. As the inotropic state of the ventricle increases, for example with catecholamine infusion, E_{max} increases, and with a negative inotropic effect such as a reduction in coronary artery pressure, it decreases. Clinically, the ESPVR has been found to be a sensitive index of functional impairment in disorders such as ischemic heart disease, and has become a useful way of quantifying the improvement in ventricular pumping produced by reperfusion techniques such as thrombolytic therapy for dissolving clots or coronary balloon angioplasty for mechanically dilating atherosclerotic coronary arteries.

Another, perhaps more fundamental, property of the ventricular pressure-volume loop is that its area is a measure of the external work done by the ventricle on the ejecting blood as it contracts. If we plot this "stroke work" of the ventricle against a suitable measure of the "preload" or filling of the ventricle, we obtain the well known "ventricular function" curve, which illustrates the single most important intrinsic mechanical property of the heart pump. In 1914, Patterson and Starling [55] performed detailed experiments on the canine heart-lung preparation, and in the famous Linacre lecture [64] Ernest Starling summarized their results with his "Law of the Heart," stating that the work of the heart increases with the ventricular volume. Therefore, the pressure-volume relation conveys not only the effect of changes in afterload, but also the increased stroke work with increased end-diastolic pressure or volume.

This so-called Frank-Starling mechanism is now well recognized to be an intrinsic mechanical property of cardiac muscle. In striated muscle, according to the sliding filament theory, the tension generated in an isometric contraction is determined by the amount of overlap of the interdigitating myofilaments and hence by the length of the basic contractile units (sarcomeres). But unlike skeletal muscle, for which there is an optimal sarcomere length for maximal active force generation, the peak twitch tension developed in isolated cardiac muscle continues to rise with increased sarcomere length in the physiological range. Early evidence for a descending limb of the cardiac muscle isometric length-tension curve has been found to be caused by shortening in the central region of the isolated muscle at the expense of stretching at the damaged ends where the specimen was tethered to the test apparatus. If the muscle length is controlled so that sarcomere length in the undamaged part of the muscle is indeed constant, or if the developed tension is plotted against the instantaneous sarcomere length rather than the muscle length, the descending limb is eliminated [67]. Thus, the increase with chamber volume of end-systolic pressure and stroke work is reflected in isolated muscle as an increase in tension development with sarcomere length or fiber stretch.

5.2.2   THE END-DIASTOLIC PRESSURE-VOLUME RELATION

The exact configuration of the left ventricle at end-diastole therefore directly determines systolic work. The end-diastolic pressure-volume relationship and end-diastolic configuration are determined by a complex interaction of various factors during the filling phase of the cardiac cycle. A recent review of these factors has been given by Gilbert and Glantz [20]. Considerable effort has been devoted to the study of the diastolic pressure-volume relation and the various clinical conditions that may cause acute or chronic alterations such as myocardial ischemia or hypertrophy [25].

The primary determinants of the diastolic ventricular filling curve are the material properties of diastolic myocardium, geometric factors such as chamber dimensions and wall thickness, and the boundary conditions at the epicardium, endocardium, and valve ring [21]. Even simple mathematical models show the direct relationship of geometry and elasticity to pressure-volume behavior. However, other factors such as active relaxation, mechanical interaction between the left and right ventricles, pericardial properties, diastolic suction, coronary vascular tone, myocardial perfusion pressure, and viscoelasticity may also play important roles. These factors affect the diastolic ventricle at different times during filling [20]. Factors that mainly influence the mechanics early in diastole, such as the rate of relaxation [79] and diastolic suction [27], probably play a minimal role in determining the end-diastolic pressure-volume relationship. Near end-diastole, important factors include the extent of relaxation, ventricular interaction and pericardial constraints, and coronary vascular engorgement. In the following paragraphs, those physiological factors that alter the end-diastolic portion of the ventricular filling relation are briefly discussed.

Incomplete Relaxation

When the energy-dependent process of cardiac muscle relaxation is complete, the extent of the relaxation will affect the end-diastolic pressure-volume relationship. Incomplete relaxation may occur at high heart rates [47] and can also be seen with clinical complications such as acute myocardial ischemia [25]. If the ventricle has not completely relaxed when filling begins, then a given influx of blood during filling will increase the end-diastolic pressure, hence an upward shift in the curve will be seen. The rate at which the relaxation process occurs does not usually affect the end-diastolic state directly because, in general, the relaxation process is complete by the time diastole ends, even if the relaxation rate has slowed [20].

Ventricular Interaction

The intraventricular septum separates the right and left ventricles and can transmit forces from one to the other. For example, an increase in right ventricular volume may increase the left ventricular pressure by deformation of the septum. This direct interaction is particularly prominent during diastole, when ventricular pressure is low [32]. Thus, the material properties and boundary conditions in the septum are important since they determine how the septum deforms [21]. Through septal interaction, the end-diastolic pressure-volume relationship of the left ventricle may be directly affected by changes in the hemodynamic loading conditions of the right ventricle. The ventricles may also interact indirectly, since the output of the right ventricle is returned as the input to the left ventricle through the pulmonary circulation. Although conflicting results have been given for the relative effects of direct and indirect interaction, Slinker and Glantz

[62] looked at changes in pressures and dimensions after pulmonary artery and venae caval occlusions for direct (immediate) and indirect (delayed) interaction transients, and concluded that the direct interaction is about half as significant as the indirect coupling. The relative contributions of direct and indirect interaction may also depend on the diastolic stiffness of the ventricles.

The Pericardium

The pericardium provides a low friction mechanical enclosure for the beating heart. It is thought that the pericardium may restrain ventricular extension [46], as well as prevent gross movement of the heart within the chest. A volume increase in one ventricle may decrease the expansion capability of the other ventricle because of the restrictive action of the pericardium. Since the pericardium has quite different material properties from the ventricles [37], direct pericardial interaction and the pericardial pressure [72] will alter the diastolic pressure-volume relationship, especially near end-diastole when the ventricular volume is greatest and the dimensions of the heart are greatest. The constraint of the pericardium also increases the mechanical coupling between the atria and ventricles. Since atrial contraction occurs during late diastole and is responsible for increased ventricular filling, the surrounding pericardium will influence the pressure-volume characteristics near end-diastole by augmenting atrial-ventricular coupling [40]. Although many investigators have shown changes in ventricular diastolic pressure-volume relationships modulated by the pericardium with acute cardiac dilation, there is probably only a small effect at normal diastolic pressures [72].

Coronary Perfusion

Increasing coronary perfusion pressure has been seen to increase the slope of the diastolic pressure-volume relation. The changes in diastolic ventricular compliance that occur with perfusion were first clearly described by Salisbury and colleagues [60] who observed that mean left ventricular end-diastolic pressure in isovolumically contracting canine hearts more than doubled when coronary artery pressure was raised from 36 to 136 mm Hg. Although they did not measure the left ventricular pressure-volume relationship directly, they concluded that increased coronary perfusion pressure caused a decrease in ventricular compliance by increasing coronary vascular volume and passively distending the elastic coronary artery tree, which then restrained the ventricular myocardium (an "erectile" effect).

In the potassium-arrested blood-perfused canine heart, Olsen and colleagues [50] measured a significant steepening in the relation between left ventricular pressure and minor-axis dimension when coronary artery pressures were increased from 40 to 120 mm Hg. However, although adenosine increased myocardial blood flow by up to 147%, it produced no significant change in the left ventricular pressure-dimension relationships at the same perfusion pressure. Therefore, these investigators concurred with Salisbury and co-workers on the mechanism and significance of the effect, but they added that coronary perfusion pressure and not flow was the most direct determinant of left ventricular diastolic properties.

An important determinant of global ventricular filling is wall thickness [20,22]. Since Morgenstern and colleagues [48] reported that increasing coronary perfusion pressure from 70 to 170 mm Hg increased intracoronary volume from a mean 11.0% to 17.8% of the myocardium, other workers have concluded that the principal mechanism of ventricular stiffening during perfusion is thickening of the ventricular wall, although they disagree on whether coronary pressure [18,19] or flow [75] is the main determinant of wall thickness. The greatest difficulty in determining the purely mechanical effects of coronary perfusion on diastolic ventricular properties is to separate these effects from the metabolic changes associated with altered coronary blood flow. Vogel and colleagues [75] measured left ventricular diastolic pressure, volume change, epicardial circumference, and wall thickness together with coronary flow rate in normoxic, hypoxic, and ischemic isolated rabbit hearts and found that there was a substantial erectile effect of coronary perfusion independent of changes in oxygenation. Bouchard and colleagues [9] found no significant changes in ATP, creatine phosphate, or inorganic phosphate levels in normal and cardiomyopathic hamster hearts freeze-clamped 10 sec after coronary perfusion pressure had been decreased from 140 to 0 cm H_2O. Total wall volume decreased acutely by 21% in the normal hearts and 15% in the cardiomyopathic hearts following the reduction in perfusion pressure.

Some investigators, however, have not seen alterations in ventricular diastolic compliance with changes in coronary perfusion [1,53,63,66]. In isovolumically contracting canine hearts, Abel and Reis [1] found no change in left ventricular end-diastolic pressure when coronary artery pressure was increased from 75 to 150 mm Hg. Templeton and colleagues [66] and Spadaro and coworkers [63] also reported no effect on the end-diastolic pressure-volume relation of alterations in coronary artery pressure and flow. However, the maximum end-diastolic pressures in these studies were only about 6 mm Hg. Therefore, whereas the erectile effect is fairly well recognized, it is probably not significant for normal physiological changes in coronary blood flow. However, it may play a role in acute myocardial ischemia [73].

5.3   Regional Ventricular Function

Although ventricular pressures and volumes have been valuable for assessing the global pumping performance of the heart as a whole, we also need suitable measures of the *regional* function of the ventricles to understand the underlying structural basis of ventricular mechanics. The need for regional information is especially important in pathological conditions, such as myocardial ischemia and infarct, where profound mechanical changes may occur in a localized portion of the myocardium without necessarily affecting the global function.

In the context of engineering mechanics, the appropriate regional indices of myocardial function are the stresses and strains in the heart wall. The measurement of stress in the intact myocardium involves resolving the local forces acting on defined planes in the heart wall. The few attempts to measure local forces [12,28] have had questionable success because of the relatively large deformations of the myocardium and uncertainties about the nature of the mechanical coupling between the transducer elements and the tissue. Efforts to measure intramyocardial pressures using miniature pressure transducers implanted in the ventricular wall have also raised controversy over whether the force transducer is measuring force generated by muscle-fiber contraction or changes in the pressure of the interstitial fluid [4]. Therefore, by far the most common approach to assessing myocardial stress distributions is the use of mathematical models based on the laws of continuum mechanics. In Section 5.3.4, we describe an analytical model of the end-diastolic left ventricle employing a cylindrical geometric approximation. However, models must also be validated with experimental measurements. Since the measurement of myocardial stress is not yet reliable, the best experimental data for model validation are measurements of strains in the ventricular wall.

The strain in a deforming body is a complete description of the local shape-change at each point in the body. In a continuum, shape changes arise from changes in length, and therefore to study regional function, researchers have used length-transducers (strain gauges) in various animal preparations. Since the strains in the myocardium are different in different parts of the ventricular wall, the objective should be to measure length changes over distances that are as small as possible without sacrificing accuracy. In practice, transducers oriented in the plane of the muscle fibers must be close enough together that the curvature of the wall does not become a significant factor. In the transmural direction, the incompressibility of the myocardial tissue can lead to significant gradients of strain. Moreover the orientation of the myofibers changes through almost 180° between epicardium and endocardium (see Chapter 2). Therefore, techniques for measuring changes in wall thickness through the cardiac cycle can only tell us about *average* radial strains. The earliest myocardial strain gauges were made from rubber tubes filled with mercury and sealed at the ends [58]. These transducers were sutured to the epicardial surface, and the resistance of the mercury column increased with the length of the silicone rubber tube in a manner that is close to linear for the range of ventricular strains.

Localized measurement of segment length change has been accomplished with the sonomicrometer, a pair of piezoelectric crystals for transmitting and receiving ultrasonic signals. The intercrystal distances are output on-line by determining the transit time of ultrasound traveling between the crystals. These crystals will function for several months in a chronic preparation. Ultrasonic dimension gauges have been used extensively for measuring overall ventricular dimensions as well as length changes of segments aligned circumferentially, longitudinally, or transmurally in the wall. A new extension to this technique uses a single epicardial crystal oriented toward the endocardium to measure relative intramural velocities from the Doppler shift in the signal reflected from a point in the wall on which the transducer is focused (range-gated Doppler ultrasound) [27].

A major limitation of all of these methods lies in the fact that the ventricular myocardium is a three-dimensional continuum, and the strain at a point is fully defined by a symmetric tensor with six independent components—three "normal" strains describing the stretching or shortening along three mutually orthogonal coordinate axes, and three "shear" strains related to the angle change in each of the three coordinate planes. As described by Walden in Chapter 7 of this book, the eigenvalues of the strain tensor are the principal strains and the eigenvectors are the corresponding principal axes. With respect to the principal axes, strains are purely extensional and the shear strains are zero, which serves to remind us that the strain tensor is a measure of length change. But only one component can be measured with a single length transducer. A significant improvement has been achieved by Villarreal and colleagues [74] who arranged three crystals (one transmitting, one receiving, and one transmitting and receiving) about 10 mm apart in a triangular arrangement so that three segment lengths could be measured simultaneously. This was sufficient information to compute the two normal strain components and the one shear strain that characterized the two-dimensional deformation at a point in a midwall plane parallel to the epicardium. Their findings showed that the principal axis of greatest shortening may not be aligned with the circumferential midwall fibers, and moreover that this direction can change with altered ventricular loading or contractility. Therefore, investigations based on uniaxial segment measurements may not reveal the true extent of alterations in regional function caused by an experimental intervention.

An alternative approach for measuring regional myocardial strains that may be less restrictive but is also usually less accurate than length transducers is the use of the various clinical techniques for ventricular imaging. These include single-plane or biplane contrast angiography, high-speed x-ray tomography, radionuclide imaging, magnetic resonance imaging (MRI), or two-dimensional echocardiography. However, in conventional clinical practice these techniques cannot be used to measure regional segment length changes because they do not identify the motion of distinct points in the myocardium. Instead they only produce a profile or silhouette of a surface such as the endocardium or epicardium, except in the unusual circumstance when radiopaque markers are implanted in the myocardium during cardiac surgery or transplantation [30]. However, in a recent review,

Hunter and Zerhouni [29] describe the prospects for noninvasive imaging of discrete points in the ventricular wall. The most promising methods include tracking the epicardial bifurcations of coronary arteries using contrast arteriography [36,83] as described in Chapter 8, and the exciting new possibility in MRI of selectively labeling defined regions of the myocardium by spatially modulating the radio-frequency excitation of the tissue to produce a difference between the magnetic resonance signals from different points in the wall [84].

In experimental research, the use of implantable radiopaque markers is much more practical. Meier and colleagues [44,45] placed triplets of metal markers 10 to 15 mm apart near the epicardium of the canine right ventricle and reconstructed their positions from biplane cinéradiographic recordings. It is apparent that more information than the lengths of the sides of the marker triangle is available from the coordinates of three points in a plane. In fact, two independent, two-dimensional vectors can be formed giving a total of four pieces of information, which Meier and coworkers used to compute the nonsymmetric deformation gradient tensor in the plane. Then, by polar decomposition, they obtained the two principal strains, the principal angle, and a fourth measurement, the local rotation of the region.

The use of radiopaque markers was extended to three dimensions by Fenton and colleagues [13] and Waldman and colleagues [76,77], who implanted three columns of five or six metal beads at three closely separated sites in the left ventricular free wall using a specially designed trochar as described in detail in Chapter 7. Using this technique, it is possible to find all six components of strain at a number of sites through the wall (see Chapter 7 for details).

Although these techniques for quantifying regional strains are generally applicable throughout the cardiac cycle, in the great majority of cases the reference state for length change or strain has been end-diastole, and studies have concentrated on regional ventricular function during systole. Whereas the isolated diastolic arrested heart has been a useful model for studying the passive pressure-volume relation [22], few workers have used it for studying regional mechanics. In Sections 5.3.1 through 5.3.4, we review our recent studies of regional mechanics in the isolated potassium-arrested dog and rat hearts as a means for learning more about the regional mechanics of the intact heart at end-diastole.

5.3.1   REGIONAL DEFORMATIONS IN THE PASSIVE HEART

Studies of regional wall motion in the intact heart have revealed substantial regional differences in the time-course, magnitude, and pattern of systolic myocardial deformations. Perhaps the most consistent finding has been that there is a significant longitudinal variation of systolic shortening in the ventricular free wall. Kong and colleagues [36], using coronary bifurcations as markers, found that epicardial shortening in the human left ventricle increased uniformly from the base to the apex. With ultrasonic crystal pairs implanted in the midwall and on the epicardium in conscious and open-chest dogs, LeWinter and colleagues [38] also measured significantly greater fiber shortening at the apex than at the base and midventricle. A similar longitudinal distribution of systolic shortening was reported by Meier and colleagues [44] on the canine right ventricular epicardium. However, the extent to which regional differences in the magnitudes of *systolic* shortening in these studies reflected variations in *diastolic* fiber stretches was unknown.

McCulloch and colleagues [43] examined regional variations in epicardial deformations on the left ventricular free wall of the isolated, potassium-arrested dog heart. A 45° biplane video camera arrangement, illustrated in Figure 5.3, was used to record the motion of triplets of epicardial markers during static ventricular inflation at pressures up to 25 mm Hg. Three markers, each consisting of two fine-silk threads sutured in a small cross to the epicardium, were placed 10 to 20 mm apart to form an approximately equilateral triangle. Studies were performed with marker triangles located on the anterior and posterior free walls at levels 30% (basal), 50% (midventricular), and 70% (apical) between the base and apex of the left ventricle.

The three-dimensional positions of the epicardial markers were reconstructed from the biplane video recordings to an accuracy within 0.25 mm [42]. Homogeneous strain theory was then used to obtain a description of the stretch and rotation components of the epicardial deformation gradients that was independent of the orientation of the marker triangle and suitable for regional comparison. All of the computed deformations were referred to the undeformed state at which the ventricular filling pressure was zero at the level of the aortic valve.

The analysis revealed a complex but consistent pattern of epicardial strain with significant regional variations. Figure 5.4 shows the mean midanterior principal strains as functions of ventricular volume. E_1 is the maximum (major) strain and E_2 is the mutually perpendicular minimum (minor) strain. The magnitudes of the maximum strain increased by approximately 100% from the thick-walled basal sites to the thinner apical regions both on the anterior and posterior walls. This finding suggested that the increase in systolic shortening toward the apex probably reflects an increase in end-diastolic sarcomere lengths nearer the apex rather than a decrease in end-systolic sarcomere lengths.

At five of the six regions, the principal epicardial strains were significantly different from each other, indicating that epicardial stretching was nonuniform unlike, for example, the inflation of a spherical balloon. Moreover, the direction of greatest epicardial stretch in these regions was consistently along an axis that was rotated clockwise from circumferential. This principal angle tended to increase with ventricular volume, and at the high loads it was always within 10° of the local epicardial fiber angle, which was typically about −40 to −50°. The resulting asymmetric deformation of the wall

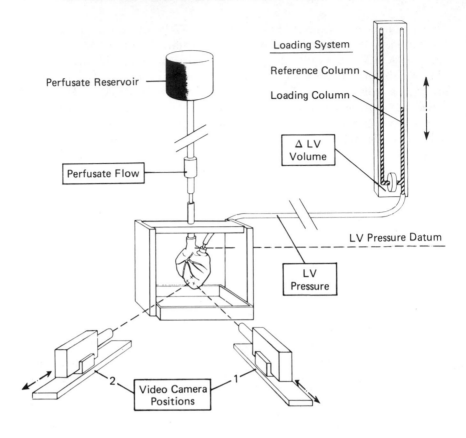

FIGURE 5.3. Schematic representation of the experimental system used by Mc-
Culloch et al. [42] showing the isolated arrested heart supported by a special
cannula assembly at the center of a biplane video camera arrangement. The left
ventricular pressure and volume were increased by raising the fluid-loading column connected to the mitral valve cannula, and a cardioplegic perfusate was
supplied from the reservoir 130 cm above the heart. The three-dimensional motion of small sutures forming markers on the epicardium was recorded on video
cassette for later analysis. (Reproduced from McCulloch et al. [42], with permission from the American Physiological Society.)

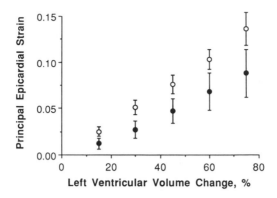

FIGURE 5.4. Principal epicardial strains, E_1 (open circles) and E_2 (closed circles), computed from Table 4 in McCulloch et al. [43]. Mean strains (± 1 standard deviation) from the left ventricular midanterior of six isolated arrested canine hearts are plotted against percent left ventricular volume change under passive loading. The mean principal axis of E_1 increased with load from about 22–55° clockwise from circumferential.

caused the ventricle to twist in a left-handed direction about a long axis pointing from base to apex. Since the base of the heart was constrained by the cannula in the mitral valve, this twist was seen to increase toward the apex consistent with a torsional deformation. A local counterclockwise rotation of the markers in the epicardial plane about the centroid of the marker triangle was also observed.

To interpret these experimental findings in the context of the helical fibrous structure of the ventricular myocardium, a simple geometric model was used to show that the measured epicardial rotations and ventricular torsion were kinematically consistent with the change in pitch and axial twist of a left-handed fiber helix. Similar analyses have been used to analyze torsional displacements during systole, which are opposite in direction to those measured in the passively filling heart [31]. The main conclusion is that the resting myocardium must be anisotropic with respect to its local fiber direction, although it is not clear from the data whether the measurements were consistent with fiber stiffness that is greater or less than the transverse stiffness. The torsional deformations resulting from this fibrous anisotropy of the passive myocardium acted to maximize end-diastolic epicardial fiber extensions.

5.3.2   EFFECTS OF CORONARY PERFUSION ON PASSIVE VENTRICULAR DEFORMATION

The isolated potassium-arrested dog heart preparation has also been used for studying the purely mechanical effects of coronary perfusion under well-defined loading conditions. Using the biplane video method, McCulloch [41] measured midanterior epicardial deformations before, during, and after perfusion of the coronary circulation with a cardioplegic buffer. During perfusion, there was a highly significant reduction in ventricular compliance; the mean cavity volume change decreased by 50% at a filling pressure of 12 mm Hg. The loss of compliance was reversible and increased with coronary artery pressure.

During perfusion, there were also highly significant reductions in the principal epicardial strains compared with those before and after perfusion. The major and minor strain both decreased with perfusion in equal proportion at all loads by a mean of 36 to 37%. However, there was no significant change in the principal angle with perfusion. The epicardial rotation and the torsional displacements also fell significantly during perfusion, but the directions of the rotation and twist were unaltered. Therefore, although the magnitudes of epicardial deformations were reduced by coronary perfusion, the characteristic patterns of deformation [43] were unchanged. This suggests that the mechanical effects of perfusion are most likely associated with the microvascular circulation or the interstitium rather than the large coronary vessels.

These results suggest that the effects of coronary perfusion cannot be neglected in an accurate continuum mechanics model of the passive left ventricle. Yet modern continuum mechanics models of passive ventricular mechanics are invariably based on uniaxial or biaxial stress-strain relationships measured in unperfused isolated myocardium [80]. Kitabatake and Suga [35] did measure the uniaxial stress-strain behavior of the intact right ventricular papillary muscle in the isolated blood-perfused canine heart. They found that the exponential coefficient of the end-diastolic length-tension relation showed a positive correlation with coronary perfusion pressure increasing from approximately 13 at 75 mm Hg coronary artery pressure to about 21 at 125 mm Hg. But simply adjusting the material parameters of a ventricular model to reflect the stiffening effects of increased perfusion pressure would provide no insight into the mechanism of altered ventricular compliance.

Several approaches for modeling the mechanical behavior of perfused tissues are possible. Bogen [7,8] has proposed biphasic elastic models of soft tissues swollen by an incompressible fluid residing in one or more fluid compartments. He showed that for a finite extension of a swollen material with nonlinear solid properties, the uniaxial stiffness was increased by swelling. When there were two fluid compartments, the stiffness was increased by a volume imbalance between the compartments. This approach may be particularly useful for understanding the decreased diastolic compliance of the left ventricle associated with myocardial edema [56]. Whether it is able to predict the effects of altered coronary perfusion pressure remains to be seen. This approach is closely related to the mechanism proposed by Salisbury and colleagues [60] that has become known as the "garden hose effect" [3].

However, it should be noted that this type of model will only give rise to increased ventricular stiffness with increased perfusion pressure if the material is nonlinear. Alternatively, if the garden hose is linearly elastic, then its stiffness in tension will be independent of the pressure in the tube.

Biphasic models have also been proposed for hydrated soft tissues in which the flow of the fluid is the important influence, such as articular cartilage [49]. However, in this type of model it is the time-dependent viscoelastic properties of the soft tissue that are affected principally by the presence of fluid.

A consequence of increased coronary perfusion pressure would appear to be an increase in ventricular wall thickness associated with increased myocardial blood volume [18,19,48,75]. However, it seems unlikely that the large effect of coronary perfusion on passive ventricular compliance will be fully accounted for by the relatively small changes in wall volume.

5.3.3 Residual Stress and Strain

To this point, we have been describing measurements of myocardial deformation referred to a zero-strain state in which the transmural pressure gradient across the ventricular wall of the passive heart was zero. However, in general, an unloaded deformable body may not be in a state of zero *stress*. "Residual stresses" are the internal forces in a body that remain after all of the external loads are removed. If the residual stresses are relieved by one or more cuts, the body adopts a new "stress-free" shape, and this shape change is described by measuring the "residual strains." The question of the stress-free state of living tissues arises because tissues are composed of cells that can proliferate, grow, or resorb. When cells are individual they grow rapidly, but growth slows when they become confluent. When cells are part of a tissue, they are no longer free to grow unrestricted. Cells in the tissue interact with each other, creating residual stresses and strains, which are borne by the intercellular structures.

The idea that an elastic body may contain stress in the absence of any external forces has been part of mechanical engineering since the mid-1800s [68]. Several early investigators studied the notion of residual stresses in materials due to plastic deformation during heat treatment processes [70]. An analytical approach to residual stress was given by B. de Saint-Venant in 1863 when he introduced the idea of residual stress into the equations of elasticity. It was soon recognized by many workers in the field of metallurgy that residual stress played an important role in many metal manufacturing processes such as welding, cold-drawing tubes, die extrusion, and cold rolling. In these mechanical engineering problems, residual stresses are left in the material because of plastic deformations. One of the first practical applications of residual stress illustrates its use in stress reduction. In 1855, R. Mallet published a paper on the construction of artillery and cannons [70]. At the time, a problem with gun and cannon barrels was their tendency to rupture when the gunpowder was set off. Mallet proposed to raise the limiting pressure at which the barrel bursts by "placing cylindrical rings of wrought-iron over each other, each new ring being shrunk on the series of rings which form its core." The idea was to create a compressive residual stress at the inner layers of the gun barrel. Hence, the stress concentration at the inner wall of the cylinder could be reduced to a value below the rupturing stress when the inner part of the cylinder was initially under compression.

Theoreticians have looked at problems in elasticity involving residual stress. Love [39] describes the analysis of an elastic hollow cylinder with residual stress. In a multiply-connected body such as the cylinder, there is the possibility of many-valued displacements. In other words, when an intact cylinder with initial stress is opened with a radial cut, the resulting deformations are different depending on the initial state of stress. This idea of many-valued displacements was first shown by G. Weingarten in 1901 [39]. Timoshenko [69] gives an example of residual stress when a curved beam with an edge dislocation is rounded up to a ring. The closed ring is in a state of pure bending, and the stress components are given in terms of the moment needed to close the gap. These classical elasticity analyses formed the basis for work with residual stress in multiply-connected biological tissues.

Recently, it was shown that the stress-free state of the passive left ventricle is not the unloaded state; residual stress exists in the left ventricle [51]. This phenomenon was first described by Y-C Fung in his book, *Biodynamics: Circulation* [15]. The initial studies looking at left ventricular residual strain used whole hearts with a longitudinal slit through the left ventricular free wall. Later, it was found that by radially resecting a cross-sectional ring of the potassium-arrested heart, more accurate measurements of residual strain could be obtained. The only quantitative experimental studies of residual strain in the ventricular myocardium have been conducted by Omens and Fung [52]. Cross-sectional equatorial slices (2–3 mm thick) were taken from potassium-arrested rat left ventricles. When the ring was cut radially, it immediately sprang open to form a curved arc when the residual stress components were relieved (Figure 5.5). Ischemic contracture was delayed with a hypothermic, high-potassium buffer containing a calcium channel blocker (nifedipine) and the calcium chelating agent EGTA. The average angle of opening measured from photographs was $45\pm10°$ in a sample of 11 rat hearts. This opening implies a nonuniform distribution of "residual strain" across the intact wall, being compressive at the endocardium and tensile at the epicardium.

To measure the strain distributions in the plane of the slice directly, several hundred reflective stainless steel microspheres were lightly pressed into the cut surface of equatorial slices, and the coordinates of the microspheres were digitized from photographs taken before and after a radial cut was made through the left ventricular free wall (Figure 5.6). The coordinates of

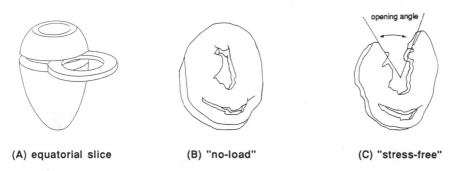

(A) equatorial slice   **(B) "no-load"**   **(C) "stress-free"**

FIGURE 5.5. (A), Equatorial slice of the left ventricle. (B), Drawing of an equatorial ring in the no-load configuration. (C), Slice shown in (B) after a radial cut was made through the left ventricular free wall, showing the definition of the "opening angle." This slice is in the stress-free configuration.

three neighboring markers in each state are sufficient to calculate the three independent residual strain components at that small region of the plane. A basic assumption of this type of analysis is that the strain in the area defined by the three points is homogeneous. Distributions of strain were thus found for several equatorial slices. Typical distributions of principal extensions along four radii are shown in Figure 5.7.

The residual strain distributions were, for the most part, consistent with the premise of simple bending to close the gap made by the radial cut, although the distributions showed consistent differences on the posterior and anterior free walls. The circumferential residual strains were compressive in the endocardial region, and either tensile or compressive with a much smaller magnitude at the epicardium. A second radial cut opposite the first produced subsequent strains that were negligibly small. Hence, a slice with one radial cut was considered to be stress-free. These studies demonstrated that the stress-free, strain-free reference state is only found when residual stresses have been removed by cutting the tissue. These simple techniques may provide a new means of studying the mechanical effects of remodeling under certain pathologic conditions.

Residual stress and strain have an important relationship to growth in an organ. Growth and change affect the stress and strain in a tissue. Conversely, stress and strain affect growth and change. Pathogenesis is often associated with growth and remodeling as seen in cardiac hypertrophy [2]. To detect and determine the spatial distribution of tissue growth and change in the whole heart is very difficult, but an effective way is offered by the measurement of residual strain and the subsequent computation of residual stress. It is known that biological tissues can adapt to external stress and strain. The remodeling of bones is an example that has been well documented [16]. It is thought that cardiac hypertrophy and growth develop in

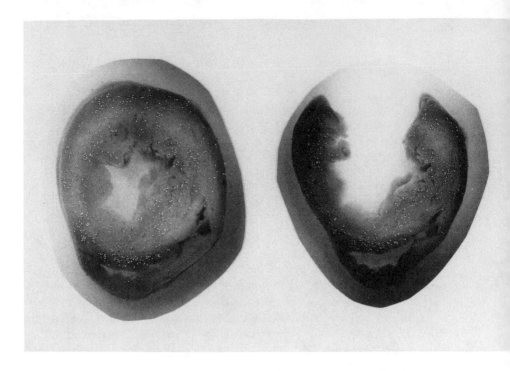

FIGURE 5.6. Photographs of an equatorial slice of a rat heart in the no-load and stress-free states. The illuminated stainless steel microspheres serve as markers for surface strain measurements.

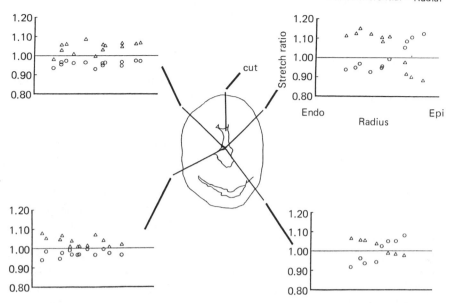

FIGURE 5.7. Calculated principal residual stretch ratios from an equatorial slice of the rat heart. Each plot shows the two principal residual stretch ratios as a function of radius at four different circumferential locations. The slice is shown in the no-load configuration. The directions of the principal axes in general corresponded to the local circumferential and radial directions. Therefore, the stretch ratios are referred to as "circumferential" and "radial." In all cases, the endocardial circumferential stretch ratios were less than unity, and a distinctive asymmetry of the distributions around the circumference was found (different stretch ratio distributions on either side of the cut).

response to changes in diastolic and systolic wall stress [24].

Residual stress and strain have also been examined in blood vessels [11]. Experimentally, the opening angle was found using a protocol similar to that described for the heart. Changes in residual strains have been shown following hypertension (pressure overload) in rats due to aortic banding [17]. The opening angle tends to increase dramatically during the early days after banding, before returning to the control value after 6 to 8 days. A similar situation may well be found in the ventricles of the heart during pressure overload hypertrophy.

The origin of residual stress from a microscopic point of view implies that residual stress is "created" in the tissue by a nonhomogeneous distribution of material. Differences in the collagen matrix may produce residual stresses. For example, it has been shown that the results of biaxial tests of ventricular myocardium are strongly dependent on whether the superficial collagenous epicardium (parietal pericardium) is intact [61]. When the thin epicardial membrane was carefully peeled off the surface of the canine heart muscle and tested it was found to be considerably stiffer than the underlying muscular tissue. Moreover, its stress-free length differed from that of the myocardium. After dissection from the muscle, the epicardium shortened to 86 to 92% of its original length. Therefore, the epicardium is pre-tensioned and hence could be a major factor in producing residual stress in the composite ventricular structure.

With a knowledge of geometry, material properties, and boundary conditions, stresses can be computed. In the analysis of cardiac mechanics, a consistent assumption has been that the unloaded passive ventricle is stress-free. As a consequence, mathematical models of ventricular mechanics—based soundly on the conservation laws of mechanics—have predicted a high endocardial stress peak at end-diastole [5,33,54,57,71]. The corresponding end-diastolic strain field implies much higher fiber extensions at the subendocardium than at the epicardium. In contrast, direct measurements of end-diastolic sarcomere lengths show a relatively uniform transmural distribution [23,82], which is much better suited to optimal systolic function. This same philosophy applies to structural engineering where residual stresses are similarly used to advantage in the manufacture of pressure vessels and prestressed concrete. In the following section, we describe a recent analytical model, that includes residual stress, and is based on our experimental measurements of strains in the isolated heart.

5.3.4   An Analytical Model of Diastolic Ventricular Mechanics

Many models of diastolic left ventricular mechanics have been proposed. Their main objective is usually to predict the distributions of stress and strain in the passive myocardium. With measurements of strain in the isolated arrested heart, it is possible to validate these models. The greatest unknown in modeling the mechanics of the diastolic heart is the constitutive equation for the material properties of the passive myocardium. Because isolated papillary muscle and trabecula preparations only permit one-dimensional stress-strain measurements, investigators have begun performing biaxial tests using isolated sheets of ventricular myocardium [81]. However, there are considerable difficulties with these measurements largely associated with the unknown effects of resecting sections of ventricular wall. Nevertheless, biaxial testing does confirm that passive myocardium exhibits the exponential stress-strain relationship, anisotropy, and strain-rate-independent hysteresis characteristic of soft biological tissues [16]. The multiaxial testing of the passive myocardium is described in detail in Chapter 3 by Humphrey and colleagues. However, a problem with existing experimental tests of isolated tissues is uncertainty about how well they describe the properties of the *intact* ventricular muscle.

Rather than review the models of passive ventricular mechanics here, we describe one recent model that uses a cylindrical geometric assumption and is based on some of our experimental measurements of epicardial deformation and residual strain described in the preceding sections. Although it is clearly a highly simplified geometric approximation, this model is able to demonstrate some important mechanical corollaries of the experimental observations, and it also provides an example of how a continuum mechanics model of the heart together with appropriate experimental measurements can be used to predict the mechanical properties of the intact myocardium.

Guccione and coworkers [26] used a thick-walled cylindrical geometry to approximate the equatorial region of the left ventricle. The passive myocardium was modeled as a nonlinear, incompressible elastic material. The components of the stress tensor were therefore defined by the gradients of a "strain-energy" potential that was a function of the components of the strain tensor. For a wide range of soft biological tissues, Fung [16] suggested the following general exponential form for the strain-energy function:

$$W = \frac{C}{2}(e^Q - 1) - \frac{1}{2}p(I_3 - 1). \tag{5.1}$$

In Equation (5.1), Q is a quadratic function of the strain components, C is a material constant, and p is a hydrostatic pressure appearing in the strain-energy function as a Lagrange multiplier that arises from the kinematic constraint that deformations must be isochoric (volume preserving). This kinematic incompressibility constraint is represented mathematically by the relation $I_3 = 1$, where I_3 is the third principal invariant of the strain tensor. The following form for Q as a function of the strain components E_{RS} gives rise to a material that is said to "transversely isotropic" about the X_1 axis:

$$Q = b_1(E_{11} + E_{22} + E_{33}) + b_2 E_{11}^2 + b_3(E_{22}^2 + E_{33}^2 + E_{23}^2 + E_{32}^2)$$
$$+ b_4(E_{12}^2 + E_{21}^2 + E_{13}^2 + E_{31}^2). \tag{5.2}$$

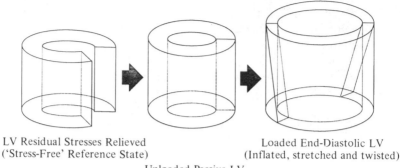

LV Residual Stresses Relieved   Loaded End-Diastolic LV
('Stress-Free' Reference State)   (Inflated, stretched and twisted)

Unloaded Passive LV
(Strain reference state)

FIGURE 5.8. The equatorial region of the canine left ventricle was modeled as a thick-walled cylinder [26]. The stress-free state of the ventricle was approximated by a cylindrical arc. A bending deformation restored the model to the unloaded, but no longer stress-free configuration. The wall deformations during passive ventricular filling were then modeled by inflation, stretch and torsion of the cylinder.

In our cylindrical model, the strains were referred to a coordinate system, $\{X_R\}$, where the axis of material symmetry X_1 was aligned with the muscle fibers in the plane of the wall, X_2 was the cross-fiber in-plane axis, and X_3 was the radial coordinate. b_1, b_2, b_3, and b_4 are positive material constants, and when only b_1 is nonzero or when $b_2 = b_3 = b_4$, the material becomes isotropic and its properties are therefore independent of the fiber orientation. The material stiffness may be increased preferentially along the fiber axis by increasing b_2 or in the transverse plane by raising b_3 from zero. The angle between the fiber axis and the circumferential axis in the model varied linearly through the wall thickness from 75° (counterclockwise about an outward normal) on the endocardium to –45° on the epicardium.

The stress-free state of the left ventricular cylinder was assumed to be a cylindrical arc based on the experimental observations described in the previous section. The unloaded state of the left ventricular cross-section was then determined by bending the arc into a closed circular ring as shown in Figure 5.8 while maintaining the normal stress on the epicardial and endocardial surfaces at zero. Thus, for given material properties the equilibrium equations and the incompressibility condition could be solved for the residual stresses and strains in the unloaded left ventricular cross-section. Consistent with the notion of bending and the experimental results in Figure 5.7, the circumferential residual stresses and strains were compressive on the endocardium and tensile on the epicardium.

To model the effects of passive ventricular filling, finite deformations were prescribed that gave rise to the epicardial strains measured in the isolated arrested dog heart as described in Section 5.3.1. At each pressure load in the

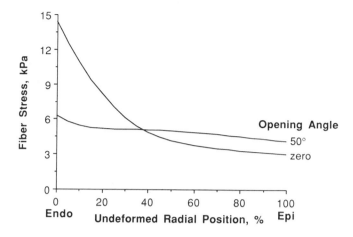

FIGURE 5.9. The effect of residual stress on the transmural distribution of fiber stress in an isotropic model at a physiological end-diastolic pressure (1.67 kPa). For an opening angle of zero there is no residual stress and a steep gradient of fiber stress at end-diastole. For a typical opening angle of 50°, a large reduction in the peak fiber stress is seen. (Left ventricular volume relative to the unloaded state = 175%; axial stretch = 11.1%; twist per unit length = $-2.25°$; $C = 0.77$ kPa; $b_1 = 4.2$.)

isolated heart, the two principal epicardial strains shown in Figure 5.4 and the corresponding principal angle were used to solve for a unique inflation, stretch, and twist of the cylindrical model. The stress distributions in the model ventricular wall were found by solving the equilibrium equations for the unknown hydrostatic pressure variable using assumed values for the material constants in Equation (5.1). In the simplest case, when the material properties are isotropic, the constants C and b_1 were chosen that gave the best fit to the known left ventricular diastolic pressure-volume relation. The influence of residual stress on the transmural distribution of fiber stress is shown in Figure 5.9 for at a pressure of 1.67 kPa (~12.5 mm Hg). When the assumed opening angle is zero and there is no residual stress, the fiber stress has a high transmural gradient with a significant stress concentration on the endocardium. For a typical opening angle of 50°, the peak stress is reduced by more than 50% and the gradient is eliminated. Increasing the torsion parameter also reduces the stress gradients in the isotropic model, although the effect is not as marked for torsion in the physiological range. A similar conclusion has been drawn from models of the actively ejecting heart that have included torsion [5,6].

Although the isotropic material model could be fitted to the observed pressure-volume and strain data, it would not twist and stretch the prescribed amount under pressure loading alone. That is, a resultant torsional moment and axial force were required to maintain the observed twisting and

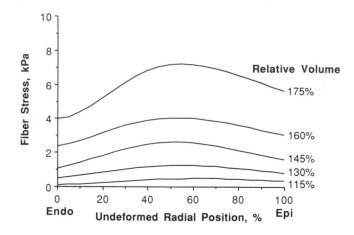

FIGURE 5.10. Transmural distributions of fiber stress in an anisotropic model at five left ventricular end-diastolic volumes defined relative to the unloaded (zero pressure) state. Notice the relatively uniform stress profiles, the combined result of residual stress, torsion, fiber orientation, and material anisotropy (see text).

stretching deformations. However, in the isolated heart, the deformations occurred in the absence of any applied forces or moments apart from the ventricular pressure. Therefore, the model was extended to include the parameters that made the material anisotropic. Guccione and colleagues [26] found that by increasing the material stiffness preferentially in the fiber direction, these resultant axial forces and moments were reduced. Hence, a least-squares parameter optimization was performed to find those values that would minimize the resultant axial force and moment subject to the known pressures, volumes, and strains. A minimum solution was found in which the predicted fiber stiffness was substantially greater than the transverse stiffness by a factor of about 3.3 (the predicted values were $C = 0.64$ kPa, $b_1 = 2.5$, $b_2 = 15.1$, $b_3 = 0.0$, and $b_4 = 10.5$). This finding is in agreement with the conclusions of most biaxial muscle testing experiments. However, it is interesting to note that these conclusions came from measured epicardial strains that were actually greatest in the fiber direction. Thus, the preferential epicardial stretch in the passively filling left ventricle along the direction of the epicardial fibers is apparently consistent with greatest muscle stiffness in the fiber direction. The corresponding torsion of the ventricular wall during filling must therefore be dominated by the endocardial fibers.

Figure 5.10 shows the transmural distribution of fiber stress predicted by the model using these optimized material parameters. Over a wide range of end-diastolic volumes, the variation in fiber stress across the wall is seen to remain small with a slight subepicardial peak. A similar flat profile of sarcomere lengths was predicted by the model and these agreed very well with direct measurements from fixed hearts [23].

Therefore, it appears that the effects of the fiber angle distribution, material anisotropy, torsion, and residual stress all combine to minimize the gradients of fiber stress and strain in the ventricular wall. In engineering, minimizing stress and strain gradients is a common objective of an optimal mechanical design. These experimental and model findings from the diastolic left ventricle lend support to the hypothesis of optimal operation in biological organs proposed by Fung [15]. Moreover, since the end-diastolic configuration determines systolic force generation, it is possible that these properties of the passive heart ultimately optimize its active pumping function as well.

Acknowledgements: The authors gratefully acknowledge the support of the NIH (through grant HL41603) and the Whitaker Foundation.

REFERENCES

[1] R.M. Abel and R.L. Reis. Effects of coronary blood flow and perfusion pressure on left ventricular contractility in dogs. *Circ. Res.*, 27:961–971, 1970.

[2] N.R. Alpert. *Cardiac Hypertrophy*. Academic Press, New York, 1971.

[3] G. Arnold, C. Morgenstern, and W. Lochner. The effect of coronary perfusion pressure on the contractility of the heart. In *Proceedings of the International Union of Physiological Sciences, XXIV International Congress*, chapters 7 and 18, 1968.

[4] T. Arts and R.S. Reneman. Interaction between intramyocardial pressure (IMP) and myocardial circulation. *J. Biomech. Eng.*, 107:51–56, 1985.

[5] T. Arts, R.S. Reneman, and P.C. Veenstra. A model of the mechanics of the left ventricle. *Ann. Biomed. Eng.*, 7:299–318, 1979.

[6] R. Beyar and S. Sideman. Effect of the twisting motion on the nonuniformities of transmyocardial fiber mechanics and energy demand— a theoretical study. *IEEE Trans. Biomed. Eng.*, BME-32:764–769, 1985.

[7] D.K. Bogen. Strain energy descriptions of biological swelling I: Single fluid compartment models. *J. Biomech. Eng.*, 109:252–256, 1987.

[8] D.K. Bogen. Strain energy descriptions of biological swelling II: Multiple fluid compartment models. *J. Biomech. Eng.*, 109:257–262, 1987.

[9] A. Bouchard, T.A. Watters, S. Wu, W.W. Parmley, R.D. Stone, E. Botvinick, R. Sievers, G. Jasmin, and J. Wikmsn-Coffelt. Effects of altered coronary perfusion pressure on function and metabolism of normal and cardiomyopathic hamster hearts. *J. Mol. Cell. Cardiol.*, 19:1011–1023, 1987.

[10] D.L. Brutsaert, F.E. Rademakers, and S.U. Sys. Triple control of relaxation: Implications in cardiac disease. *Circulation*, 69(1):190–196, 1984.

[11] C.J. Chuong and Y.C. Fung. Residual stress in arteries. In G.W. Schmid-Schoenbein, S.L-Y. Woo, and B.W. Zweifach, editors, *Frontiers in Biomechanics*, pages 118–129. Springer-Verlag, New York, 1986.

[12] E.O. Feigl, G.A. Simon, and D.L. Fry. Auxotonic and isometric cardiac force transducers. *J. Appl. Physiol.*, 23:597–600, 1967.

[13] T.R. Fenton, J.M. Cherry, and F.A. Klassen. Transmural myocardial deformation in the canine left ventricular wall. *Am. J. Physiol.*, 235:H523–H530, 1978.

[14] O. Frank. Die grundform des arteriellen pulses. *Z. Biol.*, 37:483–526, 1898.

[15] Y.C. Fung. *Biodynamics: Circulation.* Springer-Verlag, New York, 1984. QP 105. F85 1984 / 97

[16] Y.C. Fung. *Biomechanics: Mechanical Properties of Living Tissues.* Spinger-Verlag, New York, 1981.

[17] Y.C. Fung and S.Q. Liu. Change of residual strains in arteries due to hypertrophy caused by aortic constriction. *Circ. Res.*, 65:1340–1349, 1989.

[18] W.H. Gaasch and S.A. Bernard. The effect of acute changes in coronary blood flow on left ventricular end-diastolic wall thickness: An echocardiographic study. *Circulation*, 56:593–598, 1977.

[19] W.H. Gaasch, O.H.L. Bing, A. Franklin, D. Rhodes, S.A. Bernard, and R.M. Weintraub. The influence of acute alterations in coronary blood flow on left ventricular diastolic compliance and wall thickness. *Eur. J. Cardiol.*, 7(Suppl.):147–161, 1978.

[20] J.C. Gilbert and S.A. Glantz. Determinants of left ventricular filling and of the diastolic pressure-volume relation. *Circ. Res.*, 64:827–852, 1989.

[21] S.A. Glantz, G.A. Misbach, W.Y. Moores, D.G. Mathey, J. Lekuen, D.F. Stowe, W.W. Parmley, and J.V. Tyberg. The pericardium substantially affects the left ventricular diastolic pressure-volume relationship in the dog. *Circ. Res.*, 42:433–441, 1978.

[22] S.A. Glantz and W.W. Parmley. Factors which affect the diastolic pressure-volume curve. *Circ. Res.*, 42:171–180, 1978.

[23] A.F. Grimm, H.L. Lin, and B.R. Grimm. Left ventricular free wall and intraventricular pressure-sarcomere length distributions. *Am. J. Physiol.*, 239:H101–H107, 1980.

[24] W. Grossman. Cardiac hypertrophy: Useful adaptation or pathologic process? *Am. J. Med.*, 69:576–583, 1980.

[25] W. Grossman and J.T. Mann. Evidence for impaired left ventricular relaxation during acute ischemia in man. *Eur. J. Cardiol.*, 7(Suppl.): 239–249, 1978.

[26] J.M. Guccione Jr., A.D. McCulloch, and L.K. Waldman. Passive material properties of intact ventricular myocardium determined from a cylindrical model. *J. Biomech. Eng.*, 113: 1991. (in press).

[27] C.J. Hartley, L.A. Latson, L.H. Michael, C.L. Seidel, R.M. Lewis, and M.L. Entman. Doppler measurement of myocardial thickening with a single epicardial transducer. *Am. J. Physiol.*, 245:H1066–H1072, 1983.

[28] R.M. Huisman, G. Elzinga, N. Westerhof, and P. Sipkema. Measurement of left ventricular wall stress. *Cardiovasc. Res.*, 14:142–153, 1980.

[29] W.C. Hunter and E.A. Zerhouni. Imaging distinct points in left ventricular myocardium to study regional wall deformation. In J.H. Anderson, editor, *Innovations in Diagnostic Radiology*, pages 169–190. Springer-Verlag, New York, 1989.

[30] N.B. Ingels Jr., G.T. Daughters II, E.L. Stinson, and E.B. Alderman. Measurement of midwall myocardial dynamics in intact man by radiography of surgically implanted markers. *Circulation*, 52:859–867, 1975.

[31] N.B. Ingels Jr., D.E. Hansen, G.T. Daughters II, and E.B. Stinson. Relation between longitudinal, circumferential, and oblique shortening and torsional deformation in the left ventricle of the transplanted human heart. *Circ. Res.*, 64:915–927, 1989.

[32] J.S. Janicki and K.T. Weber. Factors influencing the diastolic pressure volume relation of the cardiac ventricles. *Fed. Proc.*, 39:133–140, 1980.

[33] R.F. Janz, B.R. Kubert, T.F. Moriarty, and A.F. Grimm. Deformation of the diastolic left ventricle: II. Nonlinear geometric effects. *J. Biomech.*, 7:509–516, 1974.

[34] D.A. Kass and W.L. Maughan. From e_{max} to pressure-volume relations: A broader view. *Circulation*, 77:1203–1212, 1988.

[35] A. Kitabatake and H. Suga. Diastolic stress-strain relation of non-excised blood-perfused canine papillary muscle. *Am. J. Physiol.*, 234: H416–H420, 1978.

[36] Y. Kong, J.J. Morris Jr., and H.D. McIntosh. Assessment of regional myocardial performance from biplane coronary cineangiograms. *Am. J. Cardiol.*, 27:529–537, 1971.

[37] M.C. Lee, Y.C. Fung, R. Shabetai, and M.M. LeWinter. Biaxial mechanical properties of human pericardium and canine comparisons. *Am. J. Physiol.*, 253:H75–H82, 1987.

[38] M.M. LeWinter, R.S. Kent, J.M. Kroener, T.E. Carew, and J.W. Covell. Regional differences in myocardial performance in the left ventricle of the dog. *Circ. Res.*, 37:191–199, 1975.

[39] A.E.H. Love. *A Treatise on the Mathematical Theory of Elasticity.* Dover Publications, New York, 1944.

[40] Y. Maruyama, K. Ashikawa, H. Isoyama, S. Kanatsuka, E. Ino-Oka, and T. Takishima. Mechanical interactions between the four heart chambers with and without the pericardium in canine hearts. *Circ. Res.*, 50:86–100, 1982.

[41] A.D. McCulloch. *Deformation and Stress in the Passive Heart.* PhD thesis, University of Auckland, Auckland, New Zealand, 1986.

[42] A.D. McCulloch, B.H. Smaill, and P.J. Hunter. Left ventricular epicardial deformation in isolated arrested dog heart. *Am. J. Physiol.*, 252:H233–H241, 1987.

[43] A.D. McCulloch, B.H. Smaill, and P.J. Hunter. Regional left ventricular epicardial deformation in the passive dog heart. *Circ. Res.*, 64:721–733, 1989.

[44] G.D. Meier, A.A. Bove, W.P. Santamore, and P.R. Lynch. Contractile function in canine right ventricle. *Am. J. Physiol.*, 239:H794–H804, 1980.

[45] G.D. Meier, M.C. Ziskin, W.P. Santamore, and A.A. Bove. Kinematics of the beating heart. *IEEE Trans. Biomed. Eng.*, BME-27:319–329, 1980.

[46] I. Mirsky and J.S. Rankin. The effects of geometry, elasticity, and external pressures on the diastolic pressure-volume and stiffness-stress relations: How important is the pericardium? *Circ. Res.*, 44:601–611, 1979.

[47] J.H. Mitchell, R.J. Linden, and S.J. Sarnoff. Influence of cardiac sympathetic and vagal nerve stimulation on the relation between left ventricular diastolic pressure and myocardial segment length. *Circ. Res.*, 8:1100–1107, 1960.

[48] C. Morgenstern, U. Holjes, G. Arnold, and W. Loehner. The influence of coronary pressure and coronary flow on intracoronary blood volume and geometry of the left ventricle. *Pflugers Arch.*, 340:101–111, 1973.

[49] V.C. Mow, M.K. Kwan, W.M. Lai, and M.H. Holmes. A finite deformation theory for nonlinearly permeable soft hydrated biological tissues. In S.L.Y. Woo, G.W. Schmid-Schnbein, and B.W. Zweifach, editors, *Frontiers in Biomechanics*, pages 153–179. Springer-Verlag, New York, 1986.

[50] C.O. Olsen, D.E. Attarian, R.N. Jones, R.C. Hill, J.D. Sink, K.L. Lee, and A.S. Welchsler. The coronary pressure-flow determinants of left ventricular compliance in dogs. *Circ. Res.*, 49:856–865, 1981.

[51] J.H. Omens. *Left Ventricular Strain in the No-load State due to the Existence of Residual Stress*. PhD thesis, University of California, La Jolla, CA, 1988.

[52] J.H. Omens and Y.C. Fung. Residual strain in rat left ventricle. *Circ. Res.*, 66:37–45, 1990.

[53] I. Palacios, R.A. Johnson, J.B. Newell, and W.J. Powell Jr. Left ventricular end-diastolic pressure volume relationships with experimental acute global ischemia. *Circulation*, 53:428–436, 1976.

[54] Y.C. Pao, E.L. Ritman, and E.H. Wood. Finite-element analysis of left ventricular myocardial stresses. *J. Biomech.*, 7:469–477, 1974.

[55] S.W. Patterson and E.H. Starling. On the mechanical factors which determine the output of the ventricles. *J. Physiol.*, 48:357–379, 1914.

[56] G. Pogatsa, M.Z. Koltai, and Z. Grosg. The role of myocardial water content in heart function. *Acta Physiol. Acad. Sci. Hung.*, 59:305–309, 1982.

[57] E.L. Ritman, R.M. Heethaar, R.A. Robb, and Y.C. Pao. Finite element analysis of myocardial diastolic stress and strain relationships in the intact heart. *Eur. J. Cardiol.*, 7:105–119, 1978.

[58] R.F. Rushmer. Continuous measurements of left ventricular dimensions in intact, unanesthetized dogs. *Circ. Res.*, 2:14–21, 1954.

[59] K. Sagawa, L. Maughan, H. Suga, K. Sunagawa. *Cardiac Contraction and the Pressure-Volume Relationship.* Oxford University Press, New York, 1988.

[60] P.F. Salisbury, C.E. Cross, and P.A. Rieben. Influence of coronary artery pressure upon myocardial elasticity. *Circ. Res.*, 8:794–800, 1960.

[61] A.J. Shacklock. *Biaxial Testing of Cardiac Tissue.* ME Thesis, University of Auckland, Auckland, New Zealand, 1987.

[62] B.K. Slinker and S.A. Glantz. End-systolic and end-diastolic ventricular interaction. *Am. J. Physiol.*, 251:H1062–H1075, 1986.

[63] J. Spadaro, O.H.L. Bing, P. Laraia, A. Franklin, and R.M. Weintraub. Effects of perfusion pressure on myocardial performance, metabolism, wall thickness, and compliance: Comparison of the beating and fibrillating heart. *J. Thorac. Cardiovasc. Surg.*, 84:398–405, 1982.

[64] E.H. Starling. *The Linacre Lecture on the Law of the Heart,* 1918. In C.B. Chapman and J.H. Mitchell, editors, *Starling on the heart,* pages 121–147. Dowsons, London, 1965.

[65] K. Sunagawa, K.O. Lim, D. Burkoff, and K. Sagawa. Microprocessor control of a ventricular volume servo-pump. *Ann. Biomed.*, 10(4):145–159, 1982.

[66] G.H. Templeton, K. Wildenthal, and J.H. Mitchell. Influence of coronary blood flow on left ventricular contractility and stiffness. *Am. J. Physiol.*, 223:1216–1220, 1972.

[67] HEDJ Ter Keurs, W.H. Rijnsburger, R. Van Heuningen, and M.J. Nagelsmit. Tension development and sarcomere length in rat cardiac trabeculae: Evidence of length-dependent activation. *Circ. Res.*, 46:703–713, 1980.

[68] S.P. Timoshenko. *History of Strength of Materials.* McGraw-Hill, New York, 1953.

[69] S.P. Timoshenko and J.N. Goodier. *Theory of Elasticity, 3rd Ed.* McGraw-Hill Kogakusha Ltd, Tokyo, 1970.

[70] I. Todhunter and K. Pearson. *A History of the Theory of Elasticity and of the Strength of Materials.* Dover Publications, New York, 1960.

[71] A. Tözeren. Static analysis of the left ventricle. *J. Biomech. Eng.*, 105:39–46, 1983.

[72] G.S. Tyson Jr., G.W. Maier, C.O. Olsen, J.W. Davis, and J.S. Rankin. Pericardial influences on ventricular filling in the conscious dog. an analysis based on pericardial pressure. *Circ. Res.*, 52:173–184, 1984.

[73] F.J. Villarreal, L.K. Waldman, J.W. Covell, and W.L.Y. Lew. Preferential passive reconfiguration of the endocardium in the acutely ischemic canine left ventricle. *FASEB J.*, 3(3):A841, 1989.

[74] F.J. Villarreal, L.K. Waldman, and W.Y.W. Lew. A technique for measuring regional two-dimensional finite strains in canine left ventricle. *Circ. Res.*, 62:711–721, 1988.

[75] W.M. Vogel, C.S. Apstein, L.L. Briggs, W.H. Gaasch, and J. Ahn. Acute alterations in left ventricular diastolic chamber stiffness: Role of the "erectile" effect of coronary arterial pressure and flow in normal and damaged hearts. *Circ. Res.*, 51:465–478, 1982.

[76] L.K. Waldman and J.W. Covell. Effects of ventricular pacing on finite deformation in canine left ventricles. *Am. J. Physiol.*, 252:H1023–H1030, 1987.

[77] L.K. Waldman, Y.C. Fung, and J.W. Covell. Transmural myocardial deformation in the canine left ventricle: Normal in vivo three-dimensional finite strains. *Circ. Res.*, 57:152–163, 1985.

[78] C.J. Wiggers. Studies on the consecutive phases of the cardiac cycle and criteria for their precise determination. *Am. J. Physiol.*, 56:415–438, 1921.

[79] E.L. Yellin, M. Hori, C. Yoran, E.H. Sonnenblick, S. Gabbay, and R.W.M. Frater. Left ventricular relaxation in the filling and nonfilling intact canine heart. *Am. J. Physiol.*, 250:H620–H629, 1986.

[80] F.C.P. Yin. Ventricular wall stress. *Circ. Res.*, 49:829–842, 1981.

[81] F.C.P. Yin, R.K. Strumpf, P.H. Chew, and S.L. Zeger. Quantification of the mechanical properties of noncontracting canine myocardium under simultaneous biaxial loading. *J. Biomech.*, 20:577–589, 1987.

[82] C. Yoran, J.W. Covell, and J. Ross Jr. Structural basis for the ascending limb of left ventricular function. *Circ. Res.*, 32:297–303, 1973.

[83] A.A. Young, P.J. Hunter, and B.H. Smaill. Epicardial surface estimation from coronary angiograms. *Comput. Vision Graphics Image Process.*, 47(1):111–127, 1989.

[84] E.A. Zerhouni, D.M. Parish, W.J. Rogers, A. Yang, and E.P. Shapiro. Human heart: Tagging with MR imaging—A method for noninvasive assessment of myocardial motion. *Radiology*, 169(1):59–63, 1988.

6

Finite Element Modeling of Ventricular Mechanics

Julius M. Guccione[1]
Andrew D. McCulloch[1]

ABSTRACT Computing the distributions of stress and strain in a body with the complex geometry, boundary conditions, and material properties of the heart is a difficult yet worthwhile endeavor. The most promising method for obtaining numerical solutions to this problem is the finite element method. In this chapter, we review the use of finite element analysis for modeling ventricular mechanics. And we conclude by presenting a new axisymmetric finite element model of the passive left ventricle with a realistic geometry and fibrous architecture, physiological boundary conditions, and a three-dimensional constitutive equation.

6.1 Introduction

The distributions of stress in the ventricles of the heart are determined by (1) the three-dimensional geometry and fibrous architecture of the ventricular walls, (2) the boundary conditions imposed by the ventricular cavity and pericardial pressures and structures like the fibrous valve ring skeleton at the base of the ventricles, and (3) the three-dimensional mechanical properties of the myofibers and their collagen interconnections in the relaxed and actively contracting states. Many of these determining factors have been quantified in experimental studies, some of which are described in detail in this book.

For example, Streeter and Hanna [72,73] and Nielsen and coworkers [56] have made detailed studies of the three-dimensional geometry and myocardial fiber architecture of the ventricles of the dog heart. Measurements of left and right ventricular pressures in patients are quite routine. And extensive data have been collected on the passive and active uniaxial material properties of isolated papillary muscles and trabeculae from various mammalian species [64,74]. Although fully triaxial material testing still presents significant technical difficulties, biaxial stress-strain testing of excised two-

---

[1]Department of Applied Mechanics and Engineering Sciences (Bioengineering), University of California, San Diego, La Jolla, California 92093

dimensional sheets of passive canine myocardium [14,91], epicardium [30], and pericardium [43] has been performed.

To formulate a mathematical model for predicting the distributions of wall stress in such a complex and constantly changing mechanical system is clearly very difficult, but there are important reasons to attempt this. An accurate model of the mechanics of the ventricular myocardium would provide a sound basis for interpreting the complex regional changes in cardiac function that occur in pathological conditions, such as ischemic heart disease [44,75], in terms of changes in the local properties of the tissue. Knowledge of the stress distributions in the intact myocardium would also provide valuable insight into normal ventricular function since regional coronary blood flow [34], myocardial oxygen consumption [68], hypertrophy, and remodeling [2,19] are all influenced by ventricular wall stress [90].

To keep the problem mathematically tractable, many workers have developed models of left ventricular mechanics using simple geometric approximations such as thin-walled spheres [88], thin-walled ellipsoids [16,67,86], thick-walled ellipsoids [8,21,51,52,80,87], thick-walled spheres [1,15,53,54, 69,83], thick-walled cylinders [4–6,11,17,24,34,57,58,76,79], solids of revolution [35], and noncircular cylinders [39]. However, the analyses made by these models also make other simplifying assumptions about the material behavior of the heart muscle and the governing equations of motion.

All thin-walled membrane models and even some thick-walled models [35,39] only predict the average stress across the wall thickness because they are based on a global force balance (the "Law of Laplace"); they are not based on knowledge of material properties. In order to predict transmural stress distributions using a thick-walled ellipsoidal model, either the inappropriate linear (infinitesimal strain) theory of elasticity has been assumed [21,51,52,87] or the wall has been represented by nested layers of uncoupled membranes [8,80], which neglects the tethering between adjacent layers of muscle fibers [10,66]. Some thick-walled spherical models have employed the nonlinear finite deformation theory of elasticity, which is more appropriate since experimental measurements of ventricular deformations have shown that myocardial stretches during normal diastolic filling and shortening strains during systole may be 20% or greater. However, material isotropy [15,53,54,83] or, at best, transverse isotropy with respect to the radial direction [1,69] was needed for the solution of these models so that the spherical symmetry was preserved upon inflation or ejection.

Most of the thick-walled cylindrical models, whether using concentric shells [4–6,34], finite elasticity [17,76,79], or linear elasticity [11,57,58], have incorporated a fiber-in-fluid material representation of the myocardium to study the effect of the transmural variation in muscle fiber orientation on the profiles of diastolic or systolic stress in the left ventricular wall. The fluid-fiber assumption that tension can only be borne in the muscle-fiber direction is not, however, supported by biaxial material tests of passive myocardium [14,91], which have shown that ventricular muscle can bear significant loads acting in directions transverse to the fibers. Moreover, in the intact beating heart, it has been found by Waldman and coworkers [84,85] that the direction of greatest muscle shortening does not necessarily coincide with the muscle-fiber direction, especially near the endocardium where the major axis of shortening may actually be perpendicular to the fibers. Nevertheless, unlike many other models, some of the cylindrical fluid-fiber models have been able to demonstrate the ventricular torsion [4–6,11,57,58,76] that has long been observed (see Ingels and colleagues, [33]). Interestingly, their predicted stress distributions sometimes differed significantly from those cylindrical models that did not have the freedom to twist [17,34,80].

The stresses predicted by models of ventricular mechanics cannot be directly verified because the direct measurement of local forces in the intact heart wall has not been entirely successful [29,90]. However, experimental measurements of myocardial strains have been used for model validation. Arts and coworkers [5] compared the deformations predicted by their cylindrical model of the ejecting left ventricle with systolic strains measured on the epicardium of open-chest canines using a triangular arrangement of inductance gauges. We [24] have used experimental measurements of myocardial strains to validate a thick-walled cylindrical model of the passive left ventricle that employed finite deformation theory and a three-dimensional constitutive equation referred to a system of fiber coordinates. By optimizing material parameters, the model was able to reproduce the circumferential, longitudinal, and torsional epicardial strains that had been measured in the isolated arrested dog heart, as described by McCulloch and Omens in Chapter 5 of this book.

However, cylindrical models are probably confined at best to describing the mechanics of a narrow equatorial cross-section of the left ventricular wall. Such simple models are not suitable for analyzing the nonhomogeneous effects of three-dimensional variations in the geometry, fiber orientations, and mechanical properties of the heart. Nor are factors like diastolic and systolic ventricular interactions or the motion constraints imposed by the pericardium, valve rings, and papillary muscles able to be included realistically in a cylindrical model.

Nielsen and colleagues [56] needed a three-dimensional mathematical model with a complex variation in radius and wall thickness to obtain an accurate approximation to their measurements of ventricular geometry. They also measured the distributions of muscle-fiber orientation in the canine heart and found that transmural fiber angle profiles vary significantly between the two ventricles as well as between the free walls and the interventricular septum.

Another source of heterogeneity in ventricular mechanics is the partly asynchronous activation of the myocardium, which is first activated near the apical endocardium where the wave of depolarization travels rapidly along the Purkinje fiber network toward the base and more slowly to the epicardium through muscle cell conduction with a wave-speed along the muscle fibers that is roughly two to three times that in directions transverse to the fibers [32]. In myocardial infarction, a large portion of the left ventricular wall may become stiff and scarred with no active function. Increased diastolic filling of one ventricle impedes filling of the other [18]. Direct interaction between right and left ventricles during systole has been shown to affect ventricular contraction and relaxation in the intact heart under normal conditions [70]. Diastolic ventricular interaction can be affected by the presence of the pericardium [22]. Severing the chordae tendineae of the mitral valve produces systolic bulging in the region of the papillary muscle insertions [26]. Together, these factors contribute to the regional variations in local ventricular deformations that have been observed in experimental preparations and clinical studies.

For example, passive epicardial stretches measured in the isolated arrested canine heart [48] and systolic shortening measured on the epicardium in humans [41] and at the midwall in dogs [46] were all found to increase in magnitude from the base to the apex of the left ventricular free wall. A similar longitudinal variation in shortening was also observed on the epicardium of the canine right ventricle by Meier and colleagues [49]. Lew and LeWinter [45] measured significantly greater systolic midwall shortening on the anterior free wall than on the posterior free wall of the canine left ventricle. And Kong and colleagues [41] found that systolic shortening in patients was greater on the left ventricular free wall than in the interventricular septum.

6.2   The Finite Element Method

To solve the governing equations of equilibrium for a body with such a complex geometry, boundary conditions, and material properties, computational techniques are required. The most versatile method is the finite element method in which the dependent variables are discretized by piecewise polynomial approximations over finite subdomains (elements) and expressed in terms of parameter values at interelement connection points (nodes).

As early as 1906, researchers first began suggesting the solution of continuum mechanics problems by modeling the body with a lattice of elastic bars and employing frame analysis methods (see Cook, [13]). In 1941, Courant recognized piecewise polynomial interpolation over triangular subregions as a Rayleigh-Ritz solution of a variational problem. Since there were no computers at the time, neither approach was practical and Courant's work was largely forgotten until engineers had independently developed it. By 1953, structural engineers were solving matrix stiffness equations with digital computers. The widespread use of finite element methods in engineering began with the classic papers by Turner and colleagues [78] Argyris and

Kelsey [3]. The name "finite element" was coined in 1960, and the method
began to be recognized as mathematically rigorous by 1963.

Many finite element models of ventricular mechanics have been proposed,
although most of them did not include the nonlinear kinematic terms associated with large deformations because iterative solution to the nonlinear
governing equations at each load step is required. The importance of adopt-
ing nonlinear finite deformation theory for the analysis was demonstrated
by Janz and colleagues [38]. However, their model along with a few subsequent finite element models based on finite deformation theory [25,40,92]
and most of the linear models [9,20,23,27,36,55,61,62,65] treated the myocardium as an isotropic material. The effect of the ventricular muscle-fiber
distribution has been modeled with finite elements that possess material
anisotropy with respect to a continuously varying fiber axis [7,28,31] or
more commonly by using a number of concentric elements, each with a
constant fiber direction [12,60,63,89]. The incompressibility of the heart
muscle, which is composed mostly of water, was very often accounted for
incorrectly by assuming that the myocardium has a Poisson ratio close to
0.5 [12,20,27,36,37,61,63,82,89]. More correctly, in the context of finite deformations, the hydrostatic pressure—an extra dependent variable arising
from the kinematic incompressibility constraint—is introduced as a Lagrange multiplier in the strain energy function[2] [7,25,28,31,81]. Although
a few models consider muscle activation and contraction [7,25,63], the rest
have been concerned with the accurate prediction of end-diastolic stresses
and sarcomere length distributions, a prerequisite for a realistic model of
active ventricular contraction.

Whereas a considerable body of information is available on the heterogeneity of wall motions in the beating heart, regional strain measurements
suitable for validating models of the passive left ventricle have only recently become available [48]. Only the finite element model predictions
of McCulloch [47] and Hunter and colleagues [31] have been tested with
experimental strain data. However, these comparisons were made at just
one location on the epicardial surface and the rather poor agreement was
largely attributable to the somewhat arbitrary myocardial stress-strain relation that these investigators chose and their simplified geometries, which
were each modeled with only one finite element. In this chapter, we employ the finite element method of Hunter, McCulloch and coworkers [31],
but we adopt a new constitutive equation and we extend the model to a

---

[2]It should be noted that although the "hydrostatic pressure" does have units
of stress and is added to the diagonal components of the stress tensor in the
manner of a pressure, it does not itself coincide either with the mean stress at a
point in the myocardium (the "intramyocardial pressure") or with the boundary
pressures loading the ventricular wall. It is further unsuited to direct physical
interpretation since it may not even be zero in the stress-free body.

more realistic axisymmetric geometry with more physiological boundary conditions.

6.3   An Axisymmetric Finite Element Model of the Passive Left Ventricle

Our finite element method was specifically developed for continuum analysis of the heart [31,47]. Hence, the approach includes several special features uncommon to conventional finite element methods. The Galerkin finite element equations for three-dimensional finite elasticity were derived in general tensor form and then specialized for four alternative reference coordinate systems: rectangular Cartesian, cylindrical polar, spherical polar, and prolate spheroidal. Therefore, using conforming isoparametric elements with tensor-product basis functions for each of the geometric coordinates, only one axisymmetric element with four global nodes was needed to model each of the cylindrical, spherical, and prolate spheroidal geometries illustrated in Figure 6.1. By formulating the governing equations in a coordinate system that simplifies the geometric description of the deforming body, the greater algebraic complexity is more than compensated by a significant reduction in the number of degrees of freedom required for the finite element discretization. The resulting nonlinear system of global equilibrium equations are solved, subject to appropriate boundary conditions, for the nodal displacement and hydrostatic pressure parameters using a quasi-Newton iteration scheme.

For incompressible bodies, the kinematics of the deformation are constrained by the condition that the volume of any arbitrarily small part of the wall must remain constant. For most finite element schemes, this condition can be satisfied only in the average sense for the entire finite element. However, significant improvement in the finite element stress and strain solutions was obtained using a new "isochoric" element interpolation method with which—in geometrically simple bodies such as cylinders or spheres—the kinematic incompressibility constraint was satisfied exactly throughout the element. In these isochoric elements, rather than interpolate the radial coordinate (r in a polar system), we interpolate an appropriate function of r that is proportional to the wall volume—that is, r^2 in the case of cylindrical elements, or r^3 in the case of spherical ones. In prolate spheroidal coordinates (λ, μ, θ), the function is less intuitive ($d^3 \sinh^2 \lambda \cosh \lambda$, where d is the focal length) but the idea is the same. For problems with more complex geometries the isochoric elements are no longer kinematically exact, but they still result in a considerable improvement in accuracy for incompressible finite deformations.

Since the dependent variables in the formulation of the model are the nodal displacements and the hydrostatic pressure variable that arises from

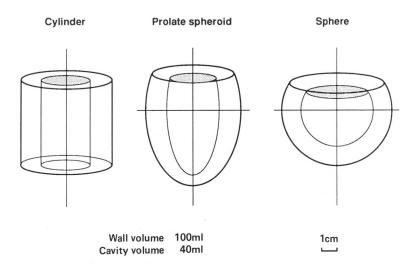

FIGURE 6.1. The cylindrical, prolate spheroidal, and spherical finite element models of the left ventricle used by Hunter et al. [31]. These geometries were each modeled exactly by a single axisymmetric finite element. All three geometric models had a wall volume of 100 ml and an undeformed cavity volume of 40 ml based on experimental measurements from the canine heart. (Reproduced from Hunter et al. [31], with permission from the ASME.)

the incompressibility condition, this finite element problem is a "mixed" formulation. With such formulations, a compatibility condition that relates the pressure and displacement approximating spaces must be satisfied to ensure that a given displacement field corresponds to a unique solution for the hydrostatic pressure [42]. In practice, this requires that for a linear nodal interpolation of the geometric parameters, such as in the elements shown in Figure 6.1, a single constant hydrostatic pressure parameter is all that can be permitted for each element to avoid numerical ill-conditioning. Since the incompressible deformation of thick-walled vessels results in a nonlinear transmural distribution of hydrostatic pressure, several finite elements each with a constant pressure are usually required through the wall thickness. However, because the ventricles of the heart are loaded by boundary pressures, it was possible to increase the hydrostatic pressure interpolation to a quadratic transmural variation by using the boundary pressures prescribed on the inner and outer surfaces of the element as additional constraints. This results in stress solutions that closely or exactly match the boundary stress field, and for simple axisymmetric problems to which closed-form solutions exist, the results obtained with a single finite element were found to be practically exact throughout the wall [47].

The material anisotropy of the heart muscle is defined by referring the element stress components to a system of embedded material coordinates, which is orthogonal in the undeformed state and has one axis aligned with the fiber direction. The fiber axis lies in the plane of the wall so that it is tangential to the boundary at the endocardium and the epicardium, but the in-plane fiber angle can change continuously through the element as defined by an appropriate nodal finite element interpolation. In the constitutive equation for stress, the myocardium is modeled as transversely isotropic with respect to the local fiber axis. These three-dimensional, curvilinear, isochoric finite elements with stress boundary constraints and embedded fiber fields meet the most important requirements for modeling passive ventricular mechanics. However, the stresses predicted by these finite elements will depend on the geometry, fiber angle distribution, material properties, displacement constraints, and loading assumed in the model.

In the study of Hunter and coworkers [31], the canine left ventricle was represented by the cylindrical, spheroidal, and prolate spheroidal finite elements shown in Figure 6.1, with a cubic variation of fiber angle across the wall from $-60°$ on the epicardium to $+60°$ on the endocardium. The myocardium was modeled both as an isotropic and as a transversely isotropic, homogeneous elastic material. All three geometric models were constrained from extending axially and rotating at the base. A uniform cavity pressure was applied to the endocardium while the epicardial pressure remained zero. For the isotropic material model, the material constants of an exponential stress-strain relation (strain energy function) were chosen to fit sarcomere length-tension measurements from isolated muscle preparations [74]. To examine the effects of transverse isotropy without altering the properties of the model material along the fiber direction, the uniaxial tensile stiffness in the transverse plane was scaled up by a constant factor chosen rather arbitrarily to be 16. The authors acknowledged that there was no firm foundation for this choice. It was based on the observation that the isotropic models were too compliant in that they significantly overestimated passive ventricular volumes at physiological pressure loads. Since the stress-strain behavior of the isotropic models had been fitted to accurate measurements of muscle sarcomere length-tension relations, the authors postulated that the additional muscle stiffness needed to produce a more physiological pressure-volume relation might be added in the transverse plane without necessarily changing the uniaxial properties of the model material in the fiber direction.

These models were evaluated by comparing predicted epicardial deformations with experimental measurements from the midanterior wall of the isolated potassium-arrested dog heart [48]. In this region, the observed principal angle of greatest stretch was consistently clockwise from circumferential, tending to increase with pressure to coincide closely with the epicardial fiber angle (mean, $-42 \pm 7°$) at high loads. This principal angle was also consistent with the torsional deformation of the ventricle that was observed. However, the principal stretches of the three isotropic models were aligned with the circumferential and longitudinal axes, so the principal angles were always zero and there was no torsion. Although the transversely isotropic models did twist upon inflation, the predicted epicardial extensions did not correspond closely with the experimental measurements. In the prolate spheroidal model, the epicardium twisted in the same direction as the experimental preparation at low pressures, but reversed direction as the load increased.

These workers had hypothesized that by making the passive myocardium most compliant in the fiber direction, the predicted epicardial extensions during filling would be greatest along this axis as they had observed experimentally. But they found that their anisotropic models were not highly sensitive to increases in the transverse stiffness, especially in the cylindrical model, which was prevented from extending longitudinally. However, in the analytical model subsequently proposed by Guccione and coworkers [24], we found that the amount a cylinder twists and extends upon inflation is indeed sensitive to the biaxial stiffness ratio and also to the fiber angle distribution. As described in Chapter 5, we obtained estimates for the parameters of a transversely isotropic material defined with an exponential strain-energy function by minimizing the differences between the experimental and model-predicted strains using a semi-inverse approach. Somewhat surprisingly, the optimal material was stiffest in the fiber direction, not the transverse direction. The fiber-to-transverse stiffness ratio[3]

---

[3] This stiffness ratio was defined as the stiffness in the fiber direction divided by

was 3.3. Here we extend the work of Hunter and coworkers [31] beyond the three simple test shapes to an axisymmetric model of the passive left ventricular free wall that incorporates the following form of the strain-energy function used by Guccione and colleagues [24]:

$$W = \frac{C}{2}(e^Q - 1) - \frac{1}{2}p(I_3 - 1) \qquad (6.1)$$

where C is a constant, p is the hydrostatic pressure variable, I_3 is the third principal strain invariant, which is unity for an incompressible deformation, and

$$Q = b_1 E_{11}^2 + b_2(E_{22}^2 + E_{33}^2 + E_{23}^2 + E_{32}^2) + b_3(E_{12}^2 + E_{21}^2 + E_{13}^2 + E_{31}^2) \quad (6.2)$$

where b_1, b_2, and b_3 are material constants and E_{RS} are Lagrangian strain components referred to the (undeformed) fiber coordinate system $\{X_R\}$, where X_1 is aligned with the muscle fibers, X_2 is the cross-fiber in-plane axis, and X_3 is the transmural coordinate. The material defined by Equations (6.1) and (6.2) possesses symmetry under the group of all rotations about the X_1-axis, and is therefore said to be "transversely isotropic" about X_1. The fiber angle used in the following models varied linearly from ϕ_i on the endocardium to ϕ_o at the epicardium, and the material constants in Equations (6.1) and (6.2) were $C = 0.88$ kPa, $b_1 = 18.5$, $b_2 = 3.58$, and $b_3 = 1.63$.

For comparison with the earlier results, we began with the one-element axisymmetric prolate spheroidal model as shown in Figure 6.2. The solid lines represent the unloaded reference state. The undeformed cavity volume was 40 ml and the volume of the ventricular myocardium was 100 ml, typical of the canine heart. The focus of the ellipsoidal surfaces was 3.75 cm and the basal plane was 30° from the equator as suggested by Streeter and Hanna [73]. At all longitudinal coordinates (μ), the transmural fiber angle distribution was defined by $\phi_i = 65°$ and $\phi_o = -55°$.

We assumed that there is no stress in the passive ventricular wall when the filling pressure is zero. However, recently it has been shown that the unloaded left ventricle is not stress-free. Omens and Fung [59] have demonstrated that when a cross-sectional ring of the freshly excised potassium-arrested rat heart is cut radially, it springs open to form an arc, as described by McCulloch and Omens in the previous chapter. Although such experiments have not yet been reported for the canine heart, the analysis of Guccione and colleagues [24] showed that the effect of this opening angle on end-diastolic stresses was significant, although the estimated material properties were not particularly sensitive to this quantity. However, it is not yet clear how to incorporate residual stresses in an axisymmetric finite element model. Since the governing equations are nonlinear, the initial

---

the stiffness in the transverse plane that would be measured in a plane equibiaxial stretch of 1.2.

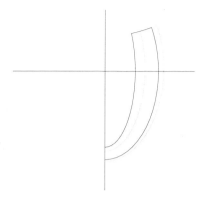

FIGURE 6.2. One-element prolate spheroidal model of the canine left ventricle. The unloaded reference state (solid lines) was bounded in the prolate spheroidal coordinate system by the surfaces defined by $\lambda = 0.43, \lambda = 0.72$, and $\mu = 120°$. Muscle-fiber angle varied linearly from 65° (endocardium) to −55° (epicardium) where zero corresponds to the circumferential direction and a positive angle is rotated counterclockwise. The dotted lines show the deformed profile for a ventricular pressure of 2 kPa (15 mm Hg).

stresses cannot be superposed in the solution. And if the residual stresses are added to the stresses in the constitutive equation, the resulting relation no longer satisfies the principal of material frame indifference. Therefore at this stage, the residual stress phenomenon is not included in our finite element models.

Using a single isochoric element, the deformed profile shown in dotted lines in Figure 6.2 was obtained for inflation to a ventricular pressure of 2.0 kPa (15 mm Hg). It shows that the deformation was greatest near the base and in the radial direction, and therefore circumferential stretch was dominant. This is shown by the dotted line in Figure 6.3, where the major principal stretch at the midanterior epicardium is plotted against the mutually perpendicular minor stretch for ventricular pressures up to 0.8 kPa. The results for the one-element model show that, in comparison with the experimental measurements, the epicardial strains were too nonuniform and the model deformed more like an inflated cylindrical tube (a vertical line) than a spherical balloon (a slope of one). Although the predicted principal angles of the major stretch were within one standard error of the experimental measurements at the lower loads (Figure 6.4, dotted line), the experimental means varied with load from −22° to −52°, whereas the predicted angles remained quite constant at −13° to −15°. Consequently, the one-element model did predict torsion in the same direction as observed experimentally at all loads; however, the amount was now exaggerated in the model (7.6° compared with 4.1° at the highest load).

To simulate the effects of the relatively stiff mitral valve ring on left ven-

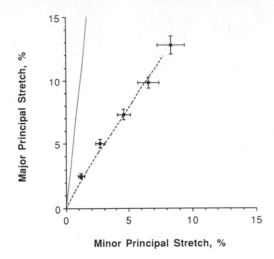

FIGURE 6.3. Major principal stretches at the midanterior epicardium plotted against the mutually perpendicular minor stretches at various left ventricular pressures. The experimental measurements (solid circles) of McCulloch et al. [48] are compared with the predictions of the one-element prolate spheroidal model (dotted line) and the 14-element prolate spheroidal model (dashed line). The error bars on the experimental measurements indicate one standard error of the mean. The five experimental points are for mean left ventricular pressures of approximately 0.2, 0.4, 0.7, 1.1, and 1.7 kPa.

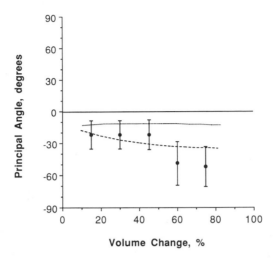

FIGURE 6.4. The principal angle of the major stretch on the midanterior epicardium plotted against left ventricular volume change. The experimental measurements (solid circles) of McCulloch et al. [48] at five left ventricular pressures are compared with the predictions of the one-element model (dotted line) and the 14-element model (dashed line). The error bars on the experimental measurements indicate one standard error of the mean.

tricular filling and make the model bulge more like a sphere, we constrained the prolate spheroidal mesh by preventing displacement of the epicardium at the base of the model. But to obtain a converged solution for this problem required considerably more degrees of freedom along the longitudinal direction than the one-element case. Therefore, the model was subdivided repeatedly into elements equally spaced along the longitudinal coordinate (μ). However, even with 16 such elements, the convergence criteria for accurate stress solutions near the base and apex were not met. To reduce the required computation, an optimum mesh layout with a minimum number of nodes was sought. To determine which portions of the mesh required refinement or optimization, the differences between the strain energy densities at neighboring Gaussian quadrature integration points within each element and across each interelement boundary were computed. A large difference indicates departure from constant strain conditions and suggests the need for mesh refinement [50,77]. The positions of the element boundaries were adjusted until these strain energy density differences were less than 10% throughout the longitude. Thus, we found that only regions near the base and apex needed refining. By having several small elements at the base and apex and a few larger elements in between, converged stress solutions were obtained for a mesh with only 14 elements (Figure 6.5).

Figure 6.5 also shows the bulging pattern of ventricular deformation in

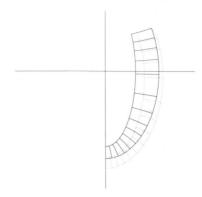

FIGURE 6.5. Fourteen-element prolate spheroidal model of the canine left ventricle. The unloaded reference shape (solid lines) and muscle-fiber angle distribution were the same as for the one-element model in Figure 6.2. The dotted lines show the deformed profile for a ventricular pressure of 2 kPa (15 mm Hg) with the base constrained from displacing at the epicardial node to simulate the effects of the stiff mitral valve ring.

contrast with the one-element model (Figure 6.2). The principal epicardial extensions near the equatorial region for ventricular pressures of 0 to 2.0 kPa are shown by the dashed line in Figure 6.3. The comparison with the one-element model and the experimental measurements shows that the effect of the basal constraint was significant and the agreement with the experimental data was excellent. Not only were the predicted principal directions of the major stretch within one standard error of the measurements, but they displayed the same variation with load as the measured means (Figure 6.4, dashed line). Consequently, the torsional displacements of the model at the midventricle were almost identical to those measured in the dog heart (4.4° compared with 4.1° for a 75% volume inflation).

The longitudinal distribution of fiber stress midway through the wall thickness of the 14-element prolate spheroidal model is shown in Figure 6.6 for a ventricular pressure of 1 kPa (7.5 mm Hg). The reasonably continuous stress distribution suggests that the convergence for this solution is good. The stresses appear to change most rapidly at the apex owing to the rapid changes of curvature there. However, the real behavior in this region may be different because the assumption that the muscle fibers lie in the plane of the endocardial and epicardial surfaces is probably not valid near the apical infundibulum [71]. Although this model provides a converged solution for base-apex stress distributions, the longitudinal variation in wall thickness and transmural fiber angle distribution of the left ventricle should also be included.

The prolate spheroidal nodal coordinates and fiber angle values of the 14-element axisymmetric mesh shown in Figure 6.7 were fitted to the anatom-

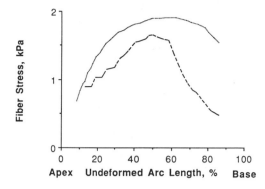

FIGURE 6.6. Longitudinal distribution of fiber stress midway through the wall thickness at a ventricular pressure of 1 kPa (7.5 mm Hg). The dotted line shows the solution for the fourteen-element prolate spheroidal model, and the dashed line corresponds to the axisymmetric model shown in Figure 6.7. Note that the stresses in the latter model were significantly lower in the basal region, and had a peak slightly nearer the apex.

ical measurements of Nielsen and coworkers at a section 36° from the anterior border of the left ventricular free wall along the same meridian on which the experimental measurements were made. The element spacing is the same as in the previous confocal prolate spheroidal mesh. Figure 6.6 shows the longitudinal distribution of fiber stress at midwall predicted by this model for a ventricular pressure of 1.0 kPa. This figure also shows that these midwall stresses were significantly lower in the basal region than those in the more idealized prolate spheroidal geometry. The deformed model shape at a load of 2.0 kPa is also shown in Figure 6.7. Again, a bulging deformation is produced by inflation with very little displacement near the fixed base.

Figure 6.8 shows the profiles from base to apex of principal epicardial strains predicted by the axisymmetric finite element model in comparison with experimental measurements at a low, medium, and high end-diastolic volume. The strains all rise from the base to a peak at about 50% along the arc length from the apex, and then again toward the apex. The model predictions appear to agree quite well with the experimental data at the basal and midventricular sites but are too low near the apex where the epicardial solutions may not be fully converged. The epicardial shear strains were negative at all volumes and fairly uniform throughout the longitude, except near the apex where they decreased rapidly to approximately zero.

The model predictions may well be improved by employing a cross-section that is thinner in the apical region. In fact, the circumferential component of epicardial strain predicted by the 14-element model shown in Figure 6.5 increased from the base to a peak much nearer the apex and

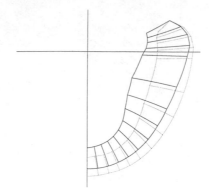

FIGURE 6.7. Fourteen-element axisymmetric model fitted by least squares to anatomical measurements from a cross-section of the anterior free wall of the canine left ventricle [56]. In the unloaded reference state (solid lines), ventricular cavity volume was 37 ml and wall volume was 125 ml. Within each element fiber angle varied linearly both in the transmural and longitudinal directions. Epicardial fiber angles varied between −43° at the base and −53° at the apex. On the endocardium, fiber angle increased continuously from 82° at the base to 97° at the apex. The dotted lines show the deformed profile for a diastolic ventricular pressure of 2 kPa (15 mm Hg) with the base constrained.

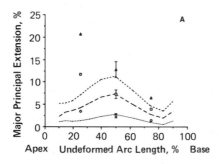
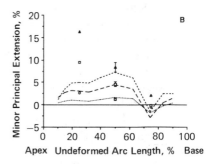

FIGURE 6.8. Longitudinal profiles of principal epicardial extensions at low, medium, and high left ventricular diastolic volumes. The lines show the predictions of the axisymmetric model shown in Figure 6.7 for increases in cavity volume of 15% (dotted line), 45% (long-dashed line), and 75% (short-dashed line). The symbols show the experimental measurements at three epicardial locations on the anterior free wall for the same three volume loads (circles, squares, and triangles, respectively). (A) Major epicardial percent extension. (B) Minor epicardial percent extension.

agreed closely with experimental results, especially at the apical and mid-ventricular sites. In that model, however, the longitudinal strain decreased slightly from base to apex in contrast to experimental observations.

A further extension to a fully three-dimensional finite element geometry is also needed. Although the regional measurements available for validating such a model remain scant, three-dimensional data will clearly be needed to understand the effects of such important factors as left and right ventricular interactions. On the other hand, since the axisymmetric model presented here is able to simulate many of the properties of the passive left ventricle, at least in significant regions of the free wall, it may be a useful starting point for a model of the active contraction of the ventricular myocardium through the cardiac cycle.

Acknowledgements: We are grateful to Dr. Peter J. Hunter, who initiated the development of our finite element models of the heart and has continued to extend the scope of the methods over the past 10 years. We would also like to thank Professor Y-C Fung, Dr. Poul M.F. Nielsen, and Dr. Lewis K. Waldman for their valuable advice and contributions to this work. This research was supported by PHS grant HL41603 and J.M. Guccione was supported by PHS Predoctoral Training Grant HL07089. This support is gratefully acknowledged.

REFERENCES

[1] H. Abe and T. Nakamura. Finite deformation model for the mechanical behavior of left ventricular wall muscles. *Math. Modeling*, 3:143–152, 1982.

[2] N.R. Alpert. *Cardiac Hypertrophy*. Academic Press, New York, 1971.

[3] J.H. Argyris and S. Kelsey. *Energy Theorems and Structural Analysis*. Butterworths, London, 1960.

[4] T. Arts and R.S. Reneman. Dynamics of left ventricular wall and mitral valve mechanics—A model study. *J. Biomech.*, 22:261–271, 1989.

[5] T. Arts, R.S. Reneman, and P.C. Veenstra. Epicardial deformation and left ventricular wall mechanics during ejection in the dog. *Am. J. Physiol.*, 243:H379–H390, 1982.

[6] T. Arts, R.S. Reneman, and P.C. Veenstra. A model of the mechanics of the left ventricle. *Ann. Biomed. Eng.*, 7:299–318, 1979.

[7] D.A. Bergel and P.J. Hunter. The mechanics of the heart. In D.R. Gross, N.H.C. Hwang and D.J. Patel, editors, *Quantitative Cardio-*

vascular Studies, pages 151–213, University Park Press, Baltimore, 1979.

[8] R. Beyar and S. Sideman. The dynamic twisting of the left ventricle: A computer study. *Ann. Biomed. Eng.*, 14:547–562, 1986.

[9] D.K. Bogen, S.A. Rabinowitz, A. Needleman, T.A. McMahon, and W.H. Abelmann. An analysis of the mechanical disadvantage of myocardial infarction in the canine left ventricle. *Circ. Res.*, 47:728–741, 1980.

[10] T.K. Borg, W.F. Ranson, and F.A. Moslehy. Structural basis of ventricular stiffness. *Lab. Invest.*, 44:49–54, 1981.

[11] R.S. Chadwick. Mechanics of the left ventricle. *J. Biophys.*, 39:279–288, 1982.

[12] C.J. Chen, B.M. Kwak, K. Rim, and H.L. Falsetti. A model for an active left ventricle deformation—Formulation of a nonlinear quasi-steady finite element analysis for orthotropic, three-dimensional myocardium. *International Conference on Finite Elements in Biomechanics*, 2:640–655, 1980.

[13] R.D. Cook. *Concepts and Applications of Finite Element Analysis.* John Wiley & Sons, New York, 1981.

[14] L.L. Demer and F.C.P. Yin. Passive biaxial mechanical properties of isolated canine myocardium. *J. Physiol. Lond.*, 339:615–630, 1983.

[15] H. Demiray. Stresses in ventricular wall. *ASME J. Appl. Mech.*, 43: 194–197, 1976.

[16] H.L. Falsetti, R.E. Mates, C. Grant, D.G. Green, and I.L. Bunnell. Left ventricular wall stress calculated from one-plane cineangiography. *Circ. Res.*, 26:71–83, 1970.

[17] T.S. Feit. Diastolic pressure-volume relations and distribution of pressure and fiber extension across the wall of a model left ventricle. *J. Biophys.*, 28:143–166, 1979.

[18] M.P. Feneley, C.O. Olsen, D.D. Glower, and J.S. Rankin. Effect of acutely increased right ventricular afterload on work output from the left ventricle in conscious dogs. *Circ. Res.*, 65:135–145, 1989.

[19] Y.C. Fung. *Biodynamics: Circulation.* Springer-Verlag, New York, 1984.

[20] D.N. Ghista and M.S. Hamid. Finite element analysis of the human left ventricle whose irregular shape is developed from single plane cineangiocardiogram. *Comp. Prog. Biomed.*, 7:219–231, 1977.

[21] D.N. Ghista and H. Sandler. An analytic elastic-viscoelastic model for the shape and the forces in the left ventricle. *J. Biomech.*, 2:35–47, 1969.

[22] S.A. Glantz, G.A. Misbach, W.Y. Moores, et al. The pericardium substantially affects the left ventricular diastolic pressure-volume relationship in the dog. *Circ. Res.*, 42:433–441, 1978.

[23] P. Gould, D. Ghista, L. Brombolich, and I. Mirsky. In vivo stresses in the human left ventricular wall: Analysis accounting for the irregular 3-dimensional geometry and comparison with idealised geometry analysis. *J. Biomech.*, 5:521–539, 1972.

[24] J.M. Guccione Jr., A.D. McCulloch, and L.K. Waldman. Passive material properties of intact ventricular myocardium determined from a cylindrical model. *ASME J. Biomech. Eng.*, 113, 1991 (in press).

[25] M.S. Hamid, H.N. Sabbah, and P.D. Stein. Determination of left ventricular wall stress during isovolumic contraction using incompressible finite elements. *Comp. Struc.*, 24:589–594, 1986.

[26] D.E. Hansen, G.E. Sarris, M.A. Niczyporuk, G.C. Derby, P.D. Cahill, and D.C. Miller. Physiologic role of the mitral apparatus in left ventricular regional mechanics, contraction synergy, and global systolic performance. *J. Thorac. Cardiovasc. Surg.*, 97:521–533, 1989.

[27] R.M. Heethaar, Y.C. Pao, and E.L. Ritman. Computer aspects of three-dimensional finite element analysis of stresses and stains in the intact heart. *Comp. Biomed. Res.*, 10:271–285, 1977.

[28] A. Horowitz, I. Sheinman, and Y. Lanir. Nonlinear incompressible finite element for simulating loading of cardiac tissue—Part II: Three dimensional formulation for thick ventricular wall segments. *ASME J. Biomech. Eng.*, 110:62–68, 1988.

[29] R.M. Huisman, G. Elzinga, N. Westerhof, and P. Sipkema. Measurement of left ventricular wall stress. *Cardiovasc. Res.*, 14:142–153, 1980.

[30] J.D. Humphrey and F.C.P. Yin. Biaxial mechanical behavior of excised epicardium. *ASME J. Biomech. Eng.*, 110:349–351, 1988.

[31] P.J. Hunter, A.D. McCulloch, P.M.F. Nielsen, and B.H. Smaill. A finite element model of passive ventricular mechanics. In R.L. Spilker and B.R. Simon, editors, *Computational Methods in Bioengineering*, pages 387–397, ASME, New York, 1988.

[32] P.J. Hunter and B.H. Smaill. The analysis of cardiac function: A continuum approach. *Prog. Biophys. Molec. Biol.*, 52:101–164, 1988.

[33] N.B. Ingels Jr., D.E. Hansen, G.T. Daughters II, E.B. Stinson, E.L. Alderman, and D.C. Miller. Relation between longitudinal, circumferential, and oblique shortening and torsional deformation in the left ventricle of the transplanted human heart. *Circ. Res.*, 64:915–927, 1989.

[34] K.M. Jan. Distribution of myocardial stress and its influence on coronary blood flow. *J. Biomech.*, 18:815–820, 1985.

[35] R.F. Janz. Estimation of local myocardial stess. *Am. J. Physiol.*, 242:H875–H881, 1982.

[36] R.F. Janz and A.F. Grimm. Deformation of the diastolic left ventricle. I. nonlinear elastic effects. *Biophys. J.*, 13:689–704, 1973.

[37] R.F. Janz and A.F. Grimm. Finite element model for the mechanical behavior of the left ventricle. *Circ. Res.*, 30:244–252, 1972.

[38] R.F. Janz, B.R. Kubert, and T.F. Moriarty. Deformation of the diastolic left ventricle. II. Nonlinear geometric effects. *J. Biomech.*, 7:509–516, 1974.

[39] R.F. Janz, S. Ozpetek, L.E. Ginzton, and M.M. Laks. Regional stress in a noncircular cylinder. *Biophys. J.*, 55:173–182, 1989.

[40] R.F. Janz and R.J. Waldron. Predicted effect of chronic apical aneurysms on the passive stiffness of the human left ventricle. *Circ. Res.*, 42:255–263, 1978.

[41] Y. Kong, J.J. Morris Jr., and H.D. McIntosh. Assessment of regional myocardial performance from biplane coronary cineangiograms. *Am. J. Cardiol.*, 27:529–537, 1971.

[42] P. Le Tallec. Compatibility condition and existence results in discrete finite incompressible elasticity. *Comp. Meth. Appl. Math. Eng.*, 27:239–259, 1981.

[43] M.C. Lee, Y.C. Fung, R. Shabetai, and M.M. LeWinter. Biaxial mechanical properties of human pericardium and canine comparisons. *Am. J. Physiol.*, 253:H75–82, 1987.

[44] W.Y.W. Lew. Influence of ischemic zone size on nonischemic area function in the canine left ventricle. *Am. J. Physiol.*, 252:H990–H997, 1987.

[45] W.Y.W. Lew and M.M. LeWinter. Regional comparison of midwall segment and area shortening in canine left ventricle. *Circ. Res.*, 58:678–691, 1986.

[46] M.M. LeWinter, R.S. Kent, J.M. Kroener, T.E. Carew, and J.W. Covell. Regional difference in myocardial performance in the left ventricle. *Circ. Res.*, 37:191–199, 1975.

[47] A.D. McCulloch. *Deformation and Stress in the Passive Heart*. Ph.D. thesis, University of Auckland, New Zealand, 1986.

[48] A.D. McCulloch, B.H. Smaill, and P.J. Hunter. Regional left ventricular epicardial deformation in the passive dog heart. *Circ. Res.*, 64:721–733, 1989.

[49] C.D. Meier, A.A. Bove, W.P. Santamore, and P.R. Lynch. Contractile function in the canine right ventricle. *Am. J. Physiol.*, 39:H794–804, 1980.

[50] R.J. Melosh and P.V. Marcal. An energy basis for mesh refinement of strectural continua. *Int. J. Num. Meth. Eng.*, 11:1083–1091, 1977.

[51] I. Mirsky. Effects of anisotropy and nonhomogeneity on left ventricular stresses in the intact heart. *Bull. Math. Biophys.*, 32:197–213, 1970.

[52] I. Mirsky. Left ventricular stresses in the intact human heart. *Biophys. J.*, 9:189–208, 1969.

[53] I. Mirsky. Ventricular and arterial wall stresses based on large deformation analyses. *Biophys. J.*, 13:1141–1159, 1973.

[54] T.F. Moriarty. Law of Laplace: Its limitations as a relation for diastolic pressure, volume or wall stress of the left ventricle. *Circ. Res.*, 46:321–331, 1980.

[55] A. Needleman, S.A. Rabinowitz, D.K. Bogen, and T.A. McMahon. A finite element model of infarcted left ventricle. *J. Biomech.*, 16:45–58, 1983.

[56] P.M.F. Nielsen, I.J. LeGrice, B.H. Smaill, and P.J. Hunter. A mathematical model of the geometry and fibrous structure of the heart. *Am. J. Physiol.*, 1991, (in press).

[57] J. Ohayon and R.S. Chadwick. Theoretical analysis of the effects of a radial activation wave and twisting motion on the mechanics of the left ventricle. *Biorheology*, 25:435–447, 1988b.

[58] J. Ohayon and R.S. Chadwick. Theoretical analysis of the effects of a radial activation wave and twisting motion on the mechanics of the left ventricle. *Biophys. J.*, 54:1077–1088, 1988a.

[59] J.H. Omens and Y.C. Fung. Residual strain in rat left ventricle. *Circ. Res.*, 66:37–45, 1990.

[60] C.S. Panda and R. Natarajan. Finite method of stress analysis in the human left ventricular layered wall structure. *Med. Biol. Eng. Comp.*, 15:67–71, 1977.

[61] Y.C. Pao, E.L. Ritman, and E.H. Wood. Finite-element analysis of left ventricular myocardial stresses. *J. Biomech.*, 7:469–477, 1974.

[62] Y.C. Pao, R.A. Robb, and E.L. Ritman. Plane-strain finite-element analysis of reconstituted diastolic left ventricular cross section. *Ann. Biomed. Eng.*, 4:232–249, 1976.

[63] M. Perl, A. Horowitz, and S. Sideman. Comprehensive model for the simulation of left ventricle mechanics. Part I. Model description and simulation procedure. *Med. Biol. Eng. Comp.*, 24:145–149, 1986.

[64] J.G. Pinto and Y.C. Fung. Mechanical properties of the heart muscle in the passive state. *J. Biomech.*, 6:597–616, 1973.

[65] E.L. Ritman, R.M. Heethaar, R.A. Robb, and Y.C. Pao. Finite element analysis of myocardial diastolic stress and strain relationships in the intact heart. *Eur. J. Cardiol.*, 7:105–119, 1978.

[66] T.F. Robinson, L. Cohen-Gould, and S.M. Factor. Skeletal framework of mammalian heart muscle: Arrangement of inter- and pericellular connective tissue structures. *Lab. Invest.*, 49:482–498, 1983.

[67] H. Sandler and H.T. Dodge. Left ventricular tension and stress in man. *Circ. Res.*, 13:91–104, 1963.

[68] S.J. Sarnoff, E. Braunwald, G.H. Welch Jr., R.B. Case, W.N. Stainsby, and R. Macruz. Hemodynamic determinants of oxygen consumption of the heart with special reference to the tension-time index. *Am. J. Physiol.*, 192:148–156, 1958.

[69] P.N. Shivakumar, C-S. Man, and S.W. Rabkin. Modelling of the heart and pericardium at end-diastole. *J. Biomech.*, 22:201–209, 1989.

[70] B.K. Slinker, Y. Goto, and M.M. LeWinter. Systolic direct ventricular interaction affects left ventricular contraction and relaxation in the intact dog circulation. *Circ. Res.*, 65:307–315, 1989.

[71] D.D. Streeter. Gross morphology and fiber geometry of the heart. In R.M. Berne, editor, *Handbook of Physiology*, pages 339–350, American Physiological Society, Bethesda, MD, 1979.

[72] D.D. Streeter Jr. and W.T. Hanna. Engineering mechanics for successive states in canine left ventricular myocardium: II. Fiber angle and sarcomere length. *Circ. Res.*, 33:656–664, 1973b.

[73] D.D. Streeter Jr. and W.T. Hanna. Engineering mechanics for successive states in canine left ventricular myocardium: I. Cavity and wall geometry. *Circ. Res.*, 33:639–655, 1973a.

[74] H.E.D.J. Ter Keurs, W.H. Rijnsburger, R. Van Heuningen, and M.J. Nagelsmit. Tension development and sarcomere length in rat cardiac trabeculae: Evidence of length-dependent activation. *Circ. Res.*, 46:703–71, 1980.

[75] P. Theroux, J. Ross Jr., D. Franklin, J.W. Covell, C.M. Bloor, and S. Sasayama. Regional myocardial function and dimensions early and late after myocardial infarction in the unanesthetized dog. *Circ. Res.*, 40:158–165, 1977.

[76] A. Tözeren. Static analysis of the left ventricle. *ASME J. Biomech. Eng.*, 105:39–46, 1983.

[77] D.J. Turcke and G.M. McNeice. Guidelines for selecting finite element grids based on an optimization study. *Comp. Struc.*, 4:499–519, 1974.

[78] M.J. Turner, R.W. Clough, H.C. Martin, and L.J. Topp. Stiffness and deflection analysis of complex structures. *J. Aero. Sci.*, 9:805–823, 1956.

[79] J.H.J.M. Van den Broek and J.J. Denier Van der Gon. A model study of isovolumic and nonisovolumic left ventricular contractions. *J. Biomech.*, 13:77–87, 1980.

[80] J.H.J.M. Van den Broek and M.H.L.M. Van den Broek. Application of an ellipsoidal heart model in studying left ventricular contractions. *J. Biomech.*, 13:493–503, 1980.

[81] D.L. Vawter. Poisson's ratio and incompressiblity. *ASME J. Biomech. Eng.*, 105:194–195, 1983.

[82] C.A. Vinsen, D.G. Gibson, and A.L. Yettram. Analysis of left ventricular behavior in diastole by means of finite element method. *Br. Heart. J.*, 41:60–67, 1979.

[83] R.P. Vito. The role of the pericardium in cardiac mechanics. *J. Biomech.*, 12:587–592, 1979.

[84] L.K. Waldman, Y.C. Fung, and J.W. Covell. Transmural myocardial deformation in the canine left ventricle: Normal in vivo three-dimensional finite strains. *Circ. Res.*, 57:152–163, 1985.

[85] L.K. Waldman, D. Nosan, F.J. Villarreal, and J.W. Covell. Relation between transmural deformation and local myofiber direction in canine left ventricle. *Circ. Res.*, 63:550–562, 1985.

[86] M.L. Walker, E.W. Hawthorne, and H. Sandler. Methods for assessing performance for the intact hypertrophied heart. In N.R. Alpert, editor, *Cardiac Hypertrophy*, pages 387–405, Academic Press, New York, 1971.

[87] A.Y.K. Wong and P.M. Rautaharju. Stress distribution within the left ventricular wall approximated as a thick ellipsoidal shell. *Am. Heart. J.*, 75:649–662, 1968.

[88] R.H. Woods. A few applications of a physical theorem to membranes in the human body in a state of tension. *J. Anat. Physiol.*, 26:362–270, 1892.

[89] A.L. Yettram, C.A. Vinson, and D.G. Gibson. Effect of myocardial fibre architecture on the behaviour of the human left ventricle in diastole. *J. Biomed. Eng.*, 5:321–328, 1983.

[90] F.C.P. Yin. Ventricular wall stress. *Circ. Res.*, 49:829–842, 1981.

[91] F.C.P. Yin, R.K. Strumpf, P.H. Chew, and S.L. Zeger. Quantification of the mechanical properities of noncontracting canine myocardium under simultaneous biaxial loading. *J. Biomech.*, 20:577–589, 1987.

[92] F.C.P. Yin, S. Zeger, R. Strumpf, L. Demer, P. Chew, and W.L. Maughan. Assessment of regional ventricular mechanics. In C. Taylor, E. Hinton, D. Owen, E. Orate, editors, *Numerical Methods for Nonlinear Problems*, Pineridge Press, Swansea, U.K., page 513, 1984.

7

Multidimensional Measurement of Regional Strains in the Intact Heart

L.K. Waldman[1]

ABSTRACT The continuum mechanics approach to characterizing the function of the heart wall uses mechanical variables such as the stress and strain tensors and the relation between them, that is, the material properties of myocardium. Difficulties with the measurement of stresses and material properties in the beating heart have led investigators to pursue the study of local deformation or strain that is more amenable to observation. However, the complexity of ventricular deformation under both normal and pathophysiologic conditions has provided the impetus for increasingly sophisticated strain measurements during the last decade. In this chapter, a brief review is given of attempts to quantify ventricular function using one-dimensional measurements including their inherent drawbacks, the analysis of deformation and strain in two and three dimensions is delineated, results and their interpretation are given for strains measured in the normal heart, and observations of abnormal strain patterns under conditions of ventricular pacing and acute ischemia are described.

7.1   Introduction

During the cardiac cycle the heart wall experiences large and complex deformations that have been partly characterized by both global and local dimensional measurements. Because the motion of the left ventricle (LV) has been identified with idealized shape changes of axisymmetric shells, many studies in cardiac mechanics have focused on variations in overall cardiac dimensions such as major and minor semi-axes of ellipsoidal representations of the LV and corresponding volume changes [23,24,38,43,45]. However, a consideration of the complex thick-walled geometry and anatomy of the heart along with more localized measurements of ventricular function indicate that these global measurements are insensitive to regional variations in ventricular function under both normal and pathophysiologic

---

[1] Division of Cardiology, Department of Medicine, University of California, San Diego, La Jolla, California 92093

conditions. The heart wall is formed from a syncytium of muscle fibers wound in a spiraling pattern about the ventricular cavities [49,52]. The fibers are embedded in an interconnected matrix of heavily cross-linked collagen [9–11,32,35,44]. Fiber orientations have been shown to vary transmurally in a continuous manner through about two radians at a number of locations on the free walls of higher mammals [48]. Recent measurements [41] have shown that these transmural distributions differ substantially at other ventricular sites such as the septum and near the base and apex, and that the ventricular surfaces have complex doubly-curved geometries; that is, both circumferential and longitudinal variations in principal curvatures preclude an assumption of axisymmetry.

Frequently, systolic function has been quantified with methods that measure uniaxial length changes at one or more ventricular sites. Accurate segmental length changes have been made with ultrasonic dimension gauges (sonomicrometers) and have revealed substantial spatial variations in function in the normal heart, indicating that one-dimensional shortening increases along the longitudinal direction from base to apex, shortening and lengthening patterns vary with ventricular site, and that endocardial shortening and thickening exceed epicardial values [28,30,39]. Some of these findings have been corroborated with the alternative technique of implanting radiopaque markers and tracking length changes between pairs with biplane cinéradiography [25–27]. Alterations in cardiac function under conditions of ischemia [1,12,15,20,21,28,47,50,51], abnormal activation [5,8], and hypertrophy [2,6,7,16,19,29,46] have been correlated with changing patterns of uniaxial strain. During occlusive ischemia, bulging or lengthening behavior observed locally, subendocardial localization of infarcts, and redistribution of transmural blood flow all provide strong motivation for fully quantifying the accompanying patterns of local deformation to help elucidate their mechanisms.

Despite high temporal and spatial resolution, uniaxial measurements made with ultrasonic dimension gauges depend on their orientation. Measurements from pairs of radiopaque markers have similar drawbacks. Therefore, to observe complex two- and three-dimensional myocardial deformations, investigations were attempted in the late 1960s and 1970s in which three or more devices were monitored simultaneously. In the studies of Dieudonné [13,14], a triad of strain gauges (strain rosette) with attached pins inserted in the canine myocardium was used to measure two-dimensional infinitesimal strains at a number of depths beneath the epicardium and at a number of ventricular sites. Dieudonné was the first to attempt to calculate two-dimensional principal strains and associated principal directions from his data. However, the strain rosette is not capable of measuring thickening and transverse shear strains that may accompany shortening. Moreover, the attached pins may create artifacts by constraining the adjacent myocardium and preventing shearing motions. In fact, a close look at his data indicates that deeper in the heart wall

the rosette seemed to be picking up large positive strains that were probably indicative of thickening rather than shortening in planes parallel to
the epicardial surface. Subsequently, Fenton and colleagues [17,18] successfully implanted columns of radiopaque markers across the canine free wall.
They acquired three-dimensional coordinate data from the markers with
biplane cinéradiography and computed displacement gradients from sets
of these coordinates. Unfortunately, the nonsymmetric displacement gradient tensor includes local rotations and cannot be diagonalized, so that
finite strains and their principal values could not be calculated. Nevertheless, Fenton and coworkers were the first to attempt such three-dimensional
local measurements.

The first successful measurement of two-dimensional finite strains in myocardium was made by Meier and coworkers [36,37]. They sewed a triplet
of radiopaque markers to the epicardial surface of the dog and tracked
them in a manner similar to Fenton although with greater temporal resolution. With the coordinate data from three closely spaced material points,
Meier correctly computed the two-dimensional deformation gradients, de-
composed them into the rotation and stretch tensors, and computed two-
dimensional principal stretches and associated directions. These studies
were further strengthened by the clever way in which a cardiac coordinate
system was computed from the three markers and two additional markers
sewn at the bifurcation of the coronary arteries and the apex.

Two-dimensional strain measurements have been pursued by others. In
particular, Arts and coworkers measured two-dimensional epicardial strains
with three closely spaced electromagnetic induction coils [3,4]. Subsequently, Prinzen and colleagues [42] implanted three closely spaced needles across
the heart wall and tracked their positions with three coils on the epicardium
in an effort to predict endocardial strains from epicardial values. As with
Dieudonné's device, the needles may have constrained the surrounding myocardium causing artifacts. In more recent studies [41] arrays of markers
have been attached to the epicardium to acquire spatial distributions of
strain across the normal left ventricular free wall and during occlusive
ischemia. Unfortunately, only two-dimensional coordinates were obtained
with a single camera, thus ignoring the potential influence of curvature.
Moreover, the strain contours computed from limited discrete measurements are of questionable accuracy. Nevertheless, these investigations are
an important first attempt at acquiring distributed two-dimensional data.

In studies of McCulloch and coworkers [33,34], two-dimensional finite deformations were measured in isolated arrested canine hearts over a range of
carefully controlled loading conditions. These studies showed that passive
principal stretches increase linearly with increasing ventricular volume, that
the principal axis associated with the largest principal stretch has an orientation similar to epicardial fibers, and the associated shear corresponds to
the torsion of a left-handed helix. In addition, these investigators extended
these experiments to account for regional variations in passive strains and

the influence of coronary perfusion on measured strains.

In the sections that follow, the use of homogeneous strain theory to determine the kinematics of multidimensional finite deformations is outlined, then measurements made by my colleagues and me in La Jolla under normal and pathophysiologic conditions are described.

7.2   Strain Analysis

Continuum kinematics is reviewed to provide the groundwork to compute two- and three-dimensional finite strains from appropriate coordinate data and oriented line segments representing material points and line segments in the heart wall. Consider a material point P in the myocardium that is identified with the position vector $\mathbf{X}$ of an implanted marker or sonomicrometer in a reference configuration (Figure 7.1). At some later time in the cardiac cycle or at a different inflation pressure for an arrested heart, point P moves to a new position Q identified by an updated position vector $\mathbf{x}$. The displacement vector $\mathbf{u}$ indicates the translation of the material point. We have

$$\mathbf{x} = \mathbf{x}(\mathbf{X}) = \mathbf{u} + \mathbf{X}. \tag{7.1}$$

As mentioned previously, many studies have used one-dimensional measures such as segmental shortening or wall thickening to quantify cardiac function. Here, an ordinary derivative suffices to relate an infinitesimal segment to a single displacement component,

$$du = \frac{du}{dX} dX. \tag{7.2}$$

This derivative is an infinitesimal normal or extensional strain that can be written in finite form to relate an updated length l to its reference length l_0:

$$e = \frac{du}{dX} \sim \frac{\Delta u}{\Delta X} = \frac{l - l_0}{l_0} = \lambda - 1. \tag{7.3}$$

The use of this strain measure presumes that the stretch ratio λ is sufficiently close to 1. Depending on the orientation of a uniaxial measurement, this strain might represent the familiar fractional shortening in normal myocardium or wall thinning during ischemia, in which case λ would be less than 1. Normal wall thickening or ischemic lengthening might be observed when λ is greater than 1. In higher dimensions the position vectors may be functions of two or three variables, that is, the coordinate components on a surface or in a volume. Then, the ordinary derivative is replaced by a partial derivative of one vector with respect to another yielding a tensor of rank two with four or nine components in two or three dimensions, respectively:

$$d\mathbf{x} = \frac{\partial \mathbf{x}}{\partial \mathbf{X}} d\mathbf{X} = \mathbf{F} d\mathbf{X} \tag{7.4}$$

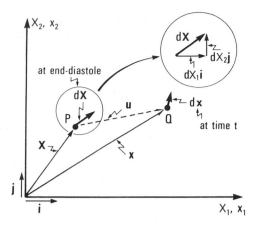

FIGURE 7.1. Continuum kinematics. A material point P of a continuous body identified by the vector $\mathbf{X}$ in a reference configuration undergoes the displacement $\mathbf{u}$ to a point Q identified by the vector $\mathbf{x}$ in an updated configuration. Infinitesimal vectors $d\mathbf{X}$ and $d\mathbf{x}$ are directed material line segments useful in deformation analysis. Their two-dimensional coordinate components are shown.

$$d\mathbf{u} = \frac{\partial \mathbf{u}}{\partial \mathbf{X}} d\mathbf{X} = \mathbf{G} d\mathbf{X}. \qquad (7.5)$$

Here, the tensors $\mathbf{F}$ and $\mathbf{G}$ are the deformation gradient tensor and the displacement gradient tensor, respectively. In earlier publications [55,57] I inadvertently interchanged the words deformation and displacement when describing these two tensors. I hope this current review serves to correct that error. $\mathbf{F}$ serves to map the contravariant vector $d\mathbf{X}$ into its updated counterpart $d\mathbf{x}$ while $\mathbf{G}$ maps $d\mathbf{X}$ into the infinitesimal displacement vector $d\mathbf{u}$. By differentiating Equation (7.1) with respect to the position vector $\mathbf{X}$, the simple relation between the deformation gradients and the displacement gradients is obtained:

$$\frac{\partial \mathbf{x}}{\partial \mathbf{X}} = \frac{\partial \mathbf{u}}{\partial \mathbf{X}} + \mathbf{I} \qquad (7.6)$$

or

$$\mathbf{F} = \mathbf{G} + \mathbf{I}. \qquad (7.7)$$

Using homogeneous strain analysis, $\mathbf{F}$ can be computed from a finite form of Equation (7.4). Two-dimensional deformation gradients can be computed as the solution of a linear algebra problem if $d\mathbf{X}$ and $d\mathbf{x}$ are known for two noncolinear vectors in a sufficiently small area; that is, an area in which strains do not vary significantly. It is convenient to obtain such data from a triangular arrangement of material points. In three dimensions, three non-coplanar vectors are required to compute $\mathbf{F}$ in a volume in which, again, strains vary little. These vectors can be any three noncoplanar edges of a tetrahedron whose vertices are now four material points. For convenience three vectors that emanate from one of the vertices might be chosen. The polar decomposition theorem can be applied to decompose $\mathbf{F}$, which is generally nonsymmetric into the orthonormal rotation tensor $\mathbf{R}$ and the symmetric stretch tensor $\mathbf{U}$:

$$\mathbf{F} = \mathbf{RU}. \tag{7.8}$$

Alternatively, an infinitesimal reference length dS can be computed from the scalar product of $d\mathbf{X}$ with itself. Similarly, the updated length ds can be computed from $d\mathbf{x}$:

$$dS^2 = d\mathbf{X}^T d\mathbf{X} \tag{7.9}$$

$$ds^2 = d\mathbf{x}^T d\mathbf{x}. \tag{7.10}$$

Substituting Equation (7.4) into Equation (7.10) and subtracting the result from Equation (7.9) gives the quadratic form that defines the Lagrangian or Green's strain tensor $\mathbf{E}$:

$$ds^2 - dS^2 = 2d\mathbf{X}^T \mathbf{E} d\mathbf{X}. \tag{7.11}$$

I made an error in writing this well-known quadratic form in earlier publications [55,57] when translating indicial notation to direct notation as currently used. Here, $\mathbf{E}$ is related to the deformation gradients $\mathbf{F}$ in the following manner:

$$\mathbf{E} = \mathbf{E}(\mathbf{X}, t) = \frac{1}{2}(\mathbf{F}^T \mathbf{F} - \mathbf{I}). \tag{7.12}$$

Observe that a simple relation exists between $\mathbf{F}$ and the symmetric stretch tensor $\mathbf{U}$ defined above:

$$\mathbf{F}^T \mathbf{F} = \mathbf{U}^2 \tag{7.13}$$

which shows how to convert the nonsymmetric $\mathbf{F}$ to symmetric stretches or strains without polar decomposition, if desired. To avoid computing $\mathbf{F}$ at all, one can use Equation (7.11) directly as has been done for both two- and three-dimensional measurements of strain [54,56,57,58]. Then, Equation (7.11) is applied repeatedly to three noncolinear line segments forming a small triangle in the two-dimensional case or to the six edges of a tetrahedron in three dimensions. Three segments are sufficient to compute three independent strain components in an area, whereas six segments are needed to compute the six independent components of a three-dimensional symmetric strain tensor. This method has the advantage that only scalar lengths need be retained connecting the updated material points. As mentioned previously for the calculation of $\mathbf{F}$, the same argument concerning the area or volume spanned by the material points applies here; that is, the strains cannot vary appreciably from point to point. If the region is small enough to justify the assumption of homogeneous strains, then additional points can be used to calculate strains in a least-squares sense. In some cases this can lead to an improvement in accuracy.

Once symmetric strain (or stretch) tensors are determined in a chosen coordinate system, they can be diagonalized; that is, an algebraic eigenvalue problem can be solved to compute the principal strains and their associated principal axes referred to that coordinate system. The principal strains or eigenvalues have the advantage that they are independent of the original x-ray or cardiac coordinates. The principal axes or eigenvectors form an orthonormal coordinate system that can be obtained from the original reference coordinates by one rotation in the two-dimensional case and by three rotations in three dimensions. Thus, in two dimensions three strains in a cardiac coordinate system [37] consist of two normal or extensional strains, such as circumferential and longitudinal strains or fiber and cross-fiber strains, and one shear strain in the plane of measurement. After diagonalization, this full strain matrix gives way to two principal strains and one angle, which serves to orient the two orthogonal principal axes relative to the cardiac coordinates. In three dimensions, a full matrix of six components has one additional normal strain (the radial strain that might indicate thickening or thinning of the heart wall locally) and two added transverse shear strains that may occur in each of the two remaining coordinate planes. The eigenpairs (principal data) can be reduced to six numbers also; that is, the three principal strains and three Euler angles that serve to reorient the cardiac coordinates along the principal axes.

Accuracy permitting, an assessment of the degree of compressibility of the myocardium between the reference and updated configurations can be made most easily from the principal stretches in the three-dimensional case. These are determined by diagonalizing the stretch tensor $\mathbf{U}$ or, if the principal strains E_i are available, as follows:

$$\lambda_i = \sqrt{2E_i + 1}, \quad i = 1, 2, 3. \tag{7.14}$$

Then, the product of these principal stretches indicates the ratio of local volume after deformation to beforehand. If $\lambda_1 \lambda_2 \lambda_3 = 1$, the deformation has resulted in no volume change. A material like myocardium consisting primarily of water might be expected to be incompressible, but the motion of fluid during systolic compression or the loss of blood supply during ischemia might lead to some compressibility. Similarly, if growth were to occur between two configurations as might be expected during normal aging or in hypertrophic disease, the product could be greater than one.

A brief comment on coordinate transformation follows. To change from x-ray coordinates to cardiac coordinates or from one set of cardiac coordinates to another (e.g., from circumferential and longitudinal coordinates to fiber and cross-fiber coordinates), one can transform the needed vectors before computing strains or other tensors. Given the orthogonal transformation matrix $\mathbf{M}$ of direction cosines relating the two coordinate systems, the initial vector components $d\mathbf{X}$ are premultiplied by this matrix to determine the components of the transformed vector $d\bar{\mathbf{X}}$:

$$d\bar{\mathbf{X}} = \mathbf{M}d\mathbf{X}. \tag{7.15}$$

If the tensor components $\mathbf{E}$ are available, coordinate transformation is performed as follows:

$$\bar{\mathbf{E}} = \mathbf{M}\mathbf{E}\mathbf{M}^T. \tag{7.16}$$

7.3   Finite Strains in the Normal Heart

Strain measurements described here are acute studies made in the anterior free wall of the left ventricle in open-chest, anesthetized dogs. After performing a thoracotomy and placing the heart in a pericardial cradle, either columns of radiopaque markers [56–58] or triplets of sonomicrometers [55] are implanted. Investigations of three-dimensional finite strains use the column technique. Usually, three columns of four or five markers each are used with an intercolumn spacing of 1 cm or less and an intermarker spacing of 2 to 3 mm (Figure 7.2). It is helpful to suture a plexiglass platform to the epicardium through which trocars are inserted perpendicular to the local surface. Sometimes an insertion is unsuccessful and a fourth column is inserted under fluoroscopy in an appropriate location to avoid confusing overlap problems. Then, additional markers are sewn above the columns and at the bifurcation of the coronary arteries and the apex to set up a local cardiac coordinate system (Figure 7.3) [37]. These coordinates are useful because the circumferential coordinate direction is close to that traditionally used to measure fiber direction while the longitudinal direction lies in the epicardial tangent plane and points from apex to base. The radial coordinate completes an orthonormal triad and provides a transmural direction with which to measure marker depths when selecting sets of markers for strain analysis. However, it should be kept in mind that any such coordinate system is inherently somewhat arbitrary in a body with such complex geometry.

After all markers are in place, biplane cinéradiography is performed to acquire the three-dimensional positions of the markers. Typically, strains are analyzed from sets of four markers available in the marker array, the tetrahedrons described previously (Figure 7.3). Errors introduced by strain variations in highly curved, thick-walled bodies like the left ventricle can be minimized if care is taken to avoid using tetrahedrons that violate the homogeneity assumption [58]; that is, in general their volumes cannot be too large. In particular, their radial legs (from the base to a fourth point along one of the columns) should not exceed 3 mm especially near the endocardium. And edges that span regions with substantial curvature (like the circumference at midventricle or probably any surface direction near the apex) must be as short as possible, say, 1 cm or less. However, the tetrahedrons cannot be too small either because of the limited spatial resolution

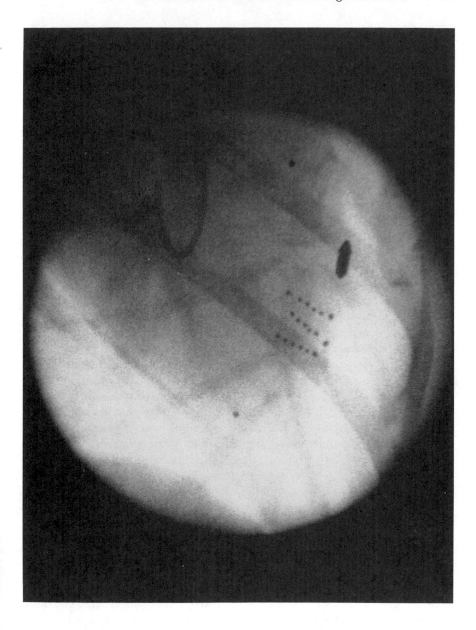

FIGURE 7.2. Sixteen-millimeter ciné photograph showing lateral view of columns of gold markers implanted chronically across the anterior free wall of the canine left ventricle. Also, a 1-cm calibration bar is visible. (Courtesy of J.H. Omens.)

of the x ray. Moreover, the radial legs cannot be too short because the volume of the tetrahedron will disappear and the six edges or three noncoplanar vectors from which deformations are calculated will no longer be independent. Moreover, when two 1-mm beads are separated by less than 2 mm from center to center, less than 1 mm of tissue lies between them. Then, one would be concerned that the markers might influence each other during deformation, thus adding artifacts to the measurement.

To remove the arbitrariness with which pairs of sonomicrometers are positioned, more elaborate arrays of these devices are needed. In recent studies [54] a triangular arrangement of these crystals has been used successfully to monitor two-dimensional finite strains at midwall ventricular sites similar to those chosen for the three-dimensional studies (Figure 7.3). The crystals are implanted so that each one can "see" the other two. The trick here is to rewire a standard sonomicrometer amplifier so that one crystal acts as a transmitter to the other two, the second both receives and transmits, and the third receives from both of the others (Figure 7.3). Because of the high sampling rates the technique still has considerably greater temporal resolution than high-speed ciné. Furthermore, the spatial resolution is vastly superior and no film or video need be processed, so that results are available virtually in real time. Unfortunately, the method has its drawbacks. Because of the size of the crystals and the wiring, only planar arrays have been tried so that only two-dimensional strains have been measured (see Section 7.2, Strain Analysis). Moreover, the orientation of one edge of the triangle is critical if a cardiac coordinate system similar to the one discussed above is desired [37,57] without imaging the reference configuration. The latter goal can be achieved by either meticulous crystal placement or biplane imaging of the reference configuration for subsequent correction with the required two-dimensional rotation. Finally, it is important to implant all three crystals at similar depths to avoid artifactual measurements of thickening when in-plane shortening and shear strains are being sought. These difficulties can be overcome with some experience.

Two- and three-dimensional finite strains measured in the normal canine heart have been reported in detail in a series of investigations [54,56,57,58]. Although the column method used to measure three-dimensional strains and some results appeared in 1985 [57], more comprehensive statistical data and a quantitative correlation between measured strains and fiber orientations are given in subsequent studies [58]. Additional two-dimensional data concerning the normal left ventricle have been acquired using the triangular array of ultrasonic dimension gauges described above. Results and their interpretation for an understanding of the normal function of the anterior free wall are given here.

Because of the complexity of strain patterns measured in the beating heart, a variety of graphical representations of the data have proven useful. As described earlier (see Section 7.2, Strain Analysis), the components of strain form a two-point tensor that may vary from one position to another

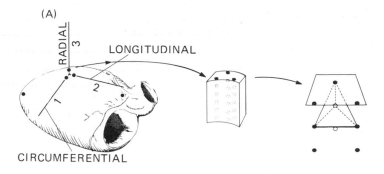

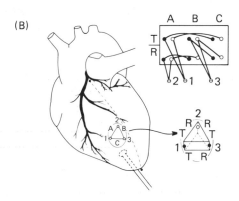

FIGURE 7.3. (A) A schematic diagram of the canine left ventricle showing a cardiac coordinate system calculated from reference markers adjacent to three implanted columns of markers and two additional reference markers at the apex and at the bifurcation of the coronary arteries. Sets of four points forming a tetrahedron are selected from the columns for homogeneous analysis of finite deformations. (B) Schematic diagram illustrating position of triangular array of ultrasonic dimension gauges (sonomicrometers) implanted in the left ventricle. One edge of the triad is oriented along an external long axis pointing from the apex to the coronary bifurcation. Dual function of the crystals is indicated by the insets that show the wiring of three channels of a transducer amplifier. As indicated in the text, each crystal either receives from or transmits to both neighboring crystals at different times during the sampling periods.

in a body and as a function of time; that is, two configurations or frames[2] are required to define these tensors. Considering the periodic dynamics of the cardiac cycle, two frames that have physiologic significance occur at end-diastole and end-systole as determined by the hemodynamics (e.g., simultaneous high-fidelity pressure measurements). Therefore, many of the results have been reported as Green's strains in which the reference frame is chosen to be end-diastole. Time series data then use subsequent frames acquired at high speeds (120 frames per sec) to update the strain tensor. To average systolic strain data from a number of experiments, it is convenient and physiologically relevant to use end-systole consistently as the updated configuration. Nevertheless, it should be kept in mind that these times in the cardiac cycle are somewhat arbitrary and that other reference frames could be used, such as a time earlier in diastole to examine filling or aortic valve opening to study ejection. It may be necessary to instrument the animals with an aortic flow probe if other reference frames with high temporal resolution are sought. With our definition of strain there can be no such thing as an end-diastolic strain unless a subsequent end-diastole is used as the updated frame. Then, unless the hemodynamic status has changed or an intervention has been performed, strains would be expected to be close to zero. Moreover, at times arbitrarily close to the reference frame, the strains are inherently small and the principal directions are indeterminate due to a multiplicity of small eigenvalues.

Typical time series of three-dimensional finite strains from a subendocardial location are shown in Figure 7.4. The six strain components of the symmetric tensor are plotted for one cardiac cycle with the initial end-diastolic frame as the reference configuration. These strains are referred to a cardiac coordinate system consisting of circumferential, longitudinal, and radial or transmural coordinate directions [37] and will be called cardiac strains. Therefore, the three normal or extensional components are circumferential, longitudinal, and radial strains. Three shear strains occur in the three respective cardiac coordinate planes; that is, shear in the circumferential-longitudinal coordinate plane (epicardial tangent plane) or in-plane shear, and two transverse shears in the circumferential-radial and the longitudinal-radial coordinate planes. Substantial negative or shortening strains are observed with circumferential strain tending to dominate longitudinal strain. At the same time a reciprocal positive or thickening strain exists radially. These strains tend to peak near end-systole, but significant magnitudes may be observed throughout ejection and even in the isovolumic periods. Simultaneously, all three components of shear occur; that is, small but consistent positive values of in-plane shear and one or both transverse shears. These transverse shears, particularly the

---

[2] Here, the term "frame" is a technical one used in continuum mechanics to describe the state of a continuous body, but for our purposes we can associate it with the film or video frames acquired to visualize this state.

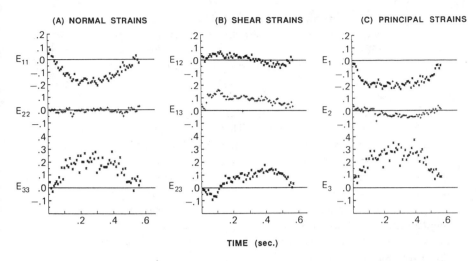

FIGURE 7.4. Time series (120 frames per sec) of three-dimensional finite strains observed during cardiac cycle near the endocardium of the ventricular wall. (A) Normal strains as a function of time referred to the cardiac coordinate system, i.e, circumferential (E_{11}), longitudinal (E_{22}), and radial strains (E_{33}). (B) Shear strains that accompany normal strains shown in (A), i.e, in-plane shear (E_{12}), circumferential-radial transverse shear (E_{13}), and longitudinal-radial transverse shear (E_{23}). (C) Principal strains computed from six strain components shown in (A) and (B). They are ranked in increasing order from top to bottom of the panel. End-diastole (ED) is the reference configuration.

longitudinal-radial component, frequently have large end-systolic magnitudes rivaling the shortening strains. Their significance is still not fully understood because of the lack of corresponding stress data and required shear moduli (material properties). Nevertheless, their frequent appearance indicates that vectors normal to the local epicardial surface at end-diastole do not remain normal during contraction; that is, Kirchhoff's hypothesis is not obeyed and even sophisticated shell models must be applied with caution. Transverse shear may influence myocardial blood flow, especially in the subendocardium.

The three principal strains computed from the cardiac strains at each time frame are shown in Figure 7.4. Notice that the first and third principal strains, which are ranked in increasing order, have larger peak values than the cardiac shortening or thickening strains. This is a result of the contribution of the shearing deformations emphasizing the need to measure all components of strain before assessing maximum or minimum values. These principal strains suggest that the myocardium at this location behaves like a "two-dimensional shortening machine"; that is, only one of the principal strains is positive so that the two negative strains tend to facilitate the large positive or thickening strain needed to eject blood. The transverse shears that come into play in the eigenvalue calculation contribute to this mechanism. Moreover, one might speculate that the reason for the large transmural variation in fiber direction is specifically to prevent lengthening along the surface directions during contraction providing for the greatest wall thickening possible. These observations are supplemented by the data shown in Figure 7.5. Here, circumferential strain and the first principal strain are plotted together along with the principal angle that serves to orient the associated principal axis in the epicardial tangent plane with respect to the circumferential direction. As alluded to previously, note that the principal strain is substantially greater than the circumferential strain over much of the cardiac cycle and that its corresponding principal direction has a fairly stable orientation that lies between circumferential and longitudinal. Thus, the peak shortening does not occur in either of the cardiac coordinate directions nor, as we shall see, does it occur in the local fiber direction deep in the anterior wall.

Another useful way to examine strain data acquired with the column technique is to select a time in the cardiac cycle, say end-systole, and to plot transmural variations or trends in various strain components. For this purpose, as many tetrahedrons as possible that meet certain selection criteria [57,58] are chosen from the marker array. Some of the selection criteria have been described in the previous section (see Section 7.2, Strain Analysis) concerning the assumption of uniform or homogeneous strains in small areas or volumes.

When comparing data from different animals, it is helpful to normalize the radial depths of the tetrahedral centroids by a wall thickness determined during a study or postmortem. Thus, strains are plotted as a function of percent depth from the epicardium. An example of the end-systolic transmural trend of the three principal strains is given in Figure 7.6. These data demonstrate the substantial variation in strain with depth that might be expected in a highly curved, thick-walled body like the left ventricle. Note that both first and third principal strains increase with depth having small values near the epicardium and considerably larger values of both shortening and lengthening near the endocardium. At end-systole the three principal strains can be negligible near the epicardium or the first (greatest shortening) and third (greatest lengthening) can be as great as 0.10 in absolute value. The greatest shortening increases by about 0.015 for each 10% depth increase (decile) while the greatest lengthening increases in a reciprocal manner by about 0.02 per decile. Near the endocardium maximum shortening may give rise to strains of -0.15 to -0.20 at these low end-diastolic pressures in the open-chest preparation while maximum lengthening ranges from 0.30 to 0.40 and sometimes even greater. Thus, the need to measure finite strains rather than using an infinitesimal approximation is evident particularly near the endocardium. The second principal strain is small throughout the wall thickness in the example shown (Figure

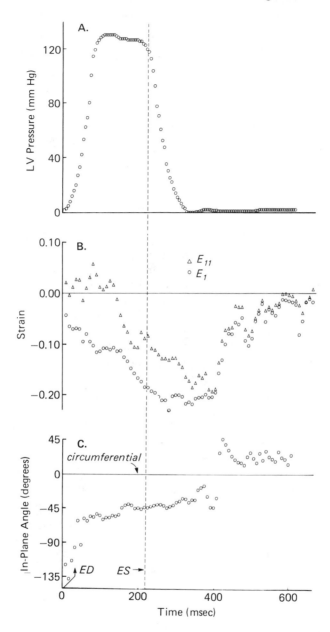

FIGURE 7.5. Simultaneous time series of left ventricular pressure (A), circumferential strain and first principal strain or greatest shortening (B), and in-plane angle (C) that orients the projection of the first principal axis on the epicardial tangent plane with respect to the circumferential coordinate direction (counterclockwise angles are positive). This angle may be defined as the first of three Euler angles that determine the orientation of the principal coordinates with respect to the cardiac coordinates. (Reproduced from Waldman et al. [58], with permission from the American Heart Association.)

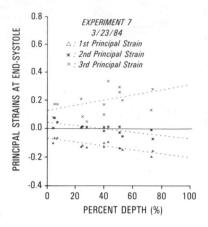

FIGURE 7.6. End-systolic principal strains as a function of percent depth beneath the epicardium (100% indicates endocardium). (Reproduced from Waldman et al. [57], with permission from the American Heart Association.)

7.6), but on average it does increase with depth yielding subendocardial magnitudes that are about half or less of the first principal strain (-0.06 to -0.08). These are still substantial negative strains supporting the concept mentioned earlier that two-dimensional shortening facilitates the large thickenings observed.

The transmural variation of both the first principal direction and fiber direction is displayed in Figure 7.7. Here, fiber orientation was measured across the marker implantation site so that a direct correlation between strain and fiber data could be made. This graph shows an increasing divergence between the direction of greatest shortening and the fiber angle as a function of increasing depth from the epicardium. Although the two directions correlate closely near the epicardium, they are virtually orthogonal to each other near the endocardium. This has been a consistent finding in this preparation, indicating that whereas fiber direction may vary by 110° to 130° across the anterior wall, the principal direction of greatest shortening typically varies by only 20° or 30°. The acquisition of fiber data throughout the region of strain measurement allows one to transform cardiac strains from circumferential and longitudinal strains to fiber and cross-fiber strains. This was done in a series of experiments [58], as indicated in Figure 7.8. Because of the difficulty in finding specific tetrahedrons at similar depths in different hearts, data from the outer or epicardial half of the ventricular wall were lumped together as were data from the inner or endocardial half to obtain statistical averages. The results reinforce the finding shown in Figure 7.7, that is, although fiber strains are somewhat greater than cross-fiber strains in the outer half (no statistically significant difference because of animal-to-animal variation), cross-fiber strains clearly dominate

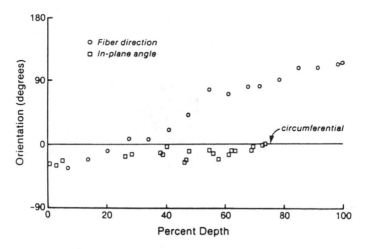

FIGURE 7.7. Transmural variation of both fiber direction and end-systolic first principal direction, as indicated by the in-plane angle (Figure 7.5C). (Reproduced from Waldman et al. [58], with permission from the American Heart Association.)

fiber strains in the inner half. This result is a corollary to the divergence between principal and fiber direction described above.

Again, without a complete knowledge of the internal forces and properties that accompany these strains, it is difficult to interpret them fully. Nevertheless, I have suggested that these observations may result from a combination of the following factors. The collagen matrix probably tethers myofibers lying at different depths so that they are closely coupled together and tend to deform together as a unit. Furthermore, fibers in the middle and outer thirds that have fiber directions ranging from circumferential to negative (i.e., clockwise) angles may dominate the inner fibers because of their relatively large volume or cross-sectional area. If corresponding stresses are as transmurally uniform as advocates of torsion and residual stress have suggested (see Chapter 5), then greater forces would indeed be generated from midwall to the epicardium. Finally, the fact that circumferential curvature is substantially greater than longitudinal curvature in the midanterior wall would suggest that circumferentially oriented (highly curved) fibers would tend to dominate as inferred by previous uniaxial measurements of substantial circumferential segment shortening near the epicardium and at midwall [20] and by some measurements showing a preponderance of hoop fibers [49]. In combination with the small in-plane shear observed locally that may be a manifestation of ventricular torsion, it is easy to see how deformations that are expected to have small or positive principal angles would tend to have more negative principal angles.

As described earlier, the triangular arrangement of sonomicrometers has advantages over radiographic methods in terms of both temporal and spa-

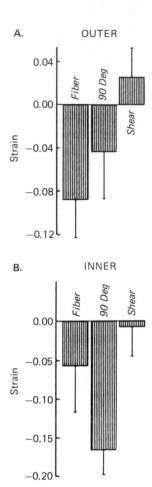

FIGURE 7.8. Bar charts of average end-systolic fiber and cross-fiber (90° from fibers in epicardial tangent plane) strains in the outer (A) and inner (B) halves of the ventricular walls in seven hearts. Measured local fiber directions were used to transform (rotate) the strain tensor into the fiber coordinate system. (Reproduced from Waldman et al. [58], with permission from the American Heart Association.)

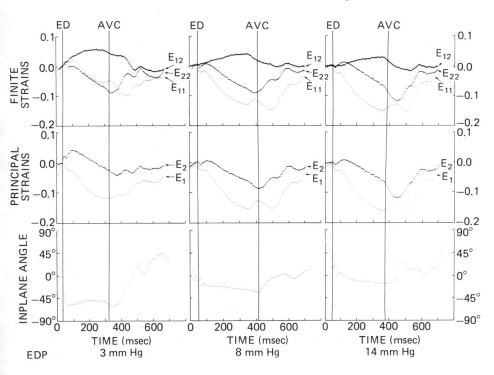

FIGURE 7.9. Times series of two-dimensional cardiac strains, principal strains, and in-plane angle acquired from an array of three ultrasonic crystals at midwall over a range of end-diastolic pressures in one experiment. Vertical lines in each panel indicate end-diastole (ED) and aortic valve closure (AVC) at the three pressures studied: 3, 8, and 14 mm Hg from left to right. (Reproduced from Villarreal et al. [54], with permission from the American Heart Association.)

tial resolution. Moreover, two-dimensional strain data can be acquired quickly without laborious film or video postprocessing. Time series of strains obtained from a midwall array of these piezoelectric crystals are shown in Figure 7.9. The array was implanted in a site on the anterior wall similar to those studied with the column technique, and the strains observed corroborate those measured with the more comprehensive technique at midwall at low end-diastolic pressure (EDP). As Figure 7.9 and the column method indicate, positive in-plane shear drives the first principal direction to an orientation that is represented by a negative principal angle of about −45°. Although at this low pressure the cardiac normal strains are almost equal over much of the cardiac cycle, the shear tends to separate the two principal strains resulting in an end-systolic first principal strain of about −0.10 with very small values of the second principal strain. It should be kept in mind that the two-dimensional method does not account for transverse shears and radial strains that accompany the three components measured. Frequently, the effect of transverse shear in the longitudinal-radial coordinate plane can be to contribute to the second principal strain, explaining why such small values are observed with the two-dimensional method. Increases in EDP result in increases in the cardiac normal strains (especially circumferential strain), increases in both principal strains, and a decrease in shear that causes the principal direction of greatest shortening to become more circumferential. On average, end-systolic circumferential strain doubles (−0.06 to −0.13) as EDP is raised from 2.3 to 17.0 mm Hg. Although the greatest shortening increases somewhat, it is the ratio of circumferential to first principal strain that changes dramatically from 0.61 to 0.87 and 0.94 as EDP rises from 2.3 to 8.0 and 17.0 mm Hg, respectively. Simultaneously, the principal direction averages −43° at the low pressure and increases (becomes more circumferential) to −26° and then −14° as the pressure rises. These findings again emphasize the need to make comprehensive strain measurements. Without all three components of strain in a plane of measurement this significant change in the direction of greatest shortening could not be observed.

7.4   Abnormal Finite Strains: Ventricular Pacing and Acute Ischemia

Although three-dimensional strains and stresses in the normal heart can be expected to vary in complex ways from one ventricular site to another and as a function of time owing to the complex geometry and properties of the cardiac microstructure, additional difficulties arise when ventricular function is compromised by disease. Abnormal conditions such as myocardial ischemia, hypertrophy, and arrhythmia alter the regional or local function of the heart. Measurements of wall strains reflect these changes in

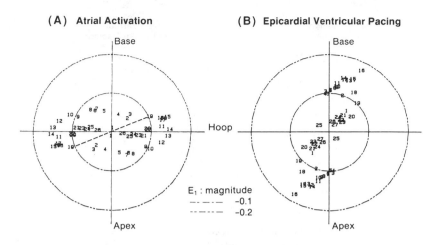

FIGURE 7.10. Time series of first principal strain and orientation of the associated principal axis in the subendocardium during atrial (A) and ventricular (B) pacing. Pairs of numbers indicate endpoints of "scaled axis" at successive times in the cardiac cycle starting at end-diastole (25-msec intervals). For example, the axis connecting the pair of 19s on the left indicates a first principal strain (greatest shortening) of −0.1 at 475 msec after end-diastole with an in-plane angle of +20°. Peak strains occur at and just after end-systole with magnitudes of about −0.16 and more circumferential orientations in this example. (Reproduced from Waldman et al. [56], with permission from the American Physiological Society.)

function and can be related to ventricular wall stress if appropriate three-dimensional material models can be applied realistically.

When the left ventricle is activated from an epicardial ventricular site rather than the normal route from right atrium through the Purkinje system, normal negative circumferential strains (associated with wall shortening) and large positive radial (or wall thickening) strains are often replaced with positive or lengthening strains in the circumferential direction and a reciprocally decreased wall thickening. Curiously, the minimum principal strain or greatest shortening reaches similar peak systolic values, but the associated principal direction is more longitudinal. This phenomenon is illustrated in Figure 7.10. The first principal strain and its corresponding principal axis are plotted simultaneously showing that the orientation of this principal axis can change markedly with time and during an episode of ventricular pacing while the local fiber orientation is fixed. Therefore, large negative strains can be measured in directions that are not aligned with local fibers, and these directions can change throughout the cardiac cycle and during a reversible intervention.

The effect of pacing on the three-dimensional strains is shown in detail at a subendocardial site at end-systole (Figure 7.11). The three-dimensional principal stretches and corresponding principal directions are shown in a

(A) Atrial Activation
(End-Systole)

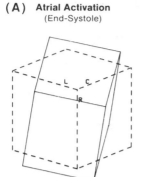

**(B)   Epicardial
Ventricular Pacing**
(End-Systole)

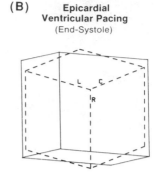

FIGURE 7.11. End-systolic principal strain data from the midwall during atrial (A) and ventricular (B) pacing. The dotted cubes indicate the orientation of the cardiac reference coordinates (C: circumferential, L: longitudinal, R: radial). The solid parallelepipeds plot the principal stretches and associated principal directions simultaneously. Three edges emanating from any vertex are scaled by the three principal stretches and oriented along the respective principal axes. (Reproduced from Waldman et al. [56], with permission from the American Physiological Society.)

three-dimensional perspective. Here, each of the three edges of a parallelepiped emanating from any vertex is scaled by a principal stretch and oriented along the associated principal axis. In this way, changes in deformation can be examined comprehensively; that is, all strain components can be observed at a given position and time. Normal end-systolic stretches give rise to an end-systolic configuration that is shorter along both principal axes in a principal plane almost parallel to the epicardial tangent plane. With ventricular pacing, this two-dimensional shortening is compromised. Instead, peak negative strains occur along with fairly circumferential positive strains in a plane tipped significantly from the tangent plane. Likewise, the third principal axis is tipped out of the tangent plane, and decreased principal thickening is apparent. Significant changes in shearing deformation in the cardiac coordinate system accompany this interesting tilt in end-systolic principal coordinates. When ventricular pacing is stopped and atrial pacing reinstated, virtually normal strains can be measured after a short recovery period.

   In experimental models of acute ischemia, coronary occlusion results in pronounced changes in patterns of regional function and accompanying local strains. Recent studies [53] in our laboratory have been performed to measure two- and three-dimensional strains in the anterior wall of the canine left ventricle during control periods and during episodes of regional ischemia induced by occlusion of the left anterior descending coronary artery

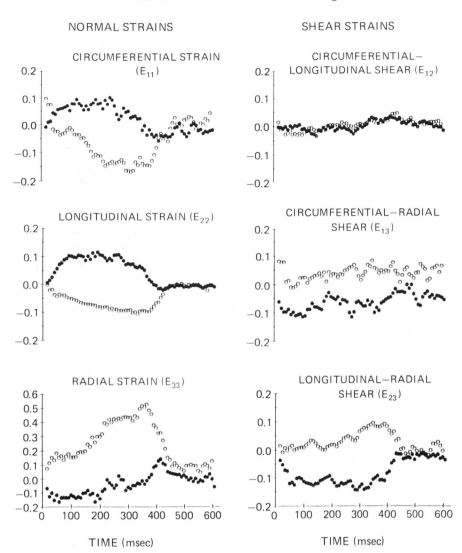

FIGURE 7.12. Time series of three-dimensional normal and shear strains at mid-wall during one cardiac cycle under control conditions (open circles) and after 10 min of occlusive ischemia (closed circles).

(LAD). The ventricular site at which strains were measured had transmural ischemia as documented by intracoronary injections of radioactive microspheres. With this protocol it is effective to study early changes in function after occlusion with the triad of ultrasonic crystals. Although only two-dimensional finite strains can be measured at, say, midwall, many of the prominent mechanical changes can be observed continuously without the collection of many video or ciné images with radiography and subsequent data reduction. Normal shortening in the plane of measurement as characterized by substantial circumferential strain (-0.10) and consistent moderate longitudinal strain (-0.06) gives way to an equibiaxial stretching after several minutes of occlusion, but the positive in-plane shear does not change significantly. If collateral circulation does not alter the hemodynamic status of the area at risk and the infarct is not too large, this bulging pattern remains fairly stable suggesting that the ischemic region is being passively stretched by surrounding more healthier muscle that continues to shorten. Thus, three-dimensional strains measured after 5 and 10 min of ischemia are equally representative of the altered function.

An example of time series of three-dimensional strain components measured in the subendocardium after 10 min of ischemia is shown along with control strains (Figure 7.12). Again, negative circumferential and longitudinal strains are replaced by positive ones but now a simultaneous change from wall thickening to wall thinning (negative radial strain) is observed. At the same time, positive transverse shears give way to negative ones, particularly in the longitudinal-radial coordinate plane. These effects are also apparent in changes in the principal data as reflected in principal coordinates that are tipped away from the cardiac coordinates and altered principal stretches. Bidirectional stretching is evident but its peak values occur in a principal plane tipped away from the epicardial tangent plane while peak thinning occurs orthogonally to that plane. During infarcts in which blood flow is compromised uniformly across the heart wall, this pattern of altered strain occurs with transmural uniformity. The combined effects of two-dimensional lengthening as opposed to shortening and remarkably consistent changes in one component of transverse shear may accompany profound changes in local myocardial material properties and local wall stresses, but a quantitative assessment awaits an accurate correlation of these kinds of data with a realistic stress analysis of the heart.

Despite recent progress in making discrete measurements of multidimensional strains in the heart wall, the assumption of uniform or homogeneous strains needs to be relaxed under conditions in which strains have substantial spatial variation. As alluded to earlier, the transmural variation of strain components in the normal heart can be expected to be increasingly pronounced deeper in the ventricular wall. In a generalization of the column technique new studies utilize a nonhomogeneous finite element analysis to compute continuous variations of strains under carefully controlled loading conditions in the isolated, arrested canine heart [31]. In addition, this

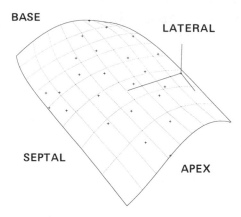

FIGURE 7.13. Plot of surface fitted to an array of 25 epicardial markers sewn on the epicardium during an acute ischemia study. Pairs of these fitted surfaces occurring at different times in the cardiac cycle and before and after an intervention that alters regional myocardial function like occlusive ischemia can be used to compute continuous distributions of two-dimensional deformation.

method is being applied to study strain gradients in the beating heart before and during volume overload hypertrophy [40]. During acute ischemia the variation of strain from the area at risk, across the border zone, and into more normal surrounding muscle is an important correlate of altered regional function. A new approach to this problem also uses the finite element method [22,59]. Here, surface fitting techniques are applied to the three-dimensional coordinates of an array of 25 epicardial markers that span the midanterior wall (Figure 7.13). The relation between a reference surface and an updated surface defines a continuous distribution of two-dimensional strains and reveals sharp gradients during ischemia (LAD occlusion) as the point of observation is moved from normally shortening myocardium into areas of abnormal lengthening (dyskinesis). Correlation of such comprehensive measurements with a realistic stress analysis may provide insights into local loading conditions and altered material properties.

Acknowledgements: The support of NIH grants HL32583 and HL17682 is gratefully acknowledged.

REFERENCES

[1] M. Akaishi, W.S. Weintraub, R.M. Schneider, L.W. Klein, J.B. Agarwal, and R.H. Helfant. Analysis of systolic bulging: Mechanical char-

acteristics of acutely ischemic myocardium in the conscious dog. *Circ Res.*, 58:209–217, 1986.

[2] N.R. Alpert. *Cardiac Hypertrophy.* Academic Press, New York, 1971.

[3] T. Arts and R.S. Reneman. Measurement of deformation of canine epicardium in vivo during cardiac cycle. *Am. J. Physiol.*, 239:H432–H437, 1980.

[4] T. Arts, R.S. Reneman, and P.C. Veenstra. Epicardial deformation and left ventricular wall mechanics during ejection in the dog. *Am. J. Physiol.*, 243:H379–H390, 1982.

[5] F.R. Badke, P. Boinay, and J.W. Covell. Effects of ventricular pacing on regional left ventricular performance in the dog. *Am. J. Physiol.*, 238:H858–H867, 1980.

[6] F.R. Badke and J.W. Covell. Early changes in left ventricular regional dimensions and function during chronic volume overloading in the conscious dog. *Circ. Res.*, 45:420–428, 1979.

[7] I. Belenkie, J.S. Baumber, and A. Rademaker. Changes in left ventricular dimensions and performance resulting from acute and chronic volume overload in the conscious dog. *Can. J. Physiol. Pharmacol.*, 61:1274–1280, 1985.

[8] R.C. Boerth and J.W. Covell. Mechanical performance and efficiency of the left ventricle during ventricular stimulation. *Am. J. Physiol.*, 221:1686–1691, 1971.

[9] T.K. Borg and J.B. Caulfield. Collagen in the heart. *Tex. Rep. Bio. Med.*, 39:321–333, 1979.

[10] T.K. Borg, W.F. Ranson, and F.A. Moslehy. Structural basis of ventricular stiffness. *Lab. Invest.*, 44:49–54, 1981.

[11] J.B. Caulfield and T.K. Borg. The collagen network of the heart. *Lab. Invest.*, 40:364–371, 1979.

[12] B. Crozatier, J. Ross Jr., D. Franklin, et al. Myocardial infarction in the baboon: Regional function and the collateral circulation. *Am. J. Physiol.*, 235:H413–H421, 1978.

[13] J.M. Dieudonné. Gradients de directions et de deformations principales dans la paroi ventriculaire gauche normale. *J. Physiol. (Paris)*, 61:305–330, 1969.

[14] J.M. Dieudonné. La determination experimentale des contraintes myocardiques. *J. Physiol. (Paris)*, 61:199–218, 1969.

[15] C.H. Edwards II, J.S. Rankin, P.A. McHale, D. Ling, and R.W. Anderson. Effects of ischemia on left ventricular regional function in the conscious dog. *Am. J. Physiol.*, 240:H413–H420, 1981.

[16] B.L. Fanburg. Experimental cardiac hypertrophy. *N. Engl. J. Med.*, 282:723–732, 1970.

[17] T.R. Fenton, J.M. Cherry, and F.A. Klassen. Transmural myocardial deformation in the canine left ventricular wall. *Am. J. Physiol.*, 235:H523–H530, 1978.

[18] T.R. Fenton, G.A. Klassen, and J.S. Outerbridge. Radiographic measurement of transmural myocardial deformation. In *Intern. Conf. Med. Biol. Eng.*, pages 714–715, 1976.

[19] W.H. Gaasch, O.H.L. Bing, and I. Mirsky. Chamber compliance and myocardial stiffness in left ventricular hypertrophy. *Eur. Heart J.*, 3(Supp A):139–145, 1982.

[20] K.P. Gallagher, G. Osakada, O.M. Hess, J.A. Koziol, W.S. Kemper, and J.J. Ross. Subepicardial segmental function during coronary stenosis and the role of myocardial fiber orientation. *Circ. Res.*, 52:352–359, 1982.

[21] K.P. Gallagher, G. Osakada, M. Matsuzaki, W.S. Kemper, and J. Ross Jr. Myocardial blood flow and function with critical coronary stenosis in excercising dogs. *Am. J. Physiol.*, 243:H698–H707, 1982.

[22] A.R. Hashima, L.K. Waldman, and A.D. McCulloch. Nonhomogeneous analysis of epicardial strain gradients in ischemic border zone. *FASEB J.*, 4(3):A432, 1990.

[23] E.W. Hawthorne. Instantaneous dimensional changes of the left ventricle in dogs. *Circ. Res.*, 9:110–119, 1961.

[24] J.E. Hinds, E.W. Hawthorne, C.B. Mullins, and J.H. Mitchell. Instantaneous changes in left ventricular lengths occurring in dogs during the cardiac cycle. *Fed. Proc.*, 28(4):1351–1357, 1969.

[25] N.B. Ingels, G.T. Daughters, and S.R. Davies. Stereo photogrammetric studies on the dynamic geometry of the canine left ventricular epicardium. *J. Biomech.*, 4:541–550, 1971.

[26] N.B. Ingels Jr., G.T. Daughters II, E.B. Stinson, and E.L. Alderman. Left ventricular midwall dynamics in the right anterior oblique projection in intact unanesthetized man. *J. Biomech.*, 14(4):221–233, 1981.

[27] N.B. Ingels Jr., G.T. Daughters II, E.L. Stinson, and E.B. Alderman. Measurement of midwall myocardial dynamics in intact man by radiography of surgically implanted markers. *Circulation*, 52:859–867, 1975.

[28] W.Y.W. Lew. Influence of ischemic zone size on nonischemic area function in the canine left ventricle. *Am. J. Physiol.*, 252:H990–H997, 1987.

[29] M.M. LeWinter, R.L. Engler, and J.S. Karliner. Enhanced left ventricular shortening during chronic volume overload in conscious dogs. *Am. J. Physiol.*, 238:H126–H133, 1980.

[30] M.M. LeWinter, R.S. Kent, J.M. Kroener, T.E. Carew, and J.W. Covell. Regional differences in myocardial performance in the left ventricle of the dog. *Circ. Res.*, 37:191–199, 1975.

[31] K.D. May, J.H. Omens, and A.D. McCulloch. Three-dimensional deformations in passive canine left ventricle. *FASEB J.*, 4(3):A432, 1990.

[32] P.E. McClain. Characteriztion of cardiac muscle collagen. *J. Biol. Chem.*, 249:2303–2311, 1974.

[33] A.D. McCulloch, B.H. Smaill, and P.J. Hunter. Left ventricular epicardial deformation in isolated arrested dog heart. *Am. J. Physiol.*, 252:H233–H241, 1987.

[34] A.D. McCulloch, B.H. Smaill, and P.J. Hunter. Regional left ventricular epicardial deformation in the passive dog heart. *Circ. Res.*, 64:721–733, 1989.

[35] I. Medugorac. Myocardial collagen in different forms of hypertrophy in the rat. *Res. Exp. Med. (Berl)*, 177:201–211, 1980.

[36] G.D. Meier, A.A. Bove, W.P. Santamore, and P.R. Lynch. Contractile function in canine right ventricle. *Am. J. Physiol.*, 239:H794–H804, 1980.

[37] G.D. Meier, M.C. Ziskin, W.P. Santamore, and A.A. Bove. Kinematics of the beating heart. *IEEE Trans. Biomed. Eng.*, BME-27:319–329, 1980.

[38] J.H. Mitchell, K. Wildenthal, and C.B. Mullins. Geometrical studies of the left ventricle utilizing biplane cinefluorography. *Fed. Proc.*, 28(4):1334–1343, 1969.

[39] J.H. Myers, M.C. Stirling, M. Choy, A.J. Buda, and K.P. Gallagher. Direct measurement of inner and outer wall thickening dynamics with epicardial echocardiography. *Circulation*, 74:164–172, 1986.

[40] J.H. Omens and J.W. Covell. Transmural variation in myocardial growth induced by volume overload hypertrophy in the dog. *FASEB J.*, 4(3):A426, 1990.

[41] F.W. Prinzen, T. Arts, G.J. Van Der Vusse, W.A. Comans, and R.S. Reneman. Gradients in fiber shortening and metabolism across the left ventricular wall. *Am. J. Physiol.*, 250:H255–H264, 1986.

[42] F.W. Prinzen, T. Arts, G.J. Van Der Vusse, and R.S. Reneman. Fiber shortening in the inner layers of the left ventricular as assessed from epicardial deformation during normoxia and ischemia. *J. Biomech.*, 17:801–811, 1984.

[43] J.S. Rankin, P.A. McHale, C.E. Arentzen, J.C. Greenfield Jr., and R.W. Anderson. The three-dimensional dynamic geometry of the left ventricle in the conscious dog. *Circ. Res.*, 39:304–313, 1976.

[44] T.F. Robinson, L. Cohen-Gould, and S.M. Factor. Skeletal framework of mammalian heart muscle: Arrangement of inter- and pericellular connective tissue structures. *Lab. Invest.*, 49:482–498, 1983.

[45] R.F. Rushmer, D.L. Franklin, and R.M. Ellis. Left ventricular dimensions recorded by sonocardiometry. *Circ. Res.*, 4:684–688, 1956.

[46] S. Sasayama, D.L. Franklin, and J. Ross Jr. Hyperfunction with normal inotropic state of the hypertrophied left ventricle. *Am. J. Physiol.*, 232:H418–H425, 1977.

[47] S. Sasayama, K.P. Gallagher, W.S. Kemper, D.L. Franklin, and J. Ross Jr. Regional left ventricular wall thickness early and late after coronary occlusion in the conscious dog. *Am. J. Physiol.*, 240:H293–H299, 1981.

[48] D.D. Streeter. Gross morphology and fiber geometry of the heart. In R.M. Berne, editor, *Handbook of Physiology, Section 2: The Cardiovascular System*. Volume 1, pages 61–112. American Physiological Society, Bethesda, MD, 1979.

[49] D.D. Streeter Jr., H.M. Spotnitz, D.P. Patel, J. Ross Jr., and E.H. Sonnenblick. Fiber orientation in the canine left ventricle during diastole and systole. *Circ. Res.*, 24:339–347, 1969.

[50] R. Tennant and C.J. Wiggers. The effect of coronary occlusion on myocardial contraction. *Am. J. Physiol.*, 112:351–361, 1935.

[51] P. Theroux, J. Ross Jr., D. Franklin, J.W. Covell, C.M. Bloor, and S. Sasayama. Regional myocardial function and dimensions early and late after myocardial infarction in the unanesthetized dog. *Circ. Res.*, 40:158–165, 1977.

[52] F. Torrent-Guasp. *The Cardiac Muscle.* Juan March Foundation, Barcelona, Spain, 1973.

[53] F.J. Villarreal, W.Y.W. Lew, L.K. Waldman, and J.W. Covell. Transmural myocardial deformation in the ischemic canine left ventricle. *Circ. Res.*, 68:, 1991. (in press).

[54] F.J. Villarreal, L.K. Waldman, and W.Y.W. Lew. A technique for measuring regional two-dimensional finite strains in canine left ventricle. *Circ. Res.*, 62:711–721, 1988.

[55] L.K. Waldman. In vivo measurement of regional strains in myocardium. In G.W. Schmid-Schönbein, S.L-Y. Woo, and B.W. Zweifach, editors, *Frontiers in Biomechanics,* pages 98–116. Springer-Verlag, New York, 1986.

[56] L.K. Waldman and J.W. Covell. Effects of ventricular pacing on finite deformation in canine left ventricles. *Am. J. Physiol.*, 252:H1023–H1030, 1987.

[57] L.K. Waldman, Y.C. Fung, and J.W. Covell. Transmural myocardial deformation in the canine left ventricle: Normal in vivo three-dimensional finite strains. *Circ. Res.*, 57:152–163, 1985.

[58] L.K. Waldman, D. Nosan, F.J. Villarreal, and J.W. Covell. Relation between transmural deformation and local myofiber direction in canine left ventricle. *Circ. Res.*, 63:550–562, 1988.

[59] A.A. Young, P.J. Hunter, and B.H. Smaill. Epicardial surface estimation from coronary angiograms. *Comput. Vision Graph. Image Process.*, 47(1):111–127, 1989.

8

Epicardial Deformation From Coronary Cinéangiograms

Alistair Young[1]

ABSTRACT A quantitative method is developed for estimating epicardial deformation in the intact heart using the motions of the coronary arteries. The method makes full use of a deformable structural model of the time-varying epicardial surface to guide the analysis of coronary cinéangiograms. A finite element model was adopted, consisting of nodal geometric parameters interpolated by bicubic Hermite basis functions in the spatial domain and sinusoidal (Fourier) basis functions in the temporal domain. The parameters of an initial static surface were fitted to the three-dimensional (3-D) locations of the coronary arteries at diastasis. The parameters of the time-varying displacement field were then fitted to the tracked displacements of the bifurcation points of the coronary arteries. The locations of the vessel centerlines were then tracked from frame to frame throughout the cycle, using the fit to the bifurcations as a reference state. Owing to the nonuniform distribution of the superficial arteries over the epicardium, weighted spline smoothness constraints were added to the error function in order to regularize the least-squares solution. The time-varying surface model provides a complete description of epicardial deformation over the entire region spanned by the ensemble basis functions. The Green's strain tensor, referred to in-plane material coordinates, was used to calculate principal (major and minor) surface strains at any point.

8.1   Introduction

A quantitative description of heart wall motion is required both for the mathematical modeling of cardiac mechanics and for the evaluation of cardiac performance in the diagnosis of disease. Traditional indices of myocardial performance include left venticular (LV) pressure-volume relationships, aortic flow, and ejection fraction. However, such global parameters provide no indication of the regional motion of the heart wall. Localized abnormalities in the pattern of myocardial deformation are clearly linked to the local dysfunction during the development of myocardial infarction and degener-

---

[1]Department of Engineering Science, School of Engineering, University of Auckland, New Zealand

ative cardiac disease (cardiomyopathy), leading to heart failure [34]. The assessment of cardiac function and performance therefore requires accurate, quantitative representations of regional wall motion.

Current efforts to quantify regional wall motion rely on data obtained by such imaging techniques as LV angiography [43], echocardiography [7], radionuclide imaging [9], and computed tomography [6]. Owing to the wide intra- and interobserver variation associated with such subjective analysis [10,52], numerous quantitative techniques have been proposed. Where few material landmarks are available, gross deformations can be approximated by length changes of diameters perpendicular to a long axis [34] or radially emanating from a centroid [42,52]. These measurements are often indexed to a coordinate system that moves with specific anatomical or derived points, in order to compensate for global translations and rotations of the heart [25]. A common problem with these methods is the lack of consistently identifiable points that can be accurately tracked throughout the cardiac cycle. This prevents an accurate description of myocardial deformation. Furthermore, evaluations of cardiac function utilizing only two-dimensional information require simplistic assumptions approximating the complex three-dimensional underlying motion and are prone to large errors [20].

One useful method of supplying identifiable points has been to implant radiopaque markers on or in the heart wall during surgery and track their movement using cinéradiography. Harrison and coworkers [15] were the first to measure cardiac dimensions using this method, demonstrating the effects of drugs and exercise on local wall motion. The reduction of myocardial shortening with hypertrophy [31] and return of contractility after coronary bypass surgery [8] have also been investigated. Rankin and colleagues [40] used implanted ultrasonic transducers to study ventricular dynamics during isovolumic contraction in dogs. Similar techniques have been used to measure regional differences in myocardial deformation, showing an increase in systolic shortening from base to apex [26,33] and from epicardium to endocardium [27,53]. In order to facilitate the use of these techniques, several automatic marker tracking systems have been developed [13,14,37]. In general, however, there are insufficient markers to describe adequately the LV geometry. A geometric model is required to interpret the measured deformations and predict the distribution of myocardial stress. Approximations to simple analytical shapes are often made, including spherical [5], cylindrical [2], ellipsoidal [22], and prolate spheroidal [30] regular geometries.

A large number of natural, well-defined, and fixed epicardial markers exist in the form of bifurcations of the coronary arteries. There are two main branches of the coronary arteries, left and right, that supply blood to the left and right sides of the heart, respectively. Both originate at the root of the aorta and branch over the epicardial surface (superficial branches) before diving into the heart wall (transmural branches). For the present application, we are more interested in the left coronary artery (LCA), because it covers most of the LV free wall. Typically, the LCA bifurcates near its origin and two main branches arise. The left anterior descending (LAD) branch approximately follows the groove between the right and left ventricles to the apex. The left circumflex (LCX) curls around the base between the left atrium and the left ventricle and supplies the posterior free wall. Several smaller branches (e.g., obtuse marginal, anterior diagonal) supply the regions in between. Bifurcations can be visualized in vivo, localized in three dimensions, and tracked through time using biplane cinéangiography. This obviates the need for marker implantation, allowing routine evaluations in a clinical environment. An analysis of regional performance is possible over a wider area, as more points can typically be defined than markers implanted. Kong and colleagues [24] were the first to demonstrate the validity of this technique, using the motion of metallic markers as a standard. Segment lengths obtained from the linear distance between opacified bifurcations correlated closely to those obtained from markers placed in corresponding locations. Although intracoronary injection of contrast medium is known to depress cardiac inotropic state, this effect was found to be insignificant within the first five cardiac cycles. Segmental lengths were plotted against time for selected sites, allowing comparison of shortening onset, rate, extent, and duration. Potel and colleagues [39] used an interactive three-dimensional graphics system incorporating animated overlays to aid the input process, including the ability to display the back projections of positions in one view across another (epipolar lines) [28]. Kim and coworkers [22] have applied homogeneous strain kinematics to the motion of bifurcations; estimations of wall stress were obtained assuming a thick-walled ellipsoidal geometry for the LV.

In an effort to interpret and understand experimental findings, it is desirable to build a mathematical model of the process based on the physical laws of the underlying continuum [19]. A knowledge of the mechanical factors that influence cardiac function and dysfunction (instantaneous myocardial stresses and strains) is central to both the understanding of contractile status and evaluation of blood/oxygen requirements within regions of the heart [29,54]. Ventricular wall stresses, which are difficult to measure directly, can be expressed as functions of pressure loading, geometrical parameters, and material properties of the myocardium. Comparison of the performance of such models with experimental observations can highlight deficiencies in the assumptions used and validate numerical values of model parameters. The static three-dimensional geometry of isolated, arrested canine hearts can be described efficiently and accurately using finite element meshes interpolating low-order polynomial geometric fields. The geometric parameters of the model can be fitted to manual geometric measurements by least-squares [36]. Models describing the spatially variant fiber field, nonlinear constitutive properties, and the electrical activation of the heart may also be incorporated; for a review of these techniques see Hunter and Smaill [19].

This chapter is concerned with the construction of an accurate and complete representation of myocardial deformation in the intact heart. Specifically, we wish to extend the above geometric model of the isolated, arrested heart to describe the entire cardiac cycle. In doing so we exploit the information inherent in biplane coronary cinéangiograms more fully than present investigations of atherosclerotic disease by using the coronary arteries as landmarks indicative of the motion of the underlying epicardial surface. The problem falls into two main parts: (1) a complete description must be obtained of the three-dimensional structure of the entire opacified arterial tree throughout the cardiac cycle, and (2) this information must be used to describe the dynamic geometry and characterize deformation in a compact and efficient manner, making explicit the contractile status of any point. The derived quantitative description of strain can then be compared with finite element models of stress, activation, and material properties in order to validate the assumptions and parameter values used. Given an accurate quantitative map of normal and abnormal regional wall motion, more informed clinical evaluations could be made as to the extent and severity of diseased portions of the heart wall.

8.2   Data Acquisition

Biplane coronary angiograms were performed at Green Lane Hospital, Auckland, in 60° right anterior oblique (RAO) and lateral (LAT) projections using a Philips biplane cinéangiocardiographic unit. Injections of 7 ml Urogratin 76% were made selectively into the left coronary artery via a catheter inserted into the right femoral artery. Images were recorded on 35-mm film at 75 frames per sec using 9-inch image intensifier fields. The two views were exposed alternately with 2-ms exposure times. Because the master/slave relationship between the cameras of each view, the phase difference between exposures of the left and right views was not constant. Breathing was suspended at full inspiration and there was no motion of the patient relative to the x-ray gantry during filming.

Two perspex discs containing a number of radiopaque markers were fixed to the faces of the image intensifiers prior to filming. Each disc had a diameter of 250 mm, a thickness of 5 mm, and contained 17 small lengths of lead-based solder wire embedded parallel to the approximate axis of the x rays in a radially symmetric pattern. These markers facilitated accurate frame to frame registration during the digitizing process. A 1-cm grid was recorded for each view to allow correction of geometric distortions. After filming, the x ray apparatus was repositioned into the same geometry as for each study and a three-dimensional cylindrical grid was imaged in both views. This grid consisted of circumferential and longitudinal lines inscribed on the surface of a length of perspex tubing. The inscribed grooves were filled with a lead oxide mixture (lithage and glycerol). This allowed the accurate estimation of each camera projection matrix using the method of MacKay and colleagues [28] and enabled the reconstruction of three-dimensional points from pairs of corresponding image points. The average error in reconstructing the three-dimensional grid points was approximately 0.35 mm in the space of the heart.

The ciné film was mounted in a device that allowed single frames to be back-lit and viewed from above by a DAGE MTI low-light vidicon camera. A slight modification was made to the camera circuitry to allow the automatic gain control to be disabled during the digitizing process. A Zeiss Tessovar zoom lens was used to focus on the 35-mm ciné frame. The camera rested on a table that allowed three-dimensional translations relative to the plane of the film. The video signal was digitized by a Data Translation DT2561 frame grabber housed within a microVax II workstation. The resolution of the digital image was 512×512 pixels with 8 bits per pixel; i.e., approximately 0.25 mm per pixel in the space of the heart.

8.3   Static Surface Estimation

8.3.1   CORONARY ARTERY TREE DEFINITION

As a prerequisite to the surface reconstruction and tracking procedures, we require a method of characterizing the locatable structure within the image. This consists not only of the branch points of the arterial tree, but also of the loci of the vessels themselves. To date, the problem of automatic vessel detection remains largely unsolved [12,17,23,38,45,51,56]. The correspondence of structure between views is especially difficult to establish. We therefore resort to the interactive approach described below.

Two frames are chosen, one from each view, during the diastasis phase of the cardiac cycle at approximately the same instant in time. The tree is constructed as an ensemble of Bézier cubics, overlay graphically on the images. Each cubic is parameterized as $\mathbf{x}(\xi)$, where $\mathbf{x} = (x, y, z)$ and is characterized by four three-dimensional points, $\mathbf{x}_i, i = 1, 2, 3, 4$. These control the curve's position and slope at either end:

$$\mathbf{x}(0) = \mathbf{x}_1; \;\; \mathbf{x}(1) = \mathbf{x}_4; \;\; \frac{d\mathbf{x}(0)}{d\xi} = 3(\mathbf{x}_2 - \mathbf{x}_1); \;\; \frac{d\mathbf{x}(1)}{d\xi} = 3(\mathbf{x}_4 - \mathbf{x}_3). \;\; (8.1)$$

Bézier polynomials are frequently used in interactive geometric modeling because the shape of the curve is easily manipulated by positioning the control points and the curve lies within the convex hull so formed. Continuity of the first derivative across interelement boundaries may be enforced by constraining the two slope points associated with the element start/end point to be colinear with that point. In the current application, the user positions a control point in one view and is immediately prompted for its corresponding location in the other. The epipolar constraint [28] is employed in this context to maintain compatibility between views.

A set of three-dimensional data points is generated at equal increments of physical arc length s, rather than the parameter ξ, in order to give a uniform distribution of data points along the vessel's length. This approach provides a complete and compact representation of the coronary artery tree with a moderate degree of operator effort. However, further automation is required before it could feasibly be applied in a routine clinical environment. A semiautomatic system would perhaps be more feasible, whereby the bifurcation points are located manually and the images are searched for the connecting paths most likely to coincide with arterial segments. Pope and colleagues [38] describe such a technique for detecting a two-dimensional structure where dynamic programming is used to carry out the search. Since the topology and material coordinates of the arteries are constant, the coronary artery tree need only be defined once for each patient. The surface representation is used to guide subsequent vessel tracking algorithms, as this requires fewer degrees of freedom and implicitly enforces the surface constraint.

8.3.2   THE STRUCTURAL MODEL

Since the model is intended primarily for use in finite element analysis, we adopt a model of the epicardial geometry in which nodal (geometric) parameters $\mathbf{u}_n$ are interpolated by low-order piecewise parametric polynomials, a technique used extensively in both computer graphics and finite element modeling [35]. The field value at a point ξ within the ensemble $\mathbf{E}$ is given by a weighted average of the element field parameters, with the weights determined by the element basis functions $\Psi_n(\xi)$:

$$\mathbf{u}(\xi) = \sum_{n=1}^{N} \Psi_n(\xi)\mathbf{u}_n. \qquad (8.2)$$

The model used in the present study is derived from Nielsen's [35,36] finite element model for the geometry of the heart. A prolate spheroidal coordinate system (λ,μ,θ) was adopted because surfaces of constant radial (λ) coordinate form good approximations to actual epicardial surfaces. Description in terms of rectangular Cartesian coordinates, for example, would require a higher order model (either higher degree basis functions or more elements) to achieve comparable accuracy. Since the radial direction is approximately perpendicular to the computed surface, radial coordinate values and derivatives provide good control over the surfaces using relatively few elements.

The element material coordinates $\xi = (\xi_1, \xi_2)$ were chosen to lie in the circumferential and longitudinal directions, respectively. The angular and azimuthal coordinate fields were thus given directly by fixed linear maps

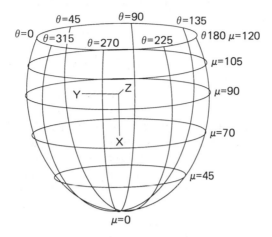

FIGURE 8.1. Initial configuration of the static surface ensemble. Both angular and azimuthal coordinates are shown in degrees. (Reproduced with permission from Young et al. [57].)

$\theta = \theta(\xi_1); \mu = \mu(\xi_2)$. In this way the data points $\mathbf{v}_d$ can be projected onto ensemble positions ξ_d simply by inverting these relations.

The central axis and focal length of the prolate coordinate system were defined so that the left main bifurcation was situated at $\lambda=1$, $\mu=120°$, $\theta=0°$ and the apex at $\lambda=1$, $\mu=0°$, $\theta=0°$. The surface defined by $\lambda=1$ was used as the initial estimate of the epicardial surface $\mathbf{u}^o$. The focal length is used as a scaling factor, allowing direct comparison of different sized hearts. The ensemble consists of 40 elements—eight circumferential by five longitudinal (Figure 8.1).

Bicubic Hermite interpolation was employed for the displacement radial field, that is, the difference between the deformed and initial configurations. Bicubic Hermite basis functions enforce first-order continuity across element boundaries and require four degrees of freedom per node to describe a single valued field. These are associated with field value, first derivatives, and cross-derivative.

8.3.3 THE FITTING PROCEDURE

The parameters of the static surface model are fitted to the Bézier coronary data by linear least squares, using the method of Young and colleagues [57]. The surface $\mathbf{u}$ is given by the sum of the initial configuration $\mathbf{u}^o$ and a displacement field $\mathbf{v}$. Given the above initial configuration in prolate coordinates (a regular ellipsoid) and ensemble data coordinates ξ_d, the radial displacement field can be denoted $\lambda(\xi_1, \xi_2)$ and the displacement data λ_d. The error function to be minimized penalizes the spatial variation of the displacement field as well as the usual Euclidean norm of the data errors:

$$\mathcal{E}(\lambda) = \mathbf{S}(\lambda(\boldsymbol{\xi})) + \sum_{d=1}^{D} \gamma_d |\, \lambda(\boldsymbol{\xi}_d) - \lambda_d\,|^2. \tag{8.3}$$

This may be regarded as a virtual work formulation of a mechanical deformation problem. The first term measures the deformation energy of the surface, that is, resistance to stretch and bending. It has the effect of regularizing the solution of the least-squares problem and is necessary in the present situation because of the nonuniform distribution of coronary data. When fitting piecewise polynomial surfaces by least squares, the density of observations must be uniformly high with respect to the density of nodal parameters, otherwise the normal equations may become ill-conditioned or singular. Restricting the solution to suitably smooth functions is a standard technique for regularizing ill-posed problems [4], notably for the approximation of surfaces to scattered data [47]. Following Terzopoulos [47], we use the weighted combination of thin plate and membrane spline kernels for the smoothness measure

$$\mathbf{S}(\lambda(\boldsymbol{\xi})) = \int_{E} \left\{ \alpha \left(\left[\frac{\partial \lambda}{\partial \xi_1} \right]^2 + \left[\frac{\partial \lambda}{\partial \xi_2} \right]^2 \right) + \beta \left(\left[\frac{\partial^2 \lambda}{\partial \xi_1^2} \right]^2 + \right.\right.$$

$$\left.\left. \left[\frac{\partial^2 \lambda}{\partial \xi_1 \partial \xi_2} \right]^2 + \left[\frac{\partial^2 \lambda}{\partial \xi_2^2} \right]^2 \right) \right\} d\xi. \tag{8.4}$$

The above smoothness functional is analogous to the small deflection energy of a thin plate under tension [47]. Such an analogy does not strictly apply in this instance as the displacement field may be quite large and is given in curvilinear coordinates. However, the relative ratios of the weights γ_d, β, and α provide intuitive control over the shape of the resulting surface. Relatively high values of γ_d will increase the fidelity to the data; α controls the tension of the surface, and β controls the degree of surface curvature. In all the cases quoted below the weights associated with the data points were:

$$\gamma_d = \begin{cases} 1 & \text{if } \mu_d \leq 120^\circ \\ 0 & \text{otherwise.} \end{cases} \tag{8.5}$$

8.3.4   FITS TO CANINE DATA

A set of 505 geometric data points taken at approximately 5-mm spacings and covering the entire epicardial surface of an isolated arrested dog heart was used for validation purposes. These data form part of the set recording the three polar coordinates and fiber angle at roughly 10,000 sites throughout the myocardium that was used by Nielsen and colleagues [36] to produce

a geometric model of the whole heart. Details of the measurement protocol can be found in Nielsen and colleagues [36]. The measurement rig gave
cylindrical polar coordinates that were referred to a central axis passing
through a point between the aortic and mitral valves and the apex. The
accuracy of the axial and radial coordinates was approximately 0.3 mm;
that of the angular coordinate was 1.2°. The data were subsequently converted to prolate spheroidal coordinates referred to the same central axis.
A set of 124 points evenly distributed at roughly 5-mm spacing along the
superficial branches of the right and left coronary arteries was acquired
concurrently.

The procedure described above was applied to 82 three-dimensional points
taken from the left coronary artery only. For convenience we call this data
set D1. Figure 8.2A shows the resulting surface (S1) using weights of $\alpha =$
0.001, $\beta = 0.01$. The geometric errors associated with D1 had a sample
mean of 0.120 mm and a sample standard deviation of 0.112 mm; the average squared error was 0.027 mm^2. (For all subsequent fits we quote the
errors in the format $0.120 \pm 0.112; 0.027$.)

These surfaces were compared to the complete epicardial data set consisting of 505 points (D2) evenly distributed around the entire epicardium.
Figure 8.2B shows the fit obtained to D2 using the 40 element mesh, the error was $0.277 \pm 0.357; 0.204$. Considering only those elements that contain
three or more points in D1 (13 out of 40), the error obtained by projecting
points in D2 onto S1 was $0.435 \pm 0.635; 0.607$. Over the six elements containing one or two coronary data points, this error was $0.923 \pm 1.506; 3.085$.
Over the remaining 21 elements (containing no coronary data points), it
was $6.13 \pm 5.31; 65.7$.

We would expect the right side of the heart to be more accuately modeled
if we included data from the right coronary artery as well as the left (D3,
124 points). The resulting surface (S3) with $\alpha = 0.01$, $\beta = 0.001$ is shown
in Figure 8.2C ($0.113 \pm 0.114; 0.0257$). Over the 19 elements containing
three or more coronary data points, the error associated with points in D2
projected onto S3 was $0.552 \pm 0.712; 0.810$. Over the 13 elements containing
one or two coronary data points, this error was $0.930 \pm 1.000; 1.86$. Over
the remaining eight elements it was $2.90 \pm 3.36; 19.5$.

8.3.5   FITS TO ANGIOGRAPHIC DATA

A set of 314 points at 2-mm spacing generated from the Bézier ensemble
defining the location of the coronary artery tree at diastasis is shown in
Figure 8.3. It comprises 54 cubic elements, defining 25 branches and 12
bifurcations of the left coronary artery. This data set (D4) is less smooth
than the canine data and the resulting surfaces are more sensitive to the
smoothing weights. In the case where $\alpha > 0$ and $\beta = 0$ the surface tries
to assume a $\lambda = $ constant geometry, the deviation of which is controlled
directly by α. In the case where $\beta > 0$ and $\alpha = 0$ the surface tends to

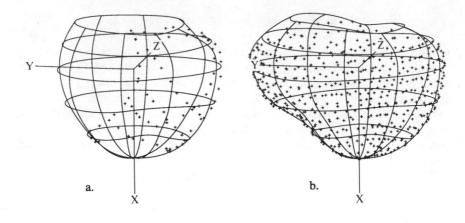

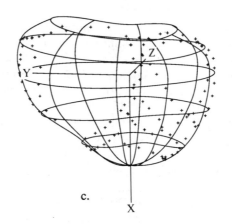

FIGURE 8.2. Fits to canine data. (A) Fit to left coronary data (82 points). (B) Fit to epicardial data (505 points). (C) Fit to left and right coronary data (124 points). (Reproduced with permission from Young et al. [57].)

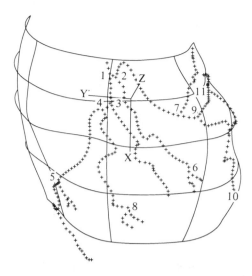

FIGURE 8.3. Surface fit to angiographic data showing location of tracked bifurcations.

maintain a linear variation of λ within each element, while conforming to the slope continuity imposed at the boundaries [57]. A combination giving acceptable data error and surface smoothness was $\alpha = 0.005, \beta = 0.001$ (Figure 8.3). The data errors were $0.518 \pm 0.454; 0.473$. This is comparable to the three-dimensional reconstruction error of 0.35 mm.

8.4 Motion from Bifurcations

The above static model is now extended to fit time-varying surfaces with the inclusion of a time-basis function. The parameters of the model are fitted to the tracked displacements of the bifurcation points of the coronary arteries, since these are readily locatable material points whose three-dimensional positions can be accurately tracked throughout the cardiac cycle.

8.4.1 BIFURCATION TRACKING PROCEDURE

The coronary data set obtained from the Bézier ensemble (Figure 8.3), which defines the location of the coronary artery tree at diastasis, also defines the initial location and structure of the bifurcation points. This information can now be used to guide the bifurcation tracking procedure. The structure and position of each bifurcation at diastasis is modeled by a three-dimensional point of intersection and a set of three-dimensional directions that define the orientations of each intersecting branch. (The term

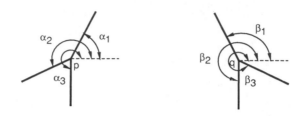

FIGURE 8.4. The three-armed bifurcation model for right and left views.

"bifurcation" is applied here very loosely: the method applies to any branch point, whether bifurcation, trifurcation, etc.) This model is then projected onto each biplane angiographic view, giving rise to two sets of intersecting line segments ("trifids") that specify the location of the bifurcation and the orientation of the branches in the two-dimensional image plane (Figure 8.4). The length of these line segments is determined a priori to be two to three times the widths of the vessels. Each bifurcation is tracked in each view separately, according to the following template matching procedure.

For each bifurcation, the view in which it is most clearly defined (the arms of the trifid are most separated) is chosen. The position of the bifurcation in the previous frame is assumed known (the first frame is given by the Bézier ensemble). Its position in the current frame is found by minimizing a cost function that measures the goodness of fit between the current image and the model. Let $F(n, \mathbf{b})$ represent such a correlation between the nth image and the trifid centered on $\mathbf{b}$. Vessels in the image are characterized by their higher gray-level intensities and by their thin, linelike structures. We therefore want to maximize the (smoothed) image intensities summed along the arms as well as some measure of image lineness. We could also include a term penalizing high edge values along the arms, since the vessel centerline points have small image gradients. The cost function is therefore a weighted average of these three terms over all points belonging to the trifid

$$F(n, \mathbf{b}) = \sum_{\mathbf{p} \in P} \{\beta_1 \text{edge}(\mathbf{p}) - \beta_2 \text{line}(\mathbf{p}) - \beta_3 \text{smooth}(\mathbf{p})\} \qquad (8.6)$$

where:
1. P is the set of image points belonging to the arms emanating from $\mathbf{b}$;
2. $\text{edge}(\mathbf{p})$ is the edge value at the image point $\mathbf{p}$ in the nth frame;
3. $\text{line}(\mathbf{p})$ is the line value at $\mathbf{p}$ in the nth frame;
4. $\text{smooth}(\mathbf{p})$ is the smoothed value at $\mathbf{p}$ in the nth frame;
5. $\beta_1, \beta_2, \beta_3$ are weights governing the relative influence of each term.
   The smoothed image values are calculated by convolution of the nth

frame with a two-dimensional gaussian function:

$$\text{smooth}(\mathbf{p}) = G(x,y) * I^n(\mathbf{p}); \ G(x,y) = exp\left(\frac{-(x^2 + y^2)}{2\sigma^2}\right). \tag{8.7}$$

Similarly, the image edge values are given by

$$\text{edge}(\mathbf{p}) = \sqrt{\left(\frac{\partial G}{\partial x} * I^n(\mathbf{p})\right)^2 + \left(\frac{\partial G}{\partial y} * I^n(\mathbf{p})\right)^2} \tag{8.8}$$

and the line values by the Laplacian of a Gaussian operator:

$$\text{line}(\mathbf{p}) = -\nabla^2 G * I^n(\mathbf{p}). \tag{8.9}$$

These operators were chosen because their properties are well understood [50] and there exist efficient algorithms for their implementation [16,44]. More sophisticated line operators, which outperform the Laplacian of a gaussian operator, are discussed in Section 8.6; however, the latter is sufficient for the present application and has the advantage of ease of implementation.

Apart from these image intrinsics, we may also consider the behavior of the bifurcation through time. Sudden jumps in position and velocity can be avoided by adding terms that penalize the point's velocity (we do not want very large displacements between frames) and acceleration (the point moves with approximately the same velocity as the previous frame). The procedure may now be written:

$$\mathbf{b}^n = \min_{\mathbf{b} \in \mathbf{I}^n} \ \{F(n, \mathbf{b}) + \beta_4 |\, \mathbf{b} - \mathbf{b}^{n-1} \,| + \beta_5 |\, \mathbf{b} - 2\mathbf{b}^{n-1} + \mathbf{b}^{n-2} \,|\} \tag{8.10}$$

where:
1. $\mathbf{b}^n$, $\mathbf{b}^{n-1}$, $\mathbf{b}^{n-2}$ are the positions of the bifurcation in the current frame, last frame and frame-before-last, respectively;
2. $\mathbf{I}^n$ is now a window of image points in the nth frame centered on $\mathbf{b}^{n-1}$;
3. $\mathbf{b}$ defines the intersection point of the model within $\mathbf{I}^n$;
4. β_4, β_5 are weights governing the model's resistance to velocity and acceleration, respectively.

The bifurcation's position in the second view is given by minimizing the same error function, with the modification that $\mathbf{I}^n$ now consists only of those points lying on the epipolar line defined by the tracked position in the first view. This assumes the views are exposed simultaneously, in contrast to the true situation in which there is a variable phase difference between views.

Eleven bifurcations were chosen from the Bézier coronary tree data (labeled in Figure 8.3). These were tracked through 60 frames, or approximately one cardiac cycle from diastasis to diastasis. Bifurcation 1 is the first bifurcation of the left coronary artery; 2, 3, and 4 are proximal. Bifurcations 7, 9, and 11 lie along the main circumflex branch, which follows the ventricular side of the groove between the left atrium and the left ventricle. Bifurcations 3, 4, and 5 lie along the left anterior descending branch, which tends to follow the groove between the right and left ventricles (anterior septum), while 8, 6, and 10 lie along anterior diagonal, obtuse marginal, and posterior lateral branches, respectively. Of the 11 bifurcations tracked, 7 are positioned near the base of the LV, the remaining 4 scattered around the midventricle to apical regions.

All bifurcations were tracked using the matching metric (8.10). Values of the weights $(\beta_1, \beta_2, \beta_3, \beta_4, \beta_5)$ giving acceptable results in the majority of cases were $(0, 0.1/C, 0.9/C, 0.1, 0.01)$ for the first view and $(0.5/C, 0, 1.0/C, 0.1, 0.01)$ for the second view, where $C = F(1, \mathbf{b}^1)$—the value of the image cost function at the initial position on the first frame. Tracked bifurcation positions can be manually corrected whenever the algorithm fails. Surprisingly, the constraint of rigid body translation is not the major cause of failure for this algorithm. Failure is more often due to the occasional presence of obscuring structure caused by overlapping vessels.

An example of the motions undergone by a single bifurcation is given in Figure 8.5, in which the coordinates of bifurcation 5, about midway down the left anterior descending branch, are plotted.

8.4.2   MODEL SPECIFICATION AND FITTING

The time-varying surface is given by the sum of an initial configuration and a displacement field

$$\mathbf{u}(\xi_1, \xi_2, t) = \mathbf{u}^1(\xi_1, \xi_2) + \mathbf{v}(\xi_1, \xi_2, t) \tag{8.11}$$

The initial surface $\mathbf{u}^1$ is defined by the static surface model that was fitted to the Bézier coronary data. Whereas the initial epicardial surface is efficiently described in prolate spheroidal coordinates, motions are best referred to a rectilinear system. Rigid body translations then give rise to uniform displacement fields and a knowledge of the motion of a central axis is not required. The tensor product representation is again invoked for the displacement field:

$$\mathbf{v}(\xi_1, \xi_2, t) = \sum_{n=1}^{N} \sum_{i=1}^{I} \mathbf{\Psi}_n(\xi_1, \xi_2) \varphi_i(t) \mathbf{v}_{in} \tag{8.12}$$

where $\mathbf{\Psi}_n$ are the two-dimensional surface basis functions, φ_i are one-dimensional time basis functions and $\mathbf{v}_{in}$ are the control vectors (nodal parameters) that specify the displacement field. The $\mathbf{\Psi}_n$ are chosen to be bicubic Hermite for each of the x, y and z coordinate fields, as these give first-order continuity across the entire ensemble. Since the motion of the

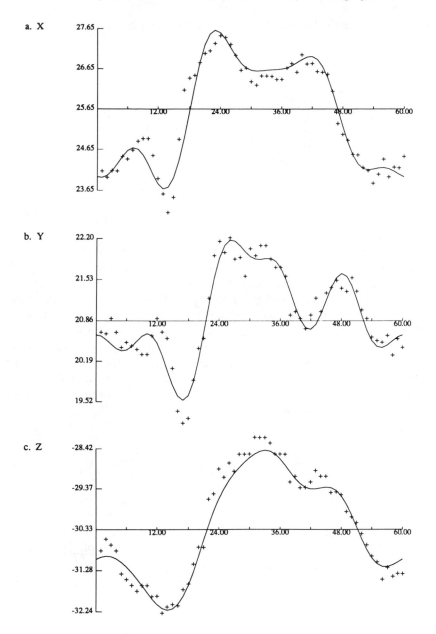

FIGURE 8.5. Tracked locations for bifucation 5 vs. frame number. Solid line represents fitted motion of the corresponding surface point.

heart is approximately periodic, the φ_i are defined to be:

$$\varphi_i = \begin{cases} 1 & , i = 1 \\ \cos\frac{\pi i t}{T} & , i \text{ even} \\ \sin\frac{\pi(i-1)t}{T} & , \text{otherwise.} \end{cases} \tag{8.13}$$

This implicitly contrains the motion to be periodic with period T. Furthermore, the magnitudes of the $\mathbf{v}_{in}, i = 1..I$, reflect the energies associated with each harmonic. The spectral components of the motions of the tracked bifurcation points can therefore guide the choice of I.

The error function to be minimized is given by

$$\mathcal{E}(\mathbf{v}) = S(\mathbf{v}(\boldsymbol{\xi}, t)) + \sum_{d=1}^{D} \gamma_d |\, \mathbf{v}(\boldsymbol{\xi}_d, t_d) - \mathbf{v}_d \,|^2. \tag{8.14}$$

The first term is a measure of the deformation energy of the surface, included to regularize the problem that is ill-posed because of the sparse distribution of the bifurcations. The second term sums the squared distances between observed and predicted data displacements. The deformation energy of the surface should ideally reflect the underlying constitutive properties of the myocardium, or alternatively be some norm of the surface strain tensor [48]. In either case the resulting deformation problem will be nonlinear in the displacement parameters. For the present we again resort to the linear spline kernel measure of surface smoothness proposed by Terzopoulos [47]. As there are enough data in the time domain, only the spatial derivatives are weighted:

$$S(\mathbf{v}) = \int_E \left\{ \alpha \left(|\frac{\partial \mathbf{v}}{\partial x}|^2 + |\frac{\partial \mathbf{v}}{\partial y}|^2 \right) + \beta \left(|\frac{\partial^2 \mathbf{v}}{\partial x^2}|^2 + \right. \right.$$
$$\left. \left. 2|\frac{\partial^2 \mathbf{v}}{\partial x \partial y}|^2 + |\frac{\partial^2 \mathbf{v}}{\partial y^2}|^2 \right) \right\} d\xi_1 d\xi_2 dt. \tag{8.15}$$

Substituting Equations (8.15), (8.12), and (8.13) into Equation (8.14) and differentiating with respect to each ensemble parameter $\mathbf{v}_{jm}$ gives a set of simultaneous linear equations that can be solved for the unknowns $\mathbf{v}_{in}$.

8.4.3   FITS TO BIFURCATION DATA

Estimated spectral components of the motion of bifurcation 5 are shown in Figure 8.6. Similar plots for each bifurcation show that the energy of motion is largely contained within the first five harmonics. We therefore choose $I=11$ in Equation (8.12). The three rectangular Cartesian displacement fields can then be fitted to the tracked bifurcations according to Equation

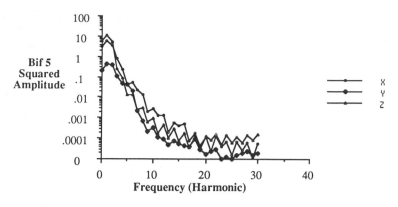

FIGURE 8.6. Estimated spectral components of the motion of bifurcation 5.

(8.14). Figure 8.7 shows a time-varying surface fitted using $\alpha = 1.0$ $\beta = 0.01$ at various stages in the cardiac cycle. The surface has 1100 parameters (25 nodes $\times$ 4 bicubic surface parameters per node $\times$ 11 Fourier coefficients per surface parameter) for each geometric field. The fit to the 660 data points (11 bifurcations $\times$ 60 frames) took 18 min on a microVax 3100 workstation. The geometric error associated with the tracked bifurcation points was $(0.657 \pm 0.364; 0.563)$. A graphical impression of the error is also possible by comparing the predicted surface location with the tracked location of bifurcations through time (Figure 8.5).

8.4.4   EPICARDIAL STRAIN

The surface deformation undergone at any point is completely described by the Lagrangian surface strain tensor $\mathbf{E}$ (see Chapter 7). The eigenvalues of this tensor are the principal (maximum and minimum) strains. Their associated eigenvectors form an orthogonal pair of directions that specify the axes of pure stretch in which they act. The principal strains at 3×3 arrays of (gaussian quadrature) points within each element are shown in Figure 8.7. These are drawn as small crosses, aligned in the principal directions, which are orthogonal at all times. The lengths of the line segments are proportional to the magnitude of the principal strains; solid lines represent positive (tensile) strains and dotted lines are negative (compressive) strains. There is substantial shortening between bifurcations 2 and 7 at frame 20 (about end-diastole), compared with lengthening between 6 and 9 at the same time, reflecting atrial contraction and ventricular expansion. The situation is reversed at end-systole (around frames 30 and 40). Here there is a noticeable contraction between bifurcations 6 and 10 (the linear distance direction is a good approximation of the direction of major principal direction) but there is also a slight lengthening in the other principal direction.

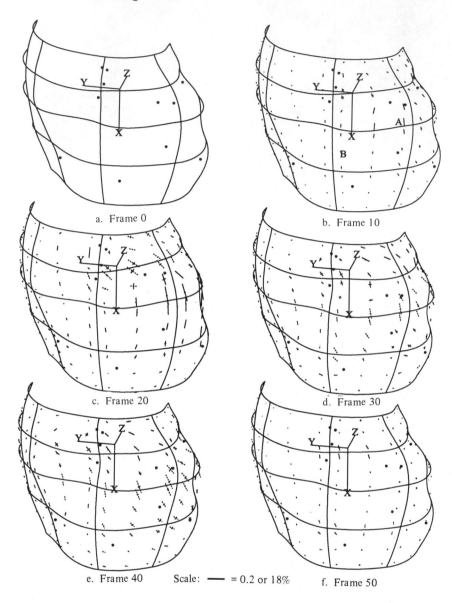

a. Frame 0

b. Frame 10

c. Frame 20

d. Frame 30

e. Frame 40   Scale: ── = 0.2 or 18%   f. Frame 50

FIGURE 8.7. Fitted dynamic surface at various stages of the cardiac cycle. Tracked bifurcations are shown as * and principal strains as crosses aligned in the principal directions. Solid lines are extensions, dotted lines are contractions.

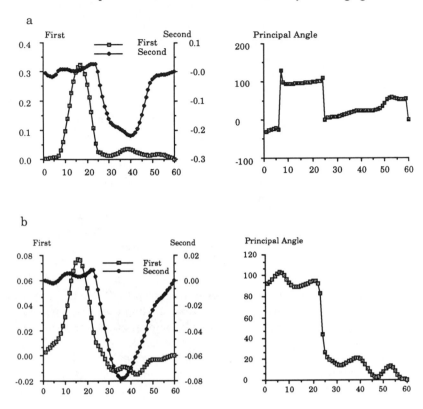

FIGURE 8.8. Principal strains vs. frame number at gauss points A and B of Figure 8.7B.

Figures 8.8A and 8.8B show time series of principal strain values for the Gauss points labeled A and B in Figure 8.7B. Here the strains are ordered so that the most positive eigenvalue is denoted "First" and the most negative eigenvalue is denoted "Second." The principal angle is defined to be the angle between the ξ_1 coordinate direction and the direction of the "First" principal stretch (counterclockwise positive). At some times during the cycle the principal angle may become ill-defined, for example, when the principal strains become vanishingly small (as at the reference state) or just numerically similar (as in equibiaxial tension or compression). Also the principal angle may undergo a sudden 90° shift when the value of the maximum strain falls below the minimum strain and the two swap roles. This happens in Figure 8.8A near frame 24: the crossover between the expansion and contraction phases, relative to diastasis at this particular gauss point. The principal angle is, however, stable around end-diastole and end-systole in all areas. Note that the direction of the largest magnitude strain is approximately longitudinal in both diastole and systole. In fact, most gauss points exhibit a preferential strain axis, along which the epicardium expands during diastole and contracts during systole. The strain in this direction is usually substantially greater than in the other principal direction, suggesting that most of the deformation appears to occur along these preferential directions in all ventricular regions. The directions of greatest strain are to some extent time dependent; however, they do appear to approximate the orientation of the epicardial fibers in all regions, varying from −40° to −90° [46].

8.5   Motion from Vessels

Although the surface fit to the bifurcations almost exactly interpolates the bifurcations themselves, significant errors can occur when predicting the loci of the interconnecting vessels. This error is governed by the order of the model (number of degrees of freedom) and the magnitude and ratio of the smoothing parameters but cannot be quantified until the vessel locations are known. This suggests that we can better utilize the information available in coronary cinéangiograms by enforcing correspondence between the predicted location of the vessels and their radiographic lumen. A more accurate description of the strain field would result, simply because the time-varying surface is better constrained. Vessel loci give slope and curvature information that would otherwise be completely determined by the smoothing parameters. Similar strain fields should then be obtained from a wide range of weights because the surface is more constrained by the data.

The surface fit to the bifurcations $\mathbf{u}^2$ forms a close approximation to the true epicardial surface $\mathbf{u}$. This approximation is a good one not only because it almost exactly interpolates the tracked bifurction positions at all times, but also because it incorporates information describing the location of the entire coronary artery tree at diastasis. At time $t = 0$ nearly all the predicted arterial points lie on the centerlines of the vessel lumen. We now treat this approximation as a reference or undeformed state and attempt to find the displacement field $\mathbf{v}$ so that $\mathbf{u} = \mathbf{u}^2 + \mathbf{v}$.

8.5.1 VESSEL TRACKING

If the point is sufficiently defined by its local gray-level distribution, as are most bifurcation points, it is often sufficient to match image tokens between frames, as above. The vessels themselves are linelike structures and hence suffer from the aperture problem [18]. Provided the displacement is small relative to the curvature of the line, local motion measures can only supply the component of displacement in the direction perpendicular to the curve. We therefore require additional constraints to define uniquely the displacement field. Most of the solution procedures make use of regularization techniques to constrain the solution [1,4]. One approach that lends itself to the current application is that of Terzopoulos and coworkers [49]. Objects are modeled as elastically deformable bodies subject to continuum mechanical laws. Constraints owing to the image data are applied as forces that deform the model from an ideal stress-free state. These mechanical models allow the integration of a wide range of intrinsic (pertaining to the model) and extrinsic (pertaining to the image and other forces) constraints into a single variational principle. The aperture problem is effectively side-stepped, since the applied structural constraints are global. Given the bifurcation fit $\mathbf{u}^2(\xi_1, \xi_2, t)$, we could apply a direct formulation of this method and attempt to find the displacement field $\mathbf{v}$ that minimizes

$$\mathcal{E}(\mathbf{v}) = \int_{\Omega} E_{internal}(\mathbf{v}) + E_{image}(\mathbf{u}^2 + \mathbf{v})d\Omega. \tag{8.16}$$

$E_{internal}$ measures the strain energy of the object, that is, its resistance to deformation from its "strain-free" or reference shape. For a surface, this can take the form of a norm of the surface strain tensor, or measure the difference between the first and second fundamental forms [48]. These measures have the advantage of being zero in the case of rigid body motion; however, they lead to nonlinear minimization problems. A linearized measure of surface deformation energy is the weighted combination of membrane and thin-plate spline kernels in Equation (8.4) [49].

E_{image} represents external potential fields derived from the images. These give rise to conservative body forces that deform the object. The image potential should measure the match between the model and the images. Since the vessels are generally manifest as long, thin structures, an appropriate measure would be given by the output of a line detector. Image lines are characterized by high curvatures in the image function or high values of the second derivative. It is possible to estimate the direction and magnitude of the maximum curvature by calculating the eigenvectors and eigenvalues of the image Hessian

$$\text{Hess}(I) = \begin{pmatrix} \dfrac{\partial^2 I}{\partial x^2} & \dfrac{\partial^2 I}{\partial x \partial y} \\[2mm] \dfrac{\partial^2 I}{\partial x \partial y} & \dfrac{\partial^2 I}{\partial y^2} \end{pmatrix}. \tag{8.17}$$

In the current implementation, the derivatives of Equation (8.17) are calculated by convolution with appropriate derivatives of a two-dimensional gaussian, and the maximal (most positive) eigenvalue is taken as a measure of image lineness. This detector offers a considerable improvement over the $\nabla^2 G$ operator, which is more susceptible to image noise and enhances image mottle.

Rather than apply the image forces to the entire surface at once, we apply a considerably simplified version, whereby each branch in the coronary artery tree is fitted separately to the image sequence. This results in a one-dimensional formulation of Equation (8.16) and allows us to apply the "snake" algorithm of Kass and colleagues [21] almost directly. The solution is more tractable than the two-dimensional case, which involves a large number of parameters in a nonlinear optimization procedure. Briefly, the procedure is as follows. The locations of the vessel centerlines $\mathbf{u}^2(s)$ are expressed as functions of arc-length s along the curve. Their undeformed or resting states are given by the surface $\mathbf{u}^2(\xi_1, \xi_2)$ as before, with the appropriate reparameterization from (ξ_1, ξ_2) to s. In the discrete formulation, the set of displacements $\mathbf{v}_i$ is found for each branch by minimizing

$$\sum_i \left(\alpha |\mathbf{v}_i - \mathbf{v}_{i-1}|^2 + \beta |\mathbf{v}_{i-1} - 2\mathbf{v}_i + \mathbf{v}_{i+1}|^2 \right) + \sum_i \text{line}(\Pi(\mathbf{u}^2{}_i + \mathbf{v}_i)). \tag{8.18}$$

The boundary conditions at each end of the branch fall into one of two categories:

1. The endpoint is a bifurcation, in which case its location, and therefore its displacement from $\mathbf{u}^2$, is known. We require one further condition and set the second derivative to zero at the boundary. If the contour is viewed as a thin beam under tension, these end conditions imply a pinned (moment free) end.

2. The endpoint is not a bifurcation but simply a point on the vessel beyond which it becomes difficult to locate the vessel reliably. In this case we apply the free end conditions of a beam, that is, the second and third derivatives of displacement are zero at the boundary.

The energy functional of Equation (8.18) is minimized using a modified gradient descent algorithm, which is implicit in the linear smoothing forces and explicit in the nonlinear image forces [21]. This procedure iterates toward a local minimum in the objective function only and relies on a sufficiently accurate starting position in order to follow it. We use the position of the tracked contour in the previous frame, linearly warped to match the locations of the tracked bifurcation endpoint(s) in the current frame, as the initial estimate of the contour's position. The initial displacement is the difference between this predicted location and the "undeformed" contour $\mathbf{u}^2$, which is also warped to match exactly the tracked bifurcations. The displacement is thereby constrained to zero at the bifurcations. This formulation allows information from the previous frame to guide the tracking process over an extended sequence while maintaining a global smoothness constraint that does not accumulate errors from frame to frame.

Most of the vessels shown in Figure 8.3 were tracked through the entire cycle using Equation (8.18) with $\alpha = 50, \beta = 500$. Each vessel could be tracked in three dimensions, using information from both views. Any branch containing fewer than six points could not be tracked by this method, owing to the pentadiagonal stiffness matrix generated by the internal forces [21]. Also, the anterior diagonal branch emanating from bifurcation 3 in Figure 8.3 could not be tracked since it became almost totally obscured in the LAT view by several overlying vessels. The tracked locations are also shown in Figure 8.9 for several frames in the cycle. The major source of error for this algorithm is the tendency of major vessels to overlap in a particular view. In the sequence under consideration the anterior descending and anterior diagonal branches overlap in the RAO view, while the circumflex and obtuse marginals overlie in the LAT view. Thus, the centerline contour could often be subject to forces arising from different vessels.

8.5.2   Fit to the Vessels

Once the vessels have been tracked, they provide exactly the same kind of four-dimensional data as the tracked bifurcations, allowing the same surface reconstruction techniques to be applied. A surface fit using $\alpha = 0.100, \beta = 0.100$ in Equation (8.14) is shown in Figure 8.9. The surface contains 1100 degrees of freedom per geometric field. The fit to the 14,580 coronary data points took 4.5 h for all three fields on a microVax 3100 workstation. The geometric error associated with the vessel data points was $0.720 \pm 0.408; 0.685$. This surface exhibits more regional variation of strain than Figure 8.7, but the overall pattern of deformation is similar. The extensions above the main circumflex branch are more substantial and are caused by the branch emanating from bifurcation 7 (Figure 8.3). This branch, however, becomes obscured in the LAT view at end-systole by other circumflex branches. The bunching of the anterior diagonal branch towards the LAD in Figure 8.9D and 8.9E is corroborated by both views, suggesting that portions of this branch may display buckling behavior that is uncoupled from the underlying myocardium. The positive extensions experienced

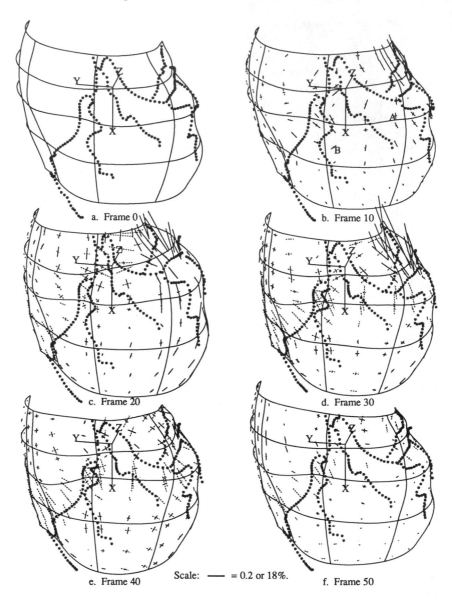

a. Frame 0

b. Frame 10

c. Frame 20

d. Frame 30

Scale: —— = 0.2 or 18%.

e. Frame 40

f. Frame 50

FIGURE 8.9. Dynamic surface fitted to tracked coronary data.

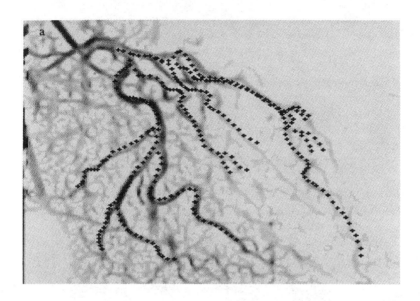

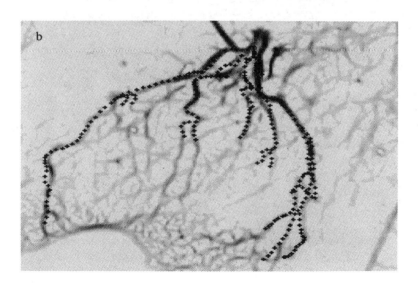

FIGURE 8.10. Surface locations of points corresponding to tracked vessel data, overlaid on line detector output at frame 40. (A) RAO view. (B) LAT view.

near bifurcation 10 in the longitudinal direction (i.e., approximately orthogonal to the basal circumflex) is in part supported by the shortening curves
between bifurcations 6 and 8, and 6 and 5, although it may be an artifact
due to the scarcity of data in the apical regions.

A wide range of smoothing parameters produces qualitatively similar
strain behavior, with slightly different magnitudes. As with the bifurcation
fits, the sensitivity of the calculated strains to the ratio $\alpha{:}\beta$ increases with
distance from the data points as the surface becomes less constrained by
the data.

A good subjective estimate of the goodness of fit of the time-varying surface is obtained by calculating the positions of the vessel material points,
as given by the ensemble coordinates of the Bézier coronary data, at particular times and displaying them in the same view as the angiograms. Figure
8.10 shows the predicted vessel points given by the surface $\alpha = 0.100$ and
$\beta = 0.100$, projected on the angiographic images at frame 40 (end-systole).
There is some discrepancy along the branch emanating from bifurcation
3 at frame 40. This branch was not tracked because it is obsured at end-
systole in the LAT view. The surface also has some trouble representing
the branches around bifurcation 4, perhaps because the basal branches
(between bifurcations 4 and 1) do not lie on the epicardium.

8.6   Discussion

The structural model of the dynamic epicardial geometry expressed by
Equations (8.11), (8.12), and (8.13) was designed to be compatible with
finite element models of stress and activation. As such it embodies a finite
element grid pattern and low-order polynomial basis functions in space.
We have chosen a Fourier series basis to interpolate the surface parameters
in time. This implicitly constrains the motion to be periodic. The Fourier
coefficients then reflect the energies associated with each harmonic. They
decrease with increasing frequency and are invariant to the number of terms
included in the series. The choice of five harmonics is consistent with Rankin
and colleagues [40], who base their estimate on global shape changes.

The advantage of fitting a time-varying surface to the data is that the
derived strain fields are temporally as well as spatially smooth. The major
and minor stretches are local deformation measures rather than average
length changes, and completely characterize the deformation at a point.
Surface strain is not constrained to be homogeneous but can vary over the
surface in a manner that depends on the order of the basis functions.

At each stage of the analysis we have used the surface model to regularize both the surface estimation and the vessel tracking problems. The
spatial variation of the displacement from an "undeformed" or target shape
was penalized in the least-squares error function. This allowed regular solutions to be obtained from sparse and nonuniform data sets. The fit to the

bifurcations was also used to guide the vessel tracking process, using Terzopoulos' [47,49] paradigm for fitting deformable models to images. This method of correspondence resolution via structural constraints has direct application to the current problem because of its clear mechanical analogy. Since we are primarily interested in epicardial deformation, the application of a deformation constraint on the solution provides a direct relationship between the type of regularization applied and the calculated strain field. A structural deformation constraint is perhaps the only method of determining the correspondence between vessel points that are not uniquely defined by their local gray-level distribution. The use of structural models to guide the image structure identification problem has been widespread in medical imaging, for example, brain atlas mapping in computed tomography [3], cell boundary detection [32], and quantitative coronary angiography [57]. Similar techniques could readily be applied to other areas of cardiac imaging where heart wall motion is of interest, for example, echocardiography, LV angiography, and magnetic resonance imaging. The main disadvantage of the currently implemented vessel tracking algorithm is that each branch is considered separately and no information is available to the model about neighboring branches, even though these branches can exert image forces on the model. Another limitation is that it is very short-sighted, that is, iterates toward local minima only in the error function. Since vessels are highly mobile and localized (fine) structures, the model can easily lose track of the vessel lumen. Extensions of the method to overcome these problems are discussed in Young [55].

Ideally, the deformation measure used in Equation (8.14) should take the form of a strain-energy function that reflects the constitutive properties of the myocardium. However, these are currently not well understood. Worse, any function that simply measures a norm of the strain tensor is inherently nonlinear. In the present study we resort to a heuristic smoothing constraint that is linear and produces regular solutions but that has no fidelity to arbitrary undeformed shapes. Small strains are therefore introduced in the case of rigid body rotations. This produces an error of about 10% in the calculated strains [55].

The choice of "optimal" smoothing weights in the surface reconstruction is guided primarily by the closeness of fit to the data points. This does not determine the weights uniquely because a range of $\alpha:\beta$ may produce similar geometric errors. For α and β small enough this ratio does not greatly affect the surface shape in regions near the points of measurement. Weights are then chosen to give a visually sensible surface in regions away from the data points. More objective criteria for the establishment of "optimal" weights may become available as our knowledge of the mechanical properties of the heart increases.

One method of determining the optimal values of smoothing parameters is Craven and Wahba's generalized cross-validation [11]. The figure of merit associated with a set of weights is given by the ability of the model to predict some of the observed points on the basis of the remaining data set. Optimal parameters are those that minimize the average error of prediction over all the observed points. An investigation of these techniques applied to vision problems has been made by Shahraray and Anderson [41]. Unfortunately, the error function is expensive to compute and nonlinear optimization techniques must be employed to locate its minima.

Geometric measurements defining the entire epicardium of several hearts, as well as their biplane angiograms, at different stages of the cardiac cycle could be used to define a set of optimum weights for the surface fit. Alternatively, a calibration study could be performed involving several marker triangles implanted between the arteries. The homogeneous strains calculated from these triangles could be compared with the surface strains derived from the arteries. The set of weights giving best agreement could be considered optimal. Similarly, the regional deformations of several normal hearts need to be examined before any statistical tests can be devised to identify pathological behavior. Although the motion of the bifurcations has been seen to correlate closely to the motion of markers implanted in nearby locations [24], it would be interesting to determine to what extent the strain within a small region between aterial branches is predicted by the current method. This would help to determine the extent to which the arteries are tethered to the actual epicardial surface.

TABLE 8.1. Average squared errors between bifurcation points and fitted surfaces for various values of the smoothing weights, α (rows) and β (columns).

$\alpha\backslash\beta$	0	0.001	0.01	0.1	1.0	10
0		7.643	6.854	6.038	5.900	6.421
0.0001		7.525				
0.001	5.439	6.971				
0.01	5.554	5.809				
0.1	5.320	5.334	5.446	5.634		
1.0	5.311	5.308	5.293	5.311		
10.0	6.454	6.455	6.458	6.489		

For the present, we can determine the best weights for the surface fit to the bifurcation points as those that best predict the location of the tracked vessels. Table 8.1 shows the results of a grid search in α and β. Each entry is the average squared error between the tracked vessel data points and their predicted location on the surface fitted to just the bifurcation data. There is a distinct, if shallow, minimum at $\alpha = 1.0, \beta = 0.01$ (hence the use of these values in Figure 8.7). These values apply only to the fit to the bifurcation points, not to the fit to the vessels themselves, since the latter is an intrinsically different data set. The surface is better constrained by the coronary data than by the bifurcation data alone, hence the smoothing weights are smaller in magnitude.

REFERENCES

[1] J.K. Aggarwal and N. Nandhakumar. On the computation of motion from sequences of images—A review. *Proceedings of the IEEE*, 76:8:917–935, 1988.

[2] T. Arts, P.C. Veenstra, and R.S. Reneman. Epicardial deformation and left ventricle wall mechanics during ejection in the dog. *Am. J. Physiol.*, 243:H379–H390, 1982.

[3] R. Bajcsy and S. Kovacic. Multiresolution elastic matching. *Comput. Vision Graph. Image Process.*, 46:1–21, 1989.

[4] M. Bertero, T.A. Poggio, and V. Torre. Ill-posed problems in early vision. *Proceedings of the IEEE*, 76:8:69–88, 1988.

[5] R. Beyar and S. Sideman. Effect of the twisting motion on the nonuniformities of transmural fibre mechanics and energy demands—A theoretical study. *IEEE Trans. Biomed. Eng.*, 32:764–769, 1983.

[6] D.P. Boyd and M.J. Lipton. Cardiac computed tomography. *Proceedings of the IEEE*, 71:298–307, 1983.

[7] L. Brevdo, S. Sideman, and R. Beyar. A simple approach to the problem of three dimensional reconstruction. *Comput. Vision Graph. Image Process.*, 37:420–427, 1987.

[8] R.W. Brower, H.T. ten Katen, and G.J. Meester. Direct method for determining regional myocardial shortening after bypass surgery from radiopaque markers in man. *Am. J. Cardiol.*, 41:1222–1229, 1978.

[9] H. Bunke, G. Sagerer, and H. Niemann. Model based analysis of scintigraphic image sequences of the human heart. In T.S. Huang, editor, *Image Sequence Processing and Dynamic Scene Analysis*, pages 725–740. Springer-Verlag, New York, 1983.

[10] B.R. Chaitman, H. Demots, J.D. Bristow, J. Rosch, and S.H. Rahimtoola. Objective and subjective analysis of left ventricular angiograms. *Circulation*, 52:420–425, 1975.

[11] P. Craven and G. Wahba. Smoothing noisy data with spline functions. *Numerische Mathematik*, 31:377–403, 1979.

[12] T. Fukui, M. Yachida, and S. Tsuji. Detection and tracking of bloodvessels in cine-angiograms. *Proceedings of the IEEE*, 1:383–385, 1980.

[13] J.J. Gerbrands, F. Booman, and J.H.C. Reiber. Comuter analysis of moving radiopaque markers from x-ray cinefilms. *Comput. Vision Graph. Image Process.*, 11:35–48, 1979.

[14] B. Hannaford and S.A. Glantz. Adaptive linear predictor tracks implanted radiopaque markers. *IEEE Trans. Biomed. Eng.*, 32:117–125, 1985.

[15] D.C. Harrison, A. Goldblatt, E. Braunwald, G. Glick, and D.T. Mason. Studies on cardiac dimensions in intact unanesthetized man. 1. Description of techniques and their validation. *Circ. Res.*, 13:448–455, 1963.

[16] M. Hashimoto and J. Sklansky. Multiple-order derivatives for detecting local image characteristics. *Comput. Vision Graph. Image Process.*, 39:28–55, 1987.

[17] G.T. Herman, L. Axel, R. Bajcsy, et al., editors. *Model Driven Visualization of Coronary Arteries.*, Proceedings of the 8th International Joint Conference on Artificial Intelligence, 1983, 1128-1131.

[18] E.C. Hildreth. Computations underlying the measurement of visual motion. *Artif. Intell.*, 23:309–354, 1984.

[19] P.J. Hunter and B.H. Smaill. The analysis of the heart: A continuum approach. *Prog. Biophys. Molec. Biol.*, 52:101–164, 1988.

[20] N.B. Ingels, G.T. Daughters, E.D. Stinson, and E.L. Alderman. Evaluation of methods for quantitating left ventricular segmental wall motion in man using myocardial markers as a standard. *Circulation*, 61:966–972, 1980.

[21] M. Kass, A. Witkin, and D. Terzopoulos. Snakes: Active contour models. *Int. J. Comput. Vision*, 1:4:321–331, 1988.

[22] H.C. Kim, B.G. Min, M.M. Lee, J.D. Seo, Y.W. Lee, and M.C. Han. Estimation of local cardiac wall deformation and regional wall stress from biplane coronary cineangiograms. *IEEE Trans. Biomed. Eng.*, 32:503–511, 1985.

[23] M. Kindelan and J. Suarez de Lezo. Artery detection and tracking in coronary angiography. In S. Levialdi, editor, *Digital Image Analysis*, pages 283–294, Pitman, New York, 1984.

[24] Y. Kong, J.J. Morris, and H.D. McIntosh. Assessment of regional myocardial performance from biplane coronary angiograms. *Am. J. Cardiol.*, 27:529–537, 1971.

[25] R.F. Leighton, S.M. Witt, and R.P. Lewis. Detection of hypokinesis by a quantitative analysis of left ventricle cineangiograms. *Circulation*, 50:121–127, 1974.

[26] M.M. LeWinter, R.S. Kent, J.M. Kroener, T.E. Carew, and J.W. Covell. Regional differences in myocardial performance in the left ventricle of the dog. *Circ. Res.*, 37:191–199, 1975.

[27] K.J. Liu, J.M. Rubin, M.J. Potel, et al. Left ventricular wall motion: Its dynamic transmural characteristics. *J. Surg. Res.*, 36:25–34, 1984.

[28] S.A. MacKay, M.J. Potel, and J.M. Rubin. Graphics methods for tracking three-dimensional heart wall motion. *Comput. Biomed. Res.*, 15:455–473, 1982.

[29] A.D. McCulloch, B.H. Smaill, and P.J. Hunter. Left ventricular epicardial deformation in isolated arrested dog heart. *Am. J. Phys.*, 252:H233–241, 1987.

[30] A.D. McCulloch, B.H. Smaill, and P.J. Hunter. Regional left ventricular epicardial deformation in the passive dog heart. *Circ. Res.*, 64:721–733, 1989.

[31] I.G. McDonald. Contraction of the hypertrophied left ventricle in man studied by cineradiography of epicardial markers. *Am. J. Cardiol.*, 30:587–594, 1972.

[32] M. McQueen. Evaluation of a generalized template matching procedure. *Australasian Phys. Eng. Sci. Med.*, 9:4:180–187, 1986.

[33] G.D. Meier, M.C. Ziskin, W.P. Santamore, and A.A. Bove. Kinematics of the beating heart. *IEEE Trans. Biomed. Eng.*, 27:319–329, 1980.

[34] Herman M.V., R.A. Heinle, and M.D. Klein. Localized disorders in myocardial contraction: Asynergy and its role in congestive heart failure. *N. Engl. J. Med.*, 277:222–232, 1967.

[35] P.M.F. Nielsen. *The Anatomy of the Heart: A Finite Element Model.* Ph.D. thesis, University of Auckland, New Zealand, 1987.

[36] P.M.F. Nielsen, I.J. Le Grice, B.H. Smaill, and P.J. Hunter. A mathematical model of the geometry and fibrous structure of the heart. *Am. J. Phys.*, 1991. (in press).

[37] C.M. Philips, J. Prenis, W.P. Santamore, and A.A. Bove. Recognition and storage of metal heart marker position from biplane X-ray images at video rates. *IEEE Trans. Biomed. Eng.*, 30:10–17, 1983.

[38] D.L. Pope, D.L Parker, D.E. Gustafson, and P.D. Clayton. Dynamic search algorithms in left ventricular border detection and analysis of coronary arteries. *Comput. Cardiol.*, September:71–75, 1984.

[39] M.J. Potel, J.M. Rubin, S.A. MacKay, A.M. Aisen, J. Al-Sadir, and R.E. Sayre. Methods for evaluating cardiac wall motion in three dimensions using bifurcation points of the coronary arterial tree. *Invest. Radiol.*, 18:47–57, 1983.

[40] J.S. Rankin, P.A. McHale, C.E. Artentzen, D. Ling, J.C. Greenfield, and R.W. Anderson. The three-dimensional dynamic geometry of the left ventricle in the conscious dog. *Circ. Res.*, 39:304–313, 1976.

[41] B. Shahraray and D.J. Anderson. Optimal estimation of contour properties by cross-validated regularization. *IEEE Trans. Patt. Anal. Mach. Intell.*, 11:600–610, 1989.

[42] T.H. Shepertycki and B.C. Morton. A computer graphic-based angiographic model for normal left ventricular contraction in man and its application to the detection of abnormalities in regional wall motion. *Circulation*, 68:1222–1230, 1983.

[43] R.W. Smalling, M.H. Skolnick, D. Myers, R. Shabetai, J.C. Cole, and D. Johnston. Digital boundary detection, volumetric and wall motion analysis of left ventricular cineangiograms. *Comput. Biol. Med.*, 6:73–85, 1976.

[44] G.E. Sotak and K.L. Boyer. The Laplacian-of-Gaussian kernel: A formal analysis and design procedure for fast, accurate full-frame output. *Comput. Vision, Graph. Image Process.*, 48:147–189, 1989.

[45] S.A. Stansfield. ANGY: A rule-based expert system for automatic segmentation of coronary vessels from digital subtracted angiograms. *IEEE Trans. Patt. Anal. Mach. Intell.*, 8:183–199, 1986.

[46] D.D. Streeter and W.T. Hanna. Engineering mechanics for successive states in the canine left ventricle myocardium. *Circ. Res.*, 33:639–664, 1973.

[47] D. Terzopoulos. Regularization of inverse problems involving discontinuities. *IEEE Trans. Patt. Anal. Mach. Intell.*, 8:413–424, 1986.

[48] D. Terzopoulos, J. Platt, A. Barr, and K. Fleischer. Elastically deformable models. *Comput. Graph.*, 22:4:279–287, 1987.

[49] D. Terzopoulos, A. Witkin, and M. Kass. Constraints on deformable models: Recovering 3D shape and non-rigid motion. *Artif. Intell.*, 36:91–123, 1988.

[50] V. Torre and T.A. Poggio. On edge detection. *IEEE Trans. Patt. Anal. Mach. Intell.*, 8:147–162, 1986.

[51] S. Tsuji and H. Nakano, editors. *Knowledge-based Identification of Artery Branches in Cine-angiograms.* Proceedings of the International Joint Conference on Artificial Intelligence, 1981, 710-718.

[52] D. Tzvioni, G. Diamond, M. Pichler, K. Stankus, R. Vas, and J. Forrester. Analysis of regional ischemic left ventricular dysfunction by quantitative cineangiography. *Circulation*, 60:1278–1283, 1979.

[53] L.K. Waldman, Y.C. Fung, and J.W. Covell. Transmural myocardial deformation in the canine left ventricle. *Circ. Res.*, 57:152–163, 1985.

[54] F.C.P. Yin. Ventricular wall stress. *Circ. Res.*, 49:829–842, 1981.

[55] A.A. Young. *Epicardial Deformation from Coronary Cinéangiograms.* Ph.D. thesis, University of Auckland, New Zealand, 1990.

[56] A.A. Young. *Image Processing of Coronary Angiograms.* M.E. thesis, University of Auckland, New Zealand., 1986.

[57] A.A. Young, P.J. Hunter, and B.H. Smaill. Epicardial surface estimation from coronary cinéangiograms. *Comput. Vision, Graph. Image Process.*, 47:111–127, 1989.

9

Functional Consequences of Regional Heterogeneity in the Left Ventricle

Wilbur Y.W. Lew[1]

ABSTRACT The left ventricle is characterized by significant regional heterogeneity in structure, electrophysiology, and function. This discussion focuses on the functional consequences of regional heterogeneity. First we examine the extent of regional heterogeneity in deformations observed under physiologic conditions. The importance of considering the *direction* as well as the magnitude of maximal deformations will be emphasized. Second, we discuss potential mechanisms for regional heterogeneity in deformations. The potential contribution of electrophysiologic, anatomic, structural, and geometric factors will be considered. Third, we examine the functional consequences of regional heterogeneity using experimental models. Regional ischemia is used to examine the mechanical interaction between ischemic and nonischemic areas, that is, the interaction between "weak" and "strong" muscles. Regional inotropic stimulation is used to produce subtle alterations in regional heterogeneity to examine the mechanical interaction between "strong" and "stronger" muscles. Finally, we correlate some of the predictions from theoretical models of acute ischemia and chronic infarction with the experimental observations.

9.1 Left Ventricular Heterogeneity Under Physiologic Conditions

The left ventricle is structurally heterogeneous with regional differences in shape, wall thickness, fiber angles [73], and a temporal dispersion in electrical activation [2,66]. Regional heterogeneity in systolic and diastolic deformations have been observed in humans and experimental animals by a variety of techniques, including contrast ventriculography [8,18,42,71], echocardiography [26,70], and ultrasonic dimension gauges (measuring segment length, wall thickness, or a ventricular diameter) [50,51,53,54,63]. There is a transmural gradient of strain with systolic shortening and wall

---

[1]Cardiology Section, Department of Medicine, University of California, San Diego, and Veterans Administration Medical Center, San Diego, California 92161

thickening increasing with depth (from the epicardium to the endocardium) [80,81]. Segment shortening and wall thickening increase from the base to the apex of the left ventricle [26,30,43,53]. Although regional differences in deformation between the anterior, lateral, and posterior walls have been found, the data are conflicting as to which region has the greatest (or least) deformation. Many of these discrepancies are largely related to differences in technique and geometric assumptions.

Many studies implicitly assume that regional left ventricular function can be evaluated by comparing regional deformations measured along a prescribed pathway. It is assumed that the orientation chosen for uniaxial measurements (e.g., the circumferential direction, a minor axis chord, or a radial pathway) is closely aligned with the direction of maximal deformation. However, different regions of the left ventricle do not necessarily deform along a common axis [36]. Furthermore, significant strains can be measured in directions orthogonal to either the predominant direction of motion or the local fiber direction [3,14,51,57,58,79–81]. Thus, unidirectional measurements of regional deformations may be severely limited for evaluating the extent of regional heterogeneity unless strains are measured in an orientation closely aligned with the direction of maximal shortening and lengthening, the contribution of shearing deformations is minimal, and the extent of anisotropy is similar in different regions.

We used ultrasonic segment length gauges to compare midwall segment shortening in the circumferential (aligned with the local fiber direction) and longitudinal directions in the anterior and posterior walls at the same midventricular level [51]. Midwall shortening in the circumferential direction was significantly greater in the anterior than lateral or posterior walls, but shortening in the longitudinal direction was similar. The calculated systolic change in midwall area did not differ between anterior and posterior walls [51]. Recently, Ingels and colleagues [38] found equal shortening in the circumferential and oblique directions in the anterior wall, with lesser shortening in the longitudinal direction. In the posterior, lateral, and septal walls, maximal shortening was in the oblique direction (in line with the subepicardial fiber direction), with significantly less shortening in the circumferential and longitudinal directions (shortening in the latter two directions did not differ). It is clear from both of these studies that interpretation of regional heterogeneity in deformations is very much dependent on the orientation of the measurements.

The direction of maximal shortening is not constant, but may change as a function of ventricular volume [79]. We developed a new technique for measuring two-dimensional finite strains using three ultrasonic crystals [79]. The three crystals were implanted in a triangular array in the anterior midwall with each crystal simultaneously focused on the other two. The three segment lengths of the crystal triangle were simultaneously measured throughout the cardiac cycle. The orientation of the crystal triangle relative to a cardiac coordinate system was measured. Figure 9.1 shows typical calculated two-dimensional finite strains for the anterior midwall at a low, mid-, and high ventricular volume, corresponding to left ventricular end-diastolic pressures (LVEDP) of 3 mm Hg, 8 mm Hg and 14 mm Hg, respectively. End-systolic circumferential and longitudinal strains were similar in magnitude at low LVEDP, but circumferential strains were significantly more negative (greater shortening) at the mid and high volumes. The end-systolic in-plane shear strain was positive and decreased with volume loading. The first principal strain (maximal shortening deformation, regardless of direction) and second principal strain (strain perpendicular to the first principal strain) became more negative (greater shortening) with volume loading. The principal axis of maximal shortening was oriented −50° (i.e., clockwise) from the circumferential direction (0°) at the low volume, but became progressively closer to the circumferential direction at higher volumes. Circumferential strain measurements adequately reflected maximal shortening deformations at mid and high volumes, but underestimated the first principal strain by approximately 40% at low volumes [79].

In a subsequent study, we compared two-dimensional finite strains in the anterior and posterior midwall [78]. With volume expansion, there was a similar increase in maximal systolic shortening (more negative first principal strain) in both regions. However, anterior wall shortening increased primarily in the circumferential direction, whereas posterior wall shortening increased more in the longitudinal direction. The orientation of the first principal axis (direction of maximal shortening) became more closely aligned with the circumferential direction in the anterior wall (with volume expansion), but did not change in the posterior wall. As a consequence, circumferential strain adequately estimated maximal shortening deformations at mid and high volumes in the anterior wall, but underestimated maximal shortening in the posterior wall by 30% at all volumes.

The positive in-plane shear strains observed in the midwall are a local reflection of torsional deformations [78,79]. The apex twists about approximately 12° to 18° relative to the base (in a counterclockwise direction as viewed from the apex) about the left ventricular axis [37,38]. Torsional deformations are greater in the lateral free wall than inferior wall, and smallest in the septal and anterior free walls [27,38]. Torsional deformations may be important for equalizing transmural fiber stresses and fiber shortening across the wall [3]. Ventricular torsion, as well as other complex three dimensional motions of the heart (e.g., cardiac translation and rigid body motion), makes it more difficult to analyze regional deformations from a cardiac silhouette or cross-sectional image (e.g., left ventriculography and echocardiography) because landmarks used for measuring motion will move in and out of the plane of measurement during the cardiac cycle.

Diastolic deformations are also regionally heterogeneous. The peak, early diastolic transmitral pressure gradient is significantly greater at the left ventricular apex than base [11,54]. Peak lengthening rates are also greater

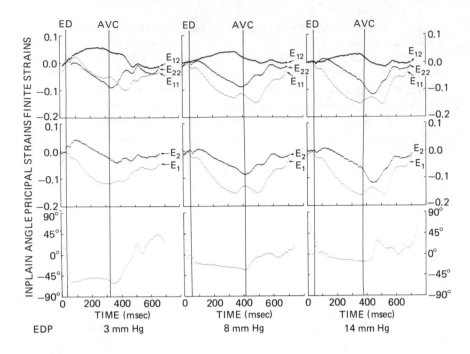

FIGURE 9.1. Two-dimensional finite strain data from the midwall of the anterior left ventricle are shown throughout a single cardiac cycle as a function of ventricular volume. The three columns (from left to right) show data at a left ventricular end-diastolic pressure (LVEDP) of 3 mm Hg, 8 mm Hg, and 14 mm Hg, respectively. In each panel the timing of end-diastole (ED) and aortic valve closure (AVC) are shown. Data for the circumferential strain (E_{11}), longitudinal strain (E_{22}), and in-plane shear strain (E_{12}) are shown in the top row. The first principal strain (E_1, maximal strain regardless of direction) and second principal strain (E_2, the strain perpendicular to E_1) are shown in the middle row. The first principal direction is shown in the bottom row.

in apical than basal segments [50,54]. Greater elastic recoil (due to greater apical than basal segment shortening) and diastolic suction effects may also contribute to greater lengthening rates at the apex than base of the left ventricle.

We examined potential regional differences in diastolic distensibility by measuring two-dimensional finite strains in the midwall of the anterior and posterior left ventricle [78]. With volume loading, the maximal stretch (regardless of direction) in the end-diastolic configuration increased to a similar extent in the anterior and posterior walls. However, the principal strain measured perpendicular to the direction of maximal stretch increased significantly more in the anterior than posterior wall, suggesting that the anterior midwall was modestly more distensible than the posterior midwall. The direction of maximal stretch (principal direction) was close to the circumferential direction in the anterior wall ($-18°$), but more longitudinally oriented in the posterior wall ($-54°$). Diastolic circumferential strain measurements adequately estimated maximal diastolic stretch in the anterior wall but underestimated maximal stretch in the posterior wall by 50% at all volumes.

McCulloch and coworkers examined regional epicardial deformations in the isolated, arrested left ventricle [57]. Volume expansion produced significantly greater epicardial stretch near the apex than base in both the anterior and posterior walls, with no significant difference in maximal epicardial stretch between anterior and posterior sites at matched levels (base, midventricle, or apex). The direction of maximal epicardial stretch at high volumes was closely aligned with the local epicardial fiber direction. Greater distensibility of the left ventricular apex than base may explain the greater segment shortening and wall thickening at apical than basal sites [26,30,43, 53].

In summary, there are substantial regional variations in systolic and diastolic deformations. Systolic strains are significantly greater (greater shortening and wall thickening) in the endocardium than epicardium, and at the apex than base of the left ventricle. Diastolic strains (peak lengthening and passive distensibility) are greater at the apex than base. Although heterogeneity of deformations have been described for the anterior, lateral, septal, and posterior walls, some of this heterogeneity is related to limitations in technique (e.g., uniaxial measurements) and geometric assumptions (e.g., regarding the principal direction), rather than a true difference in deformation. For example, circumferential strain measurements adequately reflect maximal deformations in the anterior midwall at mid and high volumes, but in the posterior midwall underestimate the maximal end-systolic strain by 30% and the maximal end-diastolic strain (maximal midwall passive stretch) by 50% at all volumes. Thus, it is important to consider the direction of maximal deformation when evaluating regional heterogeneity because this direction may vary both by region and by ventricular volume.

9.2 Potential Mechanisms for Regional Heterogeneity in Deformation

The mechanisms for the regional heterogeneity in deformation are not entirely understood. Regional heterogeneity in electrophysiologic properties, fiber anatomy, and ventricular geometry and shape may be important. Catecholamine levels are higher at the left ventricular base than apex [1,62], although the functional consequences of these differences are not known. The temporal dispersion in activation [2,66] may have important functional consequences. Regions that are activated early (e.g., the endocardium) initially contract against a lower load than regions activated late (e.g., the epicardium and base of the heart). There are regional differences in action potential duration. Monophasic action potentials are longer at early activation sites and shorter at late activation sites [15], resulting in relatively homogeneous recovery times. The action potential duration is longer at the left ventricular base than apex [82,83] and action potential amplitude and duration are greater in the endocardium than epicardium [41,55]. These electrophysiologic differences are observed both in the intact heart and in myocytes isolated from different regions. Differences in action potential amplitude and duration may influence transsarcolemmal calcium influx, contributing to the regional heterogeneity in deformations.

There are transmural differences in sarcomere lengths with longer sarcomeres in the midwall at low volumes, and more uniformly distributed sarcomere lengths at higher ventricular volumes [23,86]. This suggests a selective recruitment of endocardial and epicardial sarcomeres with volume loading. There is a transmural distribution of myocardial fiber diameters, with the fiber diameters in the epicardium larger than fiber diameters in the endocardium [33]. Sarcomere lengths are longer at the left ventricular apex than base, which can be related to the thinner wall and greater distensibility of the apex than base [46].

The orientation of fibers varies across the ventricular wall with fibers oriented near the longitudinal direction in the epicardium, becoming progressively more circumferential in the midwall, then becoming more longitudinal again in the endocardium [73]. The transmural distribution of fiber angles is not substantially different between the anterior and posterior left ventricle [22,65,74]. Although it would be predicted that fibers of different orientation may contribute to transmural differences in the principal direction (the direction of maximal shortening or lengthening), substantial deformations occur in directions away from the local fiber direction, particularly in the subendocardium [81]. Thus, local deformations must be determined by shortening along the long axis of the fiber as well significant changes in shape and/or a reorientation of the myofibers.

A transmural variation in wall stress may contribute to transmural differences in deformation. Several models predict that circumferential and longitudinal wall stresses are significantly higher in the endocardium, and decrease toward the epicardium [34]. Greater intramyocardial pressures have been measured in the endocardium than epicardium during systole, with a reversal in the transmural gradient (or no gradient) during diastole [28,72]. It is controversial whether intramyocardial pressures exceed intracavitary pressures in some regions. The validity of intramyocardial pressure measurements has been questioned because the pressure sensor unavoidably produces local distortions. It will be difficult to confirm the role of transmural differences in wall stress until acceptable techniques for directly measuring wall stress are available.

Regional differences in wall thickness may contribute to differences in wall stress. The apex is significantly thinner than the midventricle or base of the left ventricle [22,57,64,75]. The thinner wall and greater curvature contributes to a higher calculated wall stress at the apex than other regions [64]. Regional differences in wall thickness between the anterior, lateral, and posterior walls of the left ventricle are more difficult to assess because the papillary muscles and trabeculae cause considerable local variations in regional wall thickness. The anterior papillary muscle has a narrower base and is thinner than the posterior papillary muscle, which may contribute to overall differences in regional wall thickness.

Structural factors such as the valve annuli may be important. The close proximity of the posterior wall to the rigid mediastinum and the large mitral annulus may cause some local constraining effects. In contrast, the anterior wall is located farther from the smaller aortic annulus, which is attached to the more mobile aortic root. These differences may explain why the anterior wall is modestly more distensible. Regional differences in shape may play a role. The external long axis of the left ventricle is longer in the anterior than posterior wall. If the ratio of the long to short axis is also greater in the anterior wall, the anterior wall may behave more like a cylinder, whereas the posterior wall behaves more like a sphere. Passive inflation of a cylinder produces greater expansion in the direction of the minor than long axis, consistent with our findings in the anterior wall [51,79]. Greater diastolic stretch of the anterior wall in the circumferential than longitudinal direction may also explain the preferential recruitment of systolic shortening in the circumferential direction with volume loading [51,79]. In contrast, inflation of a sphere produces more equal expansion and equal recruitment of shortening in the circumferential and longitudinal directions, consistent with our experimental findings [51,78].

Regional differences in shape may be important. Regional epicardial radii of curvature and wall thickness have been measured at multiple sites and used to calculate diastolic wall stress using the law of Laplace [64]. The measurements were similar at the base and two equatorial sites. However, the radii of curvature and wall thickness were significantly smaller at the apex, leading to a calculated apical wall stress twice as great as other ventricular sites. The posterior wall is flatter than the anterior wall on left ventriculography in a 30° right anterior oblique projection [56]. Although this indicates a greater longitudinal radius of curvature in the posterior than anterior wall, any conclusions regarding regional differences in wall stress require more comprehensive data regarding regional radii of curvature in multiple directions. Furthermore, there may be complex local variations in curvature, particularly in the diaphragmatic surface, which may be convex toward the cavity [22]. In view of the current limitations in directly measuring local wall stress, radii of curvature, and wall thickness, the contribution of regional differences in wall stress to the heterogeneity in deformations remains speculative.

Finally, regional differences in extracardiac structures may be important. The proximity of the posterior wall to the rigid mediastinum has already been mentioned. The influence of the lungs and shape of the cardiac fossa on regional deformations is unknown. There are regional differences in the thickness of the pericardium and asymmetric attachments. The pericardium is attached superiorly to the root of the great vessels, posteriorly to the vertebral column, attached by ligaments anteriorly to the sternum, and attached by a tendon inferiorly to the diaphragm [69]. We found regional differences in pericardial contact pressure, which is higher over the lateral left ventricle than over the anterior or posterior left ventricle [32]. This raises the possibility that the extent of pericardial restraint may vary regionally, which may contribute to regional differences in ventricular filling and shape.

9.3   Functional Consequences of Regional Heterogeneity

Regional heterogeneity has important functional consequences under both physiologic and pathophysiologic conditions [9]. An increase in regional heterogeneity causes a redistribution of loads and an intraventricular interaction that is most apparent during isovolumic contraction and isovolumic relaxation. This section describes two experimental approaches used to examine the functional consequences of intraventricular interactions caused by an increase in left ventricular regional heterogeneity. The acute myocardial ischemia model is used to examine the intraventricular interaction between ischemic and nonischemic regions (interaction between "weak" and "strong" muscles). The regional inotropic stimulation model is used to examine the consequences of a more subtle, graded increase in both temporal and regional heterogeneity (interaction between "strong" and "stronger" muscles).

9.3.1   MECHANICAL INTERACTIONS BETWEEN ISCHEMIC AND NONISCHEMIC REGIONS

During acute myocardial ischemia, systolic shortening decreases within seconds and is replaced by paradoxical holosystolic lengthening within minutes [76]. The acutely ischemic myocardium performs little or no effective work [21], although it retains some residual contractile function [87]. Paradoxical systolic lengthening develops when the ischemic region generates tension at a reduced rate and/or level, but is unable to sustain contraction against the greater tension developed by the nonischemic areas. Isolated muscles also develop paradoxical lengthening when weak and strong muscles are placed in series [84]. Thus, an increased heterogeneity in rate and/or extent of tension development causes a mechanical interaction where function in nonischemic areas directly influences the deformation in the ischemic region.

The converse condition is also true, that is, the ischemic region influences deformations in nonischemic regions. This mechanical interaction involves nonischemic regions that are directly adjacent to (border zone) and distant from (remote areas) the ischemic region. With subendocardial ischemia, systolic strains decrease in the overlying nonischemic epicardium, even though subepicardial blood flows are normal [17]. This transmural mechanical interaction has been ascribed to a mechanical "tethering" effect, in which the abnormally contracting ischemic subendocardium imposes a mechanical disadvantage on the nonischemic subepicardium. Similarly, the ischemic zone may be mechanically "tethered" to laterally adjacent nonischemic regions, where functional impairments occur even though coronary perfusion is normal [12]. The lateral extent of the functional border zone extends approximately 10 mm from the sharply demarcated perfusion boundary, although the most severe functional impairments occur within a few mm of the ischemic region [16]. Functional impairments in the nonischemic border zone may be related to a high wall stress and constraint of motion (by the adjacent ischemic region), which are predicted by theoretical models [6,7,40].

Functional alterations also occur in remote nonischemic regions. Augmented shortening or "hyperfunction" is observed in remote nonischemic areas in patients [24,39] and experimental studies [16,20,49,77] with acute myocardial infarction. Although previously considered a "compensatory" mechanism, we have shown that augmented shortening in nonischemic regions is related to a mechanical interaction between ischemic and nonischemic regions [49]. Figure 9.2 shows the typical relationship between pressure-segment length loops in ischemic and nonischemic regions before (straight line loops) and after (loops with directional arrows) occlusion of the left anterior descending (LAD) coronary artery in anesthetized dogs. Shortening in the ischemic zone was replaced by paradoxical systolic lengthening during isovolumic contraction and akinesis during the ejection phase.

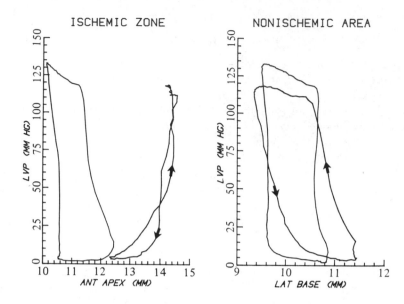

FIGURE 9.2. Regional pressure-length loops are drawn before (straight line loops) and afer (loops with directional arrows) occlusion of the left anterior descending coronary artery. The anterior (ANT) apex became ischemic, whereas the lateral (LAT) base is representative of a nonischemic area. (Reproduced from Lew et al. [49], with permission from the American Heart Association, Inc.)

Total segment shortening (from end-diastole to end-systole) increased in the nonischemic area because of an increase in end-diastolic pressure and segment length (Frank-Starling mechanism). However, shortening increased primarily during isovolumic contraction, with no "compensatory" increase in shortening during the ejection phase. Increased isovolumic shortening in the nonischemic area developed in parallel with paradoxical isovolumic lengthening in the ischemic zone, indicating a regional intraventricular unloading effect. A significant amount of shortening by the nonischemic region was "dissipated" into the paradoxically bulging ischemic zone, reducing the amount of shortening available for ventricular ejection. Thus, the ischemic zone imposed a mechanical disadvantage on the nonischemic region. This is even more apparent if acute ischemia occurs without an increase in end-diastolic pressure. If end-diastolic length does not increase in the nonischemic region (no compensatory Frank-Starling mechanism), the isovolumic shortening "wasted" in paradoxically stretching the ischemic zone causes ejection-phase shortening to decrease [48] .

The mechanical disadvantage imposed by the ischemic zone can be reduced by volume loading [48]. At higher volumes the ischemic zone is at a steeper (stiffer) position on its passive pressure-length relationship, which reduces the paradoxical isovolumic bulge and decreases regional intraventricular unloading effects (less "wasted" isovolumic shortening by nonischemic areas). Figure 9.3 shows the effects of volume loading (increase in left ventricular end-diastolic pressure from 5 mm Hg to 21 mm Hg) before (left panel) and after (right panel) LAD occlusion. During acute ischemia at a low volume, there was marked segment lengthening during isovolumic systole and akinesis during ejection in the ischemic anterior apex. An increase in volume decreased the extent of paradoxical isovolumic lengthening without improving the akinesis during ejection. When compared to control at the same end-diastolic pressure, acute ischemia was associated with a significantly greater decrease in stroke volume at low than high volumes (40% vs. 20%, respectively). Thus volume loading improves ventricular function during acute ischemia by increasing utilization of the Frank-Starling mechanism in nonischemic regions and by reducing the mechanical disadvantage imposed by the ischemic zone on nonischemic areas [48].

The mechanical disadvantage of the ischemic zone is directly related to ischemic zone size [47]. Figure 9.4 shows pressure-length loops from the ischemic zone and a nonischemic area before (C) and after occlusions of the distal (1), mid (2), and proximal (3) LAD at a matched end-diastolic pressure and nonischemic end-diastolic segment length. An increase in ischemic zone size produced a progressive increase in the amount of nonischemic area shortening "wasted" in paradoxically expanding the ischemic zone during isovolumic systole, thus progressively decreasing the amount of nonischemic area shortening during the ejection phase. Although volume loading can reduce the mechanical disadvantage imposed by a small or moderate ischemic zone size, this was less effective with a large ischemic

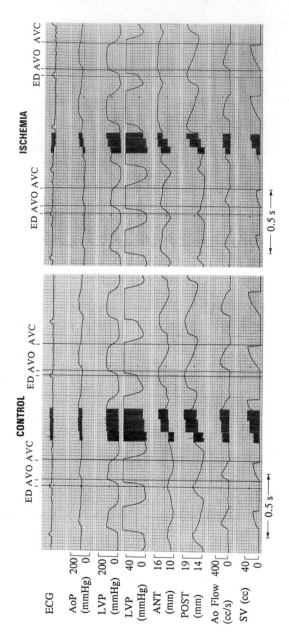

FIGURE 9.3. Typical tracing from an experiment showing (from top to bottom) the electrocardiogram (EKG), aortic (Ao) pressure, left ventricular (LV) pressure at a low and high gain, segment length signals from the anterior (ANT) apex and posterior (POST) apex, aortic flow (integrated aortic flow). The timing of end-diastole (ED), aortic valve opening (AVO), and aortic valve closure (AVC) are shown. The left panel shows control data at a low and high end-diastolic pressure (5 and 21 mmHg, respectively). The right panel shows data during ischemia (occlusion of the left anterior descending coronary artery) at matched low and high end-diastolic pressures.

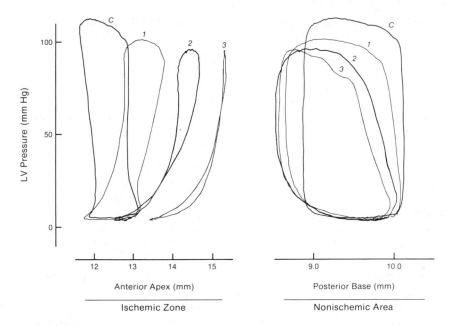

FIGURE 9.4. Pressure-length loops are shown for the ischemic zone (anterior apex) and a nonischemic area (posterior base) before (C) and after occlusion of the distal (1), mid- (2), and proximal (3) left anterior descending coronary artery, producing a small, moderate, and large ischemic zone size, respectively. (Reprinted with permission from Lew [47].)

zone size. Thus, the mechanical disadvantage imposed by the ischemic zone on nonischemic areas is directly related to ischemic zone size and inversely related to ischemic zone stiffness.

In view of the regional heterogeneity in strains, we examined whether there was a regional difference in the mechanical disadvantage of an anterior versus posterior ischemic zone [i.e., occlusion of the LAD vs. left circumflex (LCx) coronary artery, respectively] [31]. Clinical studies have found a higher morbidity and mortality for anterior than inferior wall infarctions, although this is largely related to differences in infarct size. Experimental studies have shown that for a similar extent of ischemia, ejection fraction decreases more with LAD than LCx occlusions [67,68]. We produced similar ischemic areas at risk and found that both LAD and LCx occlusions were associated with a mechanical disadvantage (i.e., increased shortening "wasted" by nonischemic areas in paradoxically stretching the ischemic zone during isovolumic systole). Ejection-phase shortening increased in nonischemic areas with LCx, but not LAD occlusions. This difference may be related to regional differences in the shape of the ischemic bed. The LAD runs longitudinally and LAD occlusions produce a wedge-shaped ischemic bed increasing in width toward the apex. The LCx runs along the minor axis and LCx occlusions produce a more rectangular-shaped ischemic bed. Since the anterior wall is slightly more distensible than the posterior wall [51,78] and the apex is more distensible than base [57], there may be greater distensibility of the ischemic anterior than posterior walls. Indeed, we found the paradoxical systolic expansion in midwall area (systolic bulge) was twice as great with LAD than LCx occlusions. Thus, an ischemic anterior wall imposed a greater mechanical disadvantage on nonischemic areas than an ischemic posterior wall of similar size [31].

Recently, Ning and colleagues also found greater regional intraventricular unloading effects with LAD than LCx occlusions in anesthetized dogs, although wall thickening in nonischemic areas did increase with both occlusions [59]. More importantly, they found that regional intraventricular unloading effects were less prominent in conscious than anesthetized dogs undergoing LCx occlusion. The ventricle operates at a higher (stiffer) position on the passive pressure-length curve in conscious animals. Thus, in conscious animals there may be less paradoxical systolic bulge, which decreases the mechanical disadvantage of the ischemic zone [48]. Ning and colleagues also found that regional intraventricular loading effects were limited to the first 3 to 5 days of acute infarction. This may reflect an increase in stiffness in the infarcted region (with reduced paradoxical systolic bulge) during the healing phase, which reduces the mechanical disadvantage imposed by the ischemic zone on nonischemic areas.

In summary, the increase in regional heterogeneity with acute myocardial ischemia produces significant regional intraventricular unloading effects, which are most apparent during isovolumic systole. There is a significant mechanical interaction between ischemic and nonischemic areas. The nonischemic area function is a major factor influencing deformations in the ischemic zone. The altered compliance and function in the ischemic zone imposes a mechanical disadvantage on nonischemic areas in direct relation to the ischemic zone size and inversely related to ischemic zone stiffness. Normal regional differences in diastolic distensibility, along with geometric differences in ischemic zone shape, contribute to differences in the mechanical disadvantage imposed by an anterior compared with posterior ischemic zone bed. The mechanical disadvantage imposed by the ischemic zone may be less severe under conscious than open-chest anesthetized conditions, and may decrease during the healing phase of an acute myocardial infarction as scar formation develops.

9.3.2   EXPERIMENTAL MODELS OF REGIONAL HETEROGENEITY

Several experimental models have been used to examine the effects of regional heterogeneity on global ventricular function, including regional ischemia, ventricular pacing, and regional inotropic stimulation or inhibition.

Regional ischemia slows the rate of left ventricular pressure fall, which has been attributed to asynchronous wall motion [10,44]. However, regional ischemia is not a useful model to study the independent effects of an increase in regional heterogeneity because ischemia can slow ventricular pressure fall directly by inhibiting myocardial relaxation or by altering loading conditions, independently of any effect on regional heterogeneity [52].

Ventricular pacing reduces global systolic function because of the asynchrony of regional contraction [4,60,85]. Ventricular pacing also slows the rate of left ventricular pressure fall, which has been attributed to asynchronous wall motion [5,29,88]. Although ventricular pacing is a useful model for increasing regional heterogeneity, the loss of atrial contraction and altered activation sequence may produce a complex regional redistribution in load, making it more difficult to isolate the independent effects of an increase in regional heterogeneity.

Regional inotropic stimulation increases regional heterogeneity and creates an imbalance of forces that impair the mechanical efficiency of ventricular ejection [25,35]. This is a useful model because the direct effects of the experimental probe (regional inotropic stimulation) should enhance, not impair, ventricular function. Thus, any decrease in mechanical efficiency can be attributed directly to an increase in regional heterogeneity, and not the direct effects of the experimental probe (in contrast to the regional ischemia model of heterogeneity). Furthermore, unlike the ventricular pacing model, the activation sequence and atrial contraction are preserved and a dose-dependent increase in regional heterogeneity can be produced.

We recently used more subtle doses of intracoronary isoproterenol to examine the effects of an increase in regional heterogeneity on the rate of left ventricular pressure fall during isovolumic relaxation [19,52]. The typical response to intracoronary (LAD) isoproterenol is shown in Figure 9.5. During the control period, all regions shorten synchronously and there are minimal segment length changes during isovolumic relaxation. After an infusion of 20 ng of isoproterenol into the LAD, there was no change in global hemodynamics as reflected by a constant left ventricular end-diastolic and peak systolic pressures and peak $+dP/dt$. However, there was a slower rate of pressure fall, as reflected by a decrease (less negative) peak $-dP/dt$. The anterior apex was directly stimulated so that segment shortening occurred at a more rapid rate, although the extent of shortening did not change. As a consequence, shortening in the anterior apex was completed earlier, and there was early, asynchronous lengthening during isovolumic relaxation. The lateral and posterior segments were not directly stimulated, but shortening increased in these regions during isovolumic relaxation, coincident with the asynchronous lengthening in the stimulated region. Thus, there was a regional intraventricular unloading effect whereby shortening by nonstimulated regions was dissipated into the stimulated region during isovolumic relaxation. As a consequence of this increased regional heterogeneity during isovolumic relaxation, the rate of left ventricular pressure

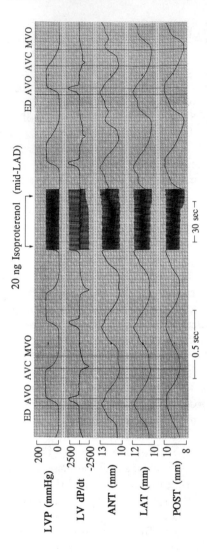

FIGURE 9.5. This tracing shows the left ventricular pressure (LVP), LV dP/dt, and three regional segment lengths (anterior apex [ANT], lateral [LAT], and posterior [post] segments) before (left panel) and after a 20-ng infusion of isoproterenol into the left anterior descending (LAD) coronary artery (right panel). The timing of end-diastole (ED), aortic valve opening (AVO), aortic valve closure (AVC), and mitral valve opening (MVO) are shown. (Reproduced with permission from Lew and Rasmussen [52].)

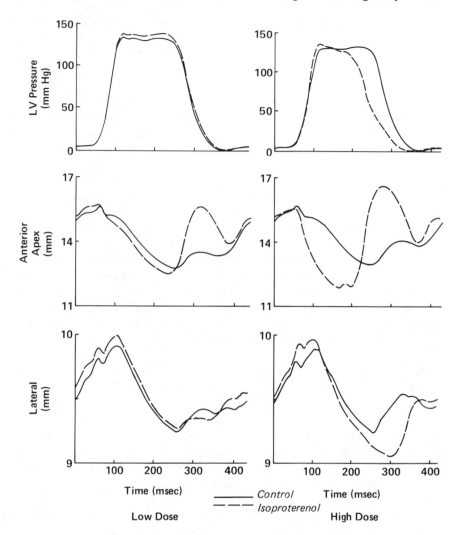

FIGURE 9.6. This tracing compares the effects of a low dose (left panel) and high dose (right panel) of isoproterenol infused into the left anterior descending coronary artery. The left ventricular (LV) pressure, anterior apex (stimulated region), and lateral (nonstimulated region) segment lengths are shown before (solid traces) and after (dotted line traces) intracoronary isoproterenol. (Reproduced with permission from Lew and Rasmussen [52].)

fall was markedly slower and the duration of isovolumic relaxation was prolonged.

Figure 9.6 demonstrates that this model can be used to produce a dose-dependent increase in regional heterogeneity. A low dose of intracoronary isoproterenol produces only a subtle decrease in the duration of anterior apex shortening, but even a slight acceleration in the onset of regional relaxation is sufficient to produce marked asynchronous segment lengthening during isovolumic contraction. There is a small but significant decrease in the rate of left ventricular pressure fall. A high dose of intracoronary isoproterenol increases the rate and extent of anterior segment shortening and produces dramatic asynchronous early relaxation. The increase in shortening in the nonstimulated region during isovolumic relaxation is also apparent. The onset of left ventricular pressure fall occurs earlier, but at a slower rate. The dose-dependent increase in regional heterogeneity produced a regional intraventricular unloading effect during isovolumic relaxation, analogous to the intraventricular unloading effect observed during isovolumic contraction with regional ischemia, as discussed earlier. These effects were dose-dependent and independent of the infusion site (i.e., regardless of whether isoproterenol was infused into the proximal, mid, or distal LAD or LCx branches). Thus, a subtle change in temporal or regional heterogeneity markedly slows the rate of left ventricular pressure fall.

The regional inotropic stimulation model can be used to separate the effects of loading conditions from the effects of regional heterogeneity [19]. Volume loading significantly slows the rate of left ventricular pressure fall, but not by increasing regional heterogeneity. Over a wide range of ventricular volumes, an increase in regional heterogeneity produces a similar decrease in the rate of left ventricular pressure fall. Thus, loading conditions and regional heterogeneity influence the rate of left ventricular pressure fall, but by largely independent mechanisms [19].

In summary, subtle changes in either regional contractile strength or duration are sufficient to produce global changes in the rate of left ventricular pressure fall during isovolumic relaxation. A slower rate of pressure fall does not necessarily reflect a pathophysiologic condition, but may merely reflect a subtle increase in regional heterogeneity and/or loading conditions. It is speculated that a small degree of regional heterogeneity is important under normal conditions for optimizing the mechanical efficiency of left ventricular contraction and relaxation.

9.4   Theoretical Models of Regional Heterogeneity

Theoretical approaches to ventricular function provide a powerful approach for analyzing the functional consequences of regional heterogeneity. The information from ventricular models and experimental data are mutually complementary. Experimental observations are used to optimize the parameters of a model to get a better fit with the experimental data, and predictions from a ventricular models can be correlated with experimental observations. The experimental data presented in the preceding sections demonstrate that under normal conditions, regional deformations are heterogeneous. The transmural heterogeneity suggests that models cannot assume that fibers deform only along their longitudinal axes (e.g., there are substantial subendocardial strains perpendicular to the fiber direction). Models need to take into account the substantial regional variation in the direction of maximal deformation. This variation cannot be simply explained by regional differences in the transmural distribution of fiber angles.

Some of the functional consequences of increased regional heterogeneity with myocardial ischemia can be predicted by appropriate models that take into account regional differences in material properties (between the ischemic and nonischemic regions), the size and shape of the ischemic zone, the extent of lateral border zones, and the severity of depressed contractility. A simple one-dimensional approach models the uniformly ischemic segment contracting in series with a normal segment [13]. The ischemic segment produces tension, although at a reduced level and/or at a reduced rate. In this model, the ischemic segment undergoes paradoxical systolic lengthening with a sudden decrease in the rate of tension development once a threshold tension level is reached [13]. This model fits well with experimental observations of weak (hypoxic) and strong (normoxic) muscles contracting in series [84], as well as segment length measurements in the intact heart during regional ischemia [49].

Our experimental observations that the ischemic zone imposes a mechanical disadvantage on nonischemic areas (due to a regional intraventricular unloading effect) in direct proportion to ischemic zone size and inversely related to ischemic zone stiffness [47–49] fits very well with the theoretical predictions of Laird and Vellekoop [45]. They modeled the left ventricle as a circular cylinder with a discrete "pie-shape" infarct area. Contraction of the normal muscle during isovolumic contraction passively stretched the infarct area, resulting in "wasted" shortening by nonischemic areas that reduced the "shortening reserve" (the amount of nonischemic area shortening available for ventricular ejection). Consistent with our experimental results, this model predicts that the "shortening reserve" decreases with an increase in infarct size and increases with increased infarct area stiffness [45].

Models have been used to examine the effects of chronic ventricular aneurysms. Early models predict that the ejection fraction decreases in direct proportion to the size of the aneurysm, and the mechanical disadvantage is greater for a compliant muscular aneurysm than for a stiff fibrous aneurysm [61]. Janz and Waldron used a finite element model to predict that the passive stiffness of the left ventricle increases with a chronic apical aneurysm, in direct relation to aneurysm size [40]. They also predicted that the apical aneurysm restrains the adjacent normal myocardium. As a consequence, the nonischemic border region has reduced end-diastolic length, less favorable utilization of the Frank-Starling mechanism, lower tension development, and less shortening than normal. The predictions from this theoretical model [40] are consistent with the experimental observation that function is impaired in nonischemic border zones that are directly adjacent ("tethered") to the ischemic zone [16].

Bogen and coworkers developed an isotropic, initially spherical membrane model to examine the effects of infarct size, stiffness, and age on ventricular performance [6,7]. During the immediate phase of infarction, the passive stiffness was not altered but there was a marked rightward shift in the end-systolic pressure-volume relationship. Passage from the acute phase (first few hours) to subacute infarction (1 week) was associated with an increase in infarct stiffness as edema and cellular infiltration occurred. A further increase in stiffness was predicted for chronic infarction (months) as fibrosis and scar formation occurred. The end-systolic pressure-volume relation was initially shifted markedly to the right with acute ischemia, but progressively shifted to the left (toward the normal relation) with a progressive increase in infarct stiffness. Stroke volume was reduced in direct relation to the infarct size, but increased in later stages with increased infarct stiffness. Normal stroke volumes were maintained by increasing ventricular volume (compensatory Frank-Starling mechanism) and/or increasing heart rates. However, these compensatory mechanisms were inadequate with large infarcts (>41% of the left ventricular mass). The effects of infarct stiffness on cardiac performance were complex, and depended on the infarct size and end-diastolic pressure. The infarct area constrained motion in the adjacent noninfarcted areas, reducing circumferential extensions during diastole, and constraining the extent of contraction during systole. A two- to fourfold increase in stress was predicted for the nonischemic region directly adjacent to the ischemic zone. The stress amplification was relatively constant for infarcts involving 15 to 41% of the left ventricle, but decreased with an increase in infarct stiffness. Stress amplification in the nonischemic border regions increased with increasing inotropic stimulation of the nonischemic myocardium [6].

These theoretical models are useful for understanding the clinical observation that ventricular function improves over time after acute myocardial infarction, a finding that has been attributed to increased stiffness in the infarct region. The stress amplification and local constraining effects in border regions provide an explanation for the functional impairment in the nonischemic border zone despite normal coronary blood flow [16]. Stress amplification in the border zone also has important clinical implications. Increased wall stress is associated with an increase in myocardial oxygen consumption, so may contribute to postinfarction angina and/or infarct extension. Stress amplification may be a pathophysiologic mechanism contributing to the arrhythmias observed during acute and chronic myocardial infarction. These concepts are supported by the clinical observation that patients with ventricular aneurysms can have an improvement in ventricular function, relief of angina, and resolution of arrhythmias after aneurysectomy.

Current models of ventricular function do not account for the more subtle effects of regional heterogeneity as examined experimentally with the regional inotropic stimulation model. In the future, more realistic models of ventricular function will need to incorporate regional differences in activation sequence, differences in contraction duration (and onset of regional lengthening), and regional differences in strain magnitude and principal directions. These theoretical models will be helpful for understanding the functional consequences of increased regional heterogeneity in common clinical conditions (e.g., ventricular pacing and bundle branch blocks) and pathophysiologic conditions (e.g., acute myocardial ischemia, chronic myocardial infarction, and hypertrophic cardiomyopathy). The experimental physiologist can measure regional deformations in only a limited number of sites. Thus, the development of models of the ventricle that are consistent with experimental data will provide important new insights into the functional consequences of regional heterogeneity.

Acknowledgements: This work was supported by the Research Service of the Veterans Administration and by National Institutes of Health grants HL-01731, HL-17682, HL-32583, and by the American Heart Association, California Affiliate Grant-in-Aid #89-167. This work was done during the tenure of a Clinical Investigator Award from the NIH and an Established Investigatorship from the American Heart Association.

REFERENCES

[1] E.T. Angelakos. Regional distribution of catecholamines in the dog heart. *Circ. Res.*, 16:39–44, 1965.

[2] G. Arisi, E. Macchi, S. Baruffi, S. Spaggiari, and B. Taccardi. Potential fields on the ventricular surface of the exposed dog heart during normal excitation. *Circ. Res.*, 52:706–715, 1983.

[3] T. Arts, P.C. Veenstra, and R.S. Reneman. Epicardial deformation and left ventricular wall mechanics during ejection in the dog. *Am. J. Physiol.*, 243:H379–H390, 1982.

[4] F.R. Badke, P. Boinay, and J.W. Covell. Effects of ventricular pacing on regional left ventricular performance in the dog. *Am. J. Physiol.*, 238:H858–H867, 1980.

[5] A.S. Blaustein and W.H. Gaasch. Myocardial relaxation VI: Effects of β-adrenergic tone and asynchrony on LV relaxation rate. *Am. J. Physiol.*, 244:H417–H422, 1983.

[6] D.K. Bogen, A. Needleman, and T.A. McMahon. An analysis of myocardial infarction: The effect of regional changes in contractility. *Circ. Res.*, 55:805–815, 1984.

[7] D.K. Bogen, S.A. Rabinowitz, A. Needleman, T.A. McMahon, and W.H. Abelmann. An analysis of the mechanical disadvantage of myocardial infarction in the canine left ventricle. *Circ. Res.*, 47:728–741, 1980.

[8] A.A. Bove, T.H. Kreulen, and J.F. Spann. Computer analysis of left ventricular dynamic geometry in man. *Am. J. Cardiol.*, 41:1239–1248, 1978.

[9] D.L. Brutsaert. Nonuniformity: A physiologic modulator of contraction and relaxation of the normal heart. *J. Am. Coll. Cardiol.*, 9:341–348, 1987.

[10] J.D. Carroll, O.M. Hess, H.O. Hirzel, and H.P. Krayenbuehl. Exercise-induced ischemia: The influence of altered relaxation on early diastolic pressures. *Circulation*, 67:521–528, 1983.

[11] M. Courtois, S.J. Kovacs Jr., and P.A. Ludbrook. Transmitral pressure-flow velocity relation: Importance of regional pressure gradients in the left ventricle during diastole. *Circulation*, 78:1459–1468, 1988.

[12] D.A. Cox and S.F. Vatner. Myocardial function in areas of heterogenous perfusion after coronary artery occlusion in conscious dogs. *Circulation*, 66:1154–1158, 1982.

[13] V.B. Elings, G.E. Jahn, and J.H.K. Vogel. A theoretical model of regionally ischemic myocardium. *Circ. Res.*, 41:722–729, 1977.

[14] T.R. Fenton, J.M. Cherry, and F.A. Klassen. Transmural myocardial deformation in the canine left ventricular wall. *Am. J. Physiol.*, 235:H523–H530, 1978.

[15] M.R. Franz, K. Bargheer, W. Rafflenbeul, A Haverich, and P.R. Lichtlen. Monophasic action potential mapping in human subjects with normal electrocardiograms: Direct evidence for the genesis of the T wave. *Circulation*, 75:379–386, 1987.

[16] K.P. Gallagher, R.A. Gerren, M.C. Stirling, et al. The distribution of functional impairment across the lateral border of acutely ischemic myocardium. *Circ. Res.*, 58:570–583, 1986.

[17] K.P. Gallagher, G. Osakada, O.M. Hess, J.A. Koziol, W.S. Kemper, and J. Ross Jr. Subepicardial segmental function during coronary stenosis and the role of myocardial fiber orientation. *Circ. Res.*, 50:352–359, 1982.

[18] H.J. Gelberg, B.H. Brundage, S. Glantz, and W.W. Parmley. Quantitative left ventricular wall motion analysis: A comparison of area, chord and radial methods. *Circulation*, 59:991–1000, 1979.

[19] T. Gillebert and W.Y.W. Lew. Nonuniformity and volume loading independently influence isovolumic relaxation rates. *Am. J. Physiol.*, 257:H1927–H1935, 1989.

[20] Y. Goto, Y. Igarashi, O. Yamada, K. Hiramori, and H. Suga. Hyperkinesis without the Frank-Starling mechanism in a nonischemic region of acutely ischemic excised canine heart. *Circulation*, 77:468–477, 1988.

[21] Y. Goto, Y. Igarashi, Y. Yasumura, et al. Integrated regional work equals total left ventricular work in regionally ischemic canine heart. *Am. J. Physiol.*, 254:H894–H904, 1988.

[22] R.A. Greenbaum, S.Y. Ho, D.G. Gibson, A.E. Becker, and R.H. Anderson. Left ventricular fibre architecture in man. *Br. Heart J.*, 45:248–263, 1981.

[23] A.F. Grimm, H.L. Lin, and B.R. Grimm. Left ventricular free wall and intraventricular pressure-sarcomere length distributions. *Am. J. Physiol.*, 239:H101–H107, 1980.

[24] C.L. Grines, E.J. Topol, R.M. Califf, et al. Prognostic implications and predictors of enhanced regional wall motion of the noninfarct zone after thrombolysis and angioplasty therapy of acute myocardial infarction. *Circulation*, 80:245–253, 1989.

[25] P.A. Gwirtz, D. Franklin, and H.J. Mass. Modulation of synchrony of left ventricular contraction by regional adrenergic stimulation in conscious dogs. *Am. J. Physiol.*, 251:H490–H495, 1986.

[26] R.V. Haendchen, H.L. Wyatt, G. Maurer, et al. Quantitation of regional cardiac function by two-dimensional echocardiography: I. Patterns of contraction in the normal left ventricle. *Circulation*, 67:1234–1244, 1983.

[27] D.E. Hansen, G.T. Daughters II, E.B. Stinson, E.L. Alderman, N.B. Ingels Jr., and D.C. Miller. Torsional deformation of the left ventricular midwall in human hearts with intramyocardial markers: Regional heterogeneity and sensitivity to the inotropic effects of abrupt rate changes. *Circ. Res.*, 62:941–952, 1988.

[28] F.W. Heineman and J. Grayson. Transmural distribution of intramyocardial pressure measured by micropipette technique. *Am. J. Physiol.*, 249:H1216–H1223, 1985.

[29] G.R. Heyndrickx, P.J. Vantrimpont, M.F. Rousseau, and H. Pouleur. Effects of asynchrony on myocardial relaxation at rest and during exercise in conscious dogs. *Am. J. Physiol.*, 254:H817–H822, 1988.

[30] L. Hittinger, B. Crozatier, J-P. Belot, and M. Pierrot. Regional ventricular segmental dynamics in normal conscious dogs. *Am. J. Physiol.*, 253:H713–H719, 1987.

[31] B.D. Hoit and W.Y.W. Lew. Functional consequences of acute anterior vs. posterior wall ischemia in canine left ventricles. *Am. J. Physiol.*, 254:H1065–H1073, 1988.

[32] B.D. Hoit, W.Y.W. Lew, and M.M. LeWinter. Regional variation in pericardial contact pressure in the canine ventricle. *Am. J. Physiol.*, 255:H1370–H1377, 1988.

[33] T. Hosino, H. Fujiwara, C. Kawai, and Y. Hamashima. Myocardial fiber diameter and regional distribution in the ventricular wall of normal adult hearts, hypertensive hearts and hearts with hypertrophic cardiomyopathy. *Circulation*, 67:1109–1116, 1983.

[34] R.M. Huisman, P. Sipkema, N. Westerhof, and G. Elzinga. Comparison of models used to calculate left ventricular wall force. *Med. Biol. Eng. Comp.*, 18:133–144, 1980.

[35] A.J. Ilebekk, J. Lekven, and F. Kiil. Left ventricular asynergy during intracoronary isoproterenol infusion in dogs. *Am. J. Physiol.*, 239:H594–H600, 1980.

[36] N.B. Ingels Jr., G.T. Daughters II, E.B. Stinson, and E.L. Alderman. Left ventricular midwall dynamics in the right anterior oblique projection in intact unanesthetized man. *J. Biomech.*, 14(4):221–233, 1981.

[37] N.B. Ingels Jr., G.T. Daughters II, E.B. Stinson, and E.L. Alderman. Measurement of midwall myocardial dynamics in intact man by radiography of surgically implanted markers. *Circulation*, 52:859–867, 1975.

[38] N.B. Ingels Jr., D.E. Hansen, G.T. Daughters II, E.B. Stinson, E.L. Alderman, and D.C. Miller. Relation between longitudinal, circumferential, and oblique shortening and torsional deformation in the left ventricle of the transplanted human heart. *Circ. Res.*, 64:915–927, 1989.

[39] W. Jaarsma, C.A. Visser, V.M.J. Eenige, et al. Prognostic implications of regional hyperkinesia and remote asynergy of noninfarcted myocardium. *Am. J. Cardiol.*, 58:394–398, 1986.

[40] R.F. Janz and R.J. Waldron. Predicted effect of chronic apical aneurysms on the passive stiffness of the human left ventricle. *Circ. Res.*, 42:255–263, 1978.

[41] S. Kimura, A.L. Bassett, T. Furukawa, J. Cuevas, and R.J. Myerburg. Electrophysiological properties and responses to stimulated ischemia in cat ventricular myocytes of endocardial and epicardial origin. *Circ. Res.*, 66:469–477, 1990.

[42] S.C. Klausner, T.J. Blair, W.F. Bulawa, G.M. Jeppson, R.L. Jensen, and P.D. Clayton. Quantitative analysis of segmental wall motion through systole and diastole in the normal human left ventricle. *Circulation*, 65:580–590, 1982.

[43] Y. Kong, J. Morris Jr., and H.D. McIntosh. Assessment of regional myocardial performance from biplane coronary cineangiograms. *Am. J. Cardiol.*, 27:529–537, 1971.

[44] T. Kumada, J.S. Karliner, H. Pouleur, K.P. Gallagher, K. Shirato, and J. Ross Jr. Effects of coronary occlusion on early ventricular diastolic events in conscious dogs. *Am. J. Physiol.*, 237:H542–H549, 1979.

[45] J.D. Laird and H.P. Vellekoop. Time course of passive elasticity of myocardial tissue following experimental infarction in rabbits and its relation to mechanical dysfunction. *Circ. Res.*, 41:715–721, 1977.

[46] M. Laks, M.J. Nisenson, and H.J.C. Swan. Myocardial cell and sarcomere lengths in the normal dog heart. *Circ. Res.*, 21:671–678, 1967.

[47] W.Y.W. Lew. Influence of ischemic zone size on nonischemic area function in the canine left ventricle. *Am. J. Physiol.*, 252:H990–H997, 1987.

[48] W.Y.W. Lew and E. Ban-Hayashi. Mechanisms of improving regional and global ventricular function by preload alterations during acute ischemia in the canine left ventricle. *Circulation*, 72:1125–1134, 1985.

[49] W.Y.W. Lew, Z. Chen, B. Guth, and J.W. Covell. Mechanisms of augmented segment shortening in nonischemic areas during acute ischemia of the canine left ventricle. *Circ. Res.*, 56:351–358, 1985.

[50] W.Y.W. Lew and M.M. LeWinter. Regional circumferential lengthening patterns in canine left ventricle. *Am. J. Physiol.*, 245:H741–H748, 1983.

[51] W.Y.W. Lew and M.M. LeWinter. Regional comparison of midwall segment and area shortening in the canine left ventricle. *Circ. Res.*, 58:678–691, 1986.

[52] W.Y.W. Lew and C.M. Rasmussen. Influence of nonuniformity on the rate of left ventricular pressure fall in the dog. *Am. J. Physiol.*, 256:H222–H232, 1989.

[53] M.M. LeWinter, R.S. Kent, J.M. Kroener, T.E. Carew, and J.W. Covell. Regional differences in myocardial performance in the left ventricle of the dog. *Circ. Res.*, 37:191–199, 1975.

[54] D. Ling, J.S. Rankin, C.H. Edwards, P.A. McHale, and R.W. Anderson. Regional diastolic mechanics of the left ventricle in the conscious dog. *Am. J. Physiol.*, 236:H323–H330, 1979.

[55] S.H. Litovsky and C. Antzelevitch. Transient outward current prominent in canine ventricular epicardium but not endocardium. *Circ. Res.*, 62:116–126, 1988.

[56] G.B.J. Mancini, S.F. DeBoe, S.B. Anselmo, M.T. LaFree, and R.A. Vogel. Quantitative regional curvature analysis: An application of shape determination for the assessment of segmental left ventricular function in man. *Am. Heart J.*, 113:326–334, 1987.

[57] A.D. McCulloch, B.H. Smaill, and P.J. Hunter. Regional left ventricular epicardial deformation in the passive dog heart. *Circ. Res.*, 64:721–733, 1989.

[58] G.D. Meier, M.C. Ziskin, W.P. Santamore, and A.A. Bove. Kinematics of the beating heart. *IEEE Trans. Biomed. Eng.*, BME-27:319–329, 1980.

[59] X-H Ning, T.N. Zweng, and K.P. Gallagher. Ejection and isovolumic contraction-phase wall thickening in nonischemic myocardium during coronary occlusion. *Am. J. Physiol.*, 258:H490–H499, 1990.

[60] R.C. Park, W.C. Little, and R.A. O'Rourke. Effect of alteration of left ventricular activation sequence on the left ventricular end systolic pressure-volume relation in closed-chest dogs. *Circ. Res.*, 57:706–717, 1985.

[61] W.W. Parmley, L. Chuck, C. Kivowitz, J.M. Matloff, and J.C. Swan. In vitro length-tension relations of human ventricular aneurysms: Relation to stiffness to mechanical disadvantage. *Am. J. Cardiol.*, 32:889–894, 1973.

[62] G.L. Pierpont, E.G. DeMaster, and J.N. Cohn. Regional differences in adrenergic function within the left ventricle. *Am. J. Physiol.*, 246:H824–H829, 1984.

[63] J.S. Rankin, P.A. McHale, C.E. Arentzen, J.C. Greenfield Jr., and R.W. Anderson. The three-dimensional dynamic geometry of the left ventricle in the conscious dog. *Circ. Res.*, 39:304–313, 1976.

[64] L. Role, D. Bogen, T.A. McMahon, and W.H. Abelmann. Regional variations in calculated diastolic wall stress in rat left ventricle. *Am. J. Physiol.*, 235:H247–H250, 1978.

[65] M.A. Ross and D.D. Streeter Jr. Nonuniform subendocardial fiber orientation in the normal macaque left ventricle. *Eur. J. Cardiol.*, 3:229–247, 1975.

[66] A.M. Scher and M.S. Spach. Cardiac depolarization and repolarization and the electrocardiogram. In R.M. Berne, N. Sperelakis, and S.R. Geiger, editors, *Handbook of Physiology, Section 2: The Cardiovascular System*, pages 357–392. American Physiological Society, Bethesda, MD, 1979.

[67] R.M. Schneider, A. Chu, M. Akaishi, W.S. Weintraub, K.G. Morris, and F.R. Cobb. Left ventricular ejection fraction after acute coronary occlusion in conscious dogs: Relation to the extent and site of myocardial infarction. *Circulation,* 72:632–638, 1985.

[68] R.M. Schneider, K.G. Morris, A. Chu, K.B. Roberts, R.E. Coleman, and F.R. Cobb. Relation between myocardial perfusion and left ventricular function following acute coronary occlusion: Disproportionate effects of anterior vs. inferior ischemia. *Circ. Res.*, 60:60–71, 1987.

[69] R. Shabetai. *The Pericardium.* Grune & Stratton, New York, 1981.

[70] E. Shapiro, D.L. Marier, M.G. St. John Sutton, and D.G. Gibson. Regional non-uniformity in wall dynamics in normal left ventricle. *Br. Heart J.*, 45:264–270, 1981.

[71] F.H. Sheehan, D.K. Stewart, H.T. Dodge, S. Mitten, E.L. Bolson, and B.G. Brown. Variability in the measurement of regional left ventricular wall motion from contrast angiograms. *Circulation,* 68:550–559, 1983.

[72] P.D. Stein, M. Marzilli, H.N. Sabbah, and T. Lee. Systolic and diastolic pressure gradients within the left ventricular wall. *Am. J. Physiol.*, 238:H625–H630, 1980.

[73] D.D. Streeter. Gross morphology and fiber geometry of the heart. In R.M. Berne, editor, *Handbook of Physiology, Section 2*, Volume 1, pages 61–112. American Physiological Society, Bethesda, MD, 1979.

[74] D.D. Streeter Jr. and D.L. Bassett. Engineering analysis of myocardial fiber orientation in pig's left ventricle in systole. *Anat. Record*, 155:503–511, 1966.

[75] D.D. Streeter Jr. and W.T. Hanna. Engineering mechanics for successive states in canine left ventricular myocardium: I. Cavity and wall geometry. *Circ. Res.*, 33:639–655, 1973.

[76] R. Tennant and C.J. Wiggers. The effect of coronary occlusion on myocardial contraction. *Am. J. Physiol.*, 112:351–361, 1935.

[77] P. Theroux, J. Ross Jr., D. Franklin, J.W. Covell, C.M. Bloor, and S. Sasayama. Regional myocardial function and dimensions early and late after myocardial infarction in the unanesthetized dog. *Circ. Res.*, 40:158–165, 1977.

[78] F.J. Villarreal and W.Y.W. Lew. Finite strains in the anterior and posterior wall of the canine left ventricle. *Am. J. Physiol.*, 259: H1409–H1418, 1990.

[79] F.J. Villarreal, L.K. Waldman, and W.Y.W. Lew. A technique for measuring regional two-dimensional finite strains in canine left ventricle. *Circ. Res.*, 62:711–721, 1988.

[80] L.K. Waldman, Y.C. Fung, and J.W. Covell. Transmural myocardial deformation in the canine left ventricle: normal in vivo three-dimensional finite strains. *Circ. Res.*, 57:152–163, 1985.

[81] L.K. Waldman, D. Nosan, F.J. Villarreal, and J.W. Covell. Relation between transmural deformation and local myofiber direction in canine left ventricle. *Circ. Res.*, 63:550–562, 1985.

[82] T. Watanabe, L.M. Delbridge, J.O. Bustamante, and T.F. McDonald. Heterogeneity of the action potential in isolated rat ventricular myocytes and tissue. *Circ. Res.*, 52:280–290, 1983.

[83] T. Watanabe, P.M. Rautaharju, and T.F. McDonald. Ventricular action potentials, ventricular extracellular potentials, and the ECG of guinea pig. *Circ. Res.*, 57:362–373, 1985.

[84] A.W. Weigner, G.J. Allen, and O.H.L. Bing. Weak and strong myocardium in series: Implications for segmental dysfunction. *Am. J. Physiol.*, 235:H776–H783, 1978.

[85] C.J. Wiggers. The muscular reactions of the mammalian ventricles to artificial surface stimuli. *Am. J. Physiol.*, 73:345–378, 1925.

[86] C. Yoran, J.W. Covell, and J. Ross Jr. Structural basis for the ascending limb of left ventricular function. *Circ. Res.*, 32:297–303, 1973.

[87] C. Yoran, E.H. Sonnenblick, and E.S. Kirk. Contractile reserve and left ventricular function in regional myocardial ischemia in the dog. *Circulation*, 66:121–128, 1982.

[88] M.R. Zile, A.S. Blaustein, G. Shimizu, and W.H. Gaasch. Right ventricular pacing reduces the rate of left ventricular relaxation and filling. *J. Am. Coll. Cardiol.*, 10:702–709, 1987.

10

Mathematical Modeling of the Electrical Activity of Cardiac Cells

Michael R. Guevara[1]

ABSTRACT We introduce the Hodgkin-Huxley (HH) formulation describing the flow of ionic currents across the membrane of a cardiac cell, paying particular attention to the central concepts of activation and inactivation. We indicate a few situations in which HH-type modeling of cardiac cells has been useful, and show that continuous models of the HH-type break down when one observes phenomena in which single-channel behavior becomes important. Finally, we show that there are some intriguing parallels between the behavior of single ionic channels, which are currently thought to be governed by stochastic processes, and the behavior of chaotic systems, which are governed not by stochastic, but rather by deterministic rules.

10.1   Introduction

Cardiac cells show evidence of electrical activity in that there is a potential difference present across the cell membrane. Changes in this transmembrane potential difference occur during the course of the cardiac cycle and are responsible for initiating contraction. These changes are mainly caused by the flow of ionic currents through a population of individual channels in the membrane of the cell. Experimental work using the patch-clamp technique, which allows the recording of the current flowing through a single channel in a small patch of membrane, reveals that there are many different types of channels present in the membrane of cardiac cells. The total current flowing across the membrane of a cell at a given point in time is the sum of the individual single-channel currents flowing through the various kinds of channels at that point in time. The channels differ in their ion selectivity (i.e., their relative permeability to ions of different species; e.g., sodium, potassium, and calcium) as well as in the way they respond to changes in the potential difference impressed across the membrane [26]. With recent developments in molecular biology, it is becoming possible to

---

[1] Department of Physiology, McGill University, Montréal, Québec H3G 1Y6

sequence the proteins of which the channel is made, and then perform selective site-directed mutagenesis, so as to sort out the biochemical origins of these differences.

Over the years, many models have been published describing in a quantitative way these flows of the various ionic currents across the cardiac cell membrane. Models of many types of cardiac cells have appeared, including Purkinje fiber [9,11,40,45], ventricular muscle [2], atrial muscle [25], sinus venosus, and the sinoatrial node [30,46] (see also Chapter 11 in this book). In working with models of cardiac cells, one must always keep in mind that there is no such thing as a "standard" ventricular cell, sinoatrial nodal cell, and so forth. For example, isolated ventricular cells show a wide range of morphologies of the action potential [60]. Since one cannot trace an isolated cell back to its site of origin in the heart, and since cells interact electrotonically in situ, one cannot say much about the spatial scale of these inhomogeneities. It is becoming increasingly clear that there is a well organized global gradient in the intrinsic electrophysiological properties of cells in the sinoatrial node as one moves from the central area of the node outward to its periphery [3,39]. The implications of this gradient in properties for the overall function of the node have not yet been explored from a theoretical point of view.

10.2   Ionic Models Using the Hodgkin-Huxley Formulation

The earliest models of cardiac tissue [45], which were developed about 30 years ago, were modifications of the Hodgkin-Huxley (HH) equations [27], which were derived from voltage-clamp studies on the giant axon of the squid. In a voltage-clamp experiment, the potential difference between the inside and outside of a cell is kept constant using electronic circuitry. In voltage-clamp work on a single isolated cardiac cell, this involves sensing the transmembrane potential difference (V) with an electrolyte-filled glass microelectrode and then injecting current into the cell using that same electrode (or a second electrode), so as to counterbalance exactly any ionic currents generated by the cell membrane itself, thus clamping V. Since the cell membrane is basically a dielectric layer of phospholipid and there is a conducting electrolytic solution both inside and outside the cell, one has a capacitor, with

$$\frac{dV}{dt} = -\sum I_i / C \qquad (10.1)$$

where C is the cell capacitance and the I_i are the various ionic currents flowing across the cell membrane at time t. By convention, an outward current (flow of positive ions out of the cell or flow of negative ions into the cell) is positive; this necessitates the negative sign in Equation (10.1), since such a current would serve to drive V in the negative direction.

In a voltage-clamp experiment, one simply clamps the membrane to some potential, say V_1, waits for a steady state in injected current to be achieved, and then clamps the membrane to some new potential, say V_2. One then observes how the current through the membrane changes as a new steady state in the injected current is eventually achieved. For a membrane containing only one kind of channel, essentially three different types of behavior will be observed: the current at the new potential V_2 will (1) instantaneously attain its new steady-state value, (2) take some time to turn on, thus gradually establishing its new steady-state value, or (3) take some time to turn on, but then gradually turn off. In the first case one speaks of a time-independent or "background" current, in the latter two cases one has a time-dependent current. The turning-on process is termed activation, the turning-off process is termed inactivation.

10.3   Background Currents

Time-independent or background currents are described by

$$I = f(V) \tag{10.2}$$

where I is the current at transmembrane potential V. Thus, the current at a potential V depends simply on V; it does not depend on the exact manner in which V was attained. If f is a linear function of V, the current is said to be linear or Ohmic; in contrast, if f is nonlinear, the current is termed a rectifying current. In many instances, the current is described by

$$I = g(V) \cdot (V - E_r) \tag{10.3}$$

where E_r is the reversal potential. If the channel is predominantly permeable to one specific ion species, E_r will be close to the Nernst potential for that ion

$$E_r = \frac{RT}{zF} \ln \frac{C_o}{C_i} \tag{10.4}$$

where T = temperature, z = charge on the ion, R = gas constant, F = the Faraday constant, and C_o and C_i are the concentrations of the ion outside and inside the cell, respectively. The function $g(V)$ in Equation (10.3) is the conductance. If it is a constant (i.e., not a function of V), the current is linear.

Perhaps the most important current in cardiac cells showing no evidence of time dependence in the physiological range of potentials is the inwardly rectifying potassium current (I_{K1}), which is responsible for maintaining the resting potential in ventricular cells. Other time-independent currents include those carried by sodium ions ($I_{Na,b}$) and calcium ions ($I_{Ca,b}$).

10.4   Activation

Currents that activate, but do not inactivate, are often described by an equation of the form

$$I = g(V) \cdot x \cdot (V - E_r) \tag{10.5}$$

where x is the activation variable ($0 \leq x \leq 1$), which satisfies the first-order kinetic equation

$$\frac{dx}{dt} = \alpha_x(V) \cdot (1 - x) - \beta_x(V) \cdot x \tag{10.6}$$

where the rate constants α_x and β_x are generally extremely complicated nonlinear functions of V. When $x = 1$, the current is said to be completely activated; when $x = 0$, there is no activation, since I will be zero from Equation (10.5), independent of V.

One can also rewrite Equation (10.6) as

$$\frac{dx}{dt} = -\frac{(x - x_\infty)}{\tau_x} \tag{10.7}$$

where

$$\tau_x = \frac{1}{(\alpha_x + \beta_x)} \tag{10.8}$$

$$x_\infty = \frac{\alpha_x}{(\alpha_x + \beta_x)}. \tag{10.9}$$

If V is held fixed, the solution of equation (10.7) is

$$x(t) = x_\infty - (x_\infty - x_o)\exp(-t/\tau_x) \tag{10.10}$$

where $x_o = x(t = 0)$. There is thus a simple exponential approach to x_∞ from x_o as $t \to \infty$. In some cases, the activation variable x is raised to some power n in Equation (10.5).

Examples of currents that only activate, and do not inactivate, include the pacemaker current (I_f, I_h), which is involved in generating spontaneous activity in some areas of the sinoatrial node and in Purkinje fiber, and the plateau potassium currents involved in repolarization of the action potential (I_K, I_{x1}, I_{x2}).

10.5   Inactivation

Currents that show time-dependent activation and inactivation processes are often described by an equation of the form

$$I = g(V) \cdot x \cdot y \cdot (V - E_r) \tag{10.11}$$

where x is the activation variable as before and y is the inactivation variable, obeying an equation identical in form to Equation (10.6), but with $x, \alpha_x(V)$, and $\beta_x(V)$ being replaced by $y, \alpha_y(V)$, and $\beta_y(V)$.

Examples of currents present in cardiac cells that inactivate as well as activate include the fast inward sodium current (I_{Na}), which is responsible for producing the upstroke phase (phase 0) of the action potential in atrial and ventricular muscle, as well as in Purkinje fiber; the transient outward potassium current (I_{to}, I_A), implicated in repolarization of atrial and ventricular muscle as well as Purkinje fiber; and the transient calcium currents (I_T, I_L) involved in raising the internal calcium concentration in the cell and so initiating mechanical contraction, as well as in generating spontaneous activity.

10.6   Pump and Exchange Currents

In addition to the currents described, which are caused by voltage-gated channels in the membrane, several pump and exchange mechanisms contribute to generation of the transmembrane potential. The sodium-potassium pump is responsible for pumping sodium ions out of the cell and potassium ions back into the cell, so as to maintain the inside-outside concentration gradients for these ions. In so doing, it generates a net outward current, which serves to hyperpolarize the membrane [9,46]. The sodium-calcium exchanger pumps calcium ions out of the cell and sodium ions into the cell, generating a net inward current [9]. The calcium pump pumps calcium ions out of the cell, generating an outward current (see Chapter 11 in this book).

10.7   Applications of Ionic Models

Ionic models are typically used in the context of voltage-clamp experiments. They have also been used in studies investigating several sorts of dynamical behaviors: for example, phase resetting of spontaneous activity [15,22,41], the response of cardiac cells to periodic stimulation [18,41] (see also Chapter 14), abolition of spontaneous activity [17,24], generation of "rotors"—circulating waves of excitation in quiescent tissue [49,54,61], block of conduction [16,34], and mutual entrainment of populations of oscillators in the sinoatrial node [33,42].

Since the rate constants α and β in Equation (10.6) are nonlinear functions of V, and g in Equations (10.3), (10.5), and (10.11) is often nonlinear, the HH-type equations used in modeling cardiac muscle are inherently a set of nonlinear ordinary differential equations. This makes possible phenomena excluded in linear systems: (1) excitability: the output is not a linear function of the input—small changes in input level can have dramatic changes in the output (virtually "all–or–none" threshold [12,13]), (2) stable spontaneous activity: this is caused by the appearance of a stable limit cycle in the phase space of the system. Small enough pertubations result in eventual relaxation back to the limit cycle trajectory [22,41], (3) harmonic and subharmonic entrainment: the output of the system becomes entrained or phase-locked to a rational multiple or submultiple of the input frequency [18,20,41,59] (see also Chapter 14 of this book)—small parameter changes will in general not affect the multiple or submultiple, and (4) "chaotic" dynamics: deterministically aperiodic or "chaotic" dynamics can occur [5,19,20,32].

10.8   Reduced Ionic Models

As time has progressed and we have acquired more knowledge about the ionic basis of electrical activity in the heart, there has been a parallel increase in the complexity of the models: the early (1962) Purkinje fiber model of Noble [45] had four variables, while the most recent (1985) model of Purkinje fiber has 15 variables [9]. As more currents, pumps, and exchangers are discovered and one begins to take into account the biochemistry of the system, one can visualize that models containing several score variables will come into existence in the not-too-distant future. It is extremely difficult to gain insight into the dynamic behavior of such very complex models. For that reason, attempts started soon after the HH equations were published to reduce the complexity of that four-dimensional system and yet retain its essential phenomenological characteristics. The most important study along these lines was that of FitzHugh [12], who demonstrated that reduction to a two-variable system (an extension of the van der Pol equations) maintained characteristic properties such as excitability, graded response, anodal-break excitation, refractoriness, and spontaneous firing in response to constant-current injection—all properties possessed by the HH equations and by the giant axon of the squid. Furthermore, it has been shown that alternans rhythms [10,19] and Wenckebach-like rhythms [47,52] can be seen in the model of FitzHugh, and Wenckebach-like rhythms can also be seen in another simple two-variable model based on the van der Pol equations [44].

More recently, we have shown that the three qualitatively different ways in which activity can be abolished in experiments on the sinoatrial node can be seen in a reduced three-variable ionic model [24]. Thus, in future work, as models become more complex it will be important, on occasion, to carry out reductions selectively on such complicated models so as to isolate causative factors that can be implicated in generating particular behaviors. In some circumstances it is possible to reduce the consideration of experimentally observed dynamics to analysis of as simple a system as a one-dimensional finite-difference equation [15,20,21,55]; it should be possible to carry out similar reductions in the corresponding modeling work. Reduced models will also be of great use in modeling situations in which propagation of the cardiac impulse occurs. This has the added advantage of dramatically reducing the otherwise rather severe computational requirements in this case. A start in this direction has already been made using the reduced equations of FitzHugh to model anisotropic propagation [35] and reentrant arrhythmias [48,58,63].

10.9 Single Channel Models

The HH equations are formulated as a set of continuous equations. However, the currents flowing through the cardiac cell membrane are caused by an effectively discontinuous process: the very quick opening and closing of single channels, each with a small but finite conductance. It is thus only in the limit of taking the averaged activity of an infinite number of channels that one obtains the continuous HH description. In fact, the HH formulation breaks down when one looks at dynamical behaviors in which the discrete nature of the single channel dynamics plays a role. We draw on two examples from our own work to illustrate this fact.

First, we have shown in space-clamped aggregates of embryonic chick ventricular cells that there is an effectively "all-or-none" threshold for obtaining an action potential [23]. Modeling with an HH-type model and a "back-of-the-envelope" calculation reveals that the opening or closing of as few as half a dozen channels should be sufficient to convert the "all" response (an action potential) into the "nothing" response (a subthreshold response) [15]. It is thus clear that an inherently stochastic model formulated as a population of single channels [6,56] would be needed to investigate the phenomenon further.

As a second example, we offer Wenckebach rhythms in single rabbit ventricular cells. In our hands, these rhythms are very unstable, with it being impossible to obtain, for example, a maintained 4:3 rhythm for more than a few cycles. However, in the HH-type equations modeling ventricular membrane, it is indeed possible to obtain such maintained rhythms (see Chapter 14 of this book). We thus believe that the irregularity of the rhythms we see experimentally is caused by "noise" that is not present in the deterministic HH-type description. This noise arises from the fact that a single cell has only a few tens of thousands of channels, and so the fluctuations in transmembrane current are quite large with respect to the mean current. Thus, a statistical description of the various transmembrane current sources is needed to replace the continuous HH description if one wishes to account for the irregular rhythms actually seen in experimental work on single cells.

10.10   Single Channel Dynamics: Stochastic or Deterministic?

"The time between bursts (i.e., the length of the stretches in the old attractor region) has a more or less random appearance when tabulated" [14].

At present it is generally held that the opening and closing of a single channel is a stochastic, rather than a deterministic, process [7,26]. However, there are several intriguing correspondences between the time series seen in recordings of the single channel current and those generated by many "chaotic" systems. Chaotic systems are systems that can generate aperiodic signals even though they are deterministic [57]. We now discuss four of these similarities.

10.10.1   CRISIS-INDUCED INTERMITTENCY

When one inspects the current record of a channel that activates but does not inactivate, one sees that after the channel spends some time in one state (open or closed), it then abruptly switches to the other state (closed or open). This shuttling back and forth between two distinct states continues indefinitely. Thus, in a mathematical model of the channel, the state-point of the system would spend most of its time in two distinct, widely separated regions of the phase space, traversing the connecting regions only during the times of rapid switching from one state to another. Similar behavior can be seen in many deterministic systems of ordinary differential equations that possess a strange attractor, for example, the Lorenz attractor [28,57] or the forced buckled beam [28]. This switching behavior often occurs when a chaotic band suddenly widens, producing "intermittent bursting," or when two disjoint chaotic bands in the "banded chaos" region merge, producing "intermittent switching" ([14,31,50] and references therein). Both of these mechanisms are forms of "crisis-induced intermittency" because they show intermittent behavior and are induced by a crisis [14].

In the case of the forced buckled beam cited above [28], the system of ordinary differential equations was obtained by reducing a partial differential equation, which is exactly the sort of equation that one might use to model the motion of a protein molecule. In many instances one can reduce the study of the dynamics of an ordinary differential equation to a one-dimensional finite-difference equation. Such relatively simple equations can also display both forms of crisis-induced intermittency [14]. It is also of interest to note that the forced buckled beam equation also describes the motion of a particle in a double-potential-well system being subjected to periodic forcing [31].

10.10.2   SENSITIVE DEPENDENCE ON INITIAL CONDITIONS

In chaotic systems, there is "sensitive dependence on initial conditions"; that is, the evolution of the system depends on exactly how the system was prepared, even though all the parameters in the system remain fixed (see, e.g., Figure 1 in Holmes [28]). This property is reminiscent of the experimental finding that repeated trials lead to different single-channel current records when the membrane is repeatedly clamped from a given holding potential to a fixed clamp potential. Patch-clampers hence resort to a statistical description of the system, computing histograms of various sorts (e.g., open time, closed time, latency to first opening). The current records of the ensemble obtained from individual trials on one channel are then summed and averaged to produce a mean current that can be compared with the macroscopic (i.e., whole-cell or multicellular) current, assuming that this current is generated by a population of identical, noninteracting, channels.

10.10.3   HISTOGRAMS AND POWER SPECTRA

The statistics of the time to switching in the single channel recordings (i.e., the open- and closed-time histograms) often yields an exponential distribution [7,26]. Similar characteristics can be found in several chaotic systems if one interprets the time spent in one or the other regions of phase space with the open- or closed-time, respectively [50,62]. This leads to a Lorenzian power spectrum [31], which is also what is found experimentally in much single channel work [7,26].

A fractal description of single channel kinetics has been shown to fit the experimental data at least as well as the usual multistate kind of Markov model in at least two cases [36,53]. The suggestion has been made that this is due to the ability of protein molecules to move on many different time scales [37]. In this context, it is interesting to note that many chaotic systems have an inherently fractal structure.

The calcium channel from ventricular cells fires in bursts; within these bursts are seen bursts on a smaller time scale, and so forth [4]. Again, this self-similar structure is reminiscent of that seen in many chaotic systems. The power spectrum produced by systems that have no characteristic time scale is often of the $1/f$ type. Simple chaotic systems demonstrating intermittency can produce $1/f$-like spectra [38,51]; such spectra have indeed been obtained from noise analyses of multichannel current records [8] and in the interspike interval of isolated snail pacemaker neurons and the interbeat interval of the human electrocardiogram (see Musha and colleagues [43] and references therein).

10.10.4 TRANSIENT CHAOS

Upon stepping the voltage from the holding to the clamp potential, channels that inactivate do not persist indefinitely in opening and closing as time proceeds. Instead, they eventually close and stay closed; indeed, on some trials, they do not open at all [4]. This is reminiscent of a behavior initially called "metastable chaos" [62], which has more recently been termed "t.ansient chaos" [14]. In systems demonstrating transient chaos, nontransient chaos can often be produced by changing a parameter. In fact, the transient chaos is in some sense a remnant or trace of the non-transient chaos. Thus, one might think that the effect of substances that remove inactivation (e.g., veratramine on the fast sodium channel) is to produce such a shift in a parameter, but in the opposite direction.

In light of the above correspondences, it is thus tempting to speculate that single channel dynamics is governed by a deterministic set of rules and not a stochastic set as is currently generally accepted. There have been at least three attempts to analyze single channel kinetics from the perspective of a non-Markovian process. The first two of these studies, which analyzed a calcium-activated potassium channel from the nodose ganglion in the neonatal rat [53] and a potassium channel from corneal epithelium [36], showed that a better fit to the data was obtained using a fractal process than a Markovian process. However, there are cases where the Markov models do better [29]. One should also keep in mind that deterministic systems of ordinary differential equations possessing two or more strange attractors can, under the influence of random "noise," display a noise-induced "hopping" motion between the two attractors, which produces, for example, $1/f$-like spectra over a not inconsiderable range of frequency (see Arrechi and Califano [1] and references therein). Thus, the possibility is raised of a combined chaotic-stochastic mechanism. Finally, in comparing things that look similar, one should always keep in mind that resemblances can be slippery things!

REFERENCES

[1] F.T. Arrechi and A. Califano. Noise-induced trapping at the boundary between two attractors: A source of 1/f spectrum and nonlinear dynamics. *Europhys. Lett.*, 3, 5–10, 1987.

[2] G.W. Beeler and H. Reuter. Reconstruction of the action potential of ventricular myocardial fibres. *J. Physiol. (Lond.)*, 268:177–210, 1977.

[3] W.K. Bleeker, A.J.C. Mackaay, M. Masson-Pévet, L.N. Bouman, and A.E. Becker. Functional and morphological organization of the rabbit sinus node. *Circ. Res.*, 46:11–22, 1980.

[4] A. Cavalié, D. Pelzer, and W. Trautwein. Fast and slow gating be-

haviour of single calcium channels in cardiac cells. *Pflüg. Arch.*, 406: 241–258, 1986.

[5] D.R. Chialvo and J. Jalife. Non-linear dynamics of cardiac excitation and impulse propagation. *Nature*, 330:749–752, 1988.

[6] J.R. Clay and L.J. DeFelice. The relationship between membrane excitability and single channel open-close kinetics. *Biophys. J.*, 42:151–157, 1983.

[7] D. Colquhoun and A.G. Hawkes. The principles of the stochastic interpretation of ion channel mechanisms. In B. Sakmann and E. Neher, editors, *Single-Channel Recording*, pages 135–175. Plenum, New York, 1983.

[8] F. Conti. Noise analysis and single-channel recordings. *Curr. Top. Memb. Transport*, 22:371–405, 1984.

[9] D. DiFrancesco and D. Noble. A model of electrical activity incorporating ionic pumps and concentration changes. *Phil. Trans. Roy. Soc. Lond.*, B 307:353–398, 1985.

[10] M. Feingold, D.L. Gonzalez, O. Piro, and H. Viturro. Phase locking, period doubling, and chaotic phenomena in externally driven excitable systems. *Physical Rev.*, A 37:4060–4063, 1988.

[11] R. Fischmeister and G. Vassort. The electrogenic Na-Ca exchange and the cardiac electrical activity. I—Simulation of Purkinje fiber action potentials. *J. Physiol. (Paris)*, 77:705–709, 1981.

[12] R. FitzHugh. Impulses and physiological states in theoretical models of nerve membrane. *Biophys J.*, 1:445–466, 1961.

[13] R. FitzHugh. Mathematical models of threshold phenomena in the nerve membrane. *Bull. Math. Biophys.*, 17:257–278, 1955.

[14] C. Grebogi, E. Ott, F. Romieras, and J.A. Yorke. Critical exponents for crisis-induced intermittency. *Physical Rev.*, 36A:5365–5380, 1987.

[15] M.R. Guevara. *Chaotic Cardiac Dynamics*. Ph.D. thesis, McGill University, Montreal, Quebec, 1984.

[16] M.R. Guevara. Spatiotemporal patterns of block in an ionic model of cardiac Purkinje fibre, In M. Markus, S.C. Müller, and G. Niclois, editors, *From Chemical to Biological Organization*, pages 273–281. Springer, Berlin, 1988.

[17] M.R. Guevara. Afterpotentials and pacemaker oscillations in an ionic model of cardiac Purkinje fibres, L. Rensing, U. an der Heiden and M.C. Mackey, editors, *Temporal Disorder in Human Oscillatory Systems*, pages 126–133. Springer, Berlin, 1987.

[18] M.R. Guevara, F. Alonso, D. Jeandupeux, and A.C.G. van Ginneken. Alternans in periodically stimulated isolated ventricular myocytes: Experiment and model, In A. Goldbeter, editor, *Cell-to-Cell Signalling: From Experiments to Theoretical Models*, pages 551–563. Academic Press, London, 1989.

[19] M.R. Guevara, L. Glass, M.C. Mackey, and S. Shrier. Chaos in neurobiology. *IEEE Trans. Syst. Man. Cybern.*, SMC-13:790–798, 1983.

[20] M.R. Guevara, L. Glass, and A. Shrier. Phase locking, period doubling bifurcations, and irregular dynamics in periodically stimulated cardiac cells. *Science*, 214:1350–1353, 1981.

[21] M.R. Guevara, D. Jeandupeux, F. Alonso, and N. Morissette. Wenckebach rhythms in isolated ventricular heart cells, In St. Pnevmatikos, T. Bountis and Sp. Pnevmatikos, editors, *Singular Behaviour and Nonlinear Dynamics*, pages 629–642. World Scientific, Singapore, 1989.

[22] M.R. Guevara and A. Shrier. Phase resetting in a model of cardiac Purkinje fibre. *Biophys. J.*, 52:165–175, 1987.

[23] M.R. Guevara, A. Shrier, and L. Glass. Phase resetting of spontaneously beating embryonic ventricular heart cell aggregates. *Am. J. Physiol.*, 251:H1298–H1305, 1986.

[24] M.R. Guevara, A.C.G. van Ginneken, and H.J. Jongsma. Patterns of activity in a reduced ionic model of a cell from the rabbit sinoatrial node, In H. Degn, A.V. Holden and L.F. Olsen, editors, *Chaos in Biological Systems*, pages 5–12. Plenum, London, 1987.

[25] D.W. Hilgemann and D. Noble. Excitation-contraction coupling and extracellular calcium transients in rabbit atrium: Reconstruction of basic cellular mechanisms. *Proc. Roy. Soc. Lond.*, B 230:163–205, 1987.

[26] B. Hille. *Ionic Channels of Excitable Membranes*. Sinauer, Sunderland, 1984.

[27] A.L. Hodgkin and A.F. Huxley. A quantitative description of membrane current and its application to conduction and excitation in nerve. *J. Physiol. (Lond.)*, 117:500–544, 1952.

[28] P. Holmes. "Strange" phenomena in dynamical systems and their physical implications. *Appl. Math. Modelling*, 1:362–366, 1977.

[29] R. Horn and S.J. Korn. Graphical discrimination of Markov and fractal models of single channel gating. *Comments Theor. Biol.*, 1:39–46, 1988.

[30] H. Irisawa and A. Noma. Pacemaker mechanisms of rabbit sinoatrial node cells, In L.N. Bouman and H.J. Jongsma, editors, *Cardiac Rate and Rhythm*, pages 35–51. Nijhoff, Amsterdam, 1982.

[31] H. Ishii, H. Fujisawa, and M. Inoue. Breakdown of chaos symmetry and intermittency in the double-well potential system. *Phys. Lett.*, 116A:257–263, 1986.

[32] J.H. Jensen, P.L. Christiansen, and A.C. Scott. Chaos in the Beeler-Reuter system for the action potential of ventricular myocardial fibers. *Physica*, 13D:269–277, 1984.

[33] R.W. Joyner and F.J.L. van Capelle. Propagation through electrically coupled cells: How a small SA node drives a large atrium. *Biophys. J.*, 50:1157–1164, 1986.

[34] R.W. Joyner, R. Veenstra, D. Rawling, and A. Chorro. Propagation through electrically coupled cells: Effects of a resistive barrier. *Biophys. J.*, 45:1017–1025, 1984.

[35] J.P. Keener. A mathematical model for the initiation of ventricular tachycardia in myocardium, In A. Goldbeter, editor, *Cell to Cell Signalling: From Experiments to Theoretical Models*, pages 589–608. Academic Press, London, 1989.

[36] L.S. Liebovitch and J.M. Sullivan. Fractal analysis of a voltage-dependent potassium channel from cultured mouse hippocampal neurons. *Biophys. J.*, 52:979–988, 1987.

[37] L.S. Liebovitch and T.I. Tóth. Fractal activity in cell membrane ion channels. *Ann. NY Acad. Sci.*, 591:375–391, 1990.

[38] P. Manneville. Intermittency, self-similarity and 1/f spectrum in dissipative dynamical systems. *J. Phys.(Paris)*, 41:1235–1243, 1980.

[39] M.A. Masson-Pévet, W.K. Bleeker, E. Besselsen, B.W. Treytel, H.J. Jongsma, and L.N. Bouman. Pacemaker cell types in the rabbit sinus node: A correlative ultrastructural and electrophysiological study. *J. Molec. Cell. Cardiol.*, 16:53–63, 1984.

[40] R.E. McAllister, D. Noble, and R.W. Tsien. Reconstruction of the electrical activity of cardiac Purkinje fibres. *J. Physiol. (Lond.)*, 251:1–59, 1975.

[41] D.C. Michaels, E.P. Matyas, and J. Jalife. A mathematical model of the effects of acetylcholine pulses on sinoatrial pacemaker activity. *Circ. Res.*, 55:89–101, 1984.

[42] D.C. Michaels, E.P. Matyas, and J. Jalife. Mechanisms of sinoatrial pacemaker synchronization: A new hypothesis. *Circ. Res.*, 61:704–714, 1987.

[43] T. Musha, H. Takeuchi, and T. Inoue. 1/f fluctuations in the spontaneous intervals of a giant snail neuron. *IEEE Trans. Biomed. Eng.*, BME-30:194–197, 1983.

[44] R.A. Nadeau, F.A. Roberge, and P. Bhéreur. The mechanism of the Wenckebach phenomenon. *Isr. J. Med. Sci.*, 5:814–818, 1969.

[45] D. Noble. A modification of the Hodgkin-Huxley equations applicable to Purkinje fibre action and pace-maker potentials. *J. Physiol. (Lond.)*, 160:317–352, 1962.

[46] D. Noble and S.J. Noble. A model of the sino-atrial node electrical acitivity based on a modification of the DiFrancesco-Noble (1984) equations. *Proc. Roy. Soc. Lond.*, B 222:295–304, 1984.

[47] H. Othmer. The dynamics of forced excitable systems. In A.V. Holden, M. Markus, H. Othmer, editors, *Nonlinear Wave Processes in Excitable Media*, Plenum, London, 1990.

[48] A.V. Panfilov and A.N. Rudenko. Two regimes of the scroll ring drift in the three-dimensional active media. *Physica*, 28D:215–218, 1987.

[49] A.A. Petrov and B.N. F'eld. Analysis of the possible mechanism of origin of the extrasystole in local ischaemia of the myocardium using a mathematical model. *Biophysics*, 18:1145–1150, 1973.

[50] T. Post, H.W. Capel, and J.P. van de Weele. New results in intermittent switching, In St. Pnevmatikos, T. Bountis, Sp. Pnevmatikos, editors, *Singular Behaviour and Nonlinear Dynamics*, pages 195–200. World Scientific, Singapore, 1989.

[51] I. Proccacia and H. Schuster. Functional renormalization group theory of universal 1/f noise in dynamical systems. *Physical Rev.*, 28A: 1210–1212, 1983.

[52] O.E. Rössler, R. Rössler, and H.D. Landahl. Arrhythmia in a periodically forced excitable system (abstract). In *Proc. Sixth Int. Biophys. Congress, Kyoto*, 1978.

[53] P. Selepova. *Single Ion Channel Dynamics*. Master's thesis, McGill University, Montreal, Quebec, 1986.

[54] A.I. Shcherbunov, N.I. Kukushkin, and M.Y. Sakson. Reverberator in a system of interrelated fibers described by the Noble equation. *Biophysics*, 18:547–554, 1973.

[55] A. Shrier, H. Dubarsky, M. Rosengarten, M.R. Guevara, S. Nattel, and L. Glass. Prediction of complex atrioventricular conduction rhythms in human beings with use of the atrioventricular nodal recovery curve. *Circulation*, 76:1196–1205, 1987.

[56] E. Skaugen. Firing behaviour in stochastic nerve membrane models with different pore densities. *Acta. Physiol. Scand.*, 108:49–60, 1980.

[57] J.M.T. Thompson and H.B. Stewart. *Nonlinear Dynamics and Chaos*. Wiley, Chichester, 1986.

[58] F.J.L. van Capelle and D. Durrer. Computer simulation of arrythmias in a network of coupled excitable elements. *Circ. Res.*, 47:454–466, 1980.

[59] B. van der Pol and J. van der Mark. The heartbeat considered as a relaxation oscillation, and an electrical model of the heart. *Phil. Mag. (Series 7)*, 6:763–775, 1928.

[60] T. Watanabe, P.M. Rautaharju, and T.F. McDonald. Ventricular action potentials, ventricular extracellular potentials, and the ECG of guinea pig. *Circ. Res.*, 57:362–373, 1985.

[61] A.T. Winfree. Electrical instability in cardiac muscle: phase singularities and rotors. *J. Theor. Biol.*, 138:353–405, 1989.

[62] J.A. Yorke and E.D. Yorke. Metastable chaos: The transition to sustained chaotic behaviour in the Lorenz model. *J. Stat. Phys.*, 21:263–277, 1979.

[63] V.S. Zykov. *Simulation of Wave Processes in Excitable Media*. Manchester University Press, Manchester, 1987.

11

Mathematical Models of Pacemaker Tissue in the Heart

J.W. Clark[1]
J.M. Shumaker[1]
C.R. Murphey[1]
W.R. Giles[2]

ABSTRACT This chapter is concerned with basic issues in modeling electrophysiological responses in pacemaker tissue in the heart. In a structural sense, this type of tissue is complex, consisting of a network of different kinds of interconnected cells. The least complex cell in this network is the primary pacemaker cell; one of the more complex in terms of its ion channel configuration is the working atrial cell from the zone bordering the pacemaker region. Transitional cells of intermediate complexity (e.g., subsidiary pacemaker cells) are interposed between these two types of cardiac cells. This study focuses primarily on the membrane dynamics of pacemaker and atrial cells, the two extreme cases regarding ion channel complexity; in addition we study both amphibian and mammalian cardiac tissue. Importantly, the study also brings in the topic of modification of the electrical behavior of these cells, by the parasympathetic neurotransmitter acetylcholine (ACh). In general, conduction in a network of interconnected cells of different types depends on: (1) the distribution of resistive coupling properties between cells, (2) the distribution of cellular membrane properties in the network, and (3) the distribution of autonomic neural influences that may dramatically change the properties of the component cells of the network.

11.1   Introduction

The microelectrode studies that first described the intracellular events underlying cardiac pacemaker activity were conducted nearly 40 years ago [19,63]. However, detailed investigations of the ionic mechanisms governing pacing were not done until the mid-1960s along with the early development

---

[1]Department of Electrical and Computer Engineering, Rice University, Houston, Texas 77251-1892

[2]Departments of Medical Physiology and Medicine, University of Calgary Medical School, Calgary, Alberta T2N 4N1

of voltage clamp methods applied to the study of cardiac tissue (for review see [2,32]). Much of this work was done on secondary or subsidiary pacemaker tissue, for example, cardiac Purkinje fibers, since this preparation was easily isolated from the heart and could be induced to beat spontaneously using a number of well known techniques [e.g., lowering $[K^+]_o$ or adding norepinephrine (NE) to the bathing medium]. It was not until the late 1970s that investigators succeeded in voltage-clamping small strips of pacemaker tissue from the bullfrog heart [11] or from the rabbit sinoatrial (SA) node (for review see [10,31]). These multicellular voltage-clamp studies have yielded useful data concerning the types and relative sizes of the ionic currents underlying the pacemaker potential and action potential of the pacing cell. However, as a result of technical difficulty in obtaining these data (e.g., maintaining two separate microelectrode impalements in contracting muscle) and the uncertainties in interpretation caused, for example, by voltage-clamp nonuniformity and extracellular depletion or accumulation of K^+ [1], it has not been possible to assign functional significance unequivocally to many of these ionic currents [12,13].

Since the early 1980s, enzymatic dispersion techniques have been employed for the isolation of a variety of cardiac cells, including pacemaker cells [7,37,50,51]. Using isolated cell preparations, the kinetics of many of the ionic currents identified previously in multicellular voltage-clamp experiments may be studied in a more quantitative fashion using whole-cell suction-pipette voltage-clamp techniques applied to the isolated cell [22,46]. In some cases these data provide conclusive information concerning the functional role of these currents. More often, however, the results are only suggestive, and important quantitative aspects are still missing.

The two most commonly used cardiac pacemaker cell preparations are obtained from the bullfrog sinus venosus and the rabbit sinoatrial (SA) node. Morphologically these cells are quite different. The bullfrog sinus venosus cell is spindle-shaped. It is approximately 6 μm in diameter measured at the level of its centroid nucleus, 200 μm long, and is tapered to either end from the center region. The cell itself has no transverse tubular system and very little sarcoplasmic reticulum (SR), in contrast with mammalian pacemaking cells [26,27]. At first glance, the bullfrog sinus venosus may appear to be a uniform, homogeneous structure comprised of coupled pacemaker cells. However, Hutter and Trautwein [29] observed that the amplitude of the recorded pacemaker potential varies in different parts of the sinus. At present, detailed morphological and electrophysiological studies of the bullfrog sinus venosus are not available, and it may eventually prove necessary to characterize the sinus venosus as something other than a simple, uniform, two-dimensional syncytium of coupled pacemaker cells.

On the other hand, the rabbit SA node is a very heterogeneous structure [3,34,35,39,45]. Electrical activity originates within a central "compact" region, where "dominant" or "primary" pacemaker cells having a low density of myofilaments generate action potentials with a low upstroke velocity. Activity proceeds outward from the small central region, first radially and then preferentially, toward the upper part of the crista terminalis. Transitional cells that conduct this activity are frequently called latent or subsidiary pacemaker cells. Their action potentials have larger upstroke velocities and more negative take-off potentials; morphologically, they possess a greater density of myofilaments. Thus, investigators attempting to study single primary pacemaker cells from the SA node must pay careful attention to the source of the tissue from which the single-cell preparations are made, since the enzymatic-dispersion process applied to small slices of tissue frequently yields mixtures of primary and subsidiary pacemaker cells. Nathan [45] has shown that these single cells may be identified electrophysiologically since I_{Na} is not present in primary pacemaker cells, but it does appear in subsidiary pacemaker cells. After a day in culture, Nathan [45] was also able to identify two morphologically distinct types of beating cells. One was spindle-shaped with bipolar cytoplasmic projections that adhered strongly to the culture dish (type I); the other tended to be spheroidal in shape with a single small attachment site at its base (type II). The cells differed dramatically in their measured upstroke velocities (1.4 V/sec, type I; 20 V/sec, type II) as well as in the general appearance of their action potentials. The dimensions of these cells are as follows: type I, $17.0 \pm .5 \times 21.2 \pm 1.5 \mu M$ (n=6); type II, $17.8 \pm .9 \times 20.9 \pm 1.4 \mu M$ (n=10). In contrast with type II cells, type I cells always exhibited prominent cytoplasmic projections and contracted much more weakly. These morphological and electrophysiological differences emphasize the importance of the precise determination of anatomic origin of the SA node cells whose electrophysiological properties are to be experimentally studied and subsequently modeled mathematically.

11.2 Modeling Aspects

Equivalent circuit models of the Hodgkin-Huxley type are capable of providing a quantitative description of the electrical behavior of the sarcolemma in individual cardiac cells, including pacemaker cells [5,6,17,25,40,47,48,64]. It is well known that the particular shape of a cardiac action potential waveform is determined by the temporal interplay of certain *inward* membrane currents [sodium current (I_{Na}) and calcium current (I_{Ca})] and a number of *outward* potassium currents [e.g., the "transient outward" potassium current (I_t), the "delayed rectifier" current (I_K), the calcium activated potassium current ($I_{K,Ca}$), and the "instantaneous rectifier" current (I_{K1})]. It is also known that the spikelike inward sodium current provides the upstroke of the cardiac action potential and the smaller and slower inward calcium current supports the plateau, whereas the activation of voltage and time-dependent potassium currents initiates repolarization. Action potentials recorded from isolated cells taken from various regions of the heart

TABLE 11.1. Typical ion channel currents in cardiac membranes and their presence ($\times$) or absence ($-$) in pacemaker (FSV,RSA) and atrial (FA,RA) cell membranes. FSV = bullfrog versus sinus venosus; FA = bullfrog atrium; RSA = rabbit sinoatrial (SA) node; and RA = rabbit atrium. See text for explanation of different types of currents.

	FSV	FA	RSA	RA
I_{Na}	$-$	$\times$	$-$	$\times$
I_{Ca}	$\times$	$\times$	$\times$	$\times$
I_t	$-$	$-$	$-$	$\times$
$I_{K,Ca}$	$-$	$-$	$\times$	$\times$
I_K	$\times$	$\times$	$\times$	$\times$
I_{K1}	$-$	$\times$	$-$	$\times$
I_f	$-$	$-$	$\times$	$-$
I_{NaK}	$\times$	$\times$	$\times$	$\times$
I_{NaCa}	$\times$	$\times$	$\times$	$\times$

[e.g., the atrium, ventricle, specialized conduction system, (SA) or atrioventricular (AV) nodes] exhibit different characteristic waveshapes. In order to provide a quantitative explanation of these differences, detailed knowledge is required of each of the ionic channel currents in a given type of cell membrane. In the specific case of the primary pacemaker cell, relatively few transmembrane ion channels are involved. In this case, the inward current providing the upstroke of the action potential is the calcium channel I_{Ca}, while repolarization is initiated by the delayed rectifier current I_K.

As electrical activity propagates away from a primary pacemaking site toward the surrounding atrium, the wavefront encounters cells that are comprised of different types of membrane ion channels. Thus, the sarcolemma of the primary pacemaking cell contains the simplest complement of ion channels, while the working atrial cell has the largest number of different types of channels. Transitional cells, including subsidiary pacemaker cells, possess a configuration of channels of intermediate complexity. Table 11.1 attempts to summarize the better known membrane currents for two different cell types (primary pacemaker and atrial cells) for two different species, the bullfrog and the rabbit. These currents are discussed in the material to follow.

11.3   The Bullfrog Sinus Venosus Pacemaker Cell

The lumped equivalent circuit of the sarcolemma of a single bullfrog sinus venosus (SV) myocyte is shown in Figure 11.1A. This membrane model contains both the ion specific channels that generate the action potential as well as pumps, background, and exchanger currents that are necessary to characterize the ongoing cyclic activity of the SV pacemaker cell. Under nonpropagating space-clamp conditions, the ordinary differential equation describing changes in membrane potential (V) is:

$$\frac{dV}{dt} = -\frac{[I_{Ca} + I_K + I_{NaK} + I_{NaCa} + I_{CaP} + I_B]}{C_M} \qquad (11.1)$$

where I_{Ca} and I_K are the voltage and time-dependent inward calcium and outward potassium currents, respectively. The outward current I_K is also known as the "delayed rectifier" current; I_{NaK} is the sodium-potassium pump current; I_{NaCa} is the sodium-calcium exchanger current; I_{CaP} is the electrogenic calcium pump current; I_B is the background current consisting of a relatively large sodium component and a smaller calcium component; and C_M is the cell membrane capacitance in nanofarads. Equations for the individual membrane channel currents are given in Rasmusson and colleagues [52,53].

The mathematical expressions for the pump, exchanger, and channel currents require specification of the concentrations for several ionic species (sodium, potassium, and calcium) both inside and outside the cell. In the case of isolated SV pacemaker myocytes, the extracellular environment is the bathing medium that is assumed to have a known ionic composition. Figure 11.1B represents a fluid compartment model of the intra- and extracellular media where provision is made for the appropriate buffering of calcium within the cell. Based on a material balance for the ionic species associated with the two-compartment system, differential equations have been derived that describe the time-dependent concentration changes for sodium, potassium, and calcium in the lumped intracellular medium [52]. Of particular interest are the differential equations describing the binding of intracellular calcium to two specific sites on the cardiac troponin molecule (a calcium-specific site and a competitive calcium-magnesium site) as well as specific sites on the calmodulin molecule (three calcium ions per molecule). This model is used to describe the $[Ca^{2+}]_i$ regulatory mechanisms in the bullfrog sinus venosus and is based on the earlier work of Robertson and colleagues [55] and the more recent work of Campbell and colleagues [14].

The single bullfrog sinus venosus cell is described mathematically by a system of 11 first-order nonlinear differential equations; 4 of these differential equations pertain to the sarcolemmal portion of the model and 7 are associated with the fluid compartment model that includes internal calcium-buffering. In applying conventional numerical methods to integrate these equations, one must be aware that the time constants involved vary from .333 msec (calcium activation kinetics) to 7500 msec (occupancy of calcium on the competitive calcium-magnesium site on troponin). Since system time constants can range over four orders of magnitude, double-precision arithmetic is used for all calculations so that roundoff errors due

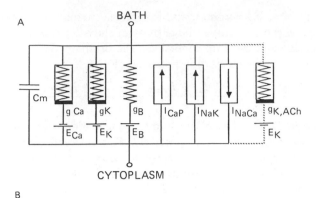

FIGURE 11.1. Model of the isolated bullfrog sinus venosus (SV) pacemaker cell. (A) Electrical equivalent circuit of the sarcolemmal ion channels, pumps and sodium-calcium exchanger. The ACh-sensitive K^+ channel (muscarinic channel) is active only in the presence of ACh and is therefore indicated as an additional (dotted) parallel branch of the circuit model. See text for explanation of notation. (B) Fluid compartment model of the sinus venosus myocyte. This model is used in writing the material balance equations for three important ionic species Na^+, K^+, and Ca^{2+}. Note that provisions are made for internal calcium ion buffering by the intracellular protein calmodulin as well as two binding sites on the protein molecule troponin: a calcium-specific site and a competitive calcium-magnesium site.

to finite precision arithmetic can be minimized. In addition, a variable step size algorithm with an appropriate error estimate should be used to minimize the time required for solution of the system of nonlinear differential equations. A variable step size Sarafyan embedding type of modification to the Runge-Kutta formulas [36] can be used for the numerical integration of the model equations using appropriate upper and lower limits on the local truncation error (10^{-4} and 10^{-6}, respectively) to change the step size.

An essential feature of any mathematical model of a primary pacemaking cell is its ability to reproduce accurately the spontaneous oscillations in transmembrane potential that are recorded under physiological conditions. Panel A of Figure 11.2 shows that typical spontaneous pacemaker and action potential waveforms can be reproduced by our sinus venosus model using the equations and nominal parameter values given in Rasmusson and coworkers [52]. The waveforms of the simulated transmembrane currents (I_{Ca}, I_K, I_B) are shown in panel B of Figure 11.2, and the pump and exchanger currents ($I_{NaK}, I_{CaP}, I_{NaCa}$) are shown in panel C.

A typical spontaneous action potential and pacemaker potential from an isolated bullfrog sinus venosus myocyte [20,57] is shown in Figure 11.3 along with a model-generated action potential. To obtain the particular model-generated waveform shown in this figure, the nominal parameter set given in Tables 1 through 5 in Rasmusson and coworkers [52] was employed, and the magnitudes of the component membrane currents were scaled appropriately to yield acceptable fits to the experimental data. The majority of the parameter values contained in the nominal parameter set for the model were derived from voltage clamp data obtained in the laboratory of W. Giles. Our model is thus able to reproduce the voltage clamp data on which it is based, and in addition is capable of providing a reasonably accurate fit to the experimental data [20,57].

11.4   The Bullfrog Atrial Cell

The membrane model for the bullfrog atrial cell is quite similar to the model for the sinus venosus pacemaker cell, as can be seen from the ordinary differential equation describing changes in membrane potential:

$$\frac{dV}{dt} = -\frac{[I_{Na} + I_{Ca} + I_K + I_{K1} + I_{NaK} + I_{NaCa} + I_{CaP} + I_B]}{C_M}. \quad (11.2)$$

The notable additions to the component membrane currents are: (1) the voltage and time-dependent inward sodium current I_{Na}, which triggers the rapid upstroke of the action potential (much faster than the upstroke velocity of the SV pacemaker cell), and (2) the voltage-dependent "instantaneous rectifier" current I_{K1}, which is important in both setting the resting potential of the cell (-85mV) and in initiating the late repolarization phase of the action potential. The pump, exchangers, and back-

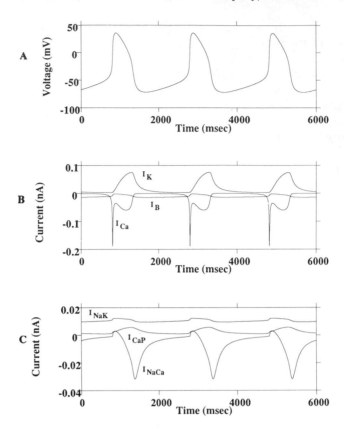

FIGURE 11.2. Model-generated bullfrog SV pacemaker cell waveforms: (A) Transmembrane potential (top), (B) the membrane ion channel currents (middle); the inward calcium current I_{Ca}, the delayed rectifier current I_K, and the background current I_B, (C) the membrane ion pump and exchanger currents, the sodium-potassium pump current I_{NaK}, the calcium extrusion pump current I_{CaP}, and the sodium-calcium exchanger current I_{NaCa} (bottom).

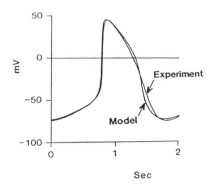

FIGURE 11.3. Comparison of model-generated and experimentally recorded action potentials from an isolated SV pacemaker cell.

ground currents are adjusted appropriately to provide a reasonable fit to action potential data from bullfrog atrial myocytes. A model-generated atrial action potential is shown in Figure 11.4A, along with selected membrane currents (I_{Ca}, I_K, I_{K1}) in panel B, and pump and exchanger currents ($I_{NaCa}, I_{Ca,P}, I_{NaCa}$) in panel C. The sodium current I_{Na} is a large spike-like inward current that coincides with the leading edge of the atrial action potential. For greater clarity, however, it was omitted from panel C, which shows the other membrane channel currents. The model-generated action potential shown in Figure 11.4A provides a reasonable fit to experimental data as shown in Rasmusson and coworkers [53].

11.5   The ACh-Sensitive K^+ Current $I_{K,ACh}$

It is generally accepted that acetylcholine (ACh) has three possible actions on cardiac muscle: (1) the activation of an ACh-sensitive potassium current $I_{K,ACh}$ that is both voltage- and time-dependent (this channel is frequently called the muscarinic channel), (2) the inhibition of the voltage- and time-dependent inward calcium current I_{Ca}, and (3) the inhibition of the hyperpolarization-activated inward current I_f, which is important in mammalian cardiac cells exhibiting pacemaking activity. Although I_f-like current components were originally observed in frog sinus venosus trabeculae [11], and continue to be observed in multicellular voltage clamp experiments [15,16], no I_f current has yet been recorded from active isolated cells from that tissue [20] One plausible interpretation of these results is that this current might be an artifact arising from ion accumulation or depletion effects in the multicellular voltage clamp preparations. Although I_f does not seem to be present in either bullfrog sinus venosus or atrial myocytes, it

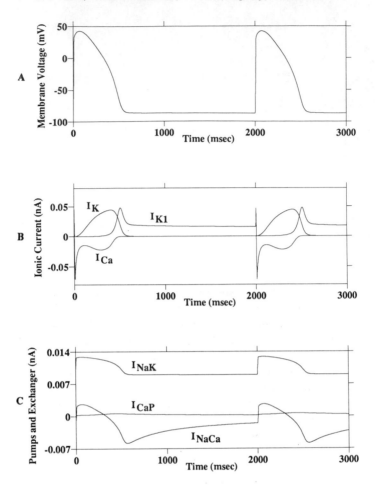

FIGURE 11.4. Model-generated bullfrog atrial cell waveforms. (A) Transmembrane potential (top), (B) the membrane ion channel currents (middle), and (C) the membrane ion pump and exchanger currents (bottom).

is an important current in mammalian cells exhibiting pacemaking activity such as the rabbit SA or AV node cell or the Purkinje fiber.

Since no I_f currents have been found in bullfrog SV myocytes, we shall assume that ACh applied to this type of cell activates an inwardly rectifying current called the muscarinic channel current $I_{K,ACh}$ and reduces the slow inward calcium current. This reduction in I_{Ca} is most likely produced by an inhibition of cyclic AMP (cAMP) production, the inhibition mediated by a GTP-binding protein (G_i), which inhibits membrane-bound adenylate cyclase activity. Reduced levels of cAMP in turn lead to reduced levels of channel protein phosphorylation by the intermediate of a cAMP-dependent protein kinase [33,54,62]. A comparison of the effects of ACh in activating $I_{K,ACh}$ and inhibiting I_{Ca} is discussed in a related modeling study by Shumaker and colleagues [59], who chose to model only the potassium-sensitive muscarinic channel mediated by a receptor-activated guanine nucleotide binding G protein (G_K). This choice was based on the relative paucity of quantitative experimental data regarding the role played by ACh in reducing I_{Ca}. Most, if not all, present-day cellular modeling studies model the effect of ACh on the activity of pacemaker cells via its effect on the muscarinic channel current $I_{K,ACh}$.

11.5.1   EFFECT OF $I_{K,ACh}$ ON ACTIVITY OF THE BULLFROG ATRIAL CELL

The muscarinic channel model developed by Shumaker and colleagues [59] for the bullfrog atrial myocyte may be incorporated into the equivalent electrical circuit of either the atrial or the sinus venosus cell model as a separate network current branch as shown via the dotted addition in Figure 11.1A. This muscarinic channel model is a modification of the well-known Osterrieder, Noma, Trautwein (ONT) model [49] of the $I_{K,ACh}$ current in the rabbit SA node cell. It differs from the ONT model in that it (1) uses a nonlinear rather than a linear ion transfer characteristic for the muscarinic channel and (2) employs an activation rate constant associated with channel opening that is both voltage- and ACh-dependent, rather than simply ACh-dependent. The parameters of the modified model are chosen to fit recent data from the literature on the muscarinic channel in bullfrog atrial myocytes [4,60]. It is incorporated into the larger mathematical models of both the bullfrog atrial and SV pacemaker cell models, which are based on quantitative whole-cell voltage clamp data. Simulations are conducted to discern the effects of applications of ACh on the muscarinic channel and subsequently the electrical behavior of the whole-cell model.

In the presence of a constant ACh bath concentration, the atrial myocyte model predicts a progressive shortening of the action potential with increasing levels of ACh, as well as an indirect influence of the $I_{K,ACh}$ current on the other currents of the atrial cell model. An example of this effect of ACh on the atrial action potential is shown in Figure 11.5. Here

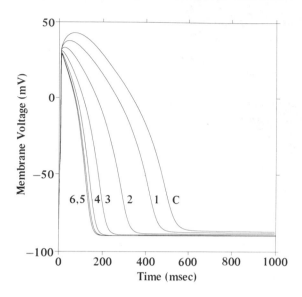

FIGURE 11.5. Model-generated atrial action potentials showing the effect produced by application of different bath concentrations of ACh. A repetitive stimulus rate of 0.5 Hz is used, which corresponds to the normal heart rate of the
bullfrog. ACh concentrations: C = control (0), 1 = 30 nM, 2 = 100 nM, 3 = 0.3
μM, 4 = 1 μM, 5 = 3 μM, 6 = 10 μM.

the fast inward sodium current is relatively unaffected by the onset of the
$I_{K,ACh}$ waveform, while the "secondary hump" and duration of the I_{Ca}
waveform are diminished, as are the amplitude and duration of the delayed
rectifier current I_K. In effect, the $I_{K,ACh}$ waveform becomes progressively
stronger as bath concentrations of ACh are increased, and can initiate a
progressively earlier repolarization of the action potential waveform. As
[ACh] is increased, the duration of the I_{Ca} waveform diminishes, which is
reflected in the diminishment of the peak height of the action potential
seen in Figure 11.5.

11.5.2   Effect of $I_{K,ACh}$ on the Activity of the Bullfrog Pacemaker Cell

The ACh-sensitive K^+ channel current has a powerful influence on the
sinus venosus pacemaker cell such that, at low to moderate bath concentrations of ACh, the pacing rate declines (period increases) according to the
relationship shown in Figure 11.6. At concentrations greater than 15nM,
however, our model predicts that electrical activity ceases. Thus, as is well
known, strong vagal stimulation can bring about cardiac arrest. A further
intriguing aspect of the effect of ACh on the activity of cardiac pacemaker

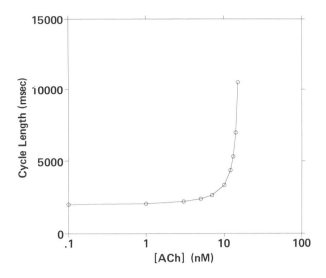

FIGURE 11.6. Dose-response curve for the steady-state cycle length of the bull-frog SV pacemaker cell action potential as a function of ACh concentration.

cells is the phase-sensitive behavior of the pacemaker cell to single, brief applications of ACh within a single cardiac cycle. On the basis of histological evidence presented by Hartzell [23] for the distribution of muscarinic ACh receptors and presynaptic nerve terminals in the amphibian heart, we consider the multiple varicosities associated with the terminal branches of parasympathetic nerves to be distributed randomly in the tissue space surrounding an individual pacemaker cell. Consequently, ACh may be considered to be essentially "bath applied" to the individual pacemaker fiber; that is, [ACh] does not vary spatially either along or around the cell, but may vary with time depending on the ACh stimulation waveform. We simulate this condition by applying an appropriate ACh concentration waveform such as one of those shown in Figure 11.7, to the lumped parameter model of the pacemaker cell (Figure 11.1), and specifically to the ACh-sensitive K^+ ion channel that produces the muscarinic channel current $I_{K,ACh}$. Figure 11.8 shows the effect of a single stimulus on the membrane potential and total membrane current using the ACh concentration waveform labeled 3 in Figure 11.7 ($k_H = 1/2000$ msec^{-1}). Note in Figure 11.8A that the stimulus is applied at the beginning of the second cycle (marked by a bar) and the cycle length of the succeeding three beats is prolonged. The temporal course of the $I_{K,ACh}$ waveform is shown in Figure 11.8B along with the total membrane current waveform. Although not shown, if the duration of the ACh stimulation waveform is of a sufficiently short duration relative to the basic cycle length and the stimulus is delivered near the upstroke of the action potential, the ACh stimulus may actually shorten (rather than

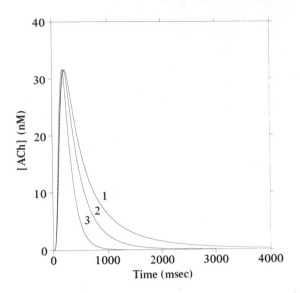

FIGURE 11.7. ACh concentration waveforms for driving the muscarinic channel shown in Figure 11.1A. The waveforms have essentially the same peak magnitude but differ in their decay characteristics. The model used to generate these [ACh] waveforms is discussed in Shumaker et al. [58] in which the decay rate is controlled by a general parameter k_H that represents the rate of hydrolysis of ACh by cholinesterase available at specific external sites on the cell sarcolemma. Notation is as follows: (1) $k_H = 1/200$ msec^{-1}, (2) $k_H = 1/617$ msec^{-1}, and (3) $k_H = 1/2000$ msec^{-1}.

lengthen) the cycle in which the stimulus is given, resulting in paradoxical acceleration of the pacemaker beat [58].

11.5.3   PHASE-SENSITIVE BEHAVIOR OF THE SV PACEMAKER CELL TO BRIEF APERIODIC ACh STIMULI

In Figure 11.9 the ACh stimulus (denoted by the triangle symbols) is delivered progressively later in the cycle with a phase delay ϕ relative to a point on the normal upstroke of the action potential (arbitrarily defined as zero time). In this figure, the basic cycle length (P_0) of the SV pacemaker cell is 2002 msec, and as the ACh stimulus is delivered progressively later in a test cycle, the length of that cycle (P_1) gradually increases. There is a point in cycle P_1, however, where the stimulus delivery will have no effect on that cycle, but rather, it will have an effect on the succeeding cycles P_2, P_3, and so forth. Similar phasic delays having no effect on cycle P_1 define a "no effect" zone labeled ρ. Thus, an ACh stimulus may be delivered

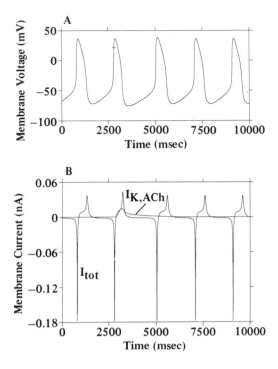

FIGURE 11.8. Model-generated potential and current waveforms for a single, brief application of ACh using the [ACh] concentration waveform labeled 3 in Figure 11.7 ($k_H = 1/2000$ msec^{-1}): (A) membrane potential waveform and (B) total membrane current I_{tot} and the muscarinic channel current $I_{K,ACh}$ as functions of time.

in a scanning fashion to a given test cardiac cycle, with a delay ϕ relative to the normal upstroke of the action potential (0 time) for ϕ in the range $-\rho \leq \phi \leq P_0 - \rho$, in order to access the phase-sensitive behavior of the SV pacemaker cell to a single, brief ACh stimulus. The data $P_{1,i}$ from multiple trials using different stimulus delays ϕ_i, $i = 1, 2, \ldots$ are gathered into plots known as phase response curves (PRCs), an example of which is shown in Figure 11.10. This figure plots the normalized change in cycle length ΔP as a function of the normalized phase Φ where

$$\Delta P \equiv \frac{P_1 - P_0}{P_0} \tag{11.3}$$

$$\Phi \equiv \frac{\phi}{P_0} \tag{11.4}$$

and Φ takes on values in the range

$$\frac{-\rho}{P_0} \leq \Phi \leq \frac{P_1 - P_0}{P_0}. \tag{11.5}$$

From Figure 11.10 it is seen that the no-effect zone ρ is approximately 10% of the control cycle length P_0. As Φ increases, there is a progressive increase in the cycle length of the pacemaker cell. Note that the phase response curve as defined here is only concerned with the first cycle (P_1) after stimulus delivery. Application of ACh clearly affects more than one cycle, but this is of no concern in the PRC. Plots that do consider the transient response of the pacemaker cell to a single, brief application of ACh for several successive cycles are called inhibition curves (ICs). Examples of such curves are well documented in the literature [8,9,18,21,38,61], and may be seen in Shumaker and colleagues [58] for the specific case of the bullfrog SV pacemaker cell.

11.5.4   PHASE-SENSITIVE BEHAVIOR OF THE SV PACEMAKER TO PERIODIC ACh STIMULATION

The steady-state response of the SV pacemaker cell to periodic stimulation is shown in Figure 11.11. The basal ACh stimulus employed in this figure has the waveshape of the ACh waveform labeled 2 in Figure 11.7 ($k_H = 1/617$ msec^{-1}). A repetitive ACh driving function is formed by choosing the repetition interval of the basal ACh stimulus and summing the component ACh waveforms to produce a simulated ACh concentration waveform at the outer membrane surface of the SV myocyte. For each point in Figure 11.11, the period of the ACh stimulus waveform was chosen, the ACh waveform produced, and stimulation of the muscarinic channel of the SV pacemaker cell model was continued for 50 pacemaker cycles in order to assure that steady-state conditions had been achieved. Subsequently, the periods of the following 20 pacemaker cycles were averaged to obtain

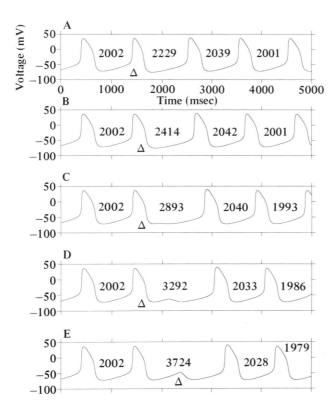

FIGURE 11.9. Transient effects of a single brief application of ACh. Model-generated action potentials for $k_H = 1/2000$ msec^{-1}. The triangle indicates the time point of ACh delivery. The phase of stimulus delivery ϕ (msec) is measured relative to a fiducial point (0 time) on the upstroke of the action potential.

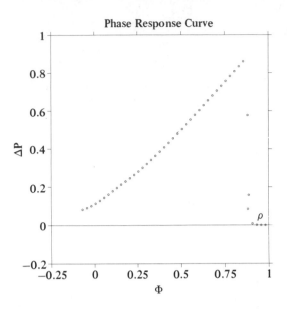

FIGURE 11.10. Model-generated phase response curve (PRC) for $k_H = 1/2000$ $msec^{-1}$. The upstroke of the action potential occurs at time $= 0$. The abcissa and ordinate are both normalized for one cardiac cycle. The no-effect zone (ρ) is denoted in the figure. See text for details.

the mean pacemaker cycle length at that ACh stimulation period. Figure 11.11 is referred to as a steady-state "entrainment" curve in that the cardiac pacemaker oscillator may become phase-locked, or entrained, to the periodic ACh driving oscillator. In certain capture regions, ACh stimulation has a strange effect on the pacemaker cycle in that the period of the pacemaker *decreases* as the period of ACh stimulation *decreases*, thus paradoxically speeding up the pacemaker. Besides the fundamental 1:1 (pacemaker period to ACh stimulation period) entrainment zone, the pacemaker oscillator is entrained at 1:2 and 2:1. The 1:1 and 2:1 entrainment regions shown in Figure 11.11 correspond to approximately 50% of the SV pacemaker cycle length. This is comparable to the experimental study of Dong and Reitz [18] and the modeling studies of Michaels and colleagues [41,42].

11.5.5   TRANSIENT RESPONSE OF THE SV PACEMAKER
TO BURST STIMULATION

In Figure 11.12A four control cycles ($P_0 = 2002$ msec) are shown prior to a single 6-sec burst of ACh stimulation at an intraburst stimulation frequency of 10 Hz. The time course of [ACh] during and after the burst is shown in Figure 11.12B. Note that the burst produces a pronounced hyperpolarizing effect of 15 mV beyond the maximum depolarization potential (MDP) level,

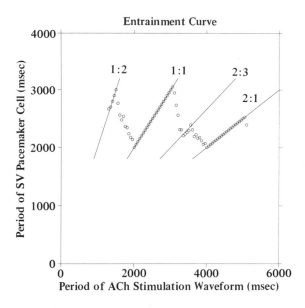

FIGURE 11.11. Model-generated entrainment curve. The period of the pacemaker cycle is measured on the ordinate and the period of the input ACh stimulus is plotted on the abcissa. Note the labeled regions of entrainment.

completely suppressing spontaneous activity in the SV pacemaker cell. After stimulation ceases, the remaining ACh is hydrolyzed rather quickly and membrane potential drifts rapidly toward the firing line. Ultimately spontaneous activity resumes. Owing to the hyperpolarization phase preceding the upstroke of the first action potential in recovery, the calcium channel is inactivated to a lesser extent than is present under control conditions. Consequently, this action potential has a higher peak magnitude. Note that if ACh were also to inhibit the calcium current I_{Ca} as mentioned previously, the enhancement of the peak magnitude of the first action potential exhibited in Figure 11.12A may not occur, and spontaneous activity may not return as quickly depending on the degree of inhibition of I_{Ca} by ACh. Figure 11.12 is intended to mimic in a loose sense the classical experiments of Hutter and Trautwein [28,29] dealing with microelectrode recordings from cells in multicellular frog sinus venosus preparations subjected to bursts of vagal stimulation.

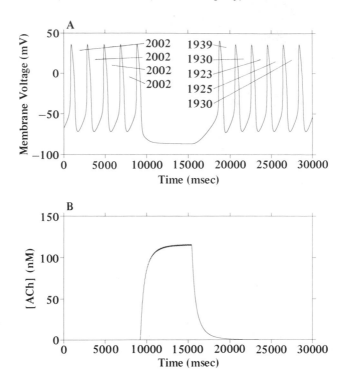

FIGURE 11.12. Model-generated response from a transient repetitive ACh delivery. Panel A shows the transmembrane voltage resulting from a 3-sec interval of ACh delivery with a repetition interval of 100 msec (10 Hz). The numbers correspond to the pacemaker cycle lengths in msec. Panel B shows the ACh waveform applied to the pacemaker cell.

11.6   Parasympathetic Control of the Rabbit SA Node Cell

We have also developed a mathematical model of the isolated rabbit SA node cell [43]. It is based on the modeling studies of DiFrancesco (DN) [17] for the Purkinje fiber, and Noble and Noble [48] for the rabbit SA node. This model contains: (1) a modified version of the calcium channel that is capable of providing a better fit to the SA node cell action potential and (2) a voltage, ACh, and time-dependent characterization of the ACh-sensitive K^+ channel. An equivalent circuit model of the membrane of this cell is given in Figure 11.13A, while the associated fluid compartment model is given in panel B of this figure. Material balance equations are written for three ionic species (sodium, potassium and calcium) distributed in the fluid compartment model. Provision is made for calcium uptake and release according to the equations provided in the DN model [17]. Note that there is no separate system provided for calcium buffering via calmodulin and troponin, as in the bullfrog SV pacemaker cell model. In Figure 11.13A, the muscarinic channel is indicated as an additional (dotted) parallel branch that is not in the circuit when ACh is absent from the bathing medium.

Figure 11.14 shows the spontaneous free-running behavior of the SA node cell model. The general waveshape of the transmembrane potential agrees very well with published data for the rabbit SA node cell. The underlying membrane currents are also shown. The slow inward current I_{si} is responsible for the leading edge of the action potential, the slight plateau region is contributed by both the exchanger current I_{NaCa} and the incomplete inactivation of the slow inward channel current that has been incorporated in our model [43], and repolarization is initiated by the delayed rectifier current I_K. The slow decrease in the tail of I_K during phase 4 is the main factor governing the rate of depolarization of membrane potential during the pacemaker potential. The time course of calcium concentration in the uptake and release compartments of the sarcoplasmic reticulum (SR) is shown in Figure 11.14B, as well as the intracellular calcium ion concentration waveform. Note that peak $[Ca]_i$ at a concentration slightly greater than $4\mu M$ coincides with the peak of the action potential.

A neural termination model of vagal activity [6,43] is employed to provide ACh concentration waveforms that are used as input to the ACh-sensitive potassium channel of this model. Figure 11.15 shows the ACh waveform presented to the outer surface of the SA node cell resulting from a single vagal volley. The numbers in the figure correspond to the number of impulses (1,3,5, or 9) contained in the vagal volley. As in the case of the bullfrog sinus venosus pacemaker cell, the rabbit SA node cell has a phase-sensitive response to ACh that is characterized by the family of phase response curves (PRCs) shown in Figure 11.16, which are parameterized by the number of vagal stimuli per volley.

A

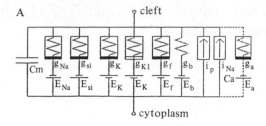

B

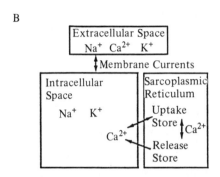

FIGURE 11.13. Model of the rabbit SA node cell: (A) equivalent network for the the cell including membrane channels, sodium/potassium pump current i_P, and sodium-calcium exchanger current i_{NaCa}, and (B) fluid compartment model.

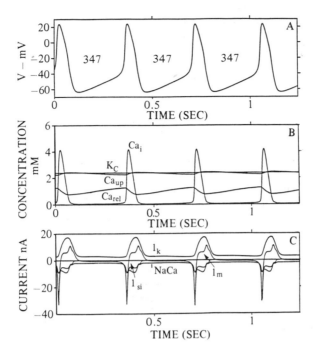

FIGURE 11.14. Free running behavior of the SA nodal cell model. The numbers seen in the figure refer to the cycle length (347 msec). Panel A (top), transmembrane potential, Panel B (middle) shows the time course of changes in the internal calcium concentration, Ca_i (μM), as well as calcium concentration in the uptake and release compartments Ca_{up} (mM) and Ca_{rel} (mM), respectively. Panel C (bottom) shows a number of currents: I_m is the total transmembrane current in nA, I_{si} is the slow inward current, I_{NaCa} is the sodium-calcium exchanger current in nA, and I_K is the outward rectified delayed potassium current. The secondary hump seen in the slow current waveform is produced by the feature of incomplete calcium inactivation that has been included in the description of the slow inward channel. (Reproduced with permission from Murphey and Clark [43].)

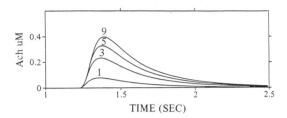

FIGURE 11.15. ACh concentration waveform seen at the surface of the SA node cell. ACh is assumed to be released from a random spatial distribution of parasympathetic varicosities near the cell. Consequently, we assume that ACh does not vary spatially either around or along the cylindrical cell, and thus varies only with time. (Reproduced with permission from Murphey and Clark [43].)

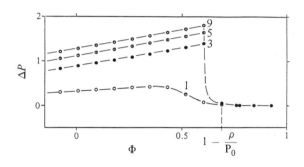

FIGURE 11.16. A family of model-generated phase reponse curves (PRCs) for the rabbit SA node cell in response to a single vagal stimulus burst. The family of PRCs is parameterized by the number of vagal ACh impulses delivered per stimulus burst (1,3,5,9). Here Φ is the normalized phase of stimulus delivery and ΔP is the normalized change in pacemaker cycle length defined according to Equation (11.3). (Reproduced with permission from Murphey and Clark [43].)

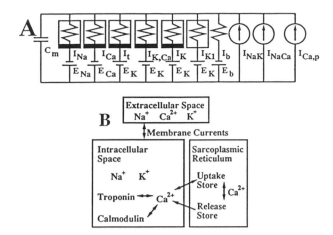

FIGURE 11.17. A mathematical model of the rabbit atrial myocyte: (A) electrical equivalent of the membrane and (B) fluid compartment model of the cell and extracellular space.

11.7 Rabbit Atrial Cell Model

We have also developed a mathematical model of the rabbit atrial myocyte [44]. Its general structure (Figure 11.17) is similar in many respects to our previous modeling work dealing with bullfrog atrial myocytes [53]. Structural differences between these models include the use of both T- and L-type calcium channels in the sarcolemma of the rabbit myocyte model as well as two transient outward potassium currents, I_t and $I_{K,Ca}$, where $I_{K,Ca}$ is a calcium-activated transient outward current. A mathematical description of the uptake and release of calcium by the SR is also included. As mentioned previously, histological investigations have revealed that the bullfrog atrial myocyte does not have a transverse tubular system and possesses very little sarcoplasmic or endoplasmic reticulum [26,27].

The rabbit atrial myocyte model (Figure 11.17) is also similar in structure to the DN model [17] for the Purkinje fiber. The structural differences between these models include: (1) the use of two calcium channels (T- and L-type) in the rabbit atrial myocyte model as opposed to one calcium channel in the DN model, (2) intracellular buffering by calmodulin and troponin are not considered in the DN model, but are included in the atrial cell model, and (3) our model uses a modified version of the SR uptake and release mechanism for calcium proposed by Hilgemann and Noble [24] rather than the SR model contained in DN [17]. In addition, this atrial cell model does not include a hyperpolarization-activated inward current I_f, since this type of current has not as yet been identified in rabbit atrial myocytes. One will also note that the DN model is based on experimental

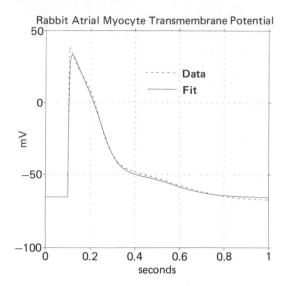

FIGURE 11.18. Comparison of model-generated and experimentally recorded rabbit atrial action potential waveforms. The data were recorded from a single atrial myocyte [30].

data from multicellular preparations. In contrast, the membrane channel descriptions used in our rabbit atrial cell model are based to a larger extent on quantitative voltage clamp data from the same species and cell-type. Figure 11.18 compares a model-generated action potential waveform using the atrial cell model with one that is experimentally recorded from a single rabbit atrial myocyte [30].

11.8   Modeling Nodal Regions

As stated earlier, the rabbit sinoatrial node region is a heterogeneous structure [3,34,35,39,45], with electrical activity originating within a central "compact" region where "dominant" or "primary" pacemaker cells generate action potentials with a very low upstroke velocity. An electrical wavefront proceeds outward from the compact central region first radially and then preferentially toward the upper part of the crista terminalis. Ultimately working fibers of the atrium as well as fibers of the specialized internodal pathways between the SA and AV nodes are activated. As it propagates, this waveform encounters a variety of cell types having different membrane properties (e.g., primary pacemaker, subsidiary pacemaker, Purkinje-like fibers of the internodal pathway, working atrial fibers). In addition, the resistive coupling properties between cells in the network are not uniformly distributed. Figure 11.19 is a crude attempt to visualize the problem of the

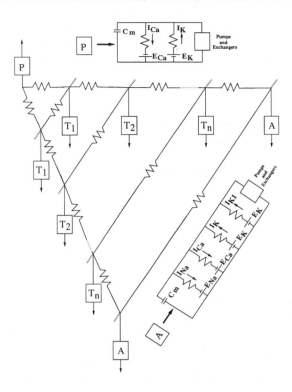

FIGURE 11.19. Schematic model of a distribution of resistively coupled cells in the region of the SA node. P = primary pacemaker cell; T_i = transitional cells, $i = 1, 2, \ldots$; A = atrial cell. See text for details.

spread of electrical activity in the nodal region from the dominant pacemaker cell region to the surrounding atrial musculature. The nodal region is considered a sheet, with the primary pacemaker region at its center. The network formed seeks to emphasize the nonuniform distribution of both coupling and membrane properties that should be included in any realistic model of conduction in the nodal region. A highly simplified spatial model would consist of a "concentric ring" model, wherein the variation in cellular properties would occur only in the radial direction in the circular sheet model. The cellular models shown in Figure 11.19 are lumped representations of cells in a given concentric ring of the sheet, or in a segment of that ring, if the cells in that ring vary in their membrane or coupling properties with angle θ. Clearly, the simplest configuration of membrane ion channels occurs in the primary pacemaker cell, and the most complex configurations occur in the Purkinje-like fibers of the internodal pathway, or in the cells of the atrium. Transitional cells of various types have complements of membrane channels of intermediate complexity. This nonuniform distribution of channels translates to: (1) a nonuniform distribution of input impedance offered by inactive cells to electrotonic current flow in the network, and (2) a nonuniform spatial distribution of currents contributed by active cardiac cells in the network at any instant of time, considered as active sources for excitatory local circuit current flow. The current strength that a given cell can contribute will largely depend on its membrane ionic channels, which in turn will influence its voltage range of operation as well as the upstroke velocity, peak height, and duration of its action potential.

An additional factor complicating the conduction of action potentials in the nodal region is the imposition of autonomic neural effects on the component cells of the network (in particular, the effects of the parasympathetic neurotransmitter ACh, which were demonstrated earlier on the bullfrog sinus venosus and atrial cells, as well as the rabbit SA node). All cells of the generalized network of Figure 11.19 are subject to the strong influence of ACh. Nonuniform distribution of ACh via parasympathetic (vagus) nerves can cause a distribution of refractory periods in the component cells of the network, which may lead to a variety of interesting effects, including the phenomenon of "pacemaker shift" and vagally induced block and delayed conduction, which may in turn serve as a mechanism of circus movement within the nodal region and its coupled surround (see Rosenshtraukh and colleagues [56]).

11.9   Summary

An equivalent network model capable of accurately representing the conduction of activity within the SA nodal region would be helpful in gaining further insight into several important and unresolved questions regarding the detailed electrophysiological behavior of nodal regions. The concentric ring network shown in Figure 11.19 is indeed highly simplified, yet it represents a reasonable starting point for investigation of nodal behavior provided that reasonably quantitative models can be developed for the component cells of the region. Inherent in this type of investigation are several factors that strongly influence conduction in a network of interconnected cells of different types, namely:

1. the distribution of resistive coupling values between cells

2. the distribution of cellular membrane properties in the network

3. the input impedance offered to current flow by the various cell types (e.g., pacemaker cells offer a relatively high input impedance while atrial cells offer a low input impedance); in general these cells also have different resting potentials

4. the distribution of autonomic innervation of the component cells of the network.

The network model of Figure 11.19 may be easily modified in both its geometry and boundary conditions to model more accurately the architecture and function of the SA node. However, to obtain meaningful results from this network, quantitative data of different types must be supplied, including geometrical data on the estimated size of the component regions of the "pacemaker or nodal region" and the locations of any anatomical barriers or boundaries. Furthermore, detailed voltage clamp and action potential data should be provided for the development of adequate models of each of the representative cell types of the nodal regions. The various cell models illustrated in this chapter indicate that such models are currently available, and thus consideration of the large-scale nodal network model problem outlined in this chapter will become progressively more feasible.

Acknowledgements: The authors gratefully acknowledge the help of Randall Rasmusson, Donald Campbell, Erwin Shibata, Robert Clark, and Y. Imaizumi. This work was supported in part by NSF Grants BNS8716568 and ECS8405435 awarded to Dr. Clark. Dr. Giles is supported as a Medical Scientist by the Alberta Heritage Foundation and receives ongoing support from the Canadian Medical Research Council and the Canadian Heart Foundation.

REFERENCES

[1] D. Atwell, D. Eisner, and I. Cohen. Voltage clamp and tracer flux data: Effects of a restricted extracellular space. *Q. Rev. Biophys.*, 12:213–263, 1979.

[2] G.W. Beeler and J.A.S McGuigan. Voltage clamping of multicellular myocardial preparations: Capabilities and limitations of existing methods. *Prog. Biophys. Mol. Biol.*, 34:219–254, 1978.

[3] W.K. Bleeker, A.J.C. Mackaay, M. Masson-Pévet, L.N. Bouman, and A.E. Becker. Functional and morphological organization of the rabbit SA node. *Circ. Res.*, 46:11–22, 1980.

[4] G.E. Breitwieser and G. Szabo. Mechanism of muscarinic receptor-induced K^+ channel activation as revealed by hydrolysis-resistant GTP analogues. *J. Gen. Physiol.*, 91:469–493, 1988.

[5] D.G. Bristow and J.W. Clark. A mathematical model of the primary pacemaking cell in the SA node of the heart. *Am. J. Physiol.*, 243:H207–H218, 1982.

[6] D.G. Bristow and J.W. Clark. A mathematical model of the vagally driven primary pacemaker. *Am. J. Physiol.*, 244:H150–H161, 1983.

[7] A.M. Brown, K.S. Lee, and T. Powell. Sodium current in single rate heart muscle cells. *J. Physiol.*, 318:479–500, 1981.

[8] G. Brown and J. Eccles. The action of a single vagal volley on the rhythm of the heart beat. *J. Physiol.*, 82:211–241, 1934.

[9] G. Brown and J. Eccles. Further experiments on vagal inhibition of the heart beat. *J. Physiol.*, 82:242–257, 1934.

[10] H.F. Brown. Electrophysiology of the sinoatrial node. *Physiol. Rev.*, 505–530, 1982.

[11] H.F. Brown, W. Giles, and S.J. Noble. Membrane currents underlying activity in frog sinus venosus. *J. Physiol.*, 271:783–816, 1977.

[12] H.F. Brown, J. Kimura, D. Noble, and S.J. Noble. The ionic currents underlying pacemaker activity in rabbit sino-atrial node: Experimental results and computer simulations. *Proc. Roy. Soc. B.*, 222:329–374, 1984.

[13] H.F. Brown, J. Kimura, D. Noble, and S.J. Noble. The slow inward current, i_{si}, in the rabbit sino-atrial node investigated by voltage clamp and computer simulation. *Proc. Roy. Soc. B.*, 222:305–328, 1984.

[14] D.L. Campbell, W.R. Giles, K. Robinson, and E.F. Shibata. Studies of the sodium-calcium exchanger in bull-frog atrial myocytes. *J. Physiol.*, 403:317–340, 1988.

[15] G. Champigny, P. Bois, and J. Lenfant. Characterization of the ionic mechanism responsible for the hyperpolarization-activated current in frog sinus venosus. *Pflügers Arch.*, 410:159–164, 1987.

[16] G. Champigny and J. Lenfant. Block and activation of the hyperpolarization-activated inward current by Ba and Cs in frog sinus venosus. *Pflügers Arch.*, 407:684–690, 1986.

[17] D. DiFrancesco and D. Noble. A model of cardiac electrical activity incorporating ionic pumps and concentration changes. *Phil. Trans. Roy. Soc. B.*, 222:353–398, 1985.

[18] E. Dong and B. Reitz. Effect of timing of vagal stimulation on heart rate in the dog. *Circ. Res.*, 27:635–646, 1970.

[19] M.H. Draper and S. Weidmann. Cardiac resting and action potentials recorded with intracellular electrodes. *J. Physiol.*, 115:74–94, 1951.

[20] W.R. Giles and E.F. Shibata. Voltage clamp of bull-frog cardiac pacemaker cells: A quantitative analysis of potassium currents. *J. Physiol.*, 368:265–292, 1985.

[21] E.C. Greco and J.W. Clark. A mathematical model of the vagally driven SA nodal pacemaker. *IEEE Trans. Biomed. Eng.*, BME–23: 192–199, 1976.

[22] O.P. Hamill, A. Marty, E. Neher, B. Sakmann, and F.J. Sigworth. Improved patch clamp techniques for high resolution current recording from cell and cell-free membrane patches. *Pflügers Arch.*, 391:85–100, 1981.

[23] H.C. Hartzell. Distribution of muscarinic acetylcholine receptors and presynaptic nerve terminals in the amphibian heart. *J. Cell Biol.*, 86:6–20, 1980.

[24] D.W. Hilgemann and D. Noble. Excitation-contraction coupling and extracellular calcium transients in rabbit atrium: Reconstruction of basic cellular mechanisms. *Proc. R. Soc. Lond. B.*, 230:163–205, 1987.

[25] A.L. Hodgkin and A.F. Huxley. A quantitative description of membrane current and its application to conduction and excitation in nerve. *J. Physiol.*, 117:500–544, 1952.

[26] J.R. Hume and W.R. Giles. Active and passive electrical properties of single bullfrog atrial cells. *J. Gen. Physiol.*, 78:19–42, 1981.

[27] J.R. Hume and W.R. Giles. Ionic currents in single isolated bullfrog atrial cells. *J. Gen. Physiol.*, 81:153–194, 1983.

[28] O.F. Hutter. Mode of action of autonomic transmitters on the heart. *Br. Med. Bull.*, 13:176–180, 1957.

[29] O.F. Hutter and W. Trautwein. Vagal and sympathetic effects on the pacemaking fibers in the sinus venosus of the heart. *J. Gen. Physiol.*, 39:715–733, 1956.

[30] Y. Imaizumi and W.R. Giles. Comparison of potassium currents in rabbit atrial and ventricular cells. *J. Physiol.*, 405:123–145, 1989.

[31] H. Irisawa. Comparative physiology of the cardiac pacemaker mechanism. *Physiol. Rev.*, 58:461–487, 1978.

[32] E.A. Johnson and M. Lieberman. Heart: Excitation and contraction. *Ann. Rev. Physiol.*, 33:417–532, 1971.

[33] M. Kameyama, F. Hofmann, and W. Trautwein. On the mechanism of β-adrenergic regulation of the Ca channel in the guinea-pig heart. *Pflügers Arch.*, 405:285–293, 1985.

[34] I. Kodama and M.R. Boyett. Regional differences in the electrical activity of the rabbit sinus node. *Pflügers Arch.*, 404:214–226, 1985.

[35] D. Kreitner. Electrophysiological study of two main pacemaker mechanisms in the rabbit sinus node. *Cardiovasc. Res.*, 19:304–318, 1985.

[36] L. Lapidus and J. Senfeld. *Numerical Solution of Ordinary Differential Equations*. Academic Press, New York, 1971.

[37] K.S. Lee, T.A. Week, R.L. Kao, N.A. Eaikee, and A.M. Brown. Sodium current in single heart muscle cells. *Nature*, 278:269–271, 1979.

[38] M.N. Levy, P.J. Martin, T. Iano, and H. Zieske. Paradoxical effect of vagus nerve stimulation on heart rate in dogs. *Circ. Res.*, 25:303–314, 1969.

[39] M. Masson-Pèvet, W.K Bleeker, L.N. Besselsen, B.W. Treytel, H.J. Jongsma, and L.N. Bouman. Pacemaker cell types in the rabbit sinus node: A correlative ultrastructural and electrophysiological study. *J. Mol. Cell. Cardiol.*, 16:53–63, 1984.

[40] R.E. McAllister, D. Noble, and R.W. Tsien. Reconstruction of the electrical activity of cardiac Purkinje fibers. *J. Physiol.*, 251:1–59, 1975.

[41] D.C. Michaels, E.P. Matyas, and J. Jalife. A mathematical model of the effects of acetylcholine pulses on sinoatrial pacemaker activity. *Circ. Res.*, 55:89–101, 1984.

[42] D.C. Michaels, V.A.J. Slenter, J.J. Salata, and J. Jalife. A model of dynamic vagus-sinoatrial node interactions. *Am. J. Physiol.*, 245: H1043–H1053, 1983.

[43] C.R. Murphey and J.W. Clark. Parasympathetic control of the SA node cell in rabbit heart: A model. In S. Sideman and R. Beyar, editors, *Activation, Metabolism and Perfusion of the Heart*, pages 41–59. Martinus Nijhoff, Boston, 1987.

[44] C.R. Murphey, J.W. Clark, W.R. Giles, Y. Imaizumi, and G.V. Naccarelli. A mathematical model of the rabbit atrial myocyte. (in preparation.)

[45] R.D. Nathan. Two electrophysiologically distinct types of cultured pacemaker cells from rabbit sinoatrial node. *Am. J. Physiol.*, 250: H325–329, 1986.

[46] E. Neher and B. Sakmann. Single channel currents recorded from membrane of denervated frog muscle fibers. *Nature*, 260:799–802, 1976.

[47] D. Noble. A modification of the Hodgkin-Huxley equations applicable to Purkinje fiber action and pacemaker potentials. *J. Physiol.*, 251:1–59, 1962.

[48] D. Noble and S. Noble. A model of the sinoatrial node electrical activity based on a modification of the DiFrancesco-Noble (1984) equations. *Proc. Roy. Soc. B.*, 222:295–304, 1984.

[49] W. Osterrieder, A. Noma, and W. Trautwein. On the kinetics of the potassium channel activated by acetylcholine in the SA node of the rabbit heart. *Pflügers Arch.*, 386:101–109, 1980.

[50] T. Powell, D.A. Terrar, and V.W. Twist. Electrical properties of individual cells isolated from adult rat ventricular myocardium. *J. Physiol.*, 302:131–153, 1980.

[51] T. Powell and V.W. Twist. A rapid technique for the isolation and purification of adult cardiac muscle cells having respiratory control and tolerance to calcium. *Biochem. Biophys. Res. Commun.*, 72:327–333, 1976.

[52] R. Rasmusson, J.W. Clark, W.R. Giles, E.F. Shibata, and D.L. Campbell. A mathematical model of a bullfrog cardiac pacemaker cell. *Am. J. Physiol.* 259:H352–H369, 1990.

[53] R. Rasmusson, J.W. Clark, W.R. Giles, et al. A mathematical model of electrophysiological activity in a bullfrog atrial cell. *Am. J. Physiol.* 259:H370–H389, 1990.

[54] H. Reuter. Calcium channel modulation by neurotransmitters, enzymes and drugs. *Nature*, 301:569–574, 1983.

[55] S. Robertson, D. Johnson, and J. Potter. The time course of Ca^{2+} exchange with calmodulin, troponin, parvalbumin, and myosin in response to transient increases in Ca^{2+}. *Biophys. J.*, 34:559–569, 1981.

[56] L.V. Rosenshtraukh, A.V. Zaitsev, V.G. Fast, A.M. Pertsov, and V.I. Krinsky. Vagally induced block and delayed conduction as a mechanism for circus movement tachychardia in frog atrial. *Circ. Res.*, 64:213–226, 1989.

[57] E.F. Shibata and W.R. Giles. Ionic currents that generate the spontaneous diastolic depolarization in individual cardiac pacemaker cells. *Proc. Natl. Acad. Sci.*, 82:7796–7800, 1985.

[58] J.M. Shumaker, J.W. Clark, and W.R. Giles. A model of the phase sensitivity of the pacemaker cell in the bullfrog heart. *J. Theoretical Biol.* (In press.)

[59] J.M. Shumaker, J.W. Clark, W.R. Giles, and G. Szabo. A model of the muscarinic receptor-induced changes in K^+-current and action potentials in the bullfrog atrial cell. *Biophys. J.*, 57:567–576, 1990.

[60] M.A. Simmons and H.C. Hartzell. A quantitative analysis of the acetylcholine-activated potassium current in single cells from frog atrium. *Pflügers Arch.*, 409:454–461, 1987.

[61] J.F. Spear, K.D. Kronhaus, E.N. Moore, and R.P. Kline. The effect of brief vagal stimulation on the isolated rabbit sinus node. *Circ. Res*, 44:75–88, 1979.

[62] W. Trautwein and M. Kameyama. Intracellular control of calcium and potassium currents in cardiac cells. *Jpn. Heart J.*, 27 Supp.:31–50, 1986.

[63] W. Trautwein and K. Zink. Über Membran-und Aktionspotentiale einzelner Myokardfasern des Kalt-und Warmblüterherzens. *Pflügers Arch.*, 256:68–84, 1952.

[64] K. Yanagihara, A. Noma, and H. Irisawa. Reconstruction of sinoatrial node pacemaker potential based on the voltage clamp experiments. *Jpn. J. Physiol.*, 30:841–857, 1980.

12

Low-Dimensional Dynamics in the Heart

Leon Glass[1]
Alvin Shrier[1]

ABSTRACT Realistic mathematical models in cardiac electrophysiology are normally expressed as complex ordinary or partial nonlinear differential equations. Yet, in a number of circumstances, simple one- (or low-) dimensional finite difference equations can be used to model the dynamics. In this chapter we consider three different circumstances: (1) the periodic forcing of spontaneously oscillating cardiac tissue, (2) stimulation of spontaneously oscillating cardiac tissue at a fixed delay after an action potential, and (3) periodic stimulation of excitable, but not spontaneously oscillating cardiac tissue. In all three circumstances, simple one-dimensional finite difference equations show qualitative similarities with experimental observations.

12.1 Introduction

The electrophysiological properties of the heart are complex. Realistic ionic models of cardiac electrical activity [7,10,39,50] (see Chapter 10 by Guevara and Chapter 11 by Clark and colleagues in this volume) are usually so complicated that the only way we can find out about their properties is by numerical simulation.

This chapter reviews the foundation for the development of theoretical models of the electrical activity of the heart based on a class of theoretical models that are one- (or low-) dimensional finite difference equations. Such a class of theoretical models is suitable to bridge the gap between realistic ionic models of action potential generation and propagation and purely phenomenological characterizations of cardiac activity that can be readily measured experimentally. Since the dynamics that can be found in low-dimensional finite difference equations have been intensively studied mathematically, it is sometimes possible to obtain theoretical insight into complex rhythms observed clinically or in the laboratory.

The use of one-dimensional finite difference equations in cardiac electrophysiology has evolved from distinguished but largely forgotten roots.

---

[1]Department of Physiology, McGill University, Montréal, Québec H3G 1Y6

Mobitz [42], one of the pioneers in cardiac electrophysiology, proposed an iterative scheme that can be written as a one-dimensional finite difference equation to study electrical propagation in the heart (see Chapter 13 in this volume). Many authors have independently developed iterative approaches to make theoretical predictions of experimentally recorded cardiac rhythms [8,36,43,44,51], but the underlying mathematical structure of these models was not studied directly.

This chapter provides an introduction to the use of one-dimensional finite difference equations in cardiac electrophysiology. Section 12.2 provides a summary of mathematical concepts and terminology that are most relevant. Section 12.3 discusses the properties of a simple mathematical model of oscillations called the Poincaré oscillator. This theoretical model provides a conceptual framework for understanding the effects of periodic stimulation of oscillators (Section 12.4) and stimulation of oscillators at a fixed delay after a marker event of the oscillation (Section 12.5). In Section 12.6 we discuss the periodic stimulation of excitable but not spontaneously oscillating cardiac tissue. Applications and limitations of the theory are discussed in Section 12.7.

12.2   Basic Concepts in Nonlinear Dynamics

This section contains a summary of well known material concerning basic concepts in nonlinear dynamics [1,9,21].

A *one-dimensional finite difference equation* is an equation of the form

$$x_{t+1} = f(x_t) \tag{12.1}$$

where x_t represents a variable at time t and f is a function that describes the evolution of the system. If $f(x_t) = \alpha x_t + \beta$ where α and β are constants then f is a linear function. Any other functional form for f is a nonlinear function and the equation is then a nonlinear finite difference equation. Such equations admit a number of qualitatively different dynamic behaviors. A *steady state* is a stationary solution for which $x_{t+1} = x_t$, and a *cycle* of period n is defined by $x_{t+n} = x_t$. When f is nonlinear, there are two other types of dynamic behavior possible in Equation (12.1). *Chaos* is aperiodic dynamics in which two initial conditions that are initially close together diverge as time proceeds. *Quasiperiodicity* is aperiodic dynamics in which two initial conditions that are close together remain close together. Quasiperiodicity is possible when f is a circle map $f:S^1 \rightarrow S^1$ (this means that each point on the circumference of a circle is mapped to a second point determined by the function, f).

A *bifurcation* is a change in the qualitative properties of dynamics as parameters vary. Examples of bifurcations include changes in which a steady state is destabilized leading to periodicity, periodic behavior changes from one period to a different period, or periodic behavior changes to nonperiodic behavior. Transitions from periodic to aperiodic dynamics often follow some recognized sequence of bifurcations [13]. For example, there can be *period-doubling bifurcations* in which the period of an oscillation doubles repeatedly as a parameter is varied, eventually leading to chaos.

A quantitative measure, the Liapunov exponent, Λ, can be used to distinguish periodicity, quasiperiodicity, and chaos [54]. The definition for the Liapunov exponent for one-dimensional maps is

$$\Lambda = \lim_{N \to \infty} \frac{1}{N} \sum_{i=1}^{N} \ln | f'(\phi_i) |, \qquad (12.2)$$

where $f'(\phi_i)$ is the slope of the function, for example, as in Equation (12.2) evaluated at subsequent iterates. Λ is negative for periodic cycles, zero for quasiperiodicity, and positive for chaos. A positive Liapunov number is often taken as a definition for chaos in one-dimensional maps and other systems of equations.

The field of nonlinear dynamics suggests an approach to study complex rhythms in cardiac physiology—to manipulate experimental stimulation parameters and to determine complex bifurcations resulting from this procedure. It is then possible to compare the experimentally observed bifurcations with bifurcations occurring in mathematical models of the physiological system. We search for experimental systems and theoretical models in which the experimental results are well approximated by comparatively simple low-dimensional finite difference equations, and in this fashion obtain theoretical insight into complex rhythms.

12.3   A Topological Model of Cardiac Oscillators

Realistic mathematical models of cardiac oscillations are extraordinarily complex high-dimensional ordinary differential equations [7,10,39,50]. However, they share the common property that they display periodic solutions that are attracting as time approaches infinity for initial conditions close to the cycle. Attracting cycles of this sort were first described by Poincaré and are called *stable limit cycle oscillations.*

A simple theoretical model of limit cycle oscillations can be used to illustrate many of the theoretical properties of limit cycles. Since the model is similar to equations originally proposed by Poincaré as a model for limit cycles, we have called the model the Poincaré oscillator [18]. The Poincaré oscillator has uncanny similarities to actual experimental data and it has been used many times as a simplified theoretical model for limit cycles [11,24,30,34,49,53].

In the Poincaré oscillator there are two variables, r and ϕ. Starting at any value of r, except $r = 0$, there is an evolution until $r = 1$. ϕ increases at a constant rate. The equations are written

$$\frac{dr}{dt} = kr(1-r), \quad \frac{d\phi}{dt} = 2\pi, \tag{12.3}$$

where k is a positive constant that regulates the rate at which the value of r approaches 1. Thus, there is a stable limit cycle oscillation at $r = 1$ that is globally attracting in the limit $t \to \infty$ for all initial conditions except the origin.

To make the connection with an experimental system we assume the period of the cycle is designated by T_0. One point of the cycle, corresponding to the upstroke of the action potential, is arbitrarily designated as zero phase, $\phi = 0$. The phase of other points on the cycle is given by $\phi = t/T_0$ (mod 1) where t measures the time elapsed since the trajectory has passed $\phi = 0$. In this notation, phase ranges from 0 to 1.

We represent the effects of electrical stimulation of the Poincaré oscillator by a horizontal translation by an amount b. This shifts from one phase, ϕ, to a new one, $g(\phi)$, as shown in Figure 12.1A. For the moment we assume there is an instantaneous relaxation back to the limit cycle along a radius of the circle. A single perturbation delivered to the cycle at phase ϕ results in a perturbed cycle length, $T(\phi)$, given by

$$\frac{T(\phi)}{T_0} = \phi + 1 - g(\phi) \tag{12.4}$$

where $g(\phi)$ is readily calculated and is given by

$$\cos(2\pi g(\phi, b)) = \frac{b + \cos 2\pi\phi}{(1 + 2b\cos 2\pi\phi + b^2)^{1/2}}. \tag{12.5}$$

The function $g(\phi)$ is called the phase transition curve (PTC). As discussed below, in experimental systems Equation (12.4) can be used to determine the PTC from the measured value of $T(\phi)/T_0$.

For the Poincaré oscillator, the perturbed cycle length and the PTC have characteristic forms depending on the strength of the perturbing stimulus [18,53]. At low stimulus amplitude the average slope of the PTC is 1 and this is called "weak" (type 1) phase resetting (Figure 12.1B). At high stimulus amplitude the average slope of the PTC is 0 and this is called "strong" (type 0) phase resetting (Figure 12.1C). For weak phase resetting both the perturbed cycle length and the PTC are continuous functions of ϕ. For strong phase resetting the perturbed cycle length is a discontinuous function of ϕ with a discontinuity of exactly one cycle at $\phi = 0.5$. However, the PTC is a continuous function in Figure 12.1C ($\phi = 1$ and $\phi = 0$ are identified with the same point).

There are qualitative similarities between the phase-resetting properties of the Poincaré oscillator and phase resetting in experimental systems. As

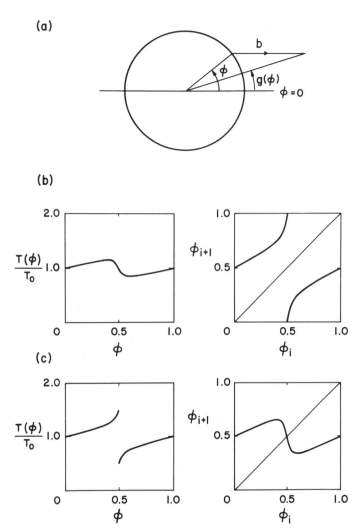

FIGURE 12.1. Schematic picture of the Poincaré oscillator. The circle represents a stable limit cycle oscillation that we assume is rapidly approached following a perturbation. (A) A horizontal perturbation with amplitude b delivered at phase ϕ gives a new phase $g(\phi)$. (B) Perturbed cycle length and phase resetting curve for the Poincaré oscillator for $b = 0.9$ corresponding to weak resetting. For convenience we plot $\phi_{i+1} = g(\phi_i) + 0.5$, see Equation (12.4). The average slope of PTC curve is equal to 1. (C) The same as in (B) except for strong resetting with the stimulus amplitude $b = 1.1$ (Reproduced with permission from Zeng et al. [56].)

an illustration, we show phase resetting in spontaneously beating aggregates of embryonic chick atrial heart cells. Figure 12.2A shows an experimental trace of transmembrane voltage as a function of time recorded with an intracellular microelectrode. A 20-msec current pulse is delivered via the same microelectrode giving rise to a large artifact at a time interval δ after the start of an action potential. The perturbed cycle length, T, is directly measured and this is used to determine the PTC using Equation (12.4). This preparation exhibits both type 1 (Figure 12.2B) and type 0 (Figure 12.2C) phase resetting.

12.4   Periodic Stimulation of Limit Cycle Oscillators

In the limit of rapid relaxation to the limit cycle following a stimulus, the effects of periodic stimulation of limit cycles can be represented by circle maps [11,12,15–17,22,24,26,28,30,31,34,45,47,55,56]. This represents an important but necessarily approximate approach to study the periodic forcing of nonlinear oscillators. We once again illustrate the basic ideas with the Poincaré oscillator, but the formulae are applicable to other limit cycle oscillators provided there is a rapid relaxation to the limit cycle.

Consider the periodic stimulation of the oscillator with a periodic train of stimuli with period t_s. Call ϕ_i the phase of the ith stimulus of a periodic train and $\tau = t_s/T_0$. Then we have

$$\phi_{i+1} = F(\phi_i, \tau) \pmod 1, \tag{12.6}$$

$$F(\phi_i, \tau) = g(\phi_i) + \tau. \tag{12.7}$$

Combining Equations (12.4), (12.6) and (12.7) we obtain

$$\phi_{i+1} = 1 + \phi_i - \frac{T(\phi_i)}{T_0} + \tau \pmod 1. \tag{12.8}$$

Equation (12.8) is the key equation for theoretical analysis.

Periodic cycles in Equation (12.8) corresponds to *phase locking* between the stimulus and the oscillator. For a periodic cycle of period N, we say there is $N{:}M$ phase locking where M counts the number of times the oscillation crosses $\phi = 0$. In the experiments, for $N{:}M$ phase locking there are N stimuli and M cardiac action potentials in each repeating cycle.

A useful quantitative measure to characterize the response of oscillations to periodic inputs is the rotation number [1,2], ρ, which is defined by

$$\rho = \lim_{N \to \infty} \frac{1}{N} \sum_{i=1}^{N} \Delta\phi_i, \tag{12.9}$$

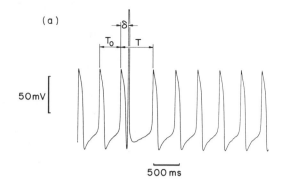

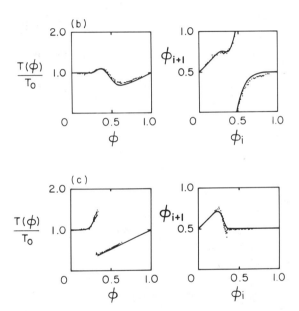

FIGURE 12.2. (A) An experimental trace showing the transmembrane potential during a stimulus and the effect of injecting a current pulse. T_0 is the basic cycle length of the preparation, T is the perturbed cycle length following a stimulus of 20-msec duration delivered at δ after the upstroke of an action potential. (B) Type 1. (C) Type 0. The right hand panels show $\phi_{i+1} = \phi_i + 0.5$. (Adapted with permission from Zeng et al. [56].)

where

$$\Delta \phi_i = F(\phi_i) - \phi_i. \tag{12.10}$$

The rotation number is rational for cycles and irrational for quasiperiodicity. The rotation number may be rational, irrational, or undefined for chaos.

If Equation (12.8) is a 1:1 invertible map (this means that for each ϕ_i there is a unique ϕ_{i+1}, and for each ϕ_{i+1} there is a unique ϕ_i), the geometrical arrangement of phase locking zones is well understood [1]. As the stimulus frequency decreases one observes $N{:}M$ phase locking where the rotation number $\rho = M/N$ increases. Theoretically, if there is $N{:}M$ phase locking at one frequency and $N'{:}M'$ at another, then over an intermediate range of frequencies there is $(N+N'){:}(M+M')$ phase locking. The rotation number is a continuous function of stimulation frequency so that as stimulation frequency varies, zones of stable phase-locked dynamics with a fixed rotation number are separated by quasiperiodic dynamics. 1:1 invertible maps are found in Equation (12.4) at low stimulation amplitudes. As the stimulation amplitude decreases to zero, the zones of stable phase locking in frequency amplitude parameter space have a "tonguelike" geometry and are often referred to as Arnold tongues [1,2,18,32]. At higher stimulation amplitudes, the functions in Equation (12.8) become noninvertible and the geometrical arrangement of phase-locking zones is much more complex [11,12,14,20,24,30,34,56].

Periodic stimulation of spontaneously beating cardiac preparations shows a large number of different regular and irregular rhythms. In Figure 12.3 we illustrate several of the regular rhythms observed during periodic stimulation of spontaneously beating aggregates of atrial heart cells. In the traces, the current pulse stimuli appear as brief, spikelike, high amplitude events, and the action potentials are broader events of lower amplitude.

A convenient way to compare the properties of different model and experimental systems subjected to periodic input is to study the plots showing the dynamics as a function of stimulus amplitude and period. This is done in Figure 12.4 for three different situations: (A) the periodically stimulated Poincaré oscillator [34], (B) theoretical computation of the entrainment in periodically stimulated atrial heart cell aggregates based on experimentally measured phase-resetting curves [56], and (C) experimental measurements of dynamics in periodically stimulated ventricular heart cell aggregates [22,27]. All three situations show common features of Arnold tongues at low stimulus amplitudes, prominent zones corresponding to entrainment in simple ratios, complex and poorly understood bifurcations at intermediate stimulus amplitudes, and somewhat simplified dynamics at the highest stimulation amplitudes.

One of the prominent features in all three cases is the appearance of aperiodic dynamics over limited ranges of stimulation parameters. These ranges are shown as stippling (corresponding to a positive Liapunov num-

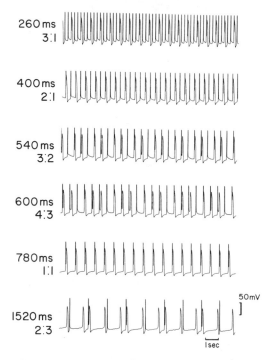

FIGURE 12.3. Regular rhythms corresponding to stable phase locking in embryonic atrial chick heart cell aggregates. 3:1 phase locking (stimulus period $t_s = 260$ ms); 2:1 phase locking ($t_s = 400$ ms); 3:2 phase locking ($t_s = 540$ ms); 4:3 phase locking ($t_s = 600$ ms); 1:1 phase locking ($t_s = 780$ ms); 2:3 phase locking ($t_s = 1520$ ms). All rhythms were obtained by stimulating for a long enough time to allow transients to pass and a steady-state rhythm to be established. (Reproduced with permission from Zeng et al. [56].)

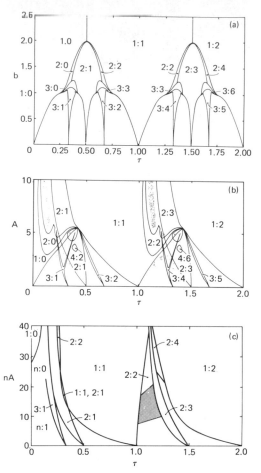

FIGURE 12.4. Phase-locking zones in stimulation amplitude–stimulation period parameter space for three different situations. (A) The Poincaré oscillator. From translational symmetry, if there is $N:M$ phase locking at τ, then there is $N:(N+M)$ phase locking at $\tau+1$. At low stimulus amplitude ($b < 1$), the phase-locking zones are well ordered and show a typical Arnold tongue structure (see text). For $b > 1$, there are complex bifurcations and bistability. (With permission from Guevara et al. [27].) (B) Computation of phase-locking zones using experimentally measured PTCs for spontaneously beating atrial aggregates. For $A < 1.5$ there is an Arnold tongue structure. For $1.5 < A < 5$ the bifurcations are typical of noninvertible type 1 circle maps. For $5 < A < 6$ the map is discontinuous and there is a complicated ordering of bifurcation zones in which some zones disappear as stimulus amplitude increases. For $A > 6$ the bifurcations are simpler, but show period doubling and chaotic dynamics over limited parameter values. The chaotic regions with positive Liapunov exponent are shown by stippling in the figure (With permission from Zeng et al. [56].) (C) Schematic composite of experimental data showing the results of periodic stimulation of spontaneously beating ventricular heart cell aggregates. Shaded area shows region of period doubling bifurcation and irregular dynamics. Unstable rhythms can also be observed over other narrow regions of parameter space. (Adapted, from Guevara [22] and Guevara et al. [27].)

ber) in Figure 12.4B, and as a shaded region in Figure 12.4C. Let us concentrate on the period-doubling bifurcations and chaotic dynamics observed at stimulation frequencies somewhat slower that the intrinsic frequency of the aggregate [26]. An experimental trace of an irregular rhythm is shown in Figure 12.5A. A plot of ϕ_{i+1} vs. ϕ_i for the rhythm strip from which the irregular rhythm in Figure 12.5A is extracted is shown in Figure 12.5B. Finally we show a plot of ϕ_{i+1} vs. ϕ_i for the Poincaré oscillator at parameter values that give chaotic dynamics ($b = 1.10$, $\tau = 1.38$) in Figure 12.5C. There are qualitative similarities in Figures 12.5B and 12.5C, reflecting the similarities in the phase-resetting curves.

One way to display the effects of parameter changes in this system is to plot successive values of ϕ_i after transients have died away for several different values of the control parameter, here either b or τ. Such a plot is called a bifurcation diagram. Figure 12.6A shows a bifurcation diagram for $\tau = 1.32$ for the Poincaré oscillator. For every value of b, starting at an initial condition of $\phi_0 = 0.9$, 100 points are generated following a transient of 350 iterations.

The Liapunov number is plotted in Figure 12.6B for the same range of stimulation amplitudes shown in Figure 12.6B. The positive values of Λ demonstrate chaos in the periodically forced Poincaré oscillator.

In summary, experimental and theoretical work have demonstrated that there are extremely rich dynamics experimentally observed as the stimulation frequency and amplitude of periodic forcing of cardiac oscillators are varied. An understandng of the origin of many of these complex rhythms can be derived from a consideration of one-dimensional finite differences equations modeling the experimental systems. However, working out the fine details of the bifurcations is a difficult task, both from an experimental and theoretical perspective.

12.5   Stimulation of the Poincaré Oscillator at a Fixed Delay after an Action Potential

In the preceding two sections we assumed that following a stimulus, the limit cycle was immediately reestablished. In this case, all stimuli delivered at a fixed delay after an action potential have exactly the same effect. However, experimental studies in both cardiac [35] and respiratory [37] physiology show that in some situations stimuli delivered at a fixed delay after some marker event in an oscillating system can have different effects. This means that the history of the previous stimulation is in some way affecting the properties of the oscillator and that the assumption that there is instantaneous reestablishment of the oscillation is not always valid. Theoretical models of stimulation of oscillators at fixed delay have been considered previously [37,40,41]. We follow the development by Lewis and colleagues [37]

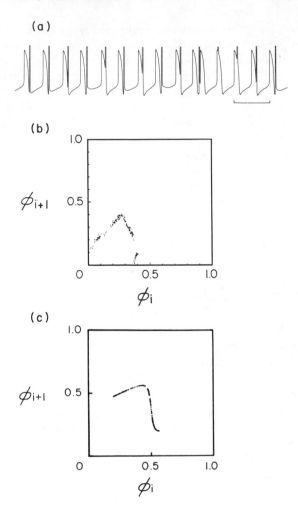

FIGURE 12.5. (A) Irregular dynamics in an aggregate of periodically stimulated ventricular heart cells. Horizontal scale represents 1 sec. (B) Plot of ϕ_{i+1} *vs.* ϕ_i derived from the data from which trace A is derived. (C) Iteration of Equation (12.8) for the Poincaré oscillator with $b = 1.05$ and $\tau = 1.38$. (Panels A and B are adapted with permission from Glass et al. [14] and panel C from Glass and Zeng [19].)

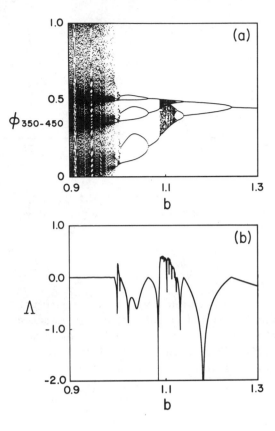

FIGURE 12.6. (A) Bifurcation diagram for the periodically forced Poincaré oscillator with $\tau = 1.32$. (B) The Liapunov exponent corresponding to the dynamics shown in panel A. (Reproduced with permission from Glass and Zeng [19].)

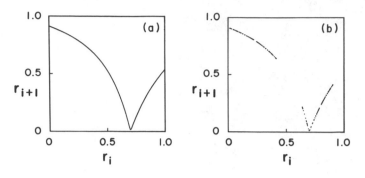

FIGURE 12.7. (A) Plot of Equation (12.13) with $\phi_0 = 0.499, b = 0.7, k = 1$. (B) Iterations for the same parameters in A. (Reproduced with permission from Glass and Zeng [19].)

and Glass and Zeng [19].

In the Poincaré oscillator the simplest way to incorporate the effects of past stimulation history is to assume that k in Equation (12.3) is finite. We consider the effects of repeated stimuli delivered at a phase ϕ_0 of the oscillation. The coordinates of the system when the ith stimulus is delivered are (r_i, ϕ_0), and this induces an immediate displacement to (r_i', ϕ'), where

$$r_i' = (r_i^2 + b^2 + 2r_i b \cos 2\pi\phi_0)^{1/2}, \qquad (12.11)$$

$$\cos 2\pi\phi' = \frac{r_i \cos 2\pi\phi_0 + b}{r_i'}. \qquad (12.12)$$

The next firing will occur after a time interval $t_i = 1 - \phi' + \phi_0$. By integrating Equation (12.3), we now compute the value r_{i+1} at the next stimulation

$$r_{i+1} = \frac{r_i'}{(1 - r_i')\exp(-kt_i) + r_i'}. \qquad (12.13)$$

A graph of this function is shown in Figure 12.7A for $\phi_0 = 0.499$, $b = 0.7$, $k = 1$. Iterating this function starting from an initial condition of $r_0 = 0.4$ is shown in Figure 12.7B. Once again, there are aperiodic, chaotic dynamics. The bifurcation diagram and the dependence of the Liapunov number as ϕ_0 varies are shown in Figures 12.8A and 12.8B, respectively.

In experimental work currently underway at McGill University, analysis is being carried out of the effects of stimulation at a fixed delay after an action potential of the heart cell aggregates. It has been observed (Zeng, Morissette, Brochu, Shrier, and Glass, unpublished) that stimuli delivered at a fixed delay after the action potential in the heart cell aggregates do not always have the same effect, and that complex rhythms between the stimulator and the intrinsic rhythm are observed. However, the preliminary

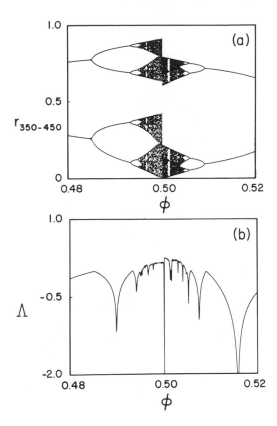

FIGURE 12.8. (A) Bifurcation diagram for Equation (12.13). (B) The Liapunov number for Equation (12.13) for the same parameters in A. (Reproduced with permission from Glass and Zeng [19].)

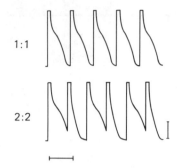

FIGURE 12.9. Intracellular recording of transmembrane potential from a periodically stimulated quiescent heart cell aggregate. Upper trace, $t_s = 300$ msec; lower trace, $t_s = 180$ msec. (Vertical calibration 50 mV; horizontal calibration 300 msec). (Reproduced with permission from Guevara et al. [29].)

analyses of the data indicate that the mechanism underlying the establishment of the complex rhythms differs somewhat from the behavior in the Poincaré oscillator with finite relaxation time to the limit cycle oscillation. In the heart cell aggregates the stimulation changes the periodicity of the beating. For example, following the termination of rapid stimulation there is a slower beating rate. This phenomenon is called overdrive suppression [46,52]. Such effects are not incorporated in the Poincaré oscillator and it will be necessary to extend the theory presented above to account for the actual mechanisms underlying the complex oscillations.

12.6 Periodic Stimulation of Excitable, Nonoscillating Cardiac Tissue

As a final example of the use of one-dimensional maps in cardiac electrophysiology, we consider the effects of periodic stimulation of nonoscillating cardiac tissue. Some years ago, a report from our group [29] described the appearance of an alternans rhythm as the stimulation frequency of a nonoscillating aggregate of chick heart cells was increased (Figure 12.9). At a stimulation period of 200 msec all action potentials had the same duration, but when stimulation period was reduced to 180 msec, the duration of the action potential alternates on subsequent stimuli. An early theoretical study provides the germ of the mechanism [44] but does not develop the mathematics.

It is well known [3] that the duration of an action potential depends on the recovery time elapsed from the end of a preceding action potential. The shorter the recovery time the shorter the action potential. However, if the

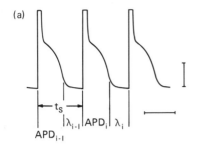

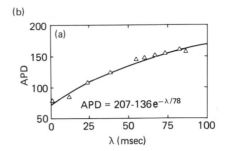

FIGURE 12.10. (A) Diagram showing definition of terms. (B) Electrical restitution curve showing the action potential duration (APD) as a function of the
recovery time (λ) derived from traces such as those shown in Figure 12.9. (Reproduced with permission from Guevara et al. [29], ©1984 IEEE.)

recovery time is too short there is no action potential.

A schematic picture of the situation is shown in Figure 12.10A. The
stimulation period is t_s, the duration of the ith action potential is APD_i,
and the recovery time before the ith action potential is λ_{i-1}. By measuring
the action potential duration and the recovery time at different stimulation frequencies it is possible to measure the restitution curve $g(\lambda)$ (Figure
12.10B), which can be approximated by

$$g(\lambda) = APD_{max} - \alpha \exp(-\lambda/\gamma), \quad \lambda > \theta \qquad (12.14)$$

where APD_{max} is the maximum action potential duration, α is a positive
constant, γ is a time constant and θ is the refractory period (the time interval following the upstroke of the action potential during which stimuli do
not elicit an action potential) of the aggregate. A finite difference equation
can be formulated based on Figure 12.10A and the electrical restitution
curve

$$APD_{i+1} = g(Nt_s - APD_i) \qquad (12.15)$$

where N is the smallest integer such that $Nt_s - APD_i > \theta$. Destabilization
of the steady state in Equation (12.15) can occur in such a way that an
alternans rhythm, such as the one shown in Figure 12.9 arises. In Figure
12.11B we show the bifurcation diagram for the restitution curve shown
in Figure 12.10B. The data points are shown as triangles. Although both
the theory and the experiment show similarities in that there is a period-
doubling bifurcation, the quantitative details do not agree. A possible reason for this is that there are time-dependent changes in the ionic properties
of the aggregate, just as there are changes in the electrical properties of
spontaneously beating aggregates due to the stimulation.

Since ventricular myocytes do not normally spontaneously oscillate, it
is important to know the effects of periodic stimulation of nonoscillating

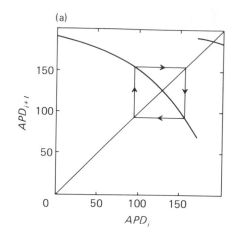

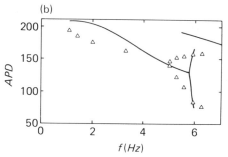

FIGURE 12.11. (A) Graphical representation of Equation (12.15) for $t_s = 170$ msec with the restitution curve in Figure 12.10B. There is a stable 2:2 rhythm and a stable steady-state corresponding to a 2:1 rhythm. (B) Bifurcation diagram showing APD as a function of stimulation frequency, f. The solid line shows theoretical results computed using the restitution curve in Figure 12.10B. Triangles give data points. (Reproduced with permission from Guevara et al. [29], ©1984 IEEE.)

cardiac tissue at rapid rates. Several studies have confirmed that there can be alternans rhythms and have also described other more complex responses [4–6,23,48]. Some progress has also been made in extending the simple finite difference equation here to take into account additional factors such as supernormality (in which a very early stimulus that falls in the refractory period and therefore would be expected to be blocked is actually conducted) and conduction effects [6].

In recent work, chaos has been discovered in Equation (12.15) [33,38] and modifications of Equation (12.15) [6]. However, the details of the bifurcations in this equation are not yet understood.

12.7   Applications and Limitations

The preceding sections have shown three different situations in which one-dimensional maps can be derived and used to describe the effects of electrical stimulation in cardiac tissue and in simple theoretical models. This work provides a foundation for further theoretical and experimental studies of complex dynamics in cardiac electrophysiology.

There are many different cardiac preparations ranging from single cells or aggregates in tissue culture, sucrose gap preparations with Purkinje fiber, in vitro preparations of the sinoatrial node or atrioventricular node, perfused heart, and the intact beating heart in vivo. Each one of these preparations can be subjected to stimulation patterns similar to those described here. On the basis of similar studies carried out in a variety of different preparations studied to date, it is clear that some of the main features described here will likewise be found in other systems. Thus, we anticipate that entrainment between a stimulator and the cardiac preparation will be readily observed resulting in rhythms in which there are simple whole number ratios between the stimulator and the cardiac action potential generation. Chaos should also be observed over more limited ranges of stimulation. It is necessary to study in detail the bifurcations in the variety of systems and to determine the extent to which bifurcations may be "universal," rather than a particular feature found only in very special circumstances. In this regard, the abstract mathematical theory described here provides a unifying feature. Even though the ionic mechanisms may differ widely in the various systems, the functional response to stimulation may be very similar. What is of most interest from a functional perspective is to characterize the different types of response to stimulation. Indeed, simulation of the responses of cardiac tissue to electrical stimulation provides a crucial way to cross-check ionic models developed based on voltage clamp techniques [7].

One-dimensional finite difference equations can be used as approximations of the dynamics in periodically stimulated Hodgkin-Huxley equations [25]. It remains to be determined whether one-dimensional finite difference equations can also be used to approximate the response of realistic ionic models of cardiac activity. If there is a rapid return to the limit cycle in these equations, the one-dimensional finite difference equations provide a simple computational tool to investigate and understand the bifurcations. However, it will be necessary to take into account time-dependent effects and to develop more complex models than the one-dimensional finite difference equations considered here.

One-dimensional finite difference equations have been used to model two different cardiac arrhythmias, atrioventricular heart block (Chapters 13 and 14) and parasystole (Chapter 15). In these situations the very simple model gives surprisingly close agreement with complex rhythms observed in clinical situations.

Acknowledgements: This research has been supported by funds from the Canadian Heart and Stroke Foundation and the Natural Sciences and Engineering Research Council of Canada. We have benefited greatly from our collaborations with Michael Guevara, Glen Ward, John Lewis, and Zeng Wan-Zhen and thank them for permission to reproduce figures published elsewhere.

REFERENCES

[1] V.I. Arnold. *Geometrical Methods in the Theory of Ordinary Differential Equations.* Springer-Verlag, New York, 1983.

[2] J. Bélair and L. Glass. Universality and self-similarity in the bifurcations of circle maps. *Physica*, 16D:143–154, 1985.

[3] M.R. Boyett and B.R. Jewel. A study of the factors responsible for the rate-dependent shortening of the action potential in mammalian ventricular muscle. *J. Physiol. Lond.*, 285:359–380, 1978.

[4] D.R Chialvo, R.F. Gilmour, and J. Jalife. Low dimensional chaos in cardiac tissue. *Nature*, 343:653–657, 1990.

[5] D.R. Chialvo and J. Jalife. Non-linear dynamics of cardiac excitation and impulse propagation. *Nature*, 330:749–752, 1987.

[6] D.R. Chialvo, D.C. Michaels, and J. Jalife. Supernormal excitability as a mechanism of chaotic dynamics of activation in cardiac Purkinje fibers. *Circ. Res.*, 66:525–545, 1990.

[7] J.R. Clay, M.R. Guevara, and A. Shrier. Phase resetting of rhythmic activity of embryonic heart cell aggregates: Experiment and theory. *Biophys. J.*, 45:699–714, 1984.

[8] G.M. Decherd and A. Ruskin. The mechanism of the Wenckebach type of A-V block. *Br. Heart J.*, 8:6–16, 1946.

[9] R.L. Devaney. *An Introduction to Chaotic Systems*. Benjamin/ Cummings, Menlo Park, 1986.

[10] D. DiFrancesco and D. Noble. A model of cardiac electrical activity incorporating ionic pumps and concentration changes. *Phil. Trans. R. Soc. Lond. B.*, 307:353–398, 1985.

[11] E.J. Ding. Analytic treatment of a driven oscillator with a limit cycle. *Phys. Rev.*, 35A:2669–2683, 1987.

[12] E.J. Ding. Structure of the parameter space for the van der Pol oscillator. *Physica Scripta*, 38:9–15, 1988.

[13] J.-P. Eckmann. Roads to turbulence in dissipative dynamical systems. *Rev. Mod. Phys.*, 53:643–654, 1981.

[14] L. Glass and J. Bélair. Continuation of Arnold tongues in mathematical models of periodically forced biological oscillators. In H.G. Othmer, editor, *Nonlinear Oscillations in Biology and Chemistry*, pages 232–243. Springer-Verlag, Berlin, 1986.

[15] L. Glass, M.R. Guevara, J. Bélair, and A. Shrier. Global bifurcations of a periodically forced biological oscillator. *Phys. Rev. A*, 29:1348–1357, 1984.

[16] L. Glass, M.R. Guevara, A. Shrier, and R. Perez. Bifurcation and chaos in a periodically stimulated cardiac oscillator. *Physica*, 7D:89–101, 1983.

[17] L. Glass and A.T. Winfree. Discontinuities in phase-resetting experiments. *Am. J. Physiol.*, 256:R251–R258, 1984.

[18] L. Glass and M.C. Mackey. *From Clocks to Chaos: The Rhythms of Life*. Princeton University Press, Princeton, 1988.

[19] L. Glass and W.-Z. Zeng. Complex bifurcations and chaos in simple theoretical models of cardiac oscillations. *Ann. N Y Acad. Sci.*, 591: 316–327, 1990.

[20] D.L. Gonzalez and O. Piro. Chaos in a nonlinear driven oscillator with exact solution. *Phys. Rev. Lett.*, 50:870–872, 1983.

[21] J. Guckenheimer and P. Holmes. *Nonlinear Oscillations, Dynamical Systems and Bifurcations of Vector Fields*. Springer-Verlag, New York, 1983.

[22] M.R. Guevara. *Chaotic Cardiac Dynamics*. Ph.D. thesis, McGill University, 1984.

[23] M.R. Guevara, F. Alonso, D. Jeandupeux, and A.C.G. van Ginneken. Alternans in periodically stimulated isolated ventricular myocytes: Experiment and model. In A. Goldbeter, editor, *Cell Signalling: From Experiments to Theoretical Models*, pages 551–563. Academic Press, London, 1989.

[24] M.R. Guevara and L. Glass. Phase locking, period doubling bifurcations, and chaos in a mathematical model of a periodically driven oscillator: A theory for the entrainment of biological oscillators and the generation of cardiac dysrhythmias. *J. Math. Biol.*, 14:1–23, 1982.

[25] M.R. Guevara, L. Glass, M.C. Mackey, and A. Shrier. Chaos in neurobiology. *IEEE Trans. Syst. Man Cybern.*, SMC-13:790–798, 1983.

[26] M.R. Guevara, L. Glass, and A. Shrier. Phase-locking, period-doubling bifurcations and irregular dynamics in periodically stimulated cardiac cells. *Science*, 214:1350–1353, 1981.

[27] M.R. Guevara, A. Shrier, and L. Glass. Chaotic and complex cardiac rhythms. In D.P. Zipes and J. Jalife, editors, *Cardiac Electrophysiology: From Cell to Bedside*, pages 192–201, W. B. Saunders, Philadelphia, 1990.

[28] M.R. Guevara, A. Shrier, and L. Glass. Phase-locked rhythms in periodically stimulated heart cell aggregates. *Am. J. Physiol.*, 254:H1–10, 1988.

[29] M.R. Guevara, G. Ward, A. Shrier, and L. Glass. Electrical alternans and period-doubling bifurcations. *IEEE Comput. in Cardiol.*, 167–170, 1984.

[30] F.C. Hoppensteadt and J. Keener. Phase locking of biological clocks. *J. Math. Biol.*, 15:339–349, 1982.

[31] N. Ikeda, N.S. Yoshizawa, and T. Sato. Difference equation model of ventricular parasystole as an interaction of pacemakers based on the phase response curve. *J. Theor. Biol.*, 103:439–465, 1983.

[32] M.H. Jensen, P. Bak, and T. Bohr. Transition to chaos by interaction of resonances in dissipative systems. I: Circle maps. *Phys. Rev.*, 30A:1960–1969, 1984.

[33] D.T. Kaplan. *The Dynamics of Cardiac Electrical Instability*. Ph.D. thesis, MIT, 1989.

[34] J. Keener and L. Glass. Global bifurcations of a periodically forced oscillator. *J. Math. Biol.*, 21:175–90, 1984.

[35] M.N. Levy, T. Iano, and H. Zieske. Effect of repetitive bursts of vagal activity on heart rate. *Circ. Res.*, 30:186–195, 1972.

[36] M.N. Levy, P.J. Martin, H. Zieske, and D. Adler. Role of positive feedback in the atrioventricular nodal Wenckebach phenomenon. *Circ. Res.*, 34:697–710, 1974.

[37] J. Lewis, M. Bachoo, L. Glass, and C. Polosa. Complex dynamics resulting from repeated stimulation of nonlinear oscillators at a fixed phase. *Phys. Lett. A*, 125:119–22, 1987.

[38] T. Lewis and M.R. Guevara. Chaotic dynamics in an ionic model of the propagated cardiac action potential. *J. Theor. Biol.*, 146:407–432, 1990.

[39] R.E. McAllister, D. Noble, and R.W. Tsien. Reconstruction of the electrical activity of cardiac Purkinje fibers. *J. Physiol. Lond.*, 251:1–59, 1975.

[40] D.C. Michaels, D.R. Chialvo, E.P. Matyas, and J. Jalife. Chaotic activity in a mathematical model of the vagally driven sinoatrial node. *Circ. Res.*, 65: 1350–60, 1989.

[41] D.C. Michaels, E.P. Matyas, and J. Jalife. A mathematical model of the effects of acetylcholine pulses on sinoatrial pacemaker activity. *Circ. Res.*, 55:89–101, 1985.

[42] W. Mobitz. Uber die unvollstandige Storung der Erregungsuberleitung zwischen Vorhof und Kammer des menschlichen Herzens. *Zeit. Exp. Med.*, 41:180–237, 1924.

[43] G.K. Moe, J. Jalife, W.J. Mueller, and B. Moe. A mathematical model of parasystole and its application to clinical arrhythmias. *Circulation*, 56:968–979, 1977.

[44] J.B. Nolasco and R.W. Dahlen. A graphic method for the study of alternation in cardiac action potentials. *J. Appl. Physiol.*, 25:191–196, 1968.

[45] T. Pavlidis. *Biological Oscillators: Their Mathematical Analysis*. Academic Press, New York, 1973.

[46] A. Pelleg, S. Vogel, L. Belardinelli, and N. Sperelakis. Overdrive suppression of automaticity in cultured chick myocardial cells. *Am. J. Physiol.*, 238:H24–30, 1983.

[47] D.H. Perkel, J.H. Schulman, T.H. Bullock, G.P. Moore, and J.P. Segundo. Pacemaker neurons: Effects of regularly spaced synaptic input. *Science*, 145:61–63, 1964.

[48] G.V. Savino, L. Romanelli, D.L. Gonzalez, O. Piro, and M. Valentinuzzi. Evidence for chaotic behavior in driven ventricles. *Biophys. J.*, 56:273–280, 1989.

[49] L. Schamroth, D.H. Martin, and M. Pachter. The extrasystolic mechanism as the entrainment of an oscillator. *Am. Heart J.*, 104:1363–1368, 1988.

[50] A. Shrier and J. Clay. Repolarization currents in embryonic chick atrial heart cell aggregates. *Biophys. J.*, 50:861–874, 1986.

[51] M.B. Simson, J.F. Spear, and E.N. Moore. Stability of an experimental atrioventricular reentrant tachycardia in dogs. *Am. J. Physiol.*, 240:H947–953, 1981.

[52] M. Vassalle. The relationship among cardiac pacemakers: Overdrive suppression. *Circ. Res.*, 41:269–277, 1977.

[53] A.T. Winfree. *The Geometry of Biological Time.* Springer-Verlag, New York, 1980.

[54] A. Wolf, J.B. Swift, H.L. Swinney, and J.A. Vastano. Determining Lyapunov numbers from a time series. *Physica*, 16D:285–317, 1985.

[55] D.L. Ypey, W.P.M. van Meerwijk, and R.L. DeHaan. Synchronization of cardiac pacemaker cells by electrical coupling. In L.N. Bouman and H.J. Jongsma, editors, *Cardiac Rate and Rhythm*, pages 363–395. Martinus Nijhoff, The Hague, 1982.

[56] W.-Z. Zeng, M. Courtemanche, L. Sehn, A. Shrier, and L. Glass. Theoretical computation of phase locking in embryonic atrial heart cell aggregates. *J. Theor. Biol.*, 145:225–244, 1990.

13

Iteration of the Human Atrioventricular (AV) Nodal Recovery Curve Predicts Many Rhythms of AV Block

Michael R. Guevara[1]

ABSTRACT The atrioventricular nodal recovery curve provides a quantitative description of how the conduction time through the atrioventricular node of a prematurely elicited atrial beat increases as the recovery time since the immediately preceding activation of the bundle of His decreases. This curve can be well approximated in human beings with normal atrioventricular nodal function by a single exponential function (the "standard" curve). Assuming that the response of the atrioventricular node to any atrial stimulus with a given recovery time during periodic pacing of the atrium is independent of the atrial rate, a simple equation ("map") can be derived, using the recovery curve. The rhythm of atrioventricular conduction expected at any atrial rate is then obtained by numerically iterating this map on a digital computer. As the atrial rate is increased, rhythms resembling normal sinus rhythm, first-degree atrioventricular block, millisecond Wenckebach, Wenckebach periodicity, reverse Wenckebach periodicity, alternating Wenckebach periodicity, and higher grades of block are successively encountered. A mathematical theorem about the map is invoked to show that these are the only rhythms of conduction permitted. In addition, the order in which the various rhythms will appear as the atrial rate is increased as well as the ordering of blocked atrial beats within a given rhythm of block are also derived. Iteration using recovery curves other than the standard one leads to rhythms in which there is atypical Wenckebach, alternation of the conduction time, or coexistence of two different conduction times. Since this work puts into a common framework many different rhythms of atrioventricular conduction, it forms the beginnings of a "unified theory" of atrioventricular block.

[1] Department of Physiology, McGill University, Montréal, Québec H3G 1Y6

13.1   Introduction

Many different forms of block of atrioventricular (AV) conduction have been described. In the normal sequence of activation of the heart, an action potential sweeps across the atria and eventually arrives in the ventricular muscle after traversing the AV node and the His-Purkinje system. During normal sinus rhythm, the AV node and the His-Purkinje network provide one output impulse to the ventricular muscle for each impulse arriving at the input of the AV node from the right atrium. In first-degree AV block, the AV conduction time is abnormally prolonged. Rhythms in which there is still one ventricular activation for each atrial activation, but with alternation of the conduction time (PR alternans [39,53]) or an apparently random, constantly fluctuating, conduction time ("flottement du PQ" [40]) are less frequently seen. In second-degree AV block, there is occasional blocking of the atrial input, leading to one of several different variants of the Wenckebach phenomenon (classical, atypical, millisecond, reverse, or alternating Wenckebach), to Mobitz type II block, or to higher grades of block. In third-degree AV block, the atrial input does not get through to the ventricular muscle. Disorders in AV conduction can be caused by disturbances in the AV node or in the His-Purkinje system.

Different mechanisms are usually invoked to explain each of the above rhythms. In this chapter I show that the existence of all of these rhythms of conduction is predicted by simple mathematical analysis using the AV nodal recovery curve. This curve gives the conduction time through the AV node as a function of the recovery time since the last successful activation of the bundle of His. The analysis involves numerical iteration on a digital computer of a function—called the "map"—that is derived from the recovery curve. Similar techniques have been previously used by several authors [10,32,37,43,54,65].

In what follows I first describe the standard AV nodal recovery curve employed for most of this study. I then explicitly state the assumptions underlying the formulation of the iterative technique before deriving the formula for the map. The results of carrying out the iterative procedure are then described and discussed—in some detail for the standard recovery curve, in lesser detail for the five other recovery curves considered. Many of the results described below were mentioned briefly in Guevara [19].

13.2   Derivation of the 1-Dimensional Map

The methods employed in this chapter are purely theoretical; new experimental findings are not described. Numerical computations using a computer are carried out and mathematical theorems are applied. The essence of the method is an iterative technique that is now derived.

13.2.1 ASSUMPTIONS

I make the following three assumptions:

Assumption (i)

The conduction time through the AV node of an impulse produced by premature atrial stimulation (i.e., the atrium-His or A_2H_2 interval of Figure 13.1A) depends only on the recovery time since the preceding successful conduction through the AV node (i.e., the preceding His-atrium or H_1A_2 interval of Figure 13.1A). It has been found that this dependence is well described by a single exponential function in 50 humans with normal AV nodal function [61]. The mean dependence of A_2H_2 on H_1A_2 for these 50 subjects can be written

$$A_2H_2 = g(H_1A_2) = a + b \ \exp\left(-H_1A_2/\tau\right), \quad \text{for } H_1A_2 \geq c, \qquad (13.1)$$

with $a = 90$ ms, $b = 780$ ms, $c = 220$ ms and $\tau = 110$ ms (see Appendix for estimation of parameters). Figure 13.1B shows the AV nodal recovery curve described by the function g of Equation (13.1). Note that the parameter a in Equation (13.1) gives the minimum conduction time through the AV node (i.e., the A_2H_2 interval in the limit $H_1A_2 \rightarrow \infty$), while τ gives the time constant of recovery and b (in combination with c and τ) influences the height of the curve above the asymptotic line $A_2H_2 = a$. For $H_1A_2 < c$, and thus for $A_1A_2 = A_1H_1 + H_1A_2 < A_1H_1 + c$, conduction is blocked through the AV node. Thus, the effective refractory period of the node is given by $A_1H_1 + c$, which is approximately equal to $a + c$, assuming that the basic cycle length (i.e., A_1A_1) is long enough to make A_1H_1 approximately equal to a. The maximum A_2H_2 interval, denoted by d, occurs for $H_1A_2 = c$ and is given by

$$d = g(c) = a + b \ \exp\left(-c/\tau\right). \qquad (13.2)$$

For the values of a, b, c, and τ given above, $d = 195.6$ msec. Figure 13.1C shows the input-output characteristics of the node—that is, $H_1H_2(= H_1A_2 + A_2H_2)$ plotted as a function of $A_1A_2(= A_1H_1 + H_1A_2)$. From Equation (13.1), again assuming $A_1H_1 = a$, one obtains

$$H_1H_2 = A_1A_2 + b \ \exp\left(-(A_1A_2 - a)/\tau\right). \qquad (13.3)$$

Note that H_1H_2 continuously falls as A_1A_2 decreases. Equation (13.1) describes the recovery curve that is used throughout most of what follows; it will be referred to henceforth as the "standard" recovery curve, since it represents the mean curve found in 50 subjects.

Assumption (ii)

The atrial input to the AV node is strictly periodic in time with period denoted by t_s.

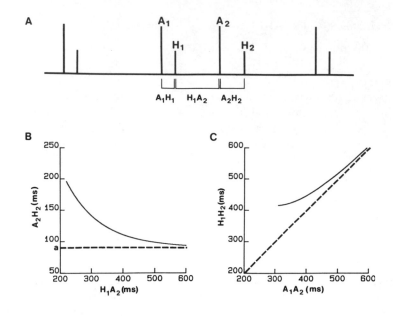

FIGURE 13.1. (A) Schematic representation of A (atrial) and H (bundle of His) deflections of His bundle electrogram (HBE) during determination of the AV nodal recovery curve. The taller spike indicates the A deflection, the shorter the H deflection. The V (ventricular) deflection on the HBE is suppressed since it is assumed that HV interval remains fixed. (B) "Standard" AV nodal recovery curve. (C) Input-output $H_1 H_2 - A_1 A_2$ curve derived from standard recovery curve.

Assumption (iii)

During periodic atrial stimulation, the AH interval of the $(i + 1)$st impulse to conduct through the AV node (denoted by AH_{i+1}) depends only on the preceding HA recovery interval (denoted by HA_i). Moreover, the dependence of AH_{i+1} on HA_i has the same functional form as that described by Equation (13.1), with A_2H_2 replaced by AH_{i+1} and H_1A_2 replaced by HA_i. Note that there may be one or more blocked atrial beats during the HA_i interval, which will be assumed to have no effect on AH_{i+1}. The above three assumptions are all approximations that eventually break down to some degree. I now show that, using the above three assumptions, one can predict the rhythm of AH conduction at any value of t_s.

13.2.2 THE MAP

Let us first consider the simplest case, in which the atrial cycle length is sufficiently long so that there are no blocked atrial beats. Then, from the construction shown in Figure 13.2A, one has

$$AH_{i+1} = g(HA_i), \tag{13.4}$$

where g is the function given in Equation (13.1). But since $HA_i = t_s - AH_i$ (Figure 13.2A), where t_s is the time between atrial stimuli, one obtains

$$AH_{i+1} = g(t_s - AH_i). \tag{13.5}$$

Substitution of the function g from Equation (13.1) leads to

$$AH_{i+1} = a + b \, \exp\left(-(t_s - AH_i)/\tau\right), \text{ for } AH_i \leq t_s - c. \tag{13.6}$$

Note that the condition $AH_i \leq t_s - c$, obtained by algebraic manipulation, is merely a restatement of our original condition of there being no blocked atrial beats, since these will occur only if the recovery time $HA_i \, (= t_s - AH_i)$ is less than c.

Consider now the situation in which there is one blocked atrial beat (Figure 13.2B). As before one assumes $AH_{i+1} = g(HA_i)$. But since $HA_i = t_s + (t_s - AH_i)$, then $AH_{i+1} = g(2t_s - AH_i)$. Substituting the function g given in Equation (13.1), one arrives at

$$AH_{i+1} = a + b \, \exp\left(-(2t_s - AH_i)/\tau\right), \text{ for } t_s - c < AH_i \leq 2t_s - c. \tag{13.7}$$

In general, for the case of $(n_i - 1)$ blocked atrial beats during the HA_i interval, one has

$$AH_{i+1} = f(AH_i) = a + b \exp(-(n_i t_s - AH_i)/\tau), \tag{13.8}$$

for

$$(n_i - 1)t_s - c < AH_i \leq n_i t_s - c. \tag{13.9}$$

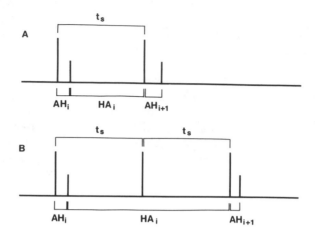

FIGURE 13.2. Schematic representation of HBE during periodic atrial stimulation with period t_s. (A) No beat skipped or dropped. (B) One beat dropped.

Note, for example, that Equation (13.7) is a special case of Equations (13.8) and (13.9) with $n_i = 2$.

Equation (13.8) is a one-dimensional finite-difference equation. For a given AV nodal recovery curve (i.e., a, b, c, and τ fixed), there is only one adjustable parameter, t_s. Given some arbitrary value AH_1, one can insert it into the right-hand side of Equation (13.8) to compute AH_2. Inserting the new value AH_2 back into Equation (13.8), one can compute AH_3, and so on. This arithmetic "iteration" of Equation (13.8) can thus be used to investigate the evolution of the AH intervals (i.e., the values $AH_1, AH_2, AH_3, ..., AH_i, AH_{i+1},...$) during ongoing atrial stimulation.

The function of Equation (13.8) can be plotted, with AH_{i+1} drawn as a function of AH_i. This curve is called the "map," since it gives the value of a variable (the AV nodal conduction time) as a function of the preceding value of that variable.

13.3   Results of Iteration of the Map

I now describe the results of iterating Equation (13.8). The atrial stimulation period t_s was changed in 5-msec steps over the range 100 msec $< t_s <$ 500 msec. Over interesting parts of this range, t_s was changed in increments of 1.0, 0.5, or even 0.1 msec to investigate the dynamics more finely. At each value of t_s, Equation (13.8) was iterated 29 times in single precision (about 6.3 decimal digits), starting with an initial condition of $AH_1 = a = 90$ msec. The values of $AH_i (1 \leq i \leq 30)$ and HA_i

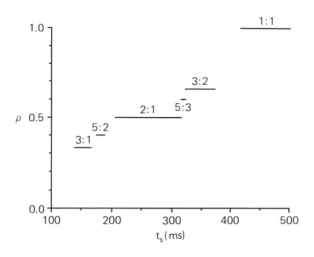

FIGURE 13.3. The asymptotic conduction ratio ρ (M/N) plotted as a function of t_s for the principal $N{:}M$ conduction rhythms. Wenckebach, Wenckebach (type X), reverse Wenckebach, Wenckebach (type Y), and alternating Wenckebach (type B) rhythms are found in the gaps between these principal zones (see text).

($1 \leq i \leq 29$) were printed out to three decimal places and inspected. A repeating rhythm in the AH_i and HA_i, corresponding to a periodic $N{:}M$ (N, M integers, $N \geq M$) AV conduction rhythm, was always found.

Figure 13.3 is a summary of the results, giving the principal $N{:}M$ conduction rhythms observed at a particular value of t_s. As t_s is decreased from 500 msec, different rhythms of AV conduction are seen. The sequence of rhythms runs as follows: normal sinus rhythm (1:1) → first degree AV block (1:1) → Wenckebach rhythms → 3:2 → Wenckebach (type X) rhythms → 5:3 → reverse Wenckebach rhythms → 2:1 → Wenckebach (type Y) rhythms → 5:2 → alternating Wenckebach (type B) rhythms → 3:1 → higher grades of block. Figure 13.4 shows the simulated His-bundle electrogram (HBE) at different values of t_s in the range 100 msec < t_s < 500 msec. The taller spikes in each trace schematically represent the atrial or A deflections; the shorter ones represent the His bundle or H deflections. The ventricular or V deflections on the HBE are suppressed for the sake of clarity of illustration. I assume that the HV interval is constant throughout; in that case, the rhythms of AV conduction are the same as the rhythms of AH conduction shown in Figure 13.4. I now discuss in some detail some of the rhythms generated by the iterative procedure. In particular, since the purely arithmetic computations shown in Figures 13.3 and 13.4 provide no insight into the dynamics, I shall examine the evolution in the geometric form of the map of Equation (13.8) that occurs as t_s is gradually decreased.

320   Michael R. Guevara

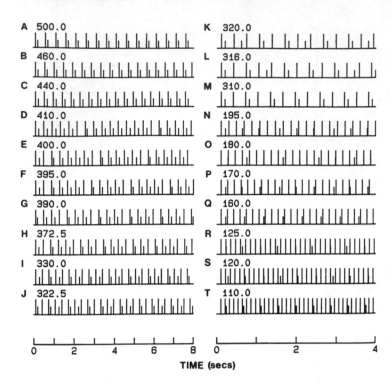

FIGURE 13.4. Schematic representation of the A and H spikes of HBE during various rhythms seen during periodic stimulation. t_s indicated above each trace. Sufficient time allowed for transients to pass. See text for further description.

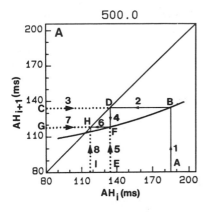
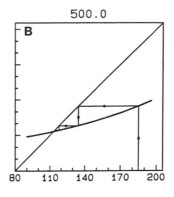

FIGURE 13.5. (A) Method of graphical iteration of the map of Equation (13.8). See text for further description. (B) Shorthand "staircase" method of iteration. In this and subsequent figures, t_s indicated over each map.

13.3.1   NORMAL SINUS RHYTHM

For $t_s \geq 457$ msec, the AV node conducts each atrial impulse, with an AH interval of < 130 msec (i.e., normal sinus rhythm). Figure 13.4A shows a rhythm in which the AH interval is about 113.2 msec ($t_s = 500$ msec). Figure 13.5A is a plot of the map $AH_{i+1} = f(AH_i)$ computed from Equation (13.8) corresponding to the rhythm shown in Figure 13.4A. Note that the range of AH_i is from the minimum conduction time a (=90.0 msec) to the maximum conduction time d (=195.6 msec). Also illustrated graphically in Figure 13.5A is the iterative process. Assume that $AH_1 = 185$ msec (i.e., start at point A). To obtain AH_2, follow the vertical line in the direction indicated by the arrow labeled 1 up to its intersection with the map (point B) and then follow arrow 2 to its intersection with the AH_{i+1}-axis (point C); this gives AH_2. Follow arrow 3 back to its intersection (point D) with the diagonal line of identity (whose equation is $AH_{i+1} = AH_i$). Then follow arrow 4 to point E on the AH_i-axis; by this "completing-the-square" construction one obtains the value $AH_i = AH_2$ at point E. Starting with the value AH_2 at point E, one can repeat the process (arrows 5,6,7,8 and points F, G, H, I), thereby obtaining AH_3. This graphical iterative process can then be repeated indefinitely to find AH_4, AH_5, and so forth.

By suppressing the dotted lines in Figure 13.5A, one obtains the clearer "staircase" method of graphical iteration shown in Figure 13.5B. Note that the iterates are attracted to the point at $AH_1^* = 113.2$ msec. This point AH_1^* is called a fixed point (or steady state or equilibrium point) of the map, since, at that point

$$AH_{i+1} = AH_i. \tag{13.10}$$

Thus, if one were to start the iterative procedure with initial condition

$AH_1 = 113.2$ msec, all succeeding iterates $AH_2, AH_3, \ldots$ would stay at the value 113.2 msec. Moreover, the fixed point in Figure 13.5B is (locally) stable. A fixed point AH_1^* is locally stable if the iterates AH_i of any initial condition AH_1 sufficiently close to the fixed point approach the fixed point as $i \to \infty$. A fixed point AH_1^* will be locally stable if the absolute value of the slope of the map at that point is less than one:

$$|\partial (AH_{i+1})/\partial (AH_i)| \, _{AH_i=AH_1^*} < 1. \tag{13.11}$$

Figure 13.4A shows the simulated HBE corresponding to the map of Figure 13.5B: there is normal sinus rhythm with an AH interval of 113.2 msec in the steady state after all transients have passed.

As t_s is decreased, the map moves vertically upward (Figure 13.6). That this must be so can be seen directly from Equation (13.6) where t_s appears in a factor of the form $[\exp\,(-t_s/\tau)]$. Thus, at any given value of AH_i, AH_{i+1} increases as t_s decreases. Decrease of t_s to 480 msec from 500 msec results in movement of the fixed point from $AH_1^* = 113.2$ msec (Figure 13.5B) to $AH_1^* = 118.0$ msec (Figure 13.6A). Further decrease to $t_s = 460$ msec results in a further increase of AH_1^* to 128.0 msec (Figure 13.6B); Figure 13.4B shows the corresponding HBE.

Figure 13.6B shows that starting from an initial condition $AH_1 = 105$ msec (arrow a), the iterates AH_i approach the same fixed point as do the iterates starting from $AH_1 = 185$ msec (arrow b). In fact, it can be proven that iterates of any initial condition AH_1 are attracted to the fixed point $AH_1^* = 128.0$ msec. The fixed point is thus not only locally stable, but is said to be globally stable since it attracts all initial conditions [33]. Therefore, although the exact nature of the particular transient seen depends on the initial condition AH_1, for example, the AH_i increase to AH_1^* for $AH_1 = 105$ msec, but decrease to AH_1^* for $AH_1 = 185$ msec in Figure 13.6B, the asymptotic (i.e., $i \to \infty$) value AH_1^* is independent of the initial value of AH_1. Either increase or decrease in the AH_i can be seen in clinical work, provided that AH_1 is set up correctly, as in Figure 13.6B, by a spontaneous extrasystole [37] or by programmed electrical stimulation [74].

13.3.2   FIRST-DEGREE AV BLOCK

For 416 msec $< t_s <$ 457 msec, there is still a 1:1 rhythm of conduction, but with an AH interval of >130 msec in the steady state (i.e., first-degree AV block [12]). Figure 13.4C gives a sample conduction rhythm ($t_s = 440$ msec); the corresponding map (Figure 13.6C) shows that the AH interval (AH_1^*) is about 142.3 msec. Again, during first-degree block, the fixed point is globally stable (Figure 13.4C). As in the case of normal sinus rhythm, decrease of t_s in first-degree AV block leads to increase in AH_1^* (Figure 13.6D). Note that since the slope of the map continually increases from left to right, the change in the AH interval produced by a fixed increment in

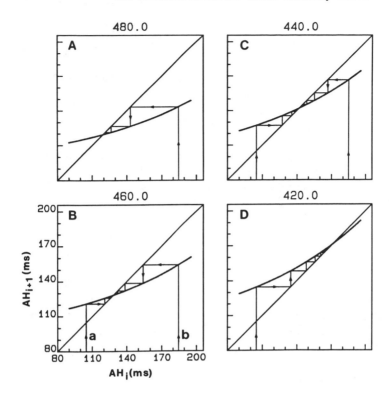

FIGURE 13.6. Period-1 orbit corresponding to normal sinus rhythm (A,B) and first-degree AV block (C,D). Iterations from different initial conditions (105 msec and 185 msec) are shown in B and C.

t_s increases as t_s decreases. The largest possible AH interval sustainable during 1:1 conduction is the maximum conduction time $d = 195.6$ msec. Substituting $AH_{i+1} = AH_i = d$ in Equation (13.6), one can calculate that the smallest value of t_s at which a 1:1 rhythm will be seen is given by $t_s = d + \tau ln\,[b/(d-a)] = 415.6$ msec.

13.3.3   WENCKEBACH RHYTHMS

For $t_s < 416$ msec, a 1:1 conduction rhythm no longer occurs. Instead, periodic rhythms in which there are blocked atrial beats are seen. These rhythms are composed of a repeating sequence of one or more Wenckebach cycles [67]. Within a Wenckebach cycle, there is a gradual increase in the AH interval, culminating in a single blocked atrial beat. Rhythms containing Wenckebach cycles are found for 323 msec $< t_s <$ 416 msec; I shall call this range of t_s the Wenckebach zone.

The usual $(n + 1){:}n$ (n is an integer, $n \geq 2$) rhythms of Wenckebach block are seen. Figures 13.4D–13.4G and 13.4I show 9:8, 6:5, 5:4, 4:3, and 3:2 rhythms respectively, while Figures 13.7A–13.7D and 13.7F show the corresponding maps. Note that the map now has two branches, a left-hand branch and a right-hand branch, separated by a point of discontinuity at $AH_i = t_s - c$. Equation (13.6) describes the left-hand branch while Equation (13.7) describes the right-hand branch. Visitation of the right-hand branch thus corresponds to a cycle of 2:1 block. Periodic orbits are shown on the maps of Figure 13.7. Only the last few of 100 iterates are shown in each case in order to suppress the transient that depends on the initial condition AH_1.

A periodic orbit of period M consists of a set of M period-M fixed points $AH_1^*, AH_2^*, ..., AH_M^*$ such that, for any j $(1 \leq j \leq M)$, one has

$$f^M(AH_j^*) = AH_j^*, \qquad (13.12)$$

but with

$$f^k(AH_j^*) \neq AH_j^*, \quad \text{for } 1 \leq k < M. \qquad (13.13)$$

The symbol f^M represents the function f applied M times, that is, M iterations of the map. Thus, each of the M periodic points of period M is taken back into itself following M (but not fewer) iterations of the map. If one starts with an initial condition on the orbit (i.e., $AH_1 = AH_j^*$ for some $1 \leq j \leq M$), one remains on the orbit (Figure 13.7). If the orbit is locally stable, starting with an initial condition AH_1 sufficiently close to one of the AH_j^* results in attraction of the iterates AH_i to the orbit in the limit $i \to \infty$. A period-M orbit is locally stable provided that

$$|[\partial^M(AH_{i+1})/\partial(AH_i)^M]_{AH_i=AH_j^*}| < 1. \qquad (13.14)$$

A stable period-M orbit on the map corresponds to a periodic $N{:}M$ conduction rhythm (for some value N). Note the Equations (13.10) and

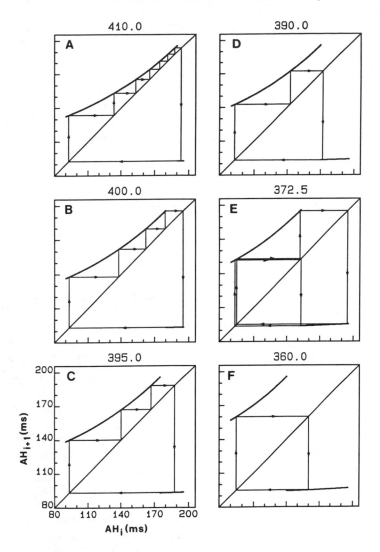

FIGURE 13.7. Period-8, 5, 4, 3, 5, and 2 orbits (A–F) corresponding to 9:8, 6:5, 5:4, 4:3, 7:5, and 3:2 Wenckebach rhythms. In this and subsequent figures transients suppressed to show steady-state orbits alone.

(13.11) given earlier are special cases of Equations (13.12) and (13.14) with $M = 1$. Thus, the fixed point AH_1^* resulting in a 1:1 rhythm (Figures 13.5, 13.6) is called a period-1 fixed point and is the only member of the period-1 orbit.

For an $N{:}M$ conduction rhythm, inspection of the periodicity of the orbit on the corresponding map gives the value of M. The integer N must also be calculated from Equation (13.8) by keeping track of the n_i, the number of atrial beats during the interval HA_i. The asymptotic ratio of ventricular to atrial beats in a conduction rhythm is called the rotation number (ρ) and is given by

$$1/\rho = \lim_{k \to \infty} \frac{1}{k} \sum_{i=1}^{k} n_i. \tag{13.15}$$

For a periodic $N{:}M$ conduction rhythm, $\rho = M/N$, so one can calculate N, once ρ and M are known. As an example of the application of the above theory, the map of Figure 13.7D admits a periodic orbit of period 3, corresponding to a 4:3 rhythm of AV conduction. The orbit can be seen to be globally stable by starting the iterative procedure with an initial condition AH_1 not on the orbit (i.e., $AH_1 \neq AH_j^*$ for $j = 1, 2, 3$); the periodic orbit is then asymptotically approached [33].

On all the maps of Figure 13.7, if AH_i is sufficiently large, AH_{i+1} falls on the right-hand branch of the map, leading to a very small value for AH_{i+1}. This occurs when AH_i is so large that the recovery time HA_i is insufficient to avoid blocking of an atrial beat (Figure 13.2B). Thus, a blocked atrial beat occurs when a transition is made from the right-hand branch of the map to the left-hand branch. The long recovery time following the blocked beat leads to rapid conduction through the AV node and a very small value of the AH interval. Moreover, since the orbit must always proceed to the left-hand branch after one iteration, at most one atrial beat can be blocked during a Wenckebach cycle.

Note that the orbits of Figure 13.7 produce "typical" Wenckebach cycles (Figures 13.4D–13.4I). In a typical Wenckebach rhythm of block, within each Wenckebach cycle there is (1) a progressive prolongation in the AH interval preceding a single blocked atrial beat, and (2) the increment in the AH interval, i.e., $(AH_{j+1}^* - AH_j^*)$ for $1 < j < M - 1$, decreases as the cycle proceeds, resulting in a progressive beat-to-beat decrease in the HH interval.

In addition to the classic $(n+1){:}n$ ($n \geq 2$) Wenckebach rhythms, one can also find more complex periodic rhythms in the Wenckebach zone. Figure 13.4H shows a 7:5 rhythm that can be regarded as an alternation between 3:2 and 4:3 cycles. It is found at a value of t_s intermediate to those at which 3:2 and 4:3 conduction rhythms are found (Figures 13.7D–13.7F). More complex rhythms composed of mixtures of various $(n + 1){:}n$ cycles have also been observed. As t_s is decreased in the Wenckebach zone, $N{:}M$ rhythms with decreasing ratios of ventricular to atrial beats (i.e., M/N)

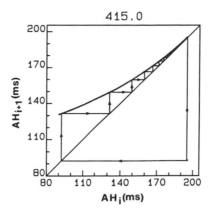

FIGURE 13.8. Period-24 orbit corresponding to a 25:24 Wenckebach rhythm.

are seen. Each $N{:}M$ rhythm is maintained over a certain interval of t_s; as t_s decreases in this interval, there is an increase in all of the conduction times $AH_j^*, 1 \leq j \leq M$. However, as the rhythm becomes more complex (i.e., as N increases), the range of t_s over which the rhythm is maintained tends to decrease. In fact, it can be proven for the class of maps studied here that there exists an interval of t_s in the Wenckebach zone over which can be found any periodic $N{:}M$ conduction rhythm, so long as N and M are relatively prime (i.e., have no common divisor) and $3/2 > N/M > 1$ [33]. The rhythm consists of a repeating unit of at most two different $(n+1){:}n$ Wenckebach cycles in which the number of blocked atrial beats differs by one; in addition, the sequence of Wenckebach cycles (i.e., the order in which atrial beats are blocked) in any given $N{:}M$ rhythm can be predicted [33].

For t_s just lower than the value at which a 1:1 rhythm is possible (t_s = 415.6 msec), $(n+1){:}n$ Wenckebach rhythms of very long period occur. Figure 13.8 shows a period-24 cycle, corresponding to 25:24 rhythm. Note that toward the end of the Wenckebach cycle the beat-to-beat increment in the AH interval is very small, producing a "millisecond Wenckebach" rhythm [13,14]. In fact, it can be proven that $(n+1){:}n$ Wenckebach cycles with n arbitrarily large can be found, provided that the parameter t_s is tuned finely enough [33].

13.3.4   REVERSE WENCKEBACH AND WENCKEBACH (TYPE X) RHYTHMS

For t_s just less than about 323 msec, a 3:2 rhythm is no longer seen; for t_s just smaller than about 316 msec, a 2:1 rhythm can be observed. Rhythms consisting of mixtures of 3:2 and 2:1 cycles occur for t_s lying between these two values (Figure 13.9, Figures 13.4I–13.4L).

For example, 5:3 and 7:4 rhythms can be seen at $t_s = 320$ msec (Figure 13.4K) and $t_s = 316$ msec (Figure 13.4L). Figures 13.9D and 13.9E show the corresponding maps. These $(2n - 1){:}n$ $(n \geq 3)$ rhythms are called "reverse Wenckebach" rhythms, because they consist of a sequence of 2:1 cycles, during which the AH intervals decrease, followed by a single 1:1 cycle [3]. Thus, the behavior is the reverse of that seen in a Wenckebach cycle, during which a sequence of 1:1 cycles with increasing AH intervals culminates in a single 2:1 cycle. The reverse Wenckebach zone is thus found between the 5:3 and 2:1 zones. For values of t_s between those at which 3:2 and 5:3 rhythms occur, one can observe cycles consisting of a sequence of 3:2 cycles terminated by a single 2:1 cycle. For example, 11:7 and 8:5 cycles occur at $t_s = 322.95$ msec (Figure 13.9B) and $t_s = 322.5$ msec (Figure 13.4J and Figure 13.9C), respectively. I have called these rhythms Wenckebach (type X) rhythms [19], to distinguish them from reverse Wenckebach rhythms, which they are not. More complex periodic rhythms, consisting of mixtures or combinations of the basic $(3n - 1){:}(2n - 1)$ $(n \geq 2)$ Wenckebach (type X) cycles can also be found. For example, a 14:9 rhythm consisting of alternating 3:2 and 11:7 Wenckebach (type X) cycles exists at $t_s = 322.99$ ms.

The 3:2 zone thus marks the border between the Wenckebach and Wenckebach (type X) zones: it may be considered as belonging to both zones. In a similar manner the 5:3 zone demarcates the border between the Wenckebach (type X) and the reverse Wenckebach zones.

13.3.5   ALTERNATING WENCKEBACH (TYPE B) AND WENCKEBACH (TYPE Y) RHYTHMS

For t_s slightly greater than about 207 msec, a 2:1 rhythm can be seen; for t_s just less than about 166 msec, a 3:1 rhythm is observed. Rhythms consisting of mixtures of 2:1 and 3:1 cycles exist for t_s lying between these two values (Figures 13.4N–13.4P and Figure 13.10).

Alternating Wenckebach (type B) rhythms can be seen. In these $(2n + 1){:}n$ $(n \geq 2)$ rhythms, there is a gradual increase in the AH interval during a sequence of 2:1 cycles that is terminated by a single 3:1 cycle, hence the original name of this rhythm—Wenckebach periods of alternate beats [27]. For example, at $t_s = 180$ msec (Figure 13.4O and Figure 13.10D) and $t_s = 195$ msec (Figure 13.4N and Figure 13.10B) are found 5:2 and 7:3 alternating Wenckebach (type B) rhythms. More complex periodic rhythms consisting of combinations of the basic $(2n + 1){:}n$ alternating Wenckebach (type B) cycles can also be found. For example, a period-5 orbit, corresponding to a 12:5 rhythm consisting of alternating 7:3 and 5:2 cycles, is found at $t_s = 190$ msec (Figure 13.10C), a value of t_s intermediate to those at which 7:3 and 5:2 cycles are found (Figures 13.10B and 13.10D, respectively). The alternating Wenckebach (type B) zone is thus found between

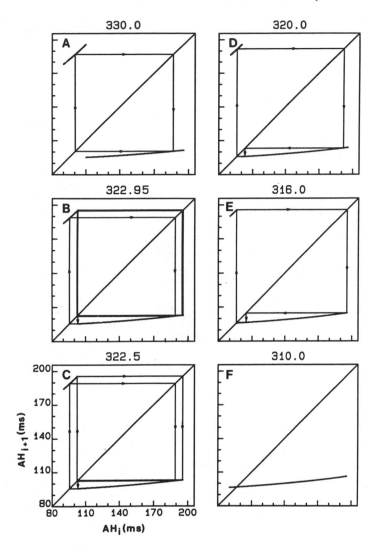

FIGURE 13.9. Period-2, 7, 5, 3, 4, and 1 orbits (A–F) corresponding to 3:2, 11:7, 8:5, 5:3, 7:4, and 2:1 rhythms. 11:7 and 8:5 rhythms are examples of Wenckebach (type X); 7:4 rhythm is example of reverse Wenckebach.

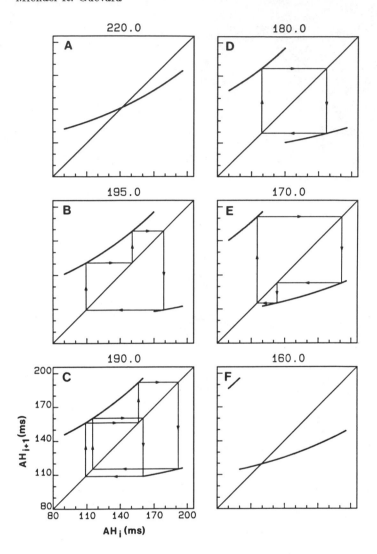

FIGURE 13.10. Period-1, 3, 5, 2, 3, and 1 orbits (A–F) corresponding to 2:1, 7:3, 12:5, 5:2, 8:3, and 3:1 rhythms, respectively. 7:3 and 12:5 rhythms are examples of alternating Wenckebach (type B); 8:3 rhythm is an example of Wenckebach (type Y). The two-branched maps in B–F are from Equation (13.8) with $n_i = 2$ and 3 for the left- and right-hand branches, respectively.

the 5:2 and 2:1 zones.

For t_s between the values at which 5:2 and 3:1 rhythms are found, one can observe $(3n-1):n$ rhythms, which I have previously called Wenckebach (type Y) rhythms [19]. For example, at $t_s = 170$ msec, an 8:3 Wenckebach (type Y) rhythm is found (Figure 13.4P and Figure 13.10E). Wenckebach (type Y) rhythms are composed of a sequence of 3:1 cycles, during which the AH interval progressively decreases, and terminate with a single 2:1 cycle. These rhythms are thus analogous to the reverse Wenckebach cycles mentioned earlier, in which an episode of consecutive 2:1 (instead of 3:1) cycles associated with gradually improving AH intervals was followed by a single 1:1 (instead of 2:1) cycle. Perhaps a better descriptive term for Wenckebach (type Y) would be *alternating reverse Wenckebach*, since one has the reverse Wenckebach phenomenon, but on alternate beats. As in the case of all the classes of rhythms previously mentioned, periodic rhythms composed of mixtures of two of the basic Wenckebach (type Y) cycles can also be seen as one changes t_s.

The 2:1 zone marks the border between the reverse Wenckebach and alternating Wenckebach (type B) zones; the 5:2 zone marks that between the alternating Wenckebach (type B) and the Wenckebach (type Y) zones.

13.3.6   HIGH-GRADE BLOCK

For t_s less than about 166 msec, $n:1$ ($n \geq 3$) rhythms of high-grade block can be seen. For example, 3:1 and 4:1 rhythms are generated at t_s=140 msec (Figure 13.4Q and Figure 13.10F) and t_s =110 msec (Figure 13.4T and Figure 13.11D), respectively. Intermediate to the values of t_s at which $n:1$ and $(n+1):n$ ($n \geq 3$) rhythms are seen, one can find complex rhythms consisting of mixtures of $n:1$ and $(n+1):1$ cycles. For example, one can find 7:2 (Figure 13.4R and Figure 13.11B) and 11:3 (Figure 13.4S and Figure 13.11C) rhythms at values of t_s intermediate to those at which 3:1 and 4:1 rhythms are found (Figure 13.4Q, Figure 13.10F and Figure 13.4T, Figure 13.11D, respectively). In general, between any two adjacent $n:1$ and $(n+1):1$ ($n \geq 3$) zones, one can see periodic rhythms containing a series of successive $n:1$ cycles in which the AH intervals gradually increase, culminating in a single $(n+1):n$ cycle (analogous to the Wenckebach and alternating Wenckebach rhythms previously described), as well as rhythms composed of consecutive $(n+1):n$ cycles in which the AH intervals decrease, followed by a single $n:1$ cycle (analogous to the reverse Wenckebach and Wenckebach (type Y) rhythms previously described). The $(2n+1):n$ zone forms the border between these two classes of rhythms.

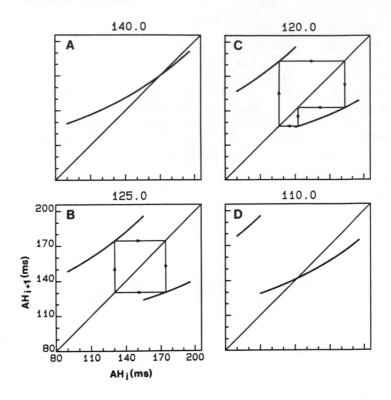

FIGURE 13.11. Period-1, 2, 3, and 1 orbits (A–D) corresponding to 3:1, 7:2, 11:3, and 4:1 rhythms respectively. These are all rhythms of high-grade block. The two branches in B and C are with $n_i = 3$ and 4 in Equation (13.8).

13.3.7   BEHAVIORS COMMON TO THE WENCKEBACH AND ALLIED RHYTHMS

Note that the various orbits corresponding to the Wenckebach, Wenckebach (type X), reverse Wenckebach, alternating Wenckebach (type B), and Wenckebach (type Y) cycles (Figures 13.7–13.10) arise naturally on the map as both the left-hand and right-hand branches move vertically upward and to the left with decreasing t_s. Iterates falling on the left-hand branch of the map correspond to a 1:1 (respectively 2:1) cycle in Figures 13.7–13.9 (respectively Figure 13.10), while iterates falling on the right-hand branch correspond to a 2:1 (respectively 3:1) cycle in Figures 13.7–13.9 (respectively Figure 13.10). The maps also graphically illustrate how the AH intervals increase during Wenckebach and alternating Wenckebach (type B) cycles, but decrease during reverse Wenckebach and Wenckebach (type Y) cycles. Furthermore, each $N{:}M$ rhythm ($1.0 < N/M < 3.0$) is maintained over an interval of t_s, which tends to grow smaller as N increases. As t_s is changed in this interval, the conduction time of the conducted beats (i.e., the $AH_j^*, 1 \le j \le M$) also changes, increasing for rhythms with conduction ratios lying in the range of $n/1 < N/M < (2n+1)/2$ and decreasing for rhythms in the range $(2n+1)/2 < N/M < (n+1)/1$ for some $n \ge 2$. Also note the similar appearance of the maps when a period-M orbit is encountered: for example, the period-2 orbits of Figures 13.9A, 13.10D, and 13.11B correspond to 3:2, 5:2 and 7:2 rhythms, the period-3 orbits of Figures 13.7D and 13.10B correspond to 4:3 and 7:3 rhythms, and the period-3 orbits of Figures 13.9D, 13.10E, and 13.11C correspond to 5:3, 8:3, and 11:3 rhythms.

13.3.8   OTHER TYPES OF AV NODAL RECOVERY CURVE

The AV nodal recovery curve shown in Figure 13.1B is the one that has been considered so far in this chapter. It was chosen because it is the "standard" curve resulting from lumping together the data from 50 different subjects with normal AV nodal function, as described in the Appendix.

This aggregate recovery curve produces the "type I" H_1H_2–A_1A_2 curve shown in Figure 13.1C (e.g., Figure 2B of Teague and colleagues [61]). However, at least five other shapes of H_1H_2–A_1A_2 input-output characteristic curves have been described. These include (1) "type II" curves, in which there is a minimum in the $H_1H_2 - A_1A_2$ curve, corresponding to a revealed functional refractory period, (2) "truncated" curves, in which there is little deviation from the line of identity $H_1H_2 = A_1A_2$ in Figure 13.1C, (3) "discontinuous" or "type III" curves, showing an abrupt jump in the H_1H_2 interval as A_1A_2 is decreased, (4) curves showing the "gap" phenomenon, in which there is an interval of A_1A_2 values over which conduction is blocked, but with conduction being successful for impulses that are either less premature or more premature, and (5) curves showing "supernormal" conduction, in which there is a range of H_1A_2 intervals over which the A_2H_2 interval paradoxically decreases with decreasing H_1A_2 in Figure 13.1B. In what follows, I will iterate the maps corresponding to each of these recovery curves, but rather than making an exhaustive study of each of these five cases (as carried out for the standard curve above), an interesting phenomenon will be illustrated in each case.

13.3.9   TYPE II RECOVERY CURVE: ATYPICAL WENCKEBACH RHYTHMS

The standard $A_2H_2–H_1A_2$ AV recovery curve of Figure 13.1B has slope everywhere more positive than -1. The slope of the $H_1H_2–A_1A_2$ curve of Figure 13.1C is thus everywhere positive, producing a type I curve, since $H_1H_2 = H_1A_2+A_2H_2$ and $A_1A_2 = A_1H_1+H_1A_2$. This standard curve also produces a map that has slope < 1 everywhere (Figures 13.5–13.11), as can be seen by differentiating Equation (13.8). However, some individuals have a type II recovery curve, resulting in a minimum in the $H_1H_2–A_1A_2$ curve of Figure 13.1C, which gives a clearly defined functional refractory period of the node (Figure 2B in [61], [72]). Such type II curves result in maps in which the slope of the left-hand branch in Figures 13.7–13.10 can increase beyond $+1$ as AH_i increases. Thus, in these cases the maximum AH interval possible during 1:1 conduction is less than the maximum AH interval (d) on the recovery curve. As t_s is decreased, the period-1 orbit corresponding to a 1:1 rhythm is then lost via a tangent bifurcation, at which point the slope of the map at the period-1 fixed point becomes equal to $+1$. At t_s just less than the value producing the bifurcation, $(n + 1){:}n$ Wenckebach rhythms with n arbitrarily long are produced, in which the beat-to-beat increment in the AH interval (i.e., $AH_{i+1} - AH_i$) *increases* progressively toward the end of the cycle, thus producing one form of atypical Wenckebach periodicity [10,13,54].

13.3.10   TRUNCATED RECOVERY CURVE: MOBITZ II BLOCK

In some individuals, the $H_1H_2–A_1A_2$ curve does not deviate very much from the line of identity $H_1H_2 = A_1A_2$ in Figure 13.1C (e.g., Figure 2C in [61]). This results in maps similar to those shown in Figure 13.7, but in which the vertical jump between the right-hand end of the left-hand branch of the map and the left-hand end of the right-hand branch is very small. The range of t_s over which Wenckebach rhythms are observed is then very small. In fact, in the limit of the $H_1H_2–A_1A_2$ curve being described by $H_1H_2 = A_1A_2$ for A_1A_2 greater than the effective refractory period, a direct transition from a 1:1 to a 2:1 rhythm occurs: Wenckebach rhythms do not exist. This situation is reminiscent of Mobitz II block, in that small fluctuations in the sinus rate or the state of conduction in the AV node would produce one or more 2:1 cycles with little or no premonitory signs of increment in AH interval, should the heart be close to its limit for 1:1 AV conduction.

13.3.11   TYPE III RECOVERY CURVE: TWO COEXISTING AH INTERVALS

Some subjects show a discontinuous AV recovery curve: as H_1A_2 is decreased in Figure 13.1B, there is a sudden discontinuous upward jump in the recovery curve [11]. For such a type III curve [61], one can obtain a map in which, over a range of values of t_s, there exist *two* stable period-1 orbits, each corresponding to a 1:1 rhythm (Figure 13.12A). The smaller AH_1^* corresponds to normal sinus rhythm, the larger to first-degree AV block. There is thus *bistability* present, with one of two different AH intervals being produced, depending on initial conditions. Note that during an episode of 1:1 conduction at either of the two possible AH intervals, injection of a premature stimulus at an appropriate HA coupling interval or dropping one stimulus of the basic drive train will flip the preexisting AH interval to the other AH interval.

13.3.12   THE GAP PHENOMENON: COEXISTING WENCKEBACH AND 1:1 RHYTHMS

Some individuals have one form or another of "gap" in their recovery curve: there is a range of H_1A_2 over which block occurs, with conduction being successful for H_1A_2 intervals smaller or greater [70]. As t_s is decreased, the map predicts that 1:1 rhythm should be replaced by Wenckebach rhythms, as in the standard case. However, with further decrease in t_s, one then obtains a three-branched map, with, for example, a 3:2 Wenckebach rhythm coexisting with a 1:1 rhythm of first-degree heart block (Figure 13.12B).

13.3.13   SUPERNORMAL CONDUCTION: ALTERNATION OF THE AH INTERVAL

Supernormal conduction is said to exist when decreasing the recovery interval H_1A_2 leads to a paradoxical improvement (i.e., decrease) in conduction time A_2H_2 over a range of H_1A_2 in Figure 13.1B [53]. The map can then possess a period-2 orbit (Figure 13.12C), in which there is one ventricular activation for each atrial activation, but with a beat-to-beat alternation in the AV conduction time. This period-2 orbit arises via a period-doubling bifurcation, which occurs when the slope of the map at the period-1 fixed point becomes equal to -1 (see e.g., Guevara and colleagues [23]). In contrast, the transition from 1:1 to 2:1 rhythm via Wenckebach, Wenckebach

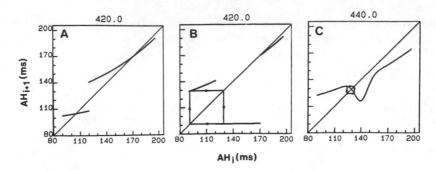

FIGURE 13.12. (A) Map resulting from type III or discontinuous AV recovery curve. The standard recovery curve of Figure 13.1B is used for shorter recovery intervals (220 msec $< H_1A_2 <$ 300 msec), while at longer recovery intervals ($H_1A_2 \geq$ 300 msec) the curve is described by Equation (13.1), but with τ decreased to 80 ms (see e.g., Figure 1C in [11]). The jump discontinuity in the map occurs at $AH_i = t_s -$ 300 msec = 120 msec. Note the existence of two stable period-1 orbits corresponding to the coexistence of two 1:1 rhythms with different AH intervals. (B) Map resulting from AV recovery curve demonstrating the gap phenomenon. The standard recovery curve of Figure 13.1B used, with a gap inserted for 350 msec $\leq H_1A_2 \leq$ 400 msec (see e.g., Figure 5 in [70]). The gap is assumed to appear in the AH interval for purposes of illustration only; in practice, it should probably be in the AV interval, since the gap phenomenon is usually infranodal. The left- and right-hand branches correspond to $n_i = 1$ in Equation (13.8), while the middle branch is for $n_i = 2$, due to the gap. Note the coexistence between a period-2 orbit, corresponding to a 3:2 rhythm, and a period-1 orbit, corresponding to 1:1 rhythm. (C) Map resulting from AV recovery curve when supernormal conduction is present. The standard recovery curve of Figure 13.1B is used, but with the term $-\alpha \exp[-(H_1A_2 - \beta)^2/\sigma^2]$ added in Equation (13.1), where $\alpha = 25$ msec, $\beta = 300$ msec, and $\sigma = 10$ msec. Thus, α controls the depth of the supernormal dip, σ controls its width, and β controls the central H_1A_2 interval about which the dip is placed.

(type X), and reverse Wenckebach rhythms (Figures 13.7, 13.9) is *not* the result of a period-doubling bifurcation, as has been previously claimed [65,68].

Bistability can also exist in this class of maps, which are nonmonotonic or noninvertible. For example, if t_s is decreased in Figure 13.12C by 10 msec to $t_s = 430$ msec, there is a coexistence between the 2:2 or alternans rhythm and a rhythm of first-degree heart block. The stable period-1 fixed point producing this latter rhythm comes in via a tangent bifurcation. If t_s is decreased to 420 msec, the stable period-doubled orbit disappears and there is a coexistence between two 1:1 rhythms. Wenckebach rhythms, in which there is a nonmonotonic trend in the increment of the AH interval, are eventually seen as t_s is reduced further.

13.4   Comparison with Clinical and Experimental Findings

13.4.1   VALIDITY OF THE ASSUMPTIONS UNDERLYING FORMULATION OF THE MAP

"The argument serves to emphasize the difficulty of treating recovery curves on a simple mathematical basis [Mobitz, 1924]; they are not necessarily simple pictures in which rest periods and consequent rates of conduction (regarded as uniform rates of conduction) are expressed" [38].

Three assumptions were made in order to derive the map of Equation (13.8). These assumptions all involved making approximations to the actual situation; like all approximations, they eventually break down.

The first assumption states that the AV nodal recovery curve can be closely approximated by a single-exponential function of the form given in Equation (13.1). This has been found to be approximately true in human beings [54,61,71] and in other species [17,29,65]. It is quite remarkable that the single-exponential fit works so well, given that the contribution to the AH interval at a given HA interval is different at different levels of the node (see Figure 4 in Billette [4]). However, depending especially on the basic cycle length at which the recovery curve is taken, deviations of varying degrees from the single-exponential fit can be seen in man [54,72] and in dogs [17]. Nevertheless, this deviation does not change the sequence of rhythms predicted to exist as t_s is decreased for the class of maps considered here [33].

The first assumption additionally states that the AH interval depends only on the preceding HA interval. This is not quite true since the AV nodal recovery curve shifts when the basic cycle length at which it is determined is changed. As the basic cycle length decreases, the curve of Figure 13.1B shifts upward at long H_1A_2 intervals because of "fatigue," and to the left at short H_1A_2 intervals because of "facilitation" [5–7,17,38,54]. A greater shift is seen if the recovery interval is defined as A_1A_2, and not H_1A_2 as used here [5], indicating that for our purpose the H_1A_2 interval is a better index of recovery time than is the A_1A_2 interval. While A_1A_2 is a better index of prematurity at the input of the node, H_1A_2 is a better indicator at the level of the output of the node. Since the AV node is a distributed structure composed of many cells of different types, with a nonconstant conduction velocity along its length, no one parameter can give the recovery time for the node as a whole.

The second assumption states that the frequency of the atrial input to the AV node is constant. While this is true in the case of artificial controlled stimulation of the atrium, it is generally not true in the natural case when the sinoatrial node acts as pacemaker to the heart. There are generally beat-to-beat changes in the rate of the sinoatrial node, often correlated with the respiratory cycle (sinus arrhythmia). These variations can sometimes cause AV block to be linked to respiration [40].

The third assumption states that the transmission time of an impulse through the AV node depends solely on the recovery time since the immediately preceding successful conduction reached the bundle of His; in addition, independently of the atrial rate the transit time is given by the same formula that describes the AV nodal recovery curve, that is, Equation (13.1). This assumption is the strongest of the three made, in that it is surely the first to break down in most instances. Perhaps the clearest example of the assumption breaking down is shown in Figure 4 of Billette and colleagues [6], where there is a beat-to-beat alternation in AH interval during stimulation at a fixed His-stimulus (HS) interval when HS is sufficiently small. Direct determination of the recovery curve on different beats of a 4:3 Wenckebach cycle reveals that the curve is not the same for all beats of the cycle [55]. A quick check of the validity of this third approximation can be obtained by constructing a "return map"—a recovery curve deduced from the AH and HA (or PR and RP) intervals during Wenckebach rhythms. The majority of such data points then lie quite close to the directly determined AV nodal recovery curve; however, some points lie at a considerable distance away from the curve [5,38,54]. These deviations are principally due to the effects of fatigue, facilitation, and concealed conduction. For example, since facilitation, which is induced by a fast nodal rate, can be dissipated by a single long cycle [6], facilitation is present to a lesser extent following the blocked beat of a Wenckebach cycle than for other beats in the cycle [75]. In addition, following a blocked beat, concealed conduction causes the next AH interval to be prolonged beyond the expected value [38,54]. . Thus, future modeling work to extend Equation (13.8) will have to take into account in a quantitative way the effects of fatigue, facilitation, and concealed conduction, along lines already started [29,30,63,71,75].

Subsequent to making the calculations detailed above, a quantitative test of the predictions of Equation (13.8) was carried out in human beings with normal AV nodal function [54]. Quite good quantitative agreement was found in all seven subjects tested, with, for example, 5:3 reverse Wenckebach and various Wenckebach rhythms being seen within about ±15% of the pacing cycle length at which they were predicted to exist based on iteration of the AV nodal recovery curve (Figure 8 of Shrier and colleagues [54]). It remains to be seen whether the agreement is as good at higher stimulation frequencies, where, for example, alternating Wenckebach rhythms are predicted to exist.

13.4.2 WENCKEBACH RHYTHMS

For t_s sufficiently large, one has a 1:1 rhythm of AV conduction (Figure 13.5). As t_s decreases, the AH interval increases from the minimum conduction time a to the maximum conduction time d, producing first-degree AV block from normal sinus rhythm (Figure 13.6). For most recovery curves, Wenckebach rhythms then ensue if t_s is decreased further.

The Wenckebach rhythm of block has been seen in almost all regions of the heart (see Guevara [19] for references), including the His-Purkinje system [1,14]. It can also be seen in other physiological (e.g., neural) systems, mathematical models, and electronic analogues of excitable membranes [19,31], as well as in smooth muscle [66]. It can be produced by periodic driving of an isopotential cardiac oscillator [23,26] and quiescent single ventricular cells (see Guevara and colleagues [24] and Chapter 14 in this volume). The Wenckebach rhythm of AV block is usually seen when the atrial rate is abnormally high or when AV conduction is compromised. I have shown above that a decrease in t_s leads to Wenckebach rhythms, keeping the parameters describing AV nodal conduction fixed. One can also produce Wenckebach block at fixed t_s by changing one of these parameters. For example, increasing the minimal conduction time a in Equation (13.1) can eventually result in Wenckebach block at fixed t_s.

Iteration of the AV nodal recovery curve has been previously used to account for the existence of the Wenckebach rhythm of AV block [10,32, 37,43,54,65]. Our contribution has been to formulate how the right-hand branch of the map is to be added in Figure 13.7, that is, Equation (13.8). In some of the above cases, the AV nodal recovery curve was not determined directly but was extracted from the electrocardiogram or HBE during spontaneous or induced Wenckebach block. Relatively simple models of excitation and conduction of the action potential incorporating threshold and refractoriness also lead to the prediction of the existence of Wenckebach patterns [2,16,34,36,44,45,50,58,59,76]. At least a dozen different mechanisms have been advanced to explain the Wenckebach rhythm of block since its initial description in the frog's heart a century ago [15]. Indeed, there may very well be more than one of these mechanisms operating in any given case; in addition, the mechanisms might differ from case to case. However, the results presented above show that the Wenckebach phenomenon is obligatory, if conduction slows continuously and monotonically with increasing prematurity of activation and eventually blocks; this is an elementary property of the class of maps considered here [33].

13.4.3   ATYPICAL WENCKEBACH RHYTHMS

The standard recovery curve, which is of type I, gives Wenckebach rhythms in which there is a gradual progressive decrease in the increment of the AH interval (Figures 13.7 and 13.8), yielding "typical" Wenckebach periodicity. For a type II curve this is not the case, and one form of atypical Wenckebach results, as previously pointed out on several occasions [10,37,43,54]. Note that for a type II curve, typical Wenckebach periodicity can be seen on the map for $(n+1):n$ rhythms if n is sufficiently small; in contrast, for n sufficiently large, the periodicity is atypical. This agrees with the findings in one clinical study in which all $(n+1):n$ cycles with $n \geq 6$ showed atypical periodicity [13]. Since the slope of the map can increase beyond $+1$ for a type II curve, as t_s is changed one can obtain cycles in which the increment in AH interval is very large—indeed even largest—for the last increment of the cycle, which again agrees with the clinical findings [13]. When there is a discontinuous (type III) recovery curve, one obtains a three-branched map in which there can be a sudden jump in the AH interval during a Wenckebach cycle, which agrees with the clinical findings [48,69]. Finally, it has been known for some time that typical Wenckebach periodicity is in fact a misnomer, since it is in fact *less* common than atypical Wenckebach periodicity in the clinic [13]. Other references are given in Guevara [19].

13.4.4   MILLISECOND WENCKEBACH AND MOBITZ II BLOCK

For t_s just less than the minimum value at which a 1:1 rhythm can be produced, one sees Wenckebach rhythms in which there is only a very occasional dropping or skipping of ventricular beats (e.g., Figure 13.8). In fact, in the absence of "noise" (uncontrolled fluctuations in the system), Wenckebach cycles of arbitrarily long period can be produced [33]. In addition, the beat-to-beat increment in the AH interval then becomes arbitrarily small.

Long Wenckebach cycles in which there is little or even no detectable increment in the AH or PR intervals preceding the dropped beat have been detected both experimentally [14] and clinically [13], and have been termed "millisecond Wenckebach" [14], which is a form of pseudo-Mobitz II block [13]. Arbitrarily long cycles can also be produced for a type II curve: however, in that instance the iterates of the map get trapped in a narrow channel between the line of identity $AH_{i+1} = AH_i$ and the region of the map where the slope is close to $+1$ (see Figure 10 of Mobitz [43]).

Note that if the Wenckebach zone of Figure 13.3 is very narrow, Wenckebach patterns would not be seen if t_s is decreased in relatively coarse steps from a value lying within the 1:1 zone. Instead, one would see a direct transition from a 1:1 to a 2:1 rhythm. In the presence of noise, one might see a rhythm resembling Mobitz II block. Indeed, as the H_1H_2–A_1A_2 curve of Figure 13.1C approaches the diagonal line $H_1H_2 = A_1A_2$, as occurs when the recovery curve is truncated, the width of the Wenckebach zone in Figure 13.3 narrows; in the limit of $H_1H_2 = A_1A_2$, the Wenckebach zone disappears completely and there is a direct transition from a 1:1 to a 2:1 rhythm. Since the recovery process in the His-Purkinje system is very rapid, that is, τ in Equation (13.1) is very small [4,60], one would then expect a very narrow Wenckebach zone; this probably explains why second-degree block is usually of the Mobitz II variety if the block is infranodal. Indeed, as t_s is decreased one can obtain a direct transition from a 1:1 to a 2:1 rhythm in Purkinje fiber [8,20], ventricular heart-cell aggregates [19,25], and isolated rabbit ventricular cells [Guevara and Jeandupeux, unpublished].

13.4.5 REVERSE WENCKEBACH RHYTHMS

Reverse Wenckebach rhythms have been described but rarely [3,41,43,47]; their existence has been predicted using the iterative method [10,43,54]. For the standard AV nodal recovery curve considered here, Equation (13.1), reverse Wenckebach rhythms (Figures 13.4 and 13.9) are found over only a comparatively narrow range of t_s (316 msec $< t_s <$ 323 msec). This fact may account for their relative rarity, especially since, as discussed below, patterns maintained over a small interval of t_s will be corrupted and destroyed by noise. Reverse Wenckebach rhythms can be more easily seen if the ability of the AV node to conduct is diminished by injecting acetylstrophanthidin into the AV nodal artery [47]. It may be that this intervention produces an AV nodal recovery curve that, upon iteration, yields a wider reverse Wenckebach zone than the one produced using the standard recovery curve.

There has been disagreement as to the mechanism of the reverse Wenckebach rhythm of AV block; it has been suggested that it might be due to AV dissociation in the presence of 2:1 AV block [41]. However, as with the Wenckebach phenomenon, the results described above strongly suggest that the reverse Wenckebach phenomenon is obligatory given that conduction through the node monotonically slows and eventually blocks with increasing prematurity of stimulation. More particularly, the reverse Wenckebach zone is to be found between the 3:2 and 5:3 zones, and not between the 3:2 and 2:1 zones as previously suggested [3,47].

Reverse Wenckebach rhythms can be seen in tissue from almost all areas of the heart, in neural tissue, in ionic models of neural and cardiac cells, and in simple mathematical models and electronic analogues of excitable systems (see Guevara [19] for references). They can also be seen in isopotential aggregates of spontaneously beating cardiac cells [26].

13.4.6 ALTERNATING WENCKEBACH RHYTHMS

The alternating Wenckebach rhythm of AV block occurs clinically in two basic forms: type A and type B. The usual mechanism invoked to account for alternating Wenckebach is block at two distinct levels of the conduction system: a 2:1 block proximally and a Wenckebach block distally produces an alternating Wenckebach (type A) rhythm, whereas a Wenckebach block proximally and a 2:1 block distally produces a type B rhythm [27,35,56,73]. An early clinical pulse tracing showing an alternating Wenckebach (type B) rhythm is shown in Figure 6 of Hay [28].

As is the case with the Wenckebach and reverse Wenckebach phenomena, the alternating Wenckebach (type B) rhythm can be seen in many types of cardiac tissue and in noncardiac systems [19], and is once again a direct consequence of the increase in AH with decrease in HA (Figure 13.1B), having been explained on this basis many years ago [10]. In contrast, the class of maps that we have considered above (Figures 13.5–13.11) do not admit alternating Wenckebach (type A) rhythms. It is conceivable that alternating Wenkebach (type A) rhythms might result in a circumstance when the AV nodal recovery curve is not the standard curve; alternatively, there might indeed be two levels of block present, thus necessitating the coupling together of two maps [32].

13.4.7 OTHER DROPPED-BEAT RHYTHMS

Iteration of the AV nodal recovery curve predicts the existence of several classes of skipped-beat rhythms different from the Wenckebach, reverse Wenckebach, and alternating Wenckebach rhythms discussed so far. These rhythms all share the feature of being composed of dropped-beat cycles in which the AH interval gradually increases or decreases. The iterations suggest the sequence of rhythms [normal sinus rhythm → first-degree AV block → Wenckebach → 3:2 → Wenckebach (type X) → 5:2 → reverse Wenckebach → 2:1 → alternating Wenckebach (type B) → 5:2 → Wenckebach (type Y) → 3:1 → · · ·] as t_s is decreased. There is both clinical and experimental evidence [19] for the two classes of rhythms that we have termed Wenckebach (type X) and Wenckebach (type Y). An early clinical example of a Wenckebach (type Y) rhythm is shown in Figure 3 of Hay [28]; a very recent clinical example of a Wenckebach (type X) rhythm is shown in Oreto and colleagues [46]. The existence of both of these rhythms was predicted by Decherd and Ruskin [10] from iterations involving the AV nodal recovery curve. Complex rhythms beyond the 3:1 rhythm with incrementing or decrementing AH intervals have also been described clinically, for example, a 7:2 rhythm was found by Hay [28] and a rhythm consisting of several consecutive 3:1 cycles with growing AH intervals terminated by a single 4:1 cycle has been documented in children [73].

13.4.8 MULTILEVEL BLOCK

Many of the more complex rhythms generated by the iterative procedure
have been interpreted as being caused by multiple levels of block within
the AV conduction system. As described above, a combination of 2:1 block
at one level and Wenckebach block at another level produces the two
types of alternating Wenckebach. Wenckebach block at both levels ("double Wenckebach") is considered to be a form of alternating Wenckebach
(type B) [35], and has been invoked on several occasions to explain various
rhythms, including Wenckebach (type X) [35,46,57]. Many other combinations of multilevel block have been proposed, usually on an ad hoc basis to
account for various rhythms, e.g., (1) a combination of 3:2 proximal block
and 2:1 distal block produces an overall 3:1 block, (2) 2:1 at both levels
produces 4:1, (3) 4:3 proximal and 2:1 distal produces 4:2, (4) 2:1 proximal
and 3:2 distal produces 6:2, (5) three levels of 2:1 produce 6:1, (6) three
levels of block with ratios 3:2, 2:1, and 3:2 from proximal to distal produce
9:2, (7) Wenckebach (type X) proximal and 2:1 distal produces a sequence
of 3:1 cycles with increasing AH intervals terminated by a single 4:1 cycle;
and (8) multilevel Wenckebach produces Wenckebach rhythms terminated
by two or more blocked beats. While there are documented cases of two or
more levels of block being involved in some instances [19], and multilevel
block most definitely can occur in an electrical model of conduction [31],
there is little or no evidence to support the majority of scenarios involving
horizontal dissociation of the conduction pathway given above. In fact, recent work on isopotential preparations shows that all eight rhythms listed
above as well as alternating Wenckebach (type B) and Wenckebach (type X)
can be seen in isopotential preparations. For example, in isopotential spontaneously beating ventricular heart-cell aggregates, one sees Wenckebach
rhythms terminated by two blocked beats [19] as well as Wenckebach (type
X), alternating Wenckebach (type B), high-grade block, and 4:2 rhythms
[25,26]; in isolated ventricular cells, one sees 4:2 and 6:2 rhythms [22].

Of course, all of the above is merely suggestive and does not prove that
multiple levels of block are not implicated in generating the various rhythms
seen clinically. This is especially so in the case of alternating Wenckebach,
since the iterative approach does not predict the existence of the type A
form, which is more common, and type A and type B are usually seen
at very different atrial rates [35]. However, the prolonged time course of
recovery seen in some instances of alternating Wenckebach (e.g., Figure 7
of Halpern and colleagues [27]) would be consistent with the predictions of
the iterative approach.

13.4.9 TOPOLOGICAL CONSIDERATIONS

By the change of variable

$$X_i = (AH_i - a)/d, \tag{13.16}$$

which simply rescales the AH interval, the map $AH_{i+1} = f(AHi)$ given by Equation (13.8) is changed into the map

$$X_{i+1} = F(X_i), \quad 0 \le X_i \le 1, \tag{13.17}$$

which is a map of the unit interval. For our standard recovery curve (Figure 13.1B), one then has a one- or two-branched map (Figures 13.5–13.11), each branch of which is monotonically increasing and moves upward with decreasing t_s. In the two-branched case, $F(0) > F(1)$ in Equation (13.17), producing a "nonoverlapping" (i.e., into, but not onto) map. One can then apply theorems due to Keener [33] which prove, among other things, that: (1) at any given value of t_s there is at most one stable periodic orbit that is globally attracting, (2) the asymptotic conduction ratio (i.e., ratio of number of ventricular beats to number of atrial beats) or rotation number defined in Equation (13.15) decreases in a monotonic fashion as t_s decreases, (3) any given rational rotation number corresponds to a periodic $N{:}M$ rhythm and is maintained over an interval of the parameter t_s, (4) when the rotation number is irrational, which occurs over a Cantor set of measure zero of the parameter t_s, the invariant set is itself a Cantor set; the measure of this latter set is zero for the standard map, which has slope < 1 at all points. Property (3) implies that should the diagram of Figure 13.3 be completely filled in, one would have a *devil's staircase*, with all rational rotation numbers generated by a Farey sequence (see left half of Figure 11 of Guevara and colleagues [26]). A similar "nonoverlapping" map occurs when one considers Wenckebach rhythms seen in isolated quiescent rabbit ventricular cells [24].

13.4.10   RHYTHMS IN SPONTANEOUSLY BEATING SYSTEMS

Many of the rhythms described above (Figure 13.4) can be seen in isopotential aggregates of spontaneously beating embryonic chick ventricular cells, for example, Wenckebach, atypical Wenckebach, millisecond Wenckebach, reverse Wenckebach, alternating Wenckebach (type B), Wenckebach (type X), and Wenckebach (type Y) [26]. One can carry out premature stimulation ("phase-resetting") experiments on an aggregate and formulate a map in a fashion similar to that described above. This map can then be iterated to predict the response of the aggregate to periodic stimulation at arbitrary t_s [23]. In this case of a spontaneously beating preparation, one obtains an invertible degree-1 map of the circle if the stimulus amplitude is sufficiently low. An invertible degree-1 circle map has properties (1)–(3) listed above for the nonoverlapping interval map. Mathematically, it is thus not surprising that similar rhythms are seen in the aggregate and the AV node. However, from a physiological point of view the two systems are not at all similar: the aggregate is a relatively homogeneous isopotential system, the AV node is a highly inhomogeneous system in which conduction occurs; the aggregate is a spontaneously firing system, and the AV node a quiescent system. In the case of the spontaneously active system, when the rotation number is irrational, one has a quasiperiodic orbit and the invariant set is not a Cantor set, as in property (4) above for an interval map. Note that the interval map of Equation (13.17) can be made into a invertible degree-1 circle map, which is both into and onto, simply by making $F(0) = F(1)$ and by letting the left-hand and right-hand limits of the slope of the map at the point of the discontinuity and at the origin be equal.

It has been known for a long time that periodic driving of an oscillator (undirectional coupling) or coupling together two oscillators (mutual coupling) can produce phase-locked rhythms resembling various forms of AV heart block [19]. For example, using an electrical analogue made from neon-bulb relaxation oscillators, van der Pol and van der Mark [64] showed that the progression $\{1{:}1 \rightarrow$ Wenckebach $\rightarrow 2{:}1 \rightarrow 5{:}2 \rightarrow 3{:}1 \rightarrow$ high-grade block $\rightarrow$ complete AV block with AV dissociation$\}$ could be mimicked as the coupling between two oscillators was reduced. Roberge and Nadeau [47] showed the additional transitions $\{$Wenckebach $\rightarrow 3{:}2 \rightarrow$ reverse Wenckebach $\rightarrow 2{:}1\}$ using a transistorized electronic analogue. Zloof and colleagues [76] considered a mathematical model of a simple mechanical "cataract" analogue and showed in addition a connection between Wenckebach and Mobitz II block. Periodic driving of a simple one-variable model oscillator produces Wenckebach, reverse Wenckebach, and alternating Wenckebach rhythms (see Guevara and colleagues [25] and Chapter 12 in this book). Studies using coupled electrical oscillators have continued down to the present day [68].

All of the above studies assume that the AV node can be treated as an oscillator being driven by a periodic supranodal input. We are not aware of any published tracings of the transmembrane potential from nodal cells showing spontaneous diastolic depolarization during Wenckebach cycles. In our view, it is a mistake to conclude that the AV node is an oscillator simply because Wenckebach and other rhythms can also be seen in forced oscillating systems. The disarming similarity in the rhythms seen in distributed quiescent systems and isopotential spontaneously active systems stems from the fact previously stated that invertible degree-1 circle maps share the above-mentioned properties (1)–(3) of nonoverlapping interval maps. One way perhaps to discriminate between the two possibilities is to note that the interval map, since it is nonoverlapping, has an intermediate range of forbidden AH intervals, which is not the case with the circle map. However, in situations in which the normally occurring overdrive suppression of the subsidiary AV nodal oscillator is released (e.g., severe sinus bradycardia) or when there is an accelerated junctional rhythm, one might then obtain a complex situation in which rhythms of AV conduction would be due to some combination of conduction block and phase locking.

13.4.11   ALTERNATION OF THE PR INTERVAL AND CHAOTIC DYNAMICS

Alternation of the PR interval, which I have called 2:2 conduction, was perhaps first documented by Lewis and Mathison [39] in a hypoxic cat. It has been seen many times in both experimental and clinical work [18,19,52] and has been attributed to the presence of dual pathways in the AV conduction system, the existence of which have been demonstrated in man [49]. In an elegant piece of experimental work, Mines showed how an alternation between two conduction pathways could be produced when such longitudinal dissociation is present (see Figure 20 of Mines [42] and associated description). However, Figure 13.12C suggests that alternans can be produced in a single pathway if supernormal conduction is present. In that case, there is a direct transition from a 1:1 to a 2:2 rhythm as t_s is decreased, which is the direct result of a period-doubling bifurcation on the map (Figure 13.12C). Supernormal conduction in the AV conduction system has been demonstrated in man [53]; in one experimental study on Purkinje fiber in which the transition $\{1{:}1 \rightarrow 2{:}2\}$ was seen, there is evidence that supernormal excitability was present in at least one of the preparations studied (Figures 2b and 2c of Chialvo and Jalife [8]). Note that should the size of the supernormal dip on the recovery curve be increased, one could then produce period-4 and other higher-order period-doubled orbits on the map, corresponding to 4:4, 8:8, ..., rhythms. As t_s is decreased further, these orbits might then progress on to "chaotic" (i.e., deterministically irregular) dynamics via a cascade of period-doubling bifurcations, in a manner analogous to that described in heart-cell aggregates [23]. The rhythms corresponding to a chaotic orbit would still show one ventricular beat for each atrial beat, but with an aperiodic wandering of the AH interval within a confined range. Although rhythms of AV block in which there is a "flottement du PQ" have been described [40], there has been no analysis showing that these rhythms are chaotic. In fact, since the "flottement" disappears when respiration is temporarily suspended, the mechanism cannot be attributed simply to the AV nodal recovery curve. However, there has been one experimental study on Purkinje fiber showing that irregular—indeed, perhaps, chaotic—conduction rhythms were produced following a period-doubling bifurcation as t_s was reduced [8]. In this regard, one should note that it is possible to produce chaotic dynamics following a single bifurcation in maps of the form shown in Figure 13.12C [62]. While one has a nonperiodic orbit when the asymptotic conduction ratio is irrational in the maps of Figures 13.7–13.11, the corresponding rhythms are not chaotic (i.e., deterministically irregular), since, among other things, sensitive dependence on initial conditions is absent from this class of nonoverlapping interval maps [33].

13.4.12   TRANSIENT ALTERNANS

Transient alternation in the AH interval has been described in the AV node
[6,9]. The discordant alternation in action potential upstroke velocity and
AH conduction time seen in N cells (Figures 4 and 6 of de Beer [9]) during
transient alternation is consistent with the hypothesis that the alternation
is being produced by the usual mechanism in which upstroke velocity determines conduction velocity. Transient alternation in the action potential
duration simultaneously occurs [9]. Transient alternation of some variable
can be accounted for if one has a map in which the slope of the map at the
stable period-1 fixed point (which corresponds to a 1:1 response) is negative (e.g., Figure 5-17A of Guevara [19]). Transient alternation of the AH_i
would occur in the map of Figure 13.12C were t_s to be increased slightly so
that the period-1 fixed point becomes stable, but lies on the negative-slope
region of the map: in fact, as t_s is decreased in Figure 13.12C one has the
sequence of rhythms {1:1 (with monotonic approach to the asymptotic AH
interval) $\rightarrow$ 1:1 (with alternating approach to the asymptotic AH interval)
$\rightarrow$ 2:2}. A similar sequence has been described in the action potential duration of quiescent (see Chapter 12 in this volume) and spontaneously beating
[25] ventricular heart-cell aggregates. Transient alternation will also be seen
in maps in which the nonmonotonicity, while still present, is not sufficiently
deep so as to produce a stable period-2 orbit as t_s is reduced, as occurs in
Figure 13.12C. This is probably the situation in the normal AV node, since
it seems that it is generally not possible to produce a sustained AH alternans simply by decreasing t_s. Since the AV nodal recovery curve does not
normally show any hint of a supernormal dip, the mechanism underlying
the alternans is probably different from that underlying generation of the
recovery curve. This conclusion is reinforced by the fact that the recovery
curve is usually obtained at a basic cycle length considerably longer than
the pacing cycle length that must be used to produce transient alternans.

13.4.13   OTHER TRANSIENTS

The transient alternans described above is superimposed on a more slowly
developing gradual increase in the AH interval. Indeed, this monotonic increase is also present when the atrial rate is not sufficiently high to generate
transient alternans. The time course of the adaptation of the AH interval
has two components: an initial fast component is superimposed on a longer
lasting, much more slowly developing component. The iterations shown in
Figure 13.6D can be interpreted as resulting from a situation in which there
is an abrupt increase in pacing rate, with a step decrease in t_s from about
540 msec to 420 msec. In that case, there is a monotonic approach to the
new steady AH_1^* at 171 msec from the preexisting value of 105 msec (see
also Figure 5a of van der Tweel and colleagues [65]). This is indeed the case
in the rabbit AV node, where there is fairly good agreement between the

AV nodal recovery curve extracted during the early part of the response to step changes in rate and the curve determined in the usual manner [5]. The AV recovery curve obtained from the responses to step changes in rate has been used to reconstruct the AV recovery curve, showing that the time constant characterizing the former also characterizes the latter [30]. In a similar vein, following an extrasystole, the postextrasystolic AH interval lies close to the AV nodal recovery curve over a wide range of coupling intervals of the extrasystole [65].

The second, much more slowly developing component ("fatigue") takes many minutes to reach its steady-state level [6]. In isopotential heart-cell aggregates, the eventual establishment in the steady state of a Wenckebach rhythm is usually preceded by higher order $(n + 1):n$ Wenckebach cycles, with n slowly decreasing to its steady-state value (Figure 4-5 of Guevara [19]). It is likely that a process such as ion accumulation or pumping is involved in producing this phenomenon as is thought to be the case with fatigue. Modeling along these lines has been carried out in neural systems, where a similar phenomenon is seen [19]. Finally, gradual adaptation of conduction time following a step change in input rate can also be seen in an electronic analogue [31].

In the 1:1 zone, one predicts that the steady-state AH interval (AH_1^*) will increase monotonically as t_s decreases (Figures 13.5 and 13.6). This can be seen directly by putting $AH_{i+1} = AH_i = AH_1^*$ in Equation (13.6). Indeed, the clinical finding is that the AV nodal recovery curve extracted by measuring AH and HA intervals during maintained 1:1 rhythm at various atrial pacing rates is quite close to the directly determined AV nodal recovery curve [5]. A close-to-exponential relationship between conduction time and t_s ("dispersion curve") is also found in a simple electronic conduction analogue [31].

13.4.14   Coexisting AH Intervals

Discontinuous breaks or jumps in the AV nodal recovery curve are quite common [11,69,72]. The discontinuity is usually attributed to the presence of "dual pathways," although there is little direct physiological or anatomical evidence for the existence of such structures. Figure 13.12A suggests that two different AH intervals ("bistability") can coexist when such a discontinuous or type III AV nodal recovery curve is present, provided that t_s is correctly chosen. In addition, it should be possible to flip from one AH interval to the other by injecting a suitably timed extrastimulus, thus changing the initial condition. While there is considerable evidence for spontaneous transitions between two AH intervals during normal sinus rhythm or incremental atrial pacing [69], as well as for the ability to flip from one AH interval to the other using programmed electrical stimulation [48,69], it is not clear that the mechanism shown in Figure 13.12A is operating in these instances. This is because there can be significant overlap between the two branches of the discontinuous AV nodal recovery curve where two coexisting AH intervals are seen (e.g., Figure 3 of Rosen and colleagues [49]). This overlap alone could account for these two phenomena. Along similar lines, the production of two PR intervals by carotid sinus massage [51] is no guarantee that bistability is present. Thus, a systematic experimental or clinical study is needed in which the range of t_s over which coexistence of two AH intervals is possible is established and compared with the results of iterating the AV nodal recovery curve obtained in the same subject when that curve demonstrates a discontinuous jump or supernormality.

Another circumstance in which coexistence occurs is shown in Figure 13.12B, where there is a bistability between a period-1 orbit corresponding to a 1:1 rhythm and a period-2 orbit corresponding to a 3:2 Wenckebach rhythm. We know of one electrocardiogram in which a 1:1 rhythm was converted into a 4:3 Wenckebach rhythm following an extrasystole (Figure 8 of Levy and colleagues [37]). However, as Levy and colleagues suggest, this might be due to a change in the state of the heart; once again, a systematic study is needed in which the results of iterating the AV nodal recovery curve would be compared with the results of pacing the right atrium at different rates. Note that while the gap phenomenon responsible for producing bistability in Figure 13.12B is usually attributed to effects at multiple levels in the heart [70], it is possible to see a gap in excitation in a space-clamped system when supernormal excitability exists, provided that the stimulus amplitude is correctly chosen (see Figure 5 of Guevara and Shrier [21]). In such a circumstance, the analysis of the response to periodic stimulation is also reduced to the study of maps in which two discontinuous jumps are present (e.g., Figure 7 (row 4) of Guevara and Shrier [21]).

13.4.15   EFFECTS OF NOISE

Noise might be defined, for our purposes, as uncontrolled variations in some parameter in the system, for example, respiration, autonomic tone, membrane noise, and so forth. When such a parameter changes, the dynamics change, with there being, for example, a change in the AH intervals, maintaining the "noise-free" rhythm, or perhaps even a change in the rhythm itself. Exactly what happens depends to a large extent on both the amplitude and frequency content of the noise. For example, high-frequency fluctuations in the excitability of the membrane due to membrane voltage noise might be enough to produce a single 4:3 cycle in what would otherwise be a maintained 3:2 rhythm in the noise-free situation. Thus, one of the major effects of noise is to break down a periodic rhythm, generally replacing it with a nonperiodic rhythm, producing a "bifurcation gap." Indeed, Wenckebach, reverse Wenckebach, and Wenckebach (type X) rhythms are usually nonperiodic in the paced human heart (Figure 8 of Shrier and colleagues [54]); the in situ dog heart must be denervated to produce a stable

4:3 Wenckebach rhythm [55]. More complex rhythms are more susceptible to the corrupting effects of noise for two reasons: (1) they are maintained over a smaller interval of the parameter t_s, and (2) the beat-to-beat change in a variable (e.g., AH interval) is smaller, and so more liable to be influenced by noise. This probably explains why reverse Wenckebach rhythms are "evanescent" [3] and why Wenckebach (type X) rhythms have been so rarely described. Even in the denervated dog heart, the 4:3 Wenckebach rhythm exists over a range of t_s that is only 4 to 8 msec wide, and higher order Wenckebach cycles are unstable [55].

13.4.16   A UNIFIED THEORY OF RHYTHMS OF AV BLOCK?

Iteration of the standard AV nodal recovery curve gives many of the rhythms of AV block that have been clinically and experimentally described. This suggests that rhythms that go by various names (e.g., Wenckebach, reverse Wenckebach, alternating Wenckebach, millisecond Wenckebach, Mobitz II) have a common origin. Several attempts have been made to unify seemingly different rhythms using analysis of clinical electrocardiograms, physiological experimentation, analogues (electrical, electronic, or mechanical), and mathematical modeling. Perhaps the most far-reaching investigation to date remains that of Decherd and Ruskin [10], who showed from analysis of the AV nodal recovery curve that rhythms such as first-degree AV block, Wenckebach, atypical Wenckebach, and Mobitz II, as well as rhythms that were subsequently termed millisecond Wenckebach, reverse Wenckebach, alternating Wenckebach (type B), Wenckebach (type X), and Wenckebach (type Y), are all expected to occur. In particular, Decherd and Ruskin suggested that Wenckebach and Mobitz II block are linked; moreover, the two rhythms are linked via what is currently termed millisecond Wenckebach: "Since all varieties of partial A-V block, with the exception of Type II, are seen to have the same fundamental explanation in terms of the auricular rate and the recovery curve of A-V conduction, it seems probable that Type II would have the same physiological basis. We have no instances available for study, but we suggest the possibility that this type of block occurs when the recovery curve is an almost horizontal straight line, from the absolute refractory period to the end of recovery. We would predict, if this assumption were true, that minute increments of the P-R interval precede the sudden dropped beat." Both of these predictions found confirmation 30 years later [13,14]. In addition, the close connection between Wenckebach and reverse Wenckebach suggested by Decherd and Ruskin has been documented in experimental work by Roberge and Nadeau [47].

The electrophysiological basis of block is not completely clear at the present time (see Chapter 14 by Jalife and Delmar in this volume). However, computer simulation of an ionic model shows that both Wenckebach and

Mobitz II block can occur in a one-dimensional strand of Purkinje fiber [20]. Even more strikingly, rhythms resembling Wenckebach [23,24,26], Mobitz II [25], and alternating Wenckebach (type B), Wenckebach (type X), Wenckebach (type Y), and reverse Wenckebach [26] can be observed in effectively isopotential cardiac preparations. While many of the above rhythms can also be seen in simple two-variable excitation-refractoriness models of a quiescent space-clamped system [2,16,36,44,45,50,58], one should be careful about drawing the conclusion that the electrophysiological mechanisms of these rhythms in the intact heart does not involve conduction, remembering the deceptive correspondence between the properties under iteration of interval and circle maps discussed above. Indeed, one obtains interval maps both for the quiescent space-clamped system (e.g., Figure 2 of Guevara and colleagues [24]) and for the quiescent distributed system (as described above), although the electrophysiology underlying the rhythms might be quite different.

Finally, we wish to mention that several patterns of conduction have been described that have not been accounted for above in the iterative approach. These include (see Guevara [19] for references): (1) alternating Wenckebach (type A) rhythms, (2) alternating Wenckebach (type B) cycles terminated by, not one, but rather two consecutive 3:1 cycles [73], (3) Wenckebach cycles terminated by two consecutive blocked atrial beats, and (4) cycles in which the blocked beat is preceded by a sequence of, not increasing, but rather decreasing AH intervals. It remains to be seen whether or not these and other rhythms can be accounted for using the iterative approach.

Acknowledgements: We thank Kate Biscomb and Diane Colizza for typing the manuscript, Robert Lamarche for photographing the figures, Dominique Jeandupeux for help with drafting, R. de Boer for help with computer graphics, and Professors L. Glass, H.J. Jongsma, and A. Shrier for use of their computing facilities. We also thank the Canadian Heart Foundation and the Natural Sciences and Engineering Research Council for pre- and postdoctoral fellowships awarded during the time this work was carried out. Supported in part by grants from the Quebec Heart Foundation and the Medical Research Council of Canada.

13.5 Appendix

Teague and colleagues [61] determined the AV recovery curve in each one of 50 subjects with normal AV nodal function. They fitted the data for each subject to the function

$$A_2 H_2 = A \exp\left[-B(A_1 A_2)\right] + C, \ \ A_1 A_2 > ERP \qquad (13.18)$$

where $A, B,$ and C are constants, and ERP is the effective refractory period of the AV node. Teague and colleagues [61] approximated C by the basic $A_1 H_1$ interval.

Note that this formula implicitly assumes that the $A_1 A_2$ interval is a good predictor of the $A_2 H_2$ interval. However, the $H_1 A_2$ interval appears to be a better determinant of the $A_2 H_2$ interval than the $A_1 A_2$ interval [5]. Use of the $H_1 A_2$ (or RP) rather than the $A_1 A_2$ (or PP) interval as an index of recovery time has a long history [37,38,43]. Therefore, I rewrite Equation (13.18) as follows:

$$
\begin{aligned}
A_2 H_2 &= A \exp\left[-B(A_1 A_2)\right] + C = A \exp\left[-B(A_1 H_1 + H_1 A_2)\right] + C \\
&= A \exp[-B(A_1 H_1)] \exp\left[-B(H_1 A_2)\right] + C, \\
&\quad H_1 A_2 > ERP - A_1 H_1.
\end{aligned}
\tag{13.19}
$$

Thus, given that $C = A_1 H_1$, with the identifications

$$
\{a = C, \quad b = A \exp\left[-B(A_1 H_1)\right], \quad c = ERP - C, \quad \tau = 1/B\}, \tag{13.20}
$$

one can obtain Equation (13.1) from Equation (13.19). Taking the mean of the 50 sets of the parameters $A, B, C,$ and ERP given in Table 1 of Teague and colleagues [61], and using Equation (13.20), one obtains $a = 90$ msec, $b = 780$ msec, $c = 220$ msec, and $\tau = 110$ msec for Equation (13.1).

REFERENCES

[1] G.H. Anderson, K. Greenspan, and C. Fisch. Electrophysiologic studies on Wenckebach structures below the atrioventricular junction. *Am. J. Cardiol.*, 30:232–236, 1972.

[2] M.B. Berkinblit. The periodic blocking of impulses in excitable tissues. In I.M. Gelfand, V.S. Gurfinkel et al., *Models of the Structural-Functional Organization of Certain Biological Systems*, pages 155–189. MIT Press, Cambridge, 1971.

[3] R. Berman. Reverse Wenckebach periods: Five cases of an unstable form of partial auriculoventricular block. *Am. Heart J.*, 50:211–217, 1955.

[4] J. Billette. Atrioventricular nodal activation during periodic premature stimulation of the atrium. *Am. J. Physiol.*, 252:H163–H177, 1987.

[5] J. Billette. Preceding His-atrial interval as a determinant of atrioventricular nodal conduction time in the human and rabbit heart. *Am. J. Cardiol.*, 38:889–896, 1976.

[6] J. Billette, R. Métayer, and M. St-Vincent. Selective functional characteristics of rate-induced fatigue in rabbit atrioventricular node. *Circ. Res.*, 62:790–799, 1988.

[7] J. Billette and M. St-Vincent. Functional origin of rate-induced changes in atrioventricular nodal conduction time of premature beats in the rabbit. *Can. J. Physiol. Pharmacol.*, 65:2329–2337, 1987.

[8] D.R. Chialvo and J. Jalife. Non-linear dynamics of cardiac excitation and impulse propagation. *Nature*, 330:749–752, 1987.

[9] E.L. de Beer. *Atrioventricular Conduction. An Experimental and Theoretical Study of Nodal Action Potentials and Propagation Times.* Ph.D. thesis, University of Utrecht, Utrecht, the Netherlands, 1977.

[10] G.M. Decherd and A. Ruskin. The mechanism of the Wenckebach type of A-V block. *Br. Heart J.*, 8:6–16, 1946.

[11] P. Denes, D. Wu, R. Dhingra, F. Amat-y-Leon, C. Wyndham, and K.M. Rosen. Dual atrioventricular nodal pathways. A common electrophysiological response. *Br. Heart J.*, 37:1069–1076, 1975.

[12] R.C. Dhingra, K.M. Rosen, and S.H. Rahimtoola. Normal conduction intervals and responses in sixty-one patients using His bundle recording and atrial pacing. *Chest*, 64:55–59, 1973.

[13] N. El-Sherif, J. Aranda, B. Befeler, and R. Lazzara. Atypical Wenckebach periodicity simulating Mobitz II AV block. *Br. Heart J.*, 40:1376–1383, 1978.

[14] N. El-Sherif, B.J. Scherlag, and R. Lazzara. Pathophysiology of second degree atrioventricular block: A unified hypothesis. *Am. J. Cardiol.*, 35:421–434, 1975.

[15] T.W. Englemann. Beobachtungen und Versuche am suspendierten Herzen. Zweite Abhandlung. Ueber die Leitung der Bewegungsreize im Herzen. *Arch. Ges. Physiol.*, 56:149–202, 1894.

[16] M. Feingold, D.L. Gonzalez, O. Piro, and H. Viturro. Phase locking, period doubling, and chaotic phenomena in externally driven excitable systems. *Phys. Rev.*, A 37:4060–4063, 1988.

[17] G.R. Ferrier and P.E. Dresel. Relationship of the functional refractory period to conduction in the atrioventricular node. *Circ. Res.*, 35:204–214, 1974.

[18] C. Fisch and E.F. Steinmetz. Supernormal phase of atrioventricular (A-V) conduction due to potassium. A-V alternans with first-degree A-V block. *Am. Heart J.*, 62:211–220, 1961.

[19] M.R. Guevara. *Chaotic Cardiac Dynamics.* Ph.D. thesis, McGill University, Montreal, 1984.

[20] M.R. Guevara. Spatiotemporal rhythms of block in an ionic model of cardiac Purkinje fibre. In M. Markus, S.C. Müller and G. Nicolis, editors, *From Chemical to Biological Organization*, pages 273–281. Springer, Berlin, 1988.

[21] M.R. Guevara and A. Shrier. Phase resetting in a model of cardiac Purkinje fiber. *Biophys. J.*, 52:165–175, 1987.

[22] M.R. Guevara, F. Alonso, D. Jeandupeux, and A.C.G. van Ginneken. Alternans in periodically stimulated isolated ventricular myocytes: Experiment and model. In A. Goldbeter, editor, *Cell to Cell Signalling: From Experiments to Theoretical Models*, pages 551–563. Academic Press, London, 1989.

[23] M.R. Guevara, L. Glass, and A. Shrier. Phase locking, period-doubling bifurcations, and irregular dynamics in periodically stimulated cardiac cells. *Science*, 214:1350–1353, 1981.

[24] M.R. Guevara, D. Jeandupeux, F. Alonso, and N. Morissette. Wenckebach rhythms in isolated ventricular heart cells. In St. Pnevmatikos, T. Bountis and Sp. Pnevmatikos, editors, *Singular Behavior and Nonlinear Dynamics*, pages 629–642. World Scientific, Singapore, 1989.

[25] M.R. Guevara, A. Shrier, and L. Glass. Chaotic and complex cardiac rhythms. In D.P. Zipes and J. Jalife, editors, *Cardiac Electrophysiology: From Cell to Bedside.*, pages 192–201. Saunders, Philadelphia, 1990.

[26] M.R. Guevara, A. Shrier, and L. Glass. Phase-locked rhythms in periodically stimulated heart cell aggregates. *Am. J. Physiol.*, 254:H1–H10, 1988.

[27] M.S. Halpern, G.J. Nau, R.J. Levi, M.V. Elizari, and M.B. Rosenbaum. Wenckebach periods of alternate beats. Clinical and experimental observations. *Circulation*, 48:41–49, 1973.

[28] J. Hay. Bradycardia and cardiac arrhythmia produced by depression of certain of the functions of the heart. *Lancet*, 1:139–143, 1906.

[29] R.M. Heethaar, R.M. de Vos Buchart, J.J.D van der Gon, and F.J. Meijler. A mathematical model of A-V conduction in the rat heart. II. Quantification of concealed conduction. *Cardiovasc. Res.*, 7:542–556, 1973.

[30] R.M. Heethaar, J.J.D. van der Gon, and F.J. Meijler. Mathematical model of A-V conduction in the rat heart. *Cardiovasc. Res.*, 7:105–114, 1973.

[31] R.M. Heethaar, J.J.D. van der Gon, and F.L. Meijler. Interpretation of some properties of A-V conduction with the help of analog simulation. *Eur. J. Cardiol.*, 1:87–93, 1973.

[32] J. Honerkamp. The heart as a system of coupled nonlinear oscillators. *J. Math. Biol.*, 18:69–88, 1983.

[33] J.P. Keener. Chaotic behavior in piecewise continuous difference equations. *Trans. Am. Math. Soc.*, 261:589–604, 1980.

[34] J.P. Keener. On cardiac arrhythmia: AV conduction block. *J. Math Biol.*, 12:215–225, 1981.

[35] B.D. Kosowsky, P. Latif, and A.M. Radoff. Multilevel atrioventricular block. *Circulation*, 54:914–921, 1976.

[36] H.D. Landahl and D. Griffeath. A mathematical model for first degree block and the Wenckebach phenomenon. *Bull. Math. Biophys.*, 33:27–38, 1971.

[37] M.N. Levy, P.J. Martin, J. Edelstein, and L.B. Goldberg. The A-V nodal Wenckebach phenomenon as a positive feedback mechanism. *Prog. Cardiovasc. Dis.*, 16:601–613, 1974.

[38] T. Lewis and A.M. Master. Observations upon conduction in the mammalian heart. A-V conduction. *Heart*, 12:210–269, 1926.

[39] T. Lewis and G.C. Mathison. Auriculo-ventricular heart block as a result of asphyxia. *Heart*, 2:47–53, 1910.

[40] S. Lo Sardo and L. Solinas. Le "flottement du PQ". II. Étude clinico-pathologénique d'une variété singulière de bloc A.V. *Arch. Mal. Coeur*, 45:824–829, 1952.

[41] H.J.L. Marriott, A.F. Schubart, and S.M. Bradley. A-V dissociation: A reappraisal. *Am. J. Cardiol.*, 2:586–605, 1958.

[42] G.R. Mines. On dynamic equilibrium in the heart. *J. Physiol. (Lond.)*, 46:349–383, 1913.

[43] W. Mobitz. Über die unvollständige Storüng der Erregungsüberleitung zwischen Vorhof und Kammer des menschlichen Herzens. *Z. ges. exp. Med.*, 41:180–237, 1924.

[44] R.A. Nadeau, F.A. Roberge, and P. Bhéreur. The mechanism of the Wenckebach phenomenon. *Isr. J. Med. Sci.*, 5:814–818, 1969.

[45] I. Nagumo and S. Sato. On a response characteristic of a mathematical neuron model. *Kybernetik*, 10:155–164, 1972.

[46] G. Oreto, F. Luzza, and G. Satullo. Progressive prolongation of the second conduction interval throughout successive 3:2 Wenckebach sequences: The double Wenckebach phenomenon. *Am. Heart J.*, 118:413–415, 1989.

[47] F.A. Roberge and R.A. Nadeau. The nature of Wenckebach cycles. *Can. J. Physiol. Pharmacol.*, 47:695–704, 1969.

[48] K.M. Rosen, P. Denes, D. Wu, and R.C. Dhingra. Electrophysiological diagnosis and manifestation of dual A-V nodal pathways. In A.J.J. Wellens, K.I. Lie, and M.J. Janse, editors, *The Conduction System of the Heart*, pages 453–466. Stenfert Kroese, Leiden, 1976.

[49] K.M. Rosen, A. Mehta, and R.A. Miller. Demonstration of dual atrioventricular nodal pathways in man. *Am. J. Cardiol.*, 33:291–294, 1974.

[50] O.E. Rössler, R. Rössler, and H.D. Landahl. Arrhythmia in a periodically forced excitable system (abstract). *Sixth Int. Biophys. Congress, Kyoto*, 1978.

[51] L. Schamroth and M.M. Perlman. Periodic variations in A-V conduction time: The "supernormal" phase of A-V conduction. A study in differential dual A-V pathway conductivity and refractoriness. *J. Electrocardiol.*, 6:81–84, 1973.

[52] M. Segers. L'alternance du temps de conduction auriculo-ventriculaire. *Arch. Mal. Coeur*, 44:525–527, 1951.

[53] M. Segers and H. Denolin. Etude de la transmission auriculo-ventriculaire. III. La phase supernormale de la conduction. *Acta Cardiol.*, 1:279–282, 1946.

[54] A. Shrier, H. Dubarsky, M. Rosengarten, M.R. Guevara, S. Nattel, and L. Glass. Prediction of complex atrioventricular conduction rhythms in humans with use of the atrioventricular nodal recovery curve. *Circulation*, 76:1196–1205, 1987.

[55] M.P. Simson, J.F. Spear, and E.N. Moore. Electrophysiological studies on atrioventricular nodal Wenckebach cycles. *Am. J. Cardiol.*, 41:244–258, 1978.

[56] R. Slama, J.F. Leclercq, M. Rosengarten, P. Coumel, and Y. Bouvrain. Multilevel block in the atrioventricular node during atrial tachycardia and flutter alternating with Wenckebach phenomenon. *Br. Heart J.*, 42:463–470, 1979.

[57] R. Slama, C. Sebag, G. Motte, and J.F. Leclercq. Double Wencke-bach phenomenon in atrioventricular node and His bundle. Electro-physiological demonstration in a case of atrial flutter. *Br. Heart J.*, 45:328–330, 1981.

[58] V.V. Smolyaninov. Analysis of the sequence of Wenckebach cycles. *Biophysics*, 11:382–390, 1966.

[59] V.V. Smolyaninov. Omission of pulses in an elementary fibre model. *Biophysics*, 13:587–599, 1968.

[60] J.F. Spear and E.N. Moore. Effect of potassium on supernormal con-duction in the bundle branch-Purkinje system of the dog. *Am. J. Cardiol.*, 40:923–928, 1977.

[61] S. Teague, S. Collins, D. Wu, P. Denes, K. Rosen, and R. Arzabaecher. A quantitative description of normal AV nodal conduction curve in man. *J. Appl. Physiol.*, 40:74–78, 1976.

[62] C. Tresser, P. Coullet, and A. Arneodo. On the existence of hysteresis in a transition to chaos after a single bifurcation. *J. Phys. (Paris) Lett.*, 41:L243–L246, 1980.

[63] F.J.L. van Capelle, J.C. du Perron, and D. Durrer. Atrioventricular conduction in isolated rat heart. *Am. J. Physiol.*, 221:284–290, 1971.

[64] B. van der Pol and J. van der Mark. The heartbeat considered as a relaxation oscillation, and an electrical model of the heart. *Phil. Mag.*, 6:763–775, 1928.

[65] I. van der Tweel, J.N. Herbschleb, C. Borst, and F.L. Meijler. Deter-ministic model of the canine atrio-ventricular node as a periodically perturbed, biological oscillator. *J. Appl. Cardiol.*, 1:157–173, 1986.

[66] W.A. van Duyl. Latent oscillating features of the smooth muscle of the ureter in relation to peristalsis, In *Oscillations in Physiological Systems: Dynamics and Control*, pages 159–164. Institute of Mea-surement and Control, London, 1984.

[67] K.F. Wenckebach. Zur Analyse des unregelmässigen Pulses. *Z. Klin. Med.*, 37:475–488, 1899.

[68] B.J. West, A.L. Goldberger, G. Rovner, and V. Bhargava. Nonlinear dynamics of the heartbeat. I. The AV junction: Passive conduit or active oscillator? *Physica*, 17D:198–206, 1985.

[69] D. Wu. Dual atrioventricular nodal pathways: A reappraisal. *PACE*, 5:72–89, 1982.

[70] D. Wu, P. Denes, R. Dhingra, and K.M. Rosen. Nature of the gap phenomenon in man. *Circ. Res.*, 34:682–692, 1974.

[71] D. Wu, P. Denes, R.C. Dhingra, C.R. Wyndham, and K.M. Rosen. Quantification of human atrioventricular nodal concealed conduction utilizing $S_1S_2S_3$ stimulation. *Circ. Res.*, 39:659–665, 1976.

[72] D.G. Wyse. Relationship between conductivity and functional refractoriness of atrioventricular node in man. *Cardiovasc. Res.*, 16:457–466, 1982.

[73] M.L. Young, H. Gelband, and G.S. Wolff. Atrial pacing-induced alternating Wenckebach periodicity and multilevel conduction block in children. *Am. J. Cardiol.*, 57:135–141, 1986.

[74] M.L. Young, G.S. Wolff, A. Castellanos, and H. Gelband. Application of the Rosenblueth hypothesis to assess atrioventricular nodal behavior. *Am. J. Cardiol.*, 57:131–134, 1986.

[75] J. Zhao and J. Billette. Modulation of recovery-dependent changes in conduction time by facilitation and fatigue during 4:3 Wenckebach cycles in rabbit atrioventricular node (abstract). *FASEB J.*, 4:A560, 1990.

[76] M. Zloof, R.M. Rosenberg, and J. Abbott. A computer model for atrio-ventricular blocks. *Math. Biosci.*, 18:87–117, 1973.

14

Ionic Basis of the Wenckebach Phenomenon

Jose Jalife[1]
Mario Delmar[1]

ABSTRACT We studied the ionic basis of slow recovery of excitability and the rate-dependent activation failure in enzymatically dissociated guinea pig ventricular myocytes and in numerical simulations using a modified version of the Beeler and Reuter model. In addition, appropriate parameters derived from biological and numerical voltage and current clamp experiments were used to devise an analytical model for diastolic recovery of excitability on the basis of the equations for current distribution in a resistive-capacitive circuit. Iteration of the analytical model equations gave rise to dynamics that closely resembled the experimentally obtained phase-locking patterns for repetitive stimulation of the ventricular cell. The results strongly suggest that slow deactivation of the delayed rectifier current (i_K) determines the time-dependent recovery of excitability during diastole while the inward rectifier (i_{K_1}) determines the amplitude and shape of depolarizations within the subthreshold range. The kinetics and voltage-dependence of both currents are responsible for the development of rate-dependent phase-locking patterns and of Wenckebach periodicity in the ventricular myocyte.

14.1 Introduction

Rate-dependent heart block was described almost 90 years ago by Wenckebach [23] and demonstrated by His [8]. In its classical manifestation, the so-called Wenckebach phenomenon is characterized by a succession of electrocardiographic complexes in which atrioventricular (AV) conduction time (the P-R interval) increases progressively in decreasing increments until transmission failure occurs. Following the dropped ventricular discharge, propagation is reinitiated and the cycle is repeated (Figure 14.1). In this chapter we give a brief and necessarily incomplete historical account of the most important developments that have led to our present-day knowledge of the dynamics and ionic mechanisms of the Wenckebach phenomenon.

[1]Department of Pharmacology, SUNY Health Science Center, Syracuse New York 13210

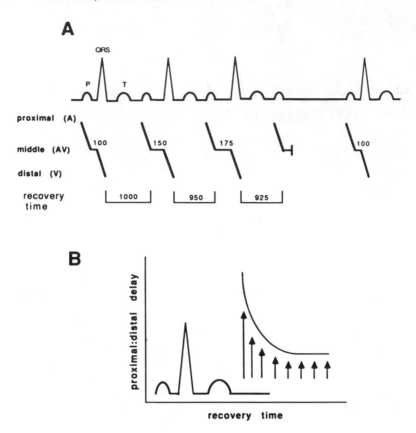

FIGURE 14.1. Schematic diagram illustrating the dynamics of Wenckebach periodicity. (A) Electrocardiogram and *Lewis diagram* of a 4:3 Wenckebach cycle. Numbers are in msec. A, atrium; AV, atrioventricular node; V, ventricle. (B) Recovery curve; the upward arrows indicate timing of atrial discharges (Modified with permission from Katz and Dick [11].)

14.2   AV Nodal Wenckebach

Several hypotheses have been put forth to account for such an interesting cyclic phenomenon and perhaps the most plausible was that originally proposed as early as 1924 by Mobitz [16]. According to this author, the progressive lengthening of the P-R interval occurs at the expense of the subsequent R-P interval (recovery time; Figure 14.1A), which shortens concomitantly after every ventricular discharge. Consequently, the pattern must be associated with a progressive shortening of the time for recovery of ventricular excitability. As the AV conduction time increases progressively on a beat-to-beat basis, the impulse will find the ventricles less and less recovered, until failure occurs (Figure 14.1B). Rosenblueth [18,19] confirmed Mobitz's contention. His experiments in the canine heart further suggested that the progressive delay and eventual failure, leading to recovery and to the start of a new cycle, were in fact the result of discontinuous propagation across the atrioventricular conduction system.

Under such a scheme, the electrical impulse initiated in the sinus node may travel throughout the atrium without interruption or decrement in its velocity. However, upon reaching the AV node, the impulse may stop altogether or resume its journey but only after a delay imposed by the functional refractory period (FRP) of the system, which Rosenblueth defined as the minimum attainable interval between two ventricular responses, both propagated from the atrium. As later demonstrated by Merideth and colleagues [15], in the mammalian AV conducting system the FRP is determined by the recovery of the excitability of the AV node tissue itself. Rosenblueth [19] concluded that the dynamics of the AV node in response to electrical impulses at high frequencies cannot be explained on the basis of smooth and uninterrupted propagation and is not compatible with the behavior of a homogeneous system. Such rate-dependent behavior could only be explained on the basis of propagation through an intrinsically inhomogeneous conducting pathway.

Microelectrode experiments in the isolated rabbit AV node by Merideth and colleagues [15] provided strong support to Mobitz's and Rosenblueth's hypotheses. Merideth and colleagues [15] demonstrated that the refractory period of the so-called "N" cells in the center of the AV node greatly outlasts the action potential duration (postrepolarization refractoriness), thus supporting Rosenblueth's contention that the FRP of the AV conducting system is determined by the time of recovery of the less excitable element in that system. Subsequently, Levy and colleagues [13] used Mobitz's idea to derive their "positive feedback" model of AV node Wenckebach periodicity. In such a model, the gradual shortening of the R-P interval that results from the progressive P-R prolongation can be considered as a gradual increase in the degree of prematurity of impulses crossing the AV node. Thus, as the R-P abbreviates progressively more, the AV node is less recovered and the P-R interval must increase with each beat, until failure occurs.

FIGURE 14.2. Example of 4:3 Wenckebach recorded from a cell in the NH region of an isolated rabbit AV node. Four superimposed traces are shown. Calibrations are: Vertical, 20 mV; horizontal, 20 msec. (Reproduced from Paes de Carvalho and de Almeida [17] with permission from the American Heart Association, Inc.)

The first microelectrode recordings of AV node Wenckebach periodicity were obtained by Paes de Carvalho and de Almeida [17]. Their classical illustration of the effects of acetylcholine on the repetitive activation of an "NH" cell in the isolated rabbit AV node is reproduced in Figure 14.2, which shows four superimposed traces obtained during a 4:3 Wenckebach cycle. The authors attributed the progressive delay in the activation process to decremental conduction [4]. However, a close look at the traces reveals that there are no significant changes in the action potential upstroke velocity that could explain the gradual increase in the activation time, and thus the idea of decremental conduction is untenable in this regard. A different interpretation of these events was proposed by Zipes and colleagues [26], who suggested that the phenomena demonstrated by Paes de Carvalho and de Almeida [17] could be explained more easily in terms of electrotonically mediated delays in the excitation of the nodal cell. In fact, it is clear from Figure 14.2 that the time course of the foot that precedes each of the superimposed active responses is essentially the same as that of the subthreshold depolarization (arrow) induced by the atrial impulse during the blocked beat. Indeed, the events in Figure 14.2 may be interpreted as follows: because of the presence of partial block for repetitive atrial stimulation, during successful propagation the electrical impulse stops momentarily at some point (the "N zone") proximal to the NH cell from which these records were obtained. However, the electrotonically mediated depolarization is sufficiently large to bring the NH cell to threshold, but only after a delay imposed by the recovery of excitability of the interposed cells. As activation is delayed progressively more, the time of recovery decreases until, finally, the electrotonically mediated depolarization is unable to reach threshold and block occurs.

Recently, Shrier and colleagues [21] used the AV nodal recovery curve, obtained by premature atrial stimulation in patients undergoing electrophysiological testing, to predict complex patterns of AV conduction induced by rapid atrial pacing. On the basis of theoretical considerations proposed by Keener [12], Shrier and colleagues [21] used a simple iterative procedure to demonstrate that, provided the recovery curve is monotonic, then if there is an $N{:}M$ conduction pattern at one pacing cycle length and an $n{:}m$ pattern at a second cycle length (where N and n are numbers of atrial stimuli and M and m are numbers of ventricular responses) there will be an $(n + N){:}(m + M)$ pattern at some intermediate cycle length. These results represent an important step forward in our understanding of the dynamics of rate-dependent heart block because they indicate that complex cyclic conduction phenomena such as Wenckebach periodicity, reverse Wenckebach, and alternating Wenckebach are demonstrable without the need of invoking complex geometrical arrangements of cell bundles or multiple levels of block.

14.3   Wenckebach in the Sucrose Gap

Although most commonly observed in the AV node, Wenckebach periodicity can also be demonstrated in the sinoatrial node [14], the subepicardial border zone [25], ischemic Purkinje fibers [7], and the Purkinje-muscle junction (R.F. Gilmour, personal communication). In fact, as has been demonstrated in 1974 by Wennemark and Bandura [24], and later by Antzelevitch and colleagues [1,2] and by Jalife and Moe [10], Wenckebach periodicity is a property of any biological model of nonhomogeneous cardiac tissue in which a discrete zone of depressed conductivity is interposed between two normal excitable zones. In Figure 14.3 we have reproduced two examples, published several years ago [9]. Superimposed transmembrane potential recordings were obtained from an isolated canine Purkinje fiber placed in a three-compartment tissue bath. A narrow zone (typically 1–2 mm) of depressed excitability was created by superfusing the central segment of the fiber with an ion-free sucrose solution. The preparation was driven by biphasic stimuli applied to the proximal segment (P; top tracing) at various basic cycle lengths (BCL) and transmission to the distal segment (D; bottom tracing) was measured for all consecutive impulses. At BCL = 2000 msec, stable 1:1 P-D locking was maintained (not shown). In panel A of Figure 14.3, the six superimposed traces represent beats 1 to 6 during a 6:5 Wenckebach cycle obtained at BCL = 1500 msec. In this case, the "typical" Wenckebach structure is clearly apparent with progressively greater P-D delays at decreasing increments. These dynamics cannot be attributed to decremental conduction. Active propagation ceases in the central segment because of the almost complete lack of ions in the external solution in the sucrose gap, but transmission to D is accomplished as a result of electrotonic spread through that segment. As in the AV node example (see Figure 14.2), the progressive delay leading to block is the result of a progressive shortening of the allowable diastolic interval in the distal tissue and of the characteristic nonlinear time course in the recovery after an active response.

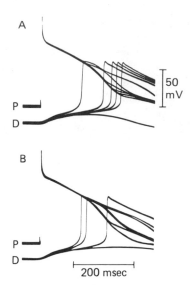

FIGURE 14.3. "Typical" (A) and "atypical" (B) Wenckebach periodicity in a dog
Purkinje fiber-sucrose gap preparation. $P =$ proximal; $D =$ distal transmembrane
potential recordings. (A) BCL = 1500 msec; six superimposed traces show a 6:5
$P{:}D$ locking pattern. (B) BCL = 1400 msec; four superimposed traces show a 4:3
pattern. (Reproduced with permission from Jalife [9].)

Abbreviating the BCL of the stimuli applied to the proximal segment led
to P-D locking phenomena that simulated atypical Wenckebach periodicity
(Figure 14.3B) with a progressive increase in the increment of the beat-by-
beat step delay. Finally, as the BCL was changed over wide ranges, other
rhythms analogous to "second degree block" also emerged, including 4:3,
5:4, and 2:1 P-D locking.

14.4   Wenckebach in the Ventricular Myocyte

The results presented in the previous section suggest that Wenckebach periodicity and post-repolarization refractoriness result from diastolic, time-
dependent changes in the excitability of cells distal to a zone of depressed
conductivity, and are therefore properties of normally polarized cardiac tissues. If this hypothesis is correct, then slow recovery of excitability and rate-
dependent activation failure should be demonstrable also in single cardiac
cells when stimulated with repetitive depolarizing current pulses of critical
amplitude and cycle lengths. Accordingly, it should be possible to determine
the ionic mechanisms of these phenomena. We used single guinea-pig ventricular myocytes [5,6], dissociated by conventional enzymatic procedures
and maintained in a HEPES-Tyrode solution, for recording of transmem-

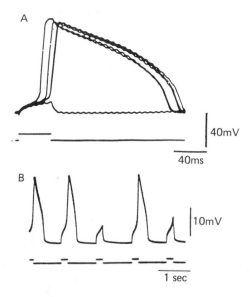

FIGURE 14.4. Wenckebach periodicity in single ventricular myocytes. (A) Five superimposed microelectrode recordings during repetitive stimulation at a BCL of 1000 msec. Note typical structure of Wenckebach periodicity with increasing delays at decreasing increments. (B) Time series showing a 5:3 pattern in a different cell. In both panels the bottom trace is current monitor.

brane potential with an intracellular microelectrode or suction pipette.

Cells were well polarized, resting potentials were always more negative than −79 mV, and gave rise to action potentials of the expected morphology. However, as shown in Figure 14.4 under conditions of repetitive stimulation with pulses of relatively low amplitude, the stimulus:response pattern varied as a function of the BCL. Panel A of Figure 14.4 was obtained from ventricular myocytes with a resting potential of −81 mV. Five superimposed traces are shown. Depolarizing current pulses, 40 msec in duration and 0.15 nA in strength, were applied at a BCL of 1000 msec. A 5:4 (stimulus:response) activation pattern was clearly manifest. In fact, the latency between the onset of the current pulse and the action potential upstroke increased in decreasing increments until failure occurred, which reproduced the typical structure of Wenckebach periodicity (see Figure 14.2). The time series in panel B of Figure 14.4 was taken from a different cell. BCL was 1100 msec, pulse amplitude 0.25 nA, and duration 200 msec. In this case, alternation between 3:2 and 2:1 locking was manifest. Moreover, as in panel A, failure always occurred several hundred milliseconds after full repolarization had been completed, which emphasizes the importance of post-repolarization refractoriness in the development of Wenckebach periodicity.

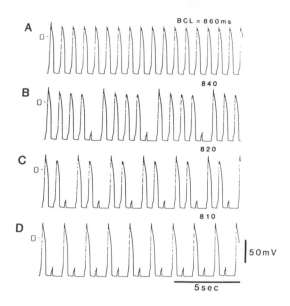

FIGURE 14.5. Wenckebach periodicity in a modified version of the Beeler and Reuter model of a ventricular cell. Action potentials were elicited by repetitive application of depolarizing current pulses ($-1.4\mu A/cm^2$; 100 msec duration) at various cycle lengths. (Reproduced from Delmar et al. [6] with permission from the American Heart Association, Inc.)

14.5   Simulating Wenckebach in the Beeler and Reuter Model

The single myocyte results have been simulated using a modified version of the Beeler and Reuter model for the ventricular cell [3], which includes four transmembrane currents: sodium, calcium, a time-independent potassium current (i_{K1}), and a time-dependent potassium current (i_K). Figure 14.5 shows the simulated action potentials recorded when current pulses were applied repetitively at various cycle lengths. Just as in the ventricular myocyte experiment [6], progressive abbreviation of the stimulus cycle length yielded patterns of frequency-dependent activation failure, including 5:4, 3:2, and 2:1.

These traces suggest that the recovery of excitability after an action potential is indeed a function of the diastolic interval, and that under certain conditions refractoriness can outlast the repolarization phase. Figure 14.6 shows additional simulations in which an S_1-S_2 protocol was used to study the conditions for recovery of excitability as a function of the diastolic interval. In panels A through C, a test pulse, S_2, of relatively long duration

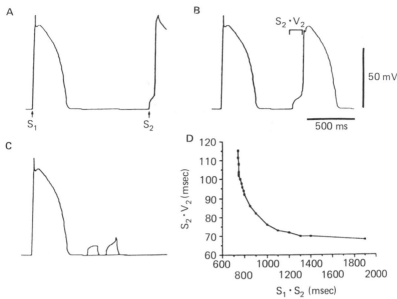

FIGURE 14.6. Post-repolarization refractoriness in the Beeler and Reuter model of a ventricular cell. A test pulse (S_2) similar to that in Figure 14.5 was used to scan the diastolic interval after an action potential induced by an S_1 stimulus. (A) At a relatively long S_1–S_2 interval, the S_2–V_2 interval (measured from the onset of the stimulus to 50% of the action potential upstroke) was brief. (B) At a shorter S_1–S_2, the S_2–V_2 was prolonged. (C) Two superimposed tracings show subthreshold responses at two different S_1–S_2 intervals. (D) S_2–V_2 as a function of S_1–S_2. (Reproduced from Delmar et al. [6] with permission from the American Heart Association, Inc.)

and low amplitude was applied at various coupling intervals after the last of a train of 10 basic S_1 stimuli applied at a cycle length of 1000 msec. Clearly, the activation delay S_2–V_2 increased as the coupling interval decreased, until activation failure occurred. In panel D, a complete plot of S_2–V_2 as a function of S_1–S_2 shows a monotonic recovery curve, which is similar to those described in multicellular preparations of AV node [15], Purkinje fibers [10], and ventricular muscle [20].

14.6 The Recovery Curve

The dynamics of a 5:4 Wenckebach period are illustrated in Figure 14.7, in terms of the recovery curve in a normally polarized myocyte (for details see Delmar and colleagues [5,6]). The white box represents the interval between the onset of the stimulus and the action potential upstroke; the black box represents the action potential duration (APD) and the dashed one, the

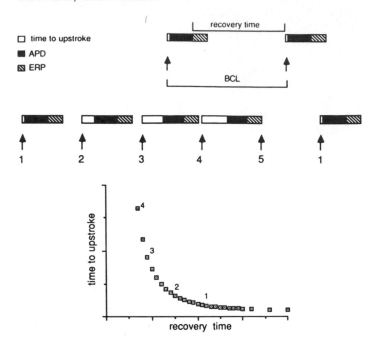

FIGURE 14.7. Dynamics of Wenckebach periodicity as derived from the recovery of excitability curve.

duration of post-repolarization refractoriness (ERP). Recovery time is the interval between the end of the preceding action potential and the occurrence of the next input. In this diagram, the first pulse (labeled one), which occurs at a relatively early interval, gives rise to an action potential whose delay is predicted by the recovery curve. The second stimulus now occurs at a briefer interval within the recovery curve, and the action potential is further delayed. Since the cycle length is constant, the next recovery time is even briefer, which leads to an even longer delay. After the fifth stimulus, the recovery time is briefer than the refractory phase and failure occurs.

14.7   Ionic Mechanisms of Wenckebach Periodicity

14.7.1   RECOVERY OF EXCITABILITY AND i_K DEACTIVATION

The ionic mechanisms of these changes were studied through a combination of voltage and current clamp techniques. Given the fact that the time course of reactivation of the sodium and calcium inward currents is too fast to account for these changes, we considered the process of deactivation

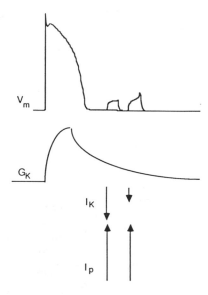

FIGURE 14.8. Diagram illustrating the proposed role of i_K in determining the time course of recovery of excitability in single myocyte. Top tracing, transmembrane potential (V_m) simulated by the Beeler and Reuter model. Bottom tracing, changes in potassium conductance (G_K) during systolic and diastolic intervals. Arrows represent relative magnitudes of i_K and current supplied internally (i_P) at two different times during i_K deactivation.

of the delayed rectifier, potassium outward current, i_K, to be an excellent candidate for controlling the changes in excitability during diastole. In addition, the nonlinear properties of the anomalous rectifying potassium current, i_{K1}, with respect to voltage can accurately explain the shape and time course of the foot potential preceding the active response as well as the subthreshold event during excitation failure. The proposed role of i_K is illustrated diagrammatically in Figure 14.8.

The top tracing represents the membrane potential; an action potential is followed by two subthreshold responses to a depolarizing current pulse of constant amplitude applied at two different intervals. The bottom tracing illustrates the time course of change in i_K conductance during the cardiac cycle and the bottom arrows show the opposing effect of i_K to the depolarizing current pulse (i_P). The conductance to potassium (G_K) increases gradually in the course of the action potential and becomes maximal during the plateau. Upon repolarization, G_K begins to decrease. Because of the slow time course of deactivation, most of the channel conductance decay occurs after repolarization has been completed. Yet, the diastolic potential remains constant because of the lack of driving force for i_K. On the other

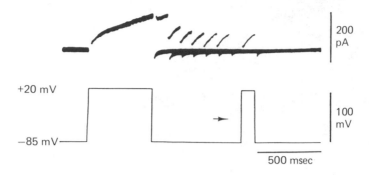

FIGURE 14.9. Envelope test to determine time course of deactivation of i_K (discontinuous single patch-electrode voltage clamp technique). Top tracing, superimposed current traces. Bottom tracing, voltage clamp protocol. Holding potential −85 mV. The presence of inwardly directed current tails on repolarization indicate that the reversal potential of i_K was less negative than HP. Reproduced from Delmar et al. [6] with permission from the American Heart Association, Inc.)

hand, if a depolarizing pulse is applied early in diastole, the membrane voltage displacement will provide sufficient driving force for i_K, which would tend to repolarize the membrane. The influence of G_K decreases at the longer diastolic intervals, which allows for an increase in the amplitude of the subthreshold response. Application of a pulse at a later interval will encounter i_K even more deactivated, and an action potential will ensue. In other words, the time course of recovery of excitability of the single myocyte should be similar to the time course of deactivation of G_K. Our experimental and theoretical work [5,6] suggests that this is indeed the case.

Using single electrode voltage clamp techniques, we have determined the time course of deactivation of i_K in single ventricular myocytes. Cobalt and TTX were superfused continuously at concentrations that were sufficient to block both sodium and calcium currents. The results are illustrated in Figure 14.9. Membrane potential was held at −85 mV. A conditioning command potential allowed for i_K activation, and was followed by a test pulse applied at various intervals. As shown by the current tracing, the amplitude of the instantaneous current jump decreased exponentially as the test interval was prolonged. Since the instantaneous jump is an indication of the degree of i_K deactivation at that particular interval, an envelope joining all such jumps revealed the time course of current decay, which was best described by a monoexponential function, with a mean time constant of 217 msec. This value corresponds very closely to the time course of recovery of excitability in our current clamp experiments [5,6].

14.7.2   THE ANOMALOUS RECTIFIER AND THE APPROACH TO THRESHOLD

Recently, Tourneur [22] demonstrated characteristic responses in membrane potentials of isolated guinea pig ventricular myocytes. Under conditions in which all inward currents had been abolished and only outward currents were operative, relatively long (150 msec) depolarizing current pulses induced membrane potential changes whose shape resembled that of the subthreshold responses observed during conduction block in nonhomogeneous multicellular preparations. Through computer simulations, this author confirmed that the mechanism of such responses is related to the existence of a negative slope region in the IV relation of i_{K1}. This current decreases upon depolarization and its slope conductance is negative for potentials positive to the potassium equilibrium potential [22]. Furthermore, since the negative slope region of this current (-20 to -30 mV positive to the resting potential) corresponds also to the threshold for the action potential, the inward rectifier must also contribute to this threshold and be a factor in the development of rate-dependent block.

To study the role of i_{K1} in determining the amplitude and time course of the action potential foot and the subthreshold local response associated with Wenckebach periodicity, both single cell experiments and computer simulations were carried out in which the inward currents, i_{Na} and i_{Ca}, were either blocked by superfusion with high concentrations of tetrodotoxin and cobalt (to block i_{Na} and i_{Ca}, respectively in the case of the experiments), or omitted from the modified Beeler and Reuter model (for details see Delmar and colleagues [5,6]).

Figure 14.10 shows superimposed tracings from one of our current clamp simulations. In panel A, only i_{K1} was operative (i.e., $i_K = 0$). After a large depolarizing conditioning pulse of 300 msec, a smaller depolarizing pulse of constant amplitude and duration was applied at a test interval that was different in each superimposed run. The voltage changes showed the characteristic shape of a subthreshold local response, but their amplitude remained constant at all coupling intervals. In panel B, when the maximum value of i_K was increased to normal, there was a time-dependent increase in the amplitude of the individual responses, with those occurring later gradually taking the shape of subthreshold local responses. These results mimic very accurately those encountered in experimental sucrose gap [20] and single cell (see Figure 14.4) preparations. Finally, panel C of the same figure shows that, also consistent with experimental results [20], when hyperpolarizing pulses are applied using similar conditions, no time-dependent changes are seen and the voltage responses are those expected from a resistance-capacitance (RC) circuit, which reflects the linearity of the i_{K1} IV relation at hyperpolarized levels of membrane potential.

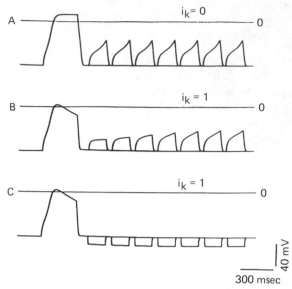

FIGURE 14.10. Superimposed tracings from a current clamp simulation. i_{Na} and i_{Ca} were omitted. See text for further details. (Reproduced from Delmar et al. [6] with permission from the American Heart Association, Inc.)

14.8   Analytical Model of Wenckebach Periodicity

To provide further support to our hypothesis, we devised a mathematical model aimed at calculating the time elapsed from the onset of the current pulse to the upstroke as a function of the preceding recovery time (t). The model was based on the assumption that variations in time to upstroke are a sole consequence of changes in the conductance of the i_K channel, which occur during the diastolic interval, and that voltage-dependent changes in the subthreshold events that precede an active response are modulated by the nonlinear characteristics of i_{K1} with respect to membrane potential.

The equations used were algebraic expressions, derived from the mathematical description of voltage response to current injection in an RC circuit (see Delmar and colleagues [5] for equations and parameters used in the formulation of the model).

Figure 14.11 shows the recovery curves. In both panels the continuous line represents the calculated time to upstroke obtained with the equations and the dots represent the experimental values recorded from the Beeler and Reuter cell (in the left panel) or from the single ventricular myocyte (in the right panel). Clearly, the experimental points fell on or very close to the theoretical curve, which strongly supports the hypothesis that the recovery of excitability during diastole is determined primarily by the time course of i_K deactivation.

As a critical test for our analytical model, we have used numerical iter-

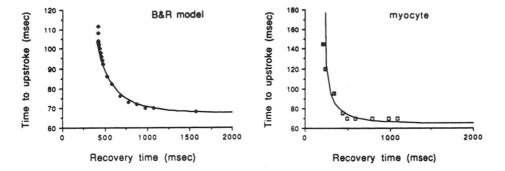

FIGURE 14.11. Predicted (continuous line) and experimental (symbols) recovery curves from the Beeler and Reuter model (left) and the ventricular myocyte (right). The analytical model equations were algebraic expressions derived from mathematical description of voltage response to current injection in a resistive capacitive current. (Modified from Delmar et al. [5] with permission from the American Heart Association, Inc.)

ation techniques to generate Wenckebach periodicities from the calculated recovery of excitability curves, and we have compared the resulting patterns with those obtained in the Beeler and Reuter cell.

A plot of the stimulus:response ratio as a function of the basic cycle lengths is presented in Figure 14.12. The response patterns of the Beeler and Reuter cell are represented by the open circles and the iteration results are represented by the closed circles. The horizontal lines represent the basic cycle length ranges at which stable patterns were maintained. A very close correspondence was demonstrated. In fact, the transition from 1:1 through various Wenckebach patterns to 2:1 occurred at very similar points in both cases as the BCL was progressively reduced, with the largest difference being less than 5 msec throughout the entire range.

14.9   Conclusion

We have come a long way in our understanding of the Wenckebach phenomenon since its original electrocardiographic description almost a century ago. Its demonstration in numerical and experimental models of isolated tissue preparations in single cells and its description in terms of ionic mechanisms provide us with a rigorous framework for better understanding heart rate-dependent conduction disturbances. Whether it be the AV node, the ischemic myocardium, or the Purkinje-muscle junction, it is now clear that the problem of Wenckebach periodicity is a consequence of the slow recovery of excitability of the distal element in a conducting pathway.

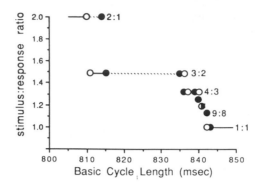

FIGURE 14.12. Steady-state stimulus:response ratios obtained with the Beeler and Reuter model (closed circles) at various basic cycle lengths are compared with those obtained by iteration of the analytical model (open circles) of cell excitability. (Modified from Delmar et al. [5] with permission from the American Heart Association, Inc.)

This contention is borne out by the demonstration of post-repolarization refractoriness and Wenckebach periodicity in normally polarized ventricular myocytes. In these cells, such phenomena are direct consequences of the long time course of deactivation of the repolarizing current i_K and of the nonlinear voltage-dependence of the anomalous rectifier, i_{K1}.

Acknowledgements: Supported in part by grants HL29439 and HL40923 from the National Heart, Lung and Blood Institute and a Grant-in-Aid from the American Heart Association, New York State Affiliate. We thank Donald C. Michaels for helpful comments, LaVerne Gilbert for typing the manuscript and Wanda Coombs for preparing the figures.

REFERENCES

[1] C. Antzelevitch, J. Jalife, and G.K. Moe. Characteristics of reflection as a mechanism of reentrant arrhythmias and its relationship to parasystole. *Circulation*, 61:182–191, 1980.

[2] C. Antzelevitch and G.K. Moe. Electrotonically mediated delayed conduction and reentry in relation to "slow" responses in mammalian ventricular conducting tissue. *Circ. Res.*, 49:1129–1139, 1981.

[3] G.W. Beeler and H. Reuter. Reconstruction of the action potential of ventricular myocardial fibers. *J. Physiol.*, 268:177–210, 1977.

[4] Hoffman B.F. and P.F. Cranefield. *Electrophysiology of the Heart*. McGraw-Hill, New York, 1960.

[5] M. Delmar, L. Glass, D.C. Michaels, and J. Jalife. Ionic bases and analytical solution of the Wenckebach phenomenon in guinea pig ventricular myocytes. *Circ. Res.*, 65:775–788, 1989.

[6] M. Delmar, D.C. Michaels, and J. Jalife. Slow recovery of excitability and the Wenckebach phenomenon in the single guinea pig ventricular myocyte. *Circ. Res.*, 65:761–774, 1989.

[7] N. El-Sherif, B.J. Scherlag, R. Lazzara, and P. Samet. Pathophysiology of tachycardia- and bradycardia-dependent block in the canine proximal His-Purkinje system after acute myocardial ischemia. *Am. J. Cardiol.*, 33:529–540, 1974.

[8] W. His. Ein Fall von Adams-Stokes'scher Krankeit mit ungleichzeitigem Schlagen der Vorhofe u. Herzkammern (Herzblock). *Deutsch Archiv. Klin. Med.*, 64:316–331, 1889.

[9] J. Jalife. The sucrose gap preparation as a model of AV nodal transmission: Are dual pathways necessary for reciprocation and AV nodal echoes? *PACE*, 6:1106–1122, 1983.

[10] J. Jalife and G.K. Moe. Excitation, conduction and reflection of impulses in isolated bovine and canine cardiac Purkinje fibers. *Circ. Res.*, 49:233–247, 1981.

[11] L.V. Katz and A. Dick. *Clinical Electrocardiography, Part I. The Arrhythmias*. Lea and Febiger, Philadelphia, 1956.

[12] J.P. Keener. On cardiac arrhythmias: AV conduction block. *J. Math. Biol.*, 12:215–225, 1981.

[13] M.N. Levy, P.J. Martin, H. Zieske, and D. Adler. Role of positive feedback in the atrioventricular nodal Wenckebach phenomenon. *Circ. Res.*, 24:697–710, 1974.

[14] S.L. Lipsius. Electrotonic interactions in delayed propagation and block within the guinea pig SA node. *Am. J. Physiol. (Heart Circ. Physiol.)*, H7-H16, 1983.

[15] J. Merideth, C. Mendez, W.J. Mueller, and G.K. Moe. Electrical excitability of atrioventricular nodal cells. *Circ. Res.*, 23:69–85, 1968.

[16] W. Mobitz. Uber die unvollstandige storung der erregungsuberleitung zwischen vorhof und kammer des menschlichen herzens. *Zeitschr. Ges. Exper. Med.*, 41:180–237, 1924.

[17] de Carvalho A. Paes and D.F. de Almeida. Spread of activity through the atrioventricular node. *Circ. Res.*, 8:801–809, 1960.

[18] A. Rosenblueth. Functional refractory period of cardiac tissues. *Am. J. Physiol.*, 194:171–183, 1958a.

[19] A. Rosenblueth. Mechanism of the Wenckebach-Luciani cycles. *Am. J. Physiol.*, 194:491–494, 1958b.

[20] G.J. Rozanski, J. Jalife, and G.K. Moe. Determinants of post-repolarization refractoriness in depressed mammalian ventricular muscle. *Circ. Res.*, 55:486–496, 1984.

[21] A. Shrier, H. Dubarsky, M. Rosengarten, M.R. Guevara, S. Nattel, and L. Glass. Prediction of complex atrioventricular conduction rhythms in humans with use of the atrioventricular nodal recovery curve. *Circulation*, 76:1196–1205, 1987.

[22] Y. Tourneur. Action potential-like responses due to the inward rectifying potassium channel. *J. Memb. Biol.*, 90:115–122, 1986.

[23] K.F. Wenckebach. Zur analyse des unregelmassigen pulses. II. Ueber den regelmassig intermittirenden puls. *Zeitschr. Klin. Med.*, 37:475–488, 1899.

[24] J.R. Wennemark and J.P. Bandura. Microelectrode study of Wenckebach periodicity in canine Purkinje fibers. *Am. J. Cardiol.*, 33:390–398, 1974.

[25] S.S. Wong, A.L. Bassett, J.S. Cameron, K. Epstein, P. Kozlovskis, and R.J. Myerburg. Dissimilarities in the electrophysiological abnormalities of lateral border and central infarct zone cells after healing of myocardial infarction. *Circ. Res.*, 51:486–493, 1982.

[26] D.P. Zipes, C. Mendez, and G.K. Moe. Some examples of Wenckebach periodicity in cardiac tissues, with an appraisal of mechanisms. In M.V. Elizari and M.B. Rosenbaum, editors, *Frontiers of Cardiac Electrophysiology*, pages 357–375. Martinus Nijhoff, Boston, 1983.

15

Parasystole and the Pacemaker Problem

Jacques Bélair[1]
Marc Courtemanche[2]
Leon Glass[3]

ABSTRACT In parasystole there is a competition between two independent cardiac pacemakers, the normal sinus rhythm and the abnormal ectopic rhythm, for the control of the heart. The physiological mechanism for parasystole has been translated into a mathematical model. Theoretical analysis of the model leads to a number of testable predictions concerning the occurrence of normal and ectopic beats in patients with parasystole. Tests of these predictions confirm the applicability of the model. The mathematical analysis uses approaches derived from number theory and nonlinear mathematics.

15.1 Introduction

The normal heart rhythm is set by the sinoatrial node. However, in some circumstances there can be pacemakers in abnormal locations, that is, ectopic foci, generating rhythms that compete with the normal sinus rhythms. The resulting rhythms are called parasystole and the identification of such a mechanism for generating cardiac arrhythmias dates to 1912 [6]. Ectopic foci generating parasystolic rhythms can be found in various parts of the heart such as the atria and ventricles, but in the following we discuss primarily ventricular foci. In the simplest situation the ectopic focus maintains a constant frequency. This situation, which we call *pure* parasystole, will be the main focus of this chapter [7]. We also briefly mention the extension of our results to *modulated* parasystole, when the sinus rhythm acts to phase reset the ectopic focus [12,13,19].

An electrocardiogram (ECG) and schematic diagram of the postulated mechanism are shown in Figure 15.1.

---

[1] Département de Mathématiques et de Statistique and Centre de Recherches Mathématiques, Université de Montréal, Québec

[2] Department of Applied Mathematics, University of Arizona, Tucson, Arizona

[3] Department of Physiology, McGill University, Québec

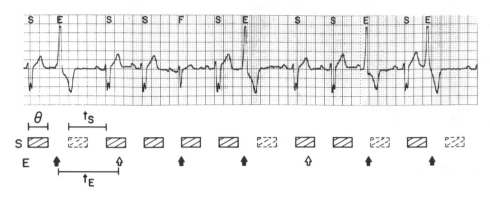

FIGURE 15.1. Electrocardiogram (ECG) from a patient with parasystole, along with a schematic representation of the model. The sinus (S), ectopic (E), and fusion (F) beats are labeled. One large box represents 0.2 sec. In the schematic diagram, the left edge of each box represents a sinus pacemaker discharge. The arrows indicate ectopic pacemaker firings, hatched boxes represent refractory periods (of length θ), and both sinus (t_S) and ectopic (t_E) periods are indicated. Filled arrows are ectopic beats falling outside the refractory period. The following sinus beats (dotted boxes) are blocked. (Reproduced with permission from Courtemanche et al. [2].)

If the depolarization from the ectopic focus falls outside the refractory period of the ventricles, it leads to an ectopic beat. Otherwise, it is blocked. Following an ectopic beat, the next sinus beat is usually blocked. This is called a compensatory pause. The morphologies of the sinus and the ectopic beats are different. When the sinus and ectopic beats fall at about the same time there is an intermediate morphology, called a fusion beat. The above description provides the classic basis for the identification of parasystole from the ECG [18]. The time intervals between ectopic beats are multiples of a common divisor, there is variable coupling between the sinus beats and the following ectopic beats, and there are fusion beats.

Hidden in this clinical and jargon-laden description of a cardiac arrhythmia is a theoretical model for ventricular parasystole. In fact, the theoretical model for parasystole that emerges naturally from the above description is equivalent to a problem in number theory, the gaps and steps problem, that attracted attention 20 to 30 years ago, but has lain dormant since. We were made aware of the connections with results in number theory from conversations with M. Waldschmidt, who directed our attention to the number theoretic literature, summarized in Slater [27] and more recently by Langevin [15]. Waldschmidt and Langevin have introduced the term "the pacemaker problem" to describe parasystole. The style of our approach to this problem is influenced by our own interest in the changes of dynamics (bifurcations) that can be observed as parameters describing physiological system vary, and the observation that the dynamics in parasystole can be thought of in the context of nonlinear dynamics. Yet the earlier results in number theory have direct implications for clinical observations. The blending of the number theoretic and nonlinear dynamic approaches poses several novel directions for research, most of which are still not explored. In the following, we try to sketch the basic mathematical structure of a theory for parasystole and give the strategy for deriving the central results. We then show how these results can be used to make clinical predictions. Finally, we mention some of the limitations of the analysis and discuss a few problems suggested by our approach. This chapter is largely based on recent papers from our group concerning the dynamics of parasystole [2,3,7–9].

15.2   The Pacemaker Problem

In this section we derive a one-dimensional map of the circle into itself to model parasystole. We also relate the biological problem to a number of problems that have been considered from a number-theoretic point of view. In our discussion we maintain a physiological flavor in describing abstract mathematical results. Those who by virtue of their training are confounded by such an approach would probably find solace in Slater [27] and Langevin [15].

We represent the sinus cycle as a circle and consider the refractory period as an arc of this circle, Figure 15.2. The angles are normalized to vary between 0 and 1. The ectopic pacemaker produces impulses characterized by the phase ϕ, in the sinus cycle at which they occur. This phase in turn corresponds to an angle on the circle. When an impulse falls inside the refractory period, it does not induce a beat. When it falls outside this period, it produces an abnormal, or ectopic, heartbeat. The pacemaker problem consists in determining rules governing the sequence of blocked and observed sinus and ectopic beats.

Mathematically, this can be set up as a study of the iterates ϕ_i of a map of the circle into itself, given by a function f as $\phi_{i+1} = f(\phi_i)$, where i takes all positive integer values. We can define a sequence of the symbols 0 and 1 associated with an orbit of the map f starting at an initial point ϕ_0: this symbolic sequence has the symbol 0 in the ith place if the ith iterate falls in the refractory period, and has the symbol 1 in that place otherwise. Counting the number of consecutive zeros between successive ones in the symbolic sequence and adding one, we obtain another sequence, which we call the *reduced* sequence. For example, for the symbolic sequence 101001000010..., the reduced sequence is 2,3,5,... The problem of determining the iterates i for which the corresponding phase ϕ_i is outside the refractory period then becomes equivalent to constructing the reduced sequence. For the case in

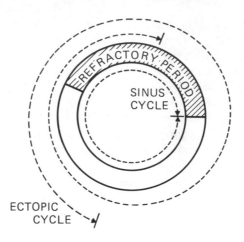

FIGURE 15.2. Schematic representation of the model. The circle represents the sinus cycle and the ectopic beat marches along. (Reproduced with permission from Gordon et al. [9].)

which the map f is a 1:1 invertible map of the circle (linear or nonlinear) the reduced sequence has a number of remarkable properties that can be summarized as follows:

Rule 1.   There are at most three integers in the reduced sequence.

Rule 2.   If two of these numbers are n and m, the third is their sum.
$$p = n + m;$$

Rule 3.   At least one of the values of m or n is odd.

Rule 4.   Exactly one of the values of m, n, or p can succeed itself in the reduced sequence.

The proofs of these results have a geometric flavor, illustrated in Figure 15.3. We assume for the moment that there are no periodic points of f. We consider the first return map to the arc of the circle, which is the complement of the refractory period of the sinus rhythm. The endpoints of this segment are denoted by α and 1, and the first return these endpoints to the segment occurs after m and n iterations, respectively. Define $r = f^{-m}(1)$ and $s = f^{-n}(\alpha)$. It is clear from Figure 15.3 and from continuity of the map f that the first return of any point in the interval $[\alpha, r)$ (respectively $[s, 1)$) to the interval $[\alpha, 1)$ is after m (respectively n) iterates. Let p be the smallest integer for which $f^p(u) \in [\alpha, 1)$, for any u in the interval $[r, s)$. Then since $\alpha = f^n(s)$, $f^m(\alpha) = f^p(s)$ and thus $f^{m+n}(s) = f^p(s)$. Since f is invertible without periodic points, this implies that $m + n = p$. This demonstrates Rules 1 and 2.

To show Rule 3, suppose that both m and n are even, so that p is also even. Consider a point t in $[\alpha, 1)$, and let $f^q(t) \in [\alpha, 1)$. Since q is a

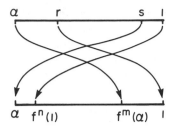

FIGURE 15.3. (Fundamental picture) First return map of the interval $[\alpha, 1)$ into itself under the action of a quasiperiodic map f. In general, there are three integers m, n, and p, corresponding to the lowest number of iterates needed to map three contigous regions in the interval back into $[\alpha, 1)$. (Reproduced with permission from Glass et al. [7].)

sum of the integers m, n, and p, all of which are supposed even, q is also even, and thus $f(t) \in [0, \alpha)$. Define v to be $f(t)$, so that there are no even iterates $f^{2j}(v)$ falling into $[\alpha, 1)$. Letting g stand for the map f^2, we obtain a transformation g without periodic points and without any iterate in the interval $[\alpha, 1)$, which is impossible. Hence, one of the integers m or n must be odd.

Finally, Rule 4 follows from geometric constructions that show that only one of the three zones $[\alpha, r), [r, s), [s, 1)$ can overlap its image [7].

When the transformation f has periodic points the above rules also apply, but it is common that there are fewer than three integers in the reduced sequence [2]. The analysis must be performed in a more combinatorial fashion, using the points of a stable periodic orbit.

The above discussion does not contain a procedure to compute the values of m, n, and p. Also, the map f does not keep track of the number of laps around the circle during the iterations, but only contains the position of successive iterates on the circle. To address these issues we need more mathematical tools. It is necessary to define the lift, the rotation number, and the Farey series.

The *lift*, $F(t)$, keeps track of the actual times that the ectopic cycle fires. Taking the sinus cycle time equal to 1, we have

$$t_i = F^i(t_0), \quad \text{and} \quad \phi_i = t_i \ (\text{mod } 1).$$

More formally, we fix an embedding of the circle in the complex plane axis, such that F is a periodic map of the real line into itself for which $e^{2\pi i F(t)} = f(e^{2\pi \phi i})$, where t projects down to [4].

Provided that the map f is monotonic and continuous, the asymptotic behavior of its iterates is well known to be completely characterized by its *rotation number*, ρ, given by the limit

$$\rho(f) = \lim_{n \to \infty} \frac{F^n(t) - t}{n}.$$

This limit is independent of the starting point t. The number ρ takes a rational value p/q if and only if the map f possesses a periodic point of period q, $F^q(u) = u + p$ for some real number u. In the case when the rotation number is irrational, then provided that the function f is smooth enough, the iterates are dense on the circle, and the dynamics are said to be *quasiperiodic*. The map is, in this case, conjugate to a rigid irrational rotation of angle ρ, that is, there exists a homomorphism h for which $f(h(s)) = h(R_\rho(s))$ for all values of s on the circle, where $R_\rho(s) = s + \rho$ defines a rigid rotation by a angle ρ.

The *Farey series* of order n, $\mathcal{F}_n$, is the set of rational numbers in the interval $[0,1]$, written in increasing order, whose denominator do not exceed n [10]. Each term in a Farey series is the mediant of its neighbors; that is, its numerator is the sum of the numerator of its neighbors and its denominator is the sum of the denominator of its neighbors. For example, $\mathcal{F}_2 = \{0, \frac{1}{2}, 1\}$ and $\mathcal{F}_3 = \{0, \frac{1}{3}, \frac{1}{2}, \frac{2}{3}, 1\}$. Each series $\mathcal{F}_n$ contains $2^n + 1$ fractions.

We now consider the properties of F imposed by the first return illustrated in Figure 15.3. It follows from the definition of the lift that there is an integer A_n for which

$$\alpha + A_n < F^n(1) \leq 1 + A_n. \tag{15.1}$$

Similarly, if the initial point is α, then we define A_m by the relationship

$$\alpha + A_m < F^m(\alpha) \leq 1 + A_m. \tag{15.2}$$

Here, m and n count the number of iterates between ectopic beats that fall outside the refractory interval, and A_m and A_n count the number of laps around the circle that occur during these iterations. The need for the lift F is apparent in the crucial role of the integers A_m and A_n, neither of which can be computed from the function f itself.

For a given map f and a fixed length α of the refractory period, the integers A_m, A_n, n, and m can be determined explicitly by using the Farey series. In the case where the mapping f defines a rigid rotation by an angle ρ, inequalities (15.1) and (15.2) can be written as

$$\frac{A_n - 1 + \alpha}{n} \leq \rho < \frac{A_n}{n} \tag{15.3}$$

and

$$\frac{A_m}{m} \leq \rho < \frac{A_m + 1 - \alpha}{m}. \tag{15.4}$$

Assume for the moment that $0 < \rho < 1$. The minimality requirements for the denominators n and m in these inequalities imply that they, along with the values of A_m and A_n, can be obtained by considering the Farey fractions. More precisely, successive entries in Farey series of increasing orders will partition the plane of the parameters α and ρ into regions associated with the ratios A_m/m and A_n/n satisfying these inequations and the minimality of the denominators.

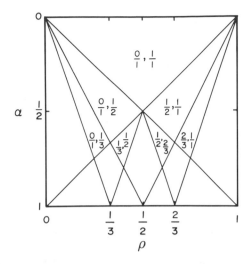

FIGURE 15.4. Combined values for A_m/m and A_n/n in the (α, ρ) plane. Each region is labeled by adjacent entries in a Farey series; the construction, shown here up to order 3, can be continued to arbitrary order. (Reproduced with permission from Glass et al. [7].)

The implementation of this procedure leads to Figure 15.4. Note that each labeled region contains two consecutive terms from the Farey series of appropriate order. In the case of a nonlinear transformation f, an explicit association of a sequence of integers with definite regions in the space of the parameters α and ρ has not been accomplished.

In analogy with the rules for the integers m, n, and p, there are rules governing the values of A_m, A_n, and A_p. These rules can be derived using a similar strategy used to derive the rules for m, n, and p [7]. For example, at least one of A_m or A_n is odd and $A_p = A_m + A_n$.

There is an intimate connection between the pacemaker problem and some well-known concepts in number theory. Although the main number-theoretic problems concern the *asymptotic* distribution of iterates [14], and thus neglect such transients properties as the first-return map defined above, there are two classical problems directly related to the pacemaker problem [27]. They are called the *gap* and the *step* problems, and they can be stated as follows. For a given ψ in the interval $(0,1)$, and j any positive integer, let $\{j\psi\}$ denote the fractional part of $j\psi$. Consider the sequence of numbers $\{j\psi\}$, all in the interval $(0,1)$. Then we have:

The gap problem. For any ξ in $(0,1)$, determine the "gaps" between the successive values of j for which $\{j\psi\} < \xi$. This is equivalent to finding the number of iterates under a rigid rotation by an angle ψ that are needed to come back in the segment $[0,\xi)$.

The step problem. For the set $\{1\psi\}, \{2\psi\}, \ldots, \{N\psi\}$ rearranged in ascending order, determine the "steps" into which the interval [0,1] is thereby partitioned. The original statement of the step problem defined the steps to be the arc lengths of the circle, but a more general formulation is to define the steps to be the differences in the index $(1, 2, \ldots, N)$ of consecutive points in the interval.

The solution to both problems is contained in our solution of the pacemaker problem. The gap problem is essentially the pacemaker problem and the equivalence to the step problem can be established using simple arguments [15]. The generalization of both problems to the case of nonlinear maps is straightforward: the rigid rotation is replaced by a nonlinear mapping f. Since the order of the iterates under an invertible circle map f of rotation number ρ is the same as that induced by a rigid rotation of angle ρ, our geometric arguments (Figure 15.3) solve both problems, even in the nonlinear case. Previous proofs ([27] and references therein) hold only for rigid rotations.

An explicit linking of these results with ours requires the consideration of the continued fraction expansion of the rotation number (see Richards [24] for a leisurely introduction to both continued fractions and Farey series).

The fundamental result is the following: if f is a circle endomorphism and the convergents (truncations of the continued fraction expansion) of its rotation number ρ are $\{p_k/q_k\}_{k \geq 1}$, then the iterates under f closest to 0 come from the set $\{q_k\}_{k \geq 1}$ and they alternate following the parity of k. The convergents p_k/q_k thus provide the desired ratios A_m/m and A_n/n: all that needs to be checked is that these iterates are close enough to the starting point (the first return map occurring outside the refractory period defining "close enough").

In fact, for a given number ρ in the interval [0,1), the successive approximations of ρ by elements in the Farey series of increasing order are exactly the convergents of the continued fraction expansion of ρ [24]. In the case when the mapping f is a rigid rotation, the algorithm described above to derive Figure 15.4 is essentially equivalent to the procedures considered in the more theoretical contexts [27].

We have considered the sinus rhythm as the reference frame and considered the number of ectopic beats with respect to it. A reversed frame of reference (i.e., the ectopic rhythm is the reference frame and the sinus rhythm marches around it), can also be used [2,3].

15.3   Dynamics of Pure Parasystole

The abstract mathematical results in the preceding section can be immediately translated into a set of concrete predictions for ventricular parasystole. It is a simple matter to count the number of interpolated beats (NIB) between ectopic beats on the ECG. This gives a sequence of integers. For example, in the record in Figure 15.1, the NIB sequence is 4,2,1. The rules

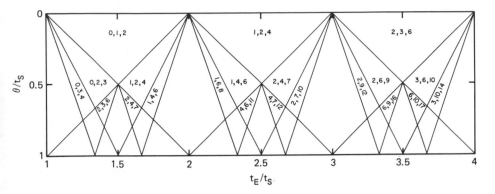

FIGURE 15.5. Allowed number of sinus beats between ectopic events, NIB, in the $(t_E/t_S, \theta/t_S)$ plane for pure parasystole. For each region the allowed values are indicated as three integers, separated by commas. To obtain the value for the reduced sequence, 1 should be added to each of the NIB values. Allowed values in the unlabeled regions can be determined from the construction described in the text. More details are in Glass et al. [7]. (Reproduced with permission from Glass et al. [7].)

were initially described for a situation in which there is an absolute independence of the sinus and ectopic cycle lengths, that is, there is no phase resetting of the ectopic pacemaker by the sinus rhythm [7]. More recently, the rules have been extended to situations in which there is a weak interaction between the sinus and ectopic rhythms [3]. The NIB values are obtained by subtracting 1 (due to the compensatory pause that blocks a sinus beat) from the values A_m, A_n, and A_p obtained in the last section.

In parasystole, the sequence of NIB values must obey the following rules:

Rule 1.   There are at most three NIB values allowed in the sequence.

Rule 2.   If three values are present, the sum of the two smaller values is one less than the larger one.

Rule 3.   If at least two values are present, one and only one of them is odd.

Rule 4.   One and only one of the allowed values can succeed itself in the sequence.

For example, in the rhythm strip shown in Figure 15.1, the three NIB values are 1,2,4, and these satisfy the first three rules.

The mathematical considerations discussed in the preceding section and translational symmetries of the zones [7] allow one to construct a diagram showing the allowed NIB values as a function of the sinus and ectopic cycle lengths and the refractory time of the ventricles. This is displayed in Figure 15.5. It is clear that as the parameters change many different combinations of NIB values are found, all of which satisfy the first three rules above.

15.4   Clinical Observations of Parasystole

Figure 15.5 makes it clear that the patterns of distribution of ectopic beats should follow definite restrictions as the sinus and ectopic cycle lengths vary. Thus, this figure can be used to make theoretical predictions concerning the dynamics that are found in certain clinical situations in which there is an independence of the sinus and ectopic rhythms. Clinical tests of these predictions are made most easily by examining circumstances in which either the sinus or ectopic rates vary independently. This may occur as a consequence of normal fluctuation of sinus or ectopic rates, fluctuation of sinus rate due to exercise, control of the sinus rate by intracardiac stimulation in the atria, or generation of an artificial ectopic site using an intraventricular fixed rate pacing.

Previous clinical studies can often be interpreted in the context of Figure 15.5. For example, several years ago, Levy and collaborators [16] observed that many patients displayed frequent ectopy in which the NIB values were either 1 or an even number. Inspection of Figure 15.5 shows that in pure parasystole these will be the observed NIB values over a large region of parameter space. Lightfoot reported on a patient observed at two different times [17]. At one time, the values of the NIB were either 1 or an even number, but somewhat later the NIB values were 2,4, or 7. Measurement of the cycle times shows that these transitions are consistent with the regions in Figure 15.5 [7].

More systematic attempts at trying to test the theoretical predictions of pure parasystole have recently been carried out. The record in Figure 15.1 is taken from a 16-year-old boy who suffered an unexplained fainting spell [2,9]. During an exercise test his sinus rate increased, but the theoretical predictions of pure parasystole were remarkably in accord with the observed rhythms. This is not to say that there was perfect agreement. Several physiological mechanisms were apparent that led to some discrepancy between predictions and observations. These included: the presence of the sinus beat that is normally blocked following the ectopic beat (this would increase the expected NIB value by one), an apparent failure to observe an ectopic beat that should have occurred outside of the refractory time of the ventricles, and the presence of a short burst of ectopic beats (two or more) that block sinus beats that would have been expected had the burst not occurred. Also, quantitative measurement of the intervals between sinus and ectopic beats indicates that there is some fluctuation. The origin of these fluctuations and their consequences on the fine details of the observed rhythms are not yet understood. However, it is clear that it is unusual to observe a situation in which there is only one NIB value that succeeds itself, that is, rule 4 of parasystole is not easily observed clinically.

By stimulating the heart directly it is possible to eliminate some of the intrinsic variability. To date, there have been two studies of parasystole in which there is a direct electrical stimulation of the heart in people. One method is to pace the atria in patients who have frequent ventricular ectopy. If the rhythm is parasystole, then the theoretical predictions of Figure 15.5 should be confirmed. In one patient, there was agreement with theoretical predictions at four out of five stimulation frequencies tested [3]. At the fifth stimulation frequency the rules of parasystole were not followed, but a new variant was observed in which there were only an odd number of sinus beats between ectopic beats. This rhythm has been called "concealed bigeminy" and previous studies have shown that it might arise as a consequence of modulation of a parasystolic focus by the sinus beat [11,13]. Because uncontrolled sinus rate fluctuation may make interpretation of ECGs difficult, direct atrial pacing may be useful clinically in situations for which it becomes important to evaluate the mechanism of the abnormal ectopic beats.

A final clinical study was carried out in patients who do not have an abnormal ectopic pacemaker, but such an artificial pacemaker can be introduced clinically by direct electrical stimulation of the ventricles during electrophysiologic testing [5]. Provided the sinus rate does not fluctuate, the first three rules of parasystole were confirmed in 16/17 episodes in eight patients. This agreement is not surprising in view of the definite nature of the mathematical theory. What is surprising is that there can even be one episode that did not give agreement, and the underlying reasons for this should be clarified. Such a procedure is potentially useful in helping to understand mechanisms leading to variability in the sinus rate and the refractory times since any discrepancies with theoretical predictions must be due to fluctuations in these parameters.

In the preceding, most of the discussion has focused on the mathematical analysis of pure parasystole in which there is an independence of the sinus and ectopic rhythms. However, it is well known that ectopic rhythms can also be reset by the sinus rhythm [25], and the resulting rhythms are called modulated parasystole [13,19]. In such situations, the classic ECG criteria are not satisfied because the intervals between ectopic beats are not multiples of a common divisor. Further, in some instances it is possible to find fixed coupling intervals from the sinus beats to the following ectopic beats. In cases of modulated parasystole, primary attention has focused on determining the modulation function based on ECG data [1,12,20–23] and understanding the resulting rhythms. The dynamics in modulated parasystole can also be understood based on one-dimensional finite difference equations [2,3,8,11,19,26,28]. The mathematical analysis involves one-dimensional maps with discontinuities. In situations in which the phase resetting is approximated by a piecewise linear function, one can obtain strong mathematical results concerning the sequences of rhythms as parameters are varied [11]. Theoretical computations also show the possibility for deterministic chaos [3]. Much more work is needed to analyze complex patterns of ectopy observed clinically to evaluate the prevalence of modulated parasystole.

Although it has been gratifying to observe the number-theoretical predictions in actual clinical data, the relevance of this for patient care is still not clear. Parasystole is normally considered to be a benign arrhythmia, and it is comparatively easy to diagnose based on classic ECG criteria. However, the occurrence of parasystole in a patient believed to be at high risk for sudden cardiac death [9] may show that a reevaluation of the risks of this arrhythmia is needed. Our analysis demonstrates that once the mathematical theory is understood, complex patterns of ectopy that previously seemed haphazard can be understood. Such detailed mathematical insight is still not possible in the great majority of patients with frequent ectopy or other cardiac arrhythmias.

15.5   Implications of Parasystole for Theory

In the preceding section we discussed new approaches to analyze arrhythmias based on the theoretical analysis. Conversely, the physiological observations suggest several directions for future mathematical studies. More prominent among these are:

1. What detailed rules govern the sequence of integers in the NIB sequence? Rule 4 gives the most basic result concerning the sequence, but a deeper analysis is still needed.

2. What is the topological structure of the boundaries of the zones in Figures 15.4 and 15.5 for nonlinear maps? Numerical results show they are very jagged (fractal?) [3].

3. Are there any simple rules that apply in more complex, physiologically plausible situations such as multiple pacemakers and noninvertible circle maps?

4. How are the rules affected by well-defined stochastic fluctuation of the parameters?

Traditionally, mathematics and quantitative approaches have had strong impact in the physical sciences, but have not had the same influence in the biological sciences or medicine. The preceding work gives one example from medical science in which mathematics can help elucidate complex rhythms. What is somewhat surprising is that the branch of mathematics that emerges naturally is number theory, which has few applications to the natural sciences. The mathematical analysis can be used to make theoretical predictions, and conversely, the clinical data suggest new avenues for mathematical research.

Acknowledgements: This research has been partially supported by funds from NSERC and the Canadian Heart and Stroke Foundation.

REFERENCES

[1] A. Castellanos, R.M. Luceri, F. Moleiro, et al. Annihilation, entrainment, and modulation of ventricular parasystolic rhythms. *Am. J. Cardiol.*, 54:317–322, 1984.

[2] M. Courtemanche, L. Glass, J. Bélair, D. Scagliotti, and D. Gordon. A circle map in a human heart. *Physica D.*, 40:299–310, 1989b.

[3] M. Courtemanche, L. Glass, M.D. Rosengarten, and A.L. Goldberger. Beyond pure parasystole: Promises and problems in modelling complex cardiac arrhythmias. *Am. J. Physiol.*, 257:H693–H706, 1989a.

[4] R.L. Devaney. *An Introduction to Chaotic Dynamical Systems.* Benjamin-Cummins, Menlo Park, 1986.

[5] P. Fernandez, A. Castellanos, G. Breuer, and A. Interian. Dynamic behavior of sinus beats during pure ventricular parasystole (abstract). *Circulation.*, 80:SII:40 (0157), 1989.

[6] G.B. Fleming. Triple rhythm of the heart due to extrasystoles. *Q. J. Med.*, 5:318–326, 1912.

[7] L. Glass, A.L. Goldberger, and J. Bélair. Dynamics of pure parasystole. *Am. J. Physiol.* 251:(*Heart Circ. Physiol.* 20):H841–H847, 1986.

[8] L. Glass, A.L. Goldberger, M. Courtemanche, and A. Shrier. Nonlinear dynamics, chaos, and complex cardiac arrhythmias. *Proc. Roy. Soc. Lond. A.*, 413:9–16, 1987.

[9] D. Gordon, D. Scagliotti, M. Courtemanche, and L. Glass. A clinical study of the dynamics of parasystole. *PACE*, 12:1412–1418, 1989.

[10] G.H. Hardy and E.M. Wright. *An Introduction to the Theory of Numbers, 4th ed.* Clarendon, Oxford, 1971.

[11] N. Ikeda, S. Yoshizawa, and T. Sato. Difference equation model of ventricular parasystole as an interaction of pacemakers based on the phase response curve. *J. Theor. Biol.*, 103:439–465, 1983.

[12] J. Jalife, C. Antzelevitch, and G.K. Moe. The case for modulated parasystole. *PACE*, 5:911–926, 1982.

[13] J. Jalife and G.K. Moe. Effect of electrotonic potentials on pacemaker activity in canine Purkinje fibers. *Circ. Res.*, 39:801–808, 1976.

[14] L. Kuipers and H. Niederreiter. *Uniform Distribution of Sequences.* John Wiley & Sons, New York, 1974.

[15] M. Langevin. Stimulateur cardiaque et suites de Farey. 1990. (in press.)

[16] M.N. Levy, N. Kerin, and I. Eisenstein. A subvariant of concealed bigeminy. *J. Electrocardiol.*, 10:225–232, 1977.

[17] P. Lightfoot. Parasystole simulating ventricular bigeminy with Wenckebach-type coupling prolongation. *J. Electrocardiol.*, 11:385–390, 1978.

[18] H.J.L. Marriot and H.J. Conover. *Advanced Concepts in Arrhythmias.* CV Mosby, St. Louis, 1983.

[19] G.K. Moe, J. Jalife, W.J. Mueller, and B. Moe. A mathematical model of parasystole and its application to clinical arrhythmias. *Circulation*, 56:968–979, 1977.

[20] G.J. Nau, A.E. Aldariz, R.S. Acunzo, et al. Modulation of parasystolic activity by non-parasystolic beats. *Circulation*, 68:462–469, 1982.

[21] G. Oreto, F. Luzza, G. Satullo, S. Coglitore, and L. Schamroth. Intermittent ventricular bigeminy as an expression of modulated parasystole. *Am. J. Cardiol.*, 55:1634–1637, 1985.

[22] G. Oreto, F. Luzza, G. Satullo, et al. The influence of sinus rhythm on a parasystolic focus. *Am. Heart J.*, 115:121–133, 1988.

[23] G. Oreto, F. Luzza, G. Satullo, and L. Schamroth. Modulated parasystole as a mechanism for concealed bigeminy. *Am. J. Cardiol.*, 58:954–958, 1986.

[24] I. Richards. Continued fractions without tears. *Math. Mag.*, 54:163–171, 1981.

[25] L. Schamroth and H.J.L. Marriot. Intermittent ventricular parasystole with observations on its relationship to extrasystolic bigeminy. *Am. J. Cardiol.*, 7:799–809, 1961.

[26] L. Schamroth, D.H. Martin, and M. Pachter. The extrasystolic mechanism as the entrainment of an oscillator. *Am. Heart J.*, 104:1363–1368, 1988.

[27] N.B. Slater. Gaps and steps for the sequence $n\theta$ mod 1. *Proc. Camb. Phil. Soc.*, 63:1115–1123, 1967.

[28] C.A. Swenne, P.A. Delang, M. Ten Hoopen, and N.M. Van Hemel. Computer simulations of ventricular arrhythmias: Ventricular bigeminy. *Comput. Cardiol.*, 295–298, 1981.

16

Electrical Propagation in Distributed Cardiac Tissue

J. Mailen Kootsey[1]

ABSTRACT Basic properties concerning electrical propagation in cardiac tissue are discussed. Electrical properties in passive and excitable tissue are described in the one-dimensional case. Realistic mathematical models in higher dimensions will have to take into account discreteness caused by finite conductance of cell couplings, anisotropy of cardiac muscle, and the finite resistance of the extracellular conductance.

16.1 Introduction

The propagation of electrical activity in cardiac muscle was first demonstrated by Engelmann [6] more than a century ago. He cut a frog ventricle into a single elongated strip, leaving one end connected to the beating auricle. The strip was then cut into many small pieces each connected to its neighbors by a thin bridge of muscle less than a millimeter wide. After an hour of recovery from the injury, the segmented strip pulsated rhythmically. Contractions began at the auricle and proceeded from segment to segment in peristaltic fashion. Engelmann found that the contractions stopped when the strip was removed from the auricle, but contractions could be initiated by an induction shock at any point along the strip and moved with equal speed in either direction. Engelmann repeated the experiment in several hearts, cutting the tissue in different directions. Since he found the same result in all the preparations, he concluded that every part of the ventricular substance was "equally endowed" and that there were no special regions for conduction.

Two years later in 1877, Marchand [19] observed propagation in whole frog hearts with an electrical galvanometer connected to the apex and base. Passing induction shocks into the tissue near one electrode and then the other, he observed "the progress of a wave of negative tension from the seat of excitation" and that the wave of excitation preceded the wave of contraction.

---

[1]National Biomedical Simulation Resource, Duke University Medical Center, Durham, North Carolina 27710

It is now generally accepted that electrical propagation in cardiac muscle proceeds from cell to cell through low-resistance intercellular connections located at the nexus. Woodbury and Crill [29] noted that low-resistance connections were necessary because of the high resistance of the normal cell membrane. By enlarging the spaces between cells, Barr and colleagues [1] were able to localize the low-resistance junctions to the nexus or tight junction.

The phenomenon of electrical propagation in cardiac muscle thus results from a large number of electrically active cells coupled together tightly enough so that an action potential in one cell may trigger an action potential in neighboring cells. The electrical properties of isolated cells and the geometry of the tissue (cell size, shape, and interconnections) have been studied extensively. The properties of the intact tissue cannot, however, be obtained by a simple combination of tissue geometry with local electrical properties. The interaction between cells is sufficiently strong to modify the cell's electrical behavior significantly from what is measured in isolation. Thus, detailed experimental measurements must be made on intact tissue (as well as on isolated cells or membrane) and theoretical models must be sufficiently detailed to include the effects of tight coupling.

This chapter presents a review of studies of the distributed electrical properties of cardiac muscle responsible for electrical propagation, with particular emphasis on the results of coupling patches of membrane to build tissue structures in one, two, or three dimensions. The goal is to understand the origins of normal activation and arrhythmias in cardiac tissue.

Because of the simple shape and large size of some nerve cells, much of the early work on propagation was done on nerve tissue and conclusions from these papers are still applied to cardiac muscle; some of these experimental and theoretical studies of nerve tissue are included below. As an example, in Hodgkin and Huxley's well-known paper on a quantitative description of the currents in the squid axon [11], the difference between the local or "membrane" action potential and the propagated action potential is apparent by comparing their Figures 13 and 15. They obtained a membrane action potential by solving the ordinary differential equations for the ion currents and transmembrane voltage at a single site. The propagated waveform was obtained by solving the partial differential equation for a one-dimensional electrical cable with distributed membrane properties, assuming a uniform propagation velocity.

16.2   Passive Properties

At low amplitudes, the equivalent circuit of a small patch of cardiac (or nerve) cell membrane is a high resistance in parallel with a low capacitance—an RC circuit. "Small" in this case means that the physical size is small enough so that the transmembrane potential is uniform. If a constant current is applied abruptly to such a circuit at time $t = 0$, the voltage across the membrane V_m rises exponentially to a constant value:

$$V_{m,\text{patch}}(t) = I_o \, r \, (1 - e^{-t/\tau}) \tag{16.1}$$

where I_o is the amplitude of the input current, r is the resistance of the membrane patch, c is the capacitance of the patch, and $\tau = rc$ is the membrane time constant.

Measuring the transmembrane potential at the point of current injection in a distributed system (one or more spatial dimensions) also produces a time-dependent waveform something like the exponential rise in Equation 16.1. The exact nature of the waveform depends not only on the local membrane properties but also on the spatial distribution of the membrane. Determining current and voltage at a single site gives "apparent" local properties, but other measurements—such as the spatial variation of voltage or current—can also be made.

The simplest extension of the RC membrane patch is the one-dimensional passive (RC) cable of infinite length. The cable equation was developed more than a century ago, but its best known application in biology was to the nerve axon by Hodgkin and Rushton in 1946 [12]. The full analytical solution developed by Hodgkin and Rushton describes the variation of transmembrane potential in time and space following the sudden application of a constant current at some point along the cable. At the current injection site, the transmembrane potential rises in a fashion similar but not identical to that across the RC patch. Instead of an exponential rise, the voltage follows the more rapidly rising *error function*:

$$V_{m,\text{cable}}(t, x = 0) = I_o \, r_{\text{input}} \, \text{erf}(\sqrt{t/\tau}) \tag{16.2}$$

where $r_{\text{input}} = \sqrt{r_m r_i}$ is the input resistance, r_m is the membrane resistance of a unit length of cable, r_i is the longitudinal or axial resistance per unit length, and τ is again the time constant of the membrane. Distributing passive membrane along one spatial dimension thus has two effects, even when the potential is measured at the current injection point as with the patch. First, the transmembrane potential from a current step rises faster in time, reaching 84% of its final value in one time constant compared to 62% for the RC patch. Second, the final value of the transmembrane potential at the injection point depends not only on the membrane resistance, but also on the longitudinal resistance.

Jack and colleagues [13] have found solutions for the RC cable equation for finite cable lengths. The finite cable might be considered intermediate in geometry between the patch and the infinite cable and the results are also intermediate, both in waveform and final value attained.

George [8] studied the input resistance (voltage divided by current at the input connection) of extended networks of linked short passive cables as approximations to a syncytium of coupled cells. He studied an *open* network where a cable splits into two after a length l, each branch again dividing and so on indefinitely, as well as a *closed* network where the branches loop back to form a hexagonal array. George showed that the input resistance depends on the length l of each cable segment. When l is small compared to the length constant $\lambda = \sqrt{r_m/r_i}$, the input resistance of the closed network depends only weakly on r_m (to the one-quarter power) and in the open network it does not depend on the membrane resistance at all!

Shiba [23,24] studied the spatial dependence of membrane potential from current injected at a point on a uniform two-dimensional sheet, calculating only steady potentials without time variations. He found a very rapid fall-off in potential with distance around the current injection point. In experimental electrophysiology, this result means that two-dimensional sheets of coupled cells cannot be stimulated successfully from a single point because the high potentials at the injection point damage the membrane. Current must be injected over a significant area to reduce the local potential to safe values. Shiba calculated a "length constant" for the sheet but had to redefine the meaning of the term in two dimensions.

16.3   Propagation in One Dimension

In a passive, linear distributed membrane system, any voltage or current disturbance falls off steadily with distance from the stimulus site. Normal cardiac membrane has nonlinear and time-dependent electrical properties that produce an action potential with constant shape and amplitude as it propagates. If the electrical properties and spatial distribution of cell membrane are known, it is possible in principle to set up equations governing current flow and potential and to solve these equations numerically to any desired degree of accuracy. In practice, solutions are impossible without substantial simplification because of the complexity of the membrane geometry and electrical description. Propagation phenomena in cardiac tissue clearly need to be understood in two or three dimensions. The cost of computer solutions, however, increases rapidly with the number of dimensions. For example, if adequate spatial resolution requires dividing a length of one-dimensional cable into 100 segments, then the same resolution in two or three dimensions would require 10^4 or 10^6 segments, respectively, with a similar increase in the total amount of numerical computation. This section summarizes some of the many studies of one-dimensional structures, where the complexities of both experiment and theory are substantially fewer than in the full three-dimensional problem.

Tasaki and Hagiwara [28] found an approximate analytical solution to the one-dimensional cable equation for the early exponential rise of the action potential undergoing uniform propagation. Substituting the exponential solution and constant velocity into the partial differential equation produced a second-order ordinary differential equation for $V_m(t)$ with the following approximate solution:

$$V_{m,\text{foot}}(t) = Ae^{r_i c_m v^2 t} \tag{16.3}$$

where v is the propagation velocity and A is a proportionality constant. Measuring the time constant of the foot of the action potential thus gives a value for the product $r_i c_m v^2$ so that if any two are known, the third can be calculated. This approximate solution has called attention to the significance of the foot of the action potential and has been used to estimate the membrane capacitance.

Lieberman and colleagues [18] solved the cable and nonlinear membrane equations for heart muscle cells on a digital computer and looked at the changes in propagation velocity of depolarization caused by changing membrane and longitudinal parameters. Propagation velocity varied most rapidly with membrane-specific capacitance, velocity being proportional to $C_m^{-.73}$. Changes in cell diameter and longitudinal resistivity had less effect on the velocity, showing proportionalities to the plus and minus 0.5 power, respectively. Least effective in changing the velocity was scaling of the time-dependent sodium and potassium currents, equivalent to changing the channel density in the membrane. The velocity was found to be proportional to the 0.25 power of the current scale factor, down to about 3% of normal where propagation failed.

Goldstein and Rall [10] tested step and flare changes in diameter and branching in a theoretical cable model using the Bonhoeffer-van der Pol equations as a simplified membrane description. Approaching a step decrease in diameter, the action potential was found to increase in both amplitude and velocity and a step increase in diameter produced the opposite result. At a branch point, the action potential shape and dimensionless velocity were unchanged if the diameters on the two sides of the branch were "matched," if they were not, the result was the same as at a step change in diameter. Tapered regions produced gradual changes in velocity.

The partial differential equation for the one-dimensional cable describes a continuous structure such as an electrical cable or a nonmyelinated nerve axon. Myelinated nerves and cardiac muscle are actually discontinuous structures, segmented by insulated wrappings from Schwann cells in the nerve and by intercellular junctions between cardiac cells. FitzHugh [7] solved the cable and membrane equations for myelinated nerves on a digital computer and found discontinuities in the transmembrane potential between a myelin sheath region and the adjacent node, but he did not see any discontinuities in the time waveform of potential at a node. Goldman and Albus [9] used a permeability description of membrane properties instead of the conductance description of Hodgkin and Huxley [11] and showed that the model accounted for the linear relationship between velocity and diameter. Brill and coworkers [2] studied the propagation velocity and amplitude of the action potential as a function of internodal spacing.

They showed that the propagation velocity is very sensitive to the internodal length when the length is small and that the amplitude and shape of the action potential also vary with the internodal length. The propagation velocity was small at a node spacing of 25 μm, increased rapidly to a broad maximum between 1 and 2 mm, then decreased slowly up to 10 mm. Peak amplitude was large at a spacing of 50 μm and decreased as the spacing was raised to 10 mm.

In myelinated nerves, extended lengths of passive, linear (RC) cable are interrupted by very small nodes with nonlinear, time-dependent properties. In cardiac muscle, the geometric model is different: most of the length is occupied by cells with nonlinear, time-dependent membrane and these cells are coupled by very small resistive junctions. Different electrical models must therefore be used to study these two types of tissue. Joyner [14] used the segmentation inherent in the numerical solution methods for partial differential equations to approximate the cellular divisions, making the spatial increment comparable to the cell length rather than the shorter length usually chosen to achieve adequate spatial resolution for representing continuous structures. Diaz and colleagues [5] and Joyner and colleagues [15] used several segments to represent each cell, with low coupling resistances between cell segments and higher resistances between cells. Both groups showed significant effects caused by the segmentation and studied the propagation velocity and waveform of the action potential as the cell length and coupling resistance were varied. They found combinations where the maximum rate of rise of the action potential ($\dot{V}_{max}$) increased as the velocity decreased, contrary to what was known from continuous structures where the two quantities increase or decrease together.

Another interaction between tissue geometry and action potential properties occurs at the ends of a one-dimensional cable—where propagation begins or ends. Whether the site of a stimulus or the end of propagation, the end of a cable is a discontinuity that alters both the velocity and waveform of the action potential. Spach and Kootsey [26] studied computer solutions of the cable equation and showed that end discontinuities alter not only the propagation velocity and time waveform of the action potential but also the time variation of the conductances generating the action potential. For example, near the end of propagation, each patch of membrane does not have as much "downstream membrane" to supply with depolarizing current. As a result, the amplitude of the action potential near the end is higher, $\dot{V}_{max}$ is higher, the peak sodium current is lower, and the areas under the time waveforms of sodium current and conductance are also smaller than in uniform propagation. Equivalent effects appear at an end where a stimulus initiates propagation.

How far do these "end effects" reach into an electrical cable or how long does a cable have to be so that no "end effects" are present over most of its length? In the passive cable, the length scale is the length constant λ defined above, approximately 1 mm in resting cardiac muscle. Since the end effects extend several λ into the cable, tissue—simulated or real—must be substantially longer than λ if the action potential is to achieve uniform propagation. In normal cardiac cells, the ratio r_m/r_i determining λ is not constant, but varies by two orders of magnitude or more during the action potential because of changes in membrane resistance. During depolarization, λ drops to about 0.1 mm. The activation pathway in atria and ventricles is much longer than 0.1 mm so an assumption of uniform propagation is reasonable for depolarization. During repolarization, λ can increase to several millimeters or more so uniform propagation cannot be assumed automatically for this phase of the action potential. There is experimental and theoretical evidence that normal repolarization is not a uniformly propagated phenomenon, but is a large-scale process involving most or all of the ventricles (see Kootsey and Johnson [16] for a review).

16.4 Propagation in Two Dimensions

After Engelmann's early experiments [6], cardiac muscle was long assumed to be uniform and isotropic in electrical properties. The first clear deviation recognized was a difference in propagation velocity with direction in the tissue. Clerc [4] observed that the velocity of propagation of depolarization along the long cell axis ("longitudinal") was three times higher than the velocity across the cells ("transverse"). Clerc attributed the difference in velocity to different internal and external resistivities in the two directions. He noticed no changes in the action potential shapes and so assumed uniform propagation in both directions.

Spach and coworkers [27] made intracellular and extracellular potential measurements on the surface of a sheet of atrial muscle, changing the stimulus site so that propagation at a measurement point could be made longitudinal or transverse. They observed not only a difference of a factor of 10 in propagation velocity in the two directions (transverse slower than longitudinal), but also differences in $\dot{V}_{max}$ of depolarization and in τ_{foot}, the time constant of the "foot" of the action potential—the early exponential charging of the membrane capacitance. Contrary to the expectations from one-dimensional theory of uniform structures, the high velocity in the longitudinal direction was associated with the lower value of $\dot{V}_{max}$ and longer time constant τ_{foot}, whereas the lower transverse velocity was associated with a higher $\dot{V}_{max}$ and shorter τ_{foot}. Spach and coworkers [27] suggested that the cause of these unexpected relationships might be an anisotropic distribution of electrical discontinuities caused by junctions between cells. Numerous junctions in the longitudinal direction provide strong electrical coupling (i.e., low resistance) and high propagation velocity while a lower density of transverse junctions (i.e., higher resistance) reduces the velocity in the transverse direction.

Another puzzling observation by Spach and colleagues [27] in the atrial tissue was directional block of early premature action potentials. As the premature stimulus was made earlier, the faster longitudinal propagation was the first to block while the slower transverse propagation continued. This is exactly the opposite of what was expected on the basis of uniform nerve models, where high velocity is associated with a fast waveform and less susceptibility to block. Spach and coworkers observed that this directional block could lead to reentry in a small region of normal tissue measuring 4 mm or less across.

The waveform at a single recording site depended, in Spach's recordings, on the direction from which the excitation approached. Thus, the directional characteristics of propagation described above cannot be caused by membrane properties but must in some way originate in the geometry of the tissue and the pattern of propagation: the size and shape of the cells, the manner in which they are packed together, the distribution of the electrical junctions coupling the cells, and the site of stimulation.

Multidimensional propagation has frequently been simplified into a set of independent one-dimensional problems, one for each orthogonal dimension. The models in the set have to be based on the same membrane description and differ only in the cell geometry and coupling pattern assumed. Such a separation is exact for plane wave propagation because in a plane wave there is coupling current only along a straight line perpendicular to the plane and there is no transverse current. A set of one-dimensional models cannot, of course, represent any curvature or other complexity in a wavefront or potential distribution.

No one has yet been able to find a set of one-dimensional models based on realistic tissue geometry that account for all the observed details of propagation. Spach and colleagues [25] studied propagation in a pair of one-dimensional, uniform cable models simulating longitudinal and transverse propagation. Published experimental values of axial resistance were used in the longitudinal model along with a fast sodium current model for depolarization derived from experimental data. In the transverse cable, the axial resistance was increased to produce the lower velocity and the membrane capacitance was cut in half. No geometric rationale could be found for reducing the capacitance in the transverse direction, but it was observed empirically that the transverse model reproduced all of the anomalous features in the experimental data, including the effects on $\dot{V}_{max}$, τ_{foot}, the peak amplitude of the action potential, and the increased resistance to block. Thus, the equivalent circuit for the cell structure somehow produces the equivalent effect of reduced membrane capacitance in the transverse direction.

Most of the detailed studies of propagation have been made on the surface of sheets of tissue. Since no satisfactory pair of one-dimensional models has been proposed, several groups have studied two-dimensional models— even though they require much more computer time to solve. Plonsey and Barr [21] studied a model consisting of two rectangular grids to represent intracellular and extracellular spaces. An arbitrary function was selected to approximate the waveform of the transmembrane potential V_m during depolarization. With isotropic resistances in both grids, propagation was a straightforward generalization of one-dimensional cable theory: longitudinal currents in intracellular and extracellular spaces were equal in magnitude and opposite in direction and the transmembrane currents were directly related to spatial second derivatives of V_m at the same site. Similar relationships held when the resistances were anisotropic with the same longitudinal-to-transverse ratios used in both grids. With unequal anisotropic ratios, the transmembrane current had to be obtained by a surface integral even though the second derivative was strongest at the measurement point.

Roberge and colleagues [22] also studied a rectangular grid model with anisotropic intracellular resistances, but assumed the extracellular resistances negligible and replaced the arbitrary V_m waveform with a more realistic membrane model. They tested a grid size small enough to give a reasonable representation of continuous, uniform tissue and also a coarser grid to approximate discrete cell connections. They found that the coarser grid produced increases in $\dot{V}_{\max}$ and peak amplitude similar to those observed experimentally in transverse propagation by Spach and colleagues [27], but the magnitudes of the changes were smaller and τ_{foot} changed in the opposite direction to what was observed in the experiments.

Kootsey and Wu [17] replaced the rectangular grid with an offset, overlapping cell pattern and added a representation of extracellular resistance. The cell pattern was chosen as a simplified and regular representation of the branching and offset structure seen in histological studies of cardiac tissue. In this pattern, a single set of cell connections produced both longitudinal and transverse coupling; no arbitrary coupling ratio was required. Kootsey and Wu found that $\dot{V}_{\max}$ increased and τ_{foot} decreased in the transverse direction (along with slower propagation), as in the experimental data of Spach and coworkers [27]. The model did not, however, show the directional difference in block seen in the tissue.

Other aspects of propagation besides the waveform of depolarization have also been studied in two-dimensional models. Michaels and coworkers [20] studied the interactions of sinoatrial pacemaker cells coupled in a two-dimensional grid. With a nonuniform distribution of intrinsic cell rates, activation originated near cells with the shortest cycle length and appeared to propagate to the rest of the cells, even though the pattern actually resulted from mutual entrainment of all cells. The apparent propagation time increased with increased coupling resistance.

Burgess and coworkers [3] measured activation and repolarization in a transmural plane and constructed a two-dimensional model to represent activity in the plane. The sheet model was anisotropic and represented the histology of the canine pulmonary conus. The model produced activation patterns similar to those recorded in the canine tissue and also showed that the nonuniform activation pattern modulated repolarization, altering the duration of the action potential over the region.

16.5 Summary

Normal electrical activation and relaxation in cardiac tissue are three-dimensional phenomena resulting from coupling in a complex network of coupled cells. The standard analytical approach to this problem has been to isolate the different factors in the problem and study them in detail: nonlinear and time-dependent transport mechanisms in the cell membrane; internal compartments and ion binding in the cell; cell size, shape, and packing; the distribution of cell couplings; the electrical characteristics of cell couplings. This detailed analysis, however, is only a part of what is required to understand propagation in intact tissue. Coupling cells together modifies their behavior and produces new behavior not described as simple combinations of their properties in isolation. Detailed measurements are thus needed on intact tissue and realistic multidimensional models are required to test hypotheses. Much can still be learned from simplified models and models of lower dimensionality, but the key is to find simplifications that preserve essential mechanisms.

Small changes in electrical waveforms may not seem significant in themselves, but they become significant when they are associated with important aspects of propagation, such as directional block or reentry. No single model yet proposed accounts for the peculiar details of propagation in cardiac muscle, but the studies described above indicate several factors that probably contribute: discreteness, caused by the finite conductance of cell couplings; the anisotropic distribution of cell couplings resulting from their location on the cells as well as the shape and packing of the cells; and the finite resistance of the extracellular conductor. Detailed models including these factors and solved on digital computers hopefully will enable us to understand electrical propagation in cardiac tissue and help us diagnose and treat the conditions that result from tissue damage and disease.

REFERENCES

[1] L. Barr, M.M. Dewey, and W. Berger. Propagation of action potentials and the structure of the nexus in cardiac muscle. *J. Gen. Physiol.*, 48:797–823, 1965.

[2] M.H. Brill, S.G. Waxman, J.W. Moore, and R.W. Joyner. Conduction velocity and spike configuration in myelinated fibres: Computed dependence on internode distance. *J. Neurol. Neurosurg. Psychiatry*, 40:769–774, 1977.

[3] M.J. Burgess, B.M. Steinhaus, K.W. Spitzer, and P.R. Ershler. Non-

uniform epicardial activation and repolarization properties of in vivo canine pulmonary conus. *Circ. Res.*, 62:233–246, 1988.

[4] L. Clerc. Directional differences of impulse spread in trabecular muscle from mammalian heart. *J. Physiol. (Lond.)*, 255:335–346, 1976.

[5] P.J. Diaz, Y. Rudy, and R. Plonsey. Intercalated discs as a cause for discontinuous propagation in cardiac muscle: A theoretical simulation. *Ann. Biomed. Eng.*, 11:177–189, 1983.

[6] T.W. Engelmann. Ueber die leitung der erregung im herzmuskel. *Pfluegers Arch. Ges. Physiol. Mensch. Tiere*, 11:465–480, 1875.

[7] R. FitzHugh. Computation of impulse initiation and saltatory conduction in a myelinated nerve fiber. *Biophys. J.*, 2:11–21, 1962.

[8] E.P. George. Resistance values in a syncytium. *Austr. J. Exp. Biol.*, 39:267–274, 1961.

[9] L. Goldman and J.S. Albus. Computation of impulse conduction in myelinated fibers: Theoretical basis of the velocity-diameter relation. *Biophys. J.*, 8:596–607, 1968.

[10] S.S. Goldstein and W. Rall. Changes of action potential shape and velocity for changing core conductor geometry. *Biophys. J.*, 14:731–757, 1974.

[11] A.L. Hodgkin and A.F. Huxley. A quantitative description of membrane current and its application to conduction and excitation in nerve. *J. Physiol. (Lond.)*, 117:500–544, 1952.

[12] A.L. Hodgkin and W.A.H. Rushton. The electrical constants of a crustacean nerve fibre. *Proc. Roy. Soc. Lond. Ser. B.*, 133:444–479, 1946.

[13] J.J.B. Jack, D. Noble, and R.W. Tsien. *Electrical Current Flow in Excitable Cells.* Oxford University Press, Oxford, 1975.

[14] R.W. Joyner. Effects of the discrete pattern of electrical coupling on propagation through an electrical syncytium. *Circ. Res.*, 50:192–200, 1982.

[15] R.W. Joyner, R. Veenstra, D. Rawling, and A. Chorro. Propagation through electrically coupled cells: Effects of a resistive barrier. *Biophys. J.*, 45:1017–1025, 1984.

[16] J.M. Kootsey and E.A. Johnson. The origin of the T-wave. *CRC Crit. Rev. Bioeng.*, 4:233–270, 1980.

[17] J.M. Kootsey and J. Wu. Models of electrophysiological activation. In S. Sideman and R. Beyar, editors, *Imaging, Measurement and Analysis of the Heart*. Freund and Kaplan, Ltd, Haifa, 1990.

[18] M. Lieberman, J.M. Kootsey, E.A. Johnson, and T. Sawanobori. Slow conduction in cardiac muscle. *Biophys. J.*, 13:37–55, 1973.

[19] R. Marchand. Beiträge zur kenntniss der reizwell und contractionswelle des herzmuskels. *Pfluegers Arch. Ges. Physiol. Mensch. Tiere*, 15:511–536, 1877.

[20] D.C. Michaels, E.P. Matyas, and J. Jalife. Mechanisms of sinoatrial pacemaker synchronization: A new hypothesis. *Circ. Res.*, 61:704–714, 1987.

[21] R. Plonsey and R.C. Barr. Current flow patterns in two-dimensional anisotropic bisyncytia with normal and extreme conductivities. *Biophys. J.*, 45:557–572, 1984.

[22] F.A. Roberge, A. Vinet, and B. Victorri. Reconstruction of propagated electrical activity with a two-dimensional model of anisotropic heart muscle. *Circ. Res.*, 58:461–475, 1986.

[23] H. Shiba. An electric model for flat epithelial cells with low resistive junctional membranes. A mathematical supplement. *Jpn. J. Appl. Physiol.*, 9:1405–1409, 1970.

[24] H. Shiba. Heaviside's "Bessel cable" as an electric model for flat simple epithelial cells with low resistivity junctional membranes. *J. Theoret. Biol.*, 30:59–68, 1971.

[25] M.S. Spach, P.C. Dolber, J.F. Heidlage, J.M. Kootsey, and E.A. Johnson. Propagating depolarization in anisotropic human and canine cardiac muscle: Apparent directional differences in membrane capacitance. *Circ. Res.*, 60:206–219, 1987.

[26] M.S. Spach and J.M. Kootsey. Relating the sodium current and conductance to the shape of the transmembrane and extracellular potentials by simulation: Effects of propagation boundaries. *IEEE Trans. Biomed. Eng.*, BME-32:743–755, 1985.

[27] M.S. Spach, W.T. Miller, D.B. Geselowitz, R.C. Barr, J.M. Kootsey, and E.A. Johnson. The discontinuous nature of propagation in normal canine cardiac muscle. *Circ. Res.*, 48:39–54, 1981.

[28] I. Tasaki and S. Hagiwara. Capacity of muscle fiber membrane. *Am. J. Physiol.*, 188:423–429, 1957.

[29] J.W. Woodbury and W.E. Crill. On the problem of impulse conduction in the atrium. In Ernst Florey, editor, *Nervous Inhibition*, pages 124–135. Pergamon Press, New York, 1961.

17

Wave Propagation in Myocardium

James P. Keener[1]

ABSTRACT I present an overview of cable theory for propagation in one-dimensional excitable cables and some modifications necessary to account for the discrete cellular nature of myocardial cells and gap junctions. The modified cable theory is shown to agree quite well with a number of recent experiments on propagation in anisotropic tissue that are unexplained by the classical continuous theory.

17.1 Introduction

One of the most important mathematical advances in a biological problem resulted from the work of Hodgkin and Huxley describing action potential activity and propagation in squid axon [13]. Their work not only won for them a well-deserved Nobel prize, but it also initiated the investigation of electrochemical propagation in physiological systems that has resulted in thousands of papers blending mathematical and biological observations.

This chapter gives an introduction to some of the consequences of this developing mathematical theory for the propagation of action potentials in the myocardium by addressing three questions, namely, "Under what conditions is propagation possible?", "Why does propagation fail?", and "What are the rules that quantify these phenomena?". One of the goals of a mathematical theory is that it should not only provide an explanation of known experimental results and make predictions about experiments that have not yet been done, but it should also be easily conceptualized, that is, one would like to obtain from a good theory *rules of thumb* that are both easy to understand and use and are simultaneously quantitatively reliable. Indeed that is one of the goals of this chapter. Along the way, however, we shall discover that there are some pitfalls to be aware of when trying to apply cable theory.

The outline of this discussion is as follows. I begin with a derivation of the most basic model of action potential propagation in an excitable medium, namely the Nagumo equation. In Section 17.3, I summarize the

---

[1] Department of Mathematics, University of Utah, Salt Lake City, Utah 84112

mathematical theory of this equation by showing that traveling waves of this equation exist, that they are stable, and that they can be initiated by superthreshold initial data. This section of the chapter contains only well-known mathematical results, which are included here as background.

The purpose of the rest of the chapter is to find simple yet quantitatively reliable rules describing propagation in myocardium. In Section 17.4 I use dimensional analysis to find the properties of the simplest cable model. This set of rules is usually referred to as continuous cable theory. Unfortunately, these rules are significantly flawed for application to myocardium, so in the remaining sections we describe corrections to cable theory necessitated by the physical structure of myocardium and summarize the experimental evidence for their validity. In the process, I give a theoretical explanation of some apparently contradictory experimental data on propagation failure in myocardium.

17.2   The Nagumo Equation

The most important electrical feature of cardiac cells is that they are excitable. Under normal conditions a cardiac cell maintains a difference in the concentration of certain ions between its interior and exterior, resulting in a voltage potential difference called the polarized state. The ionic balance is maintained by the membrane's ionic channels that are opened or closed depending on the transmembrane potential. When a current stimulus is added to the cell, the transmembrane potential can respond in one of two ways. If the stimulus is very small, there is no essential change in the conductance of the membrane and the potential returns to its equilibrium. If the stimulus is large enough, the conductance of sodium channels is affected (sodium channels are opened) followed by an inrush of sodium ions. This is an autocatalytic event because the inrush of positive sodium ions further raises the transmembrane voltage potential, which further opens sodium channels, and so on. The rapid inrush of sodium ions and subsequent increase of the transmembrane potential do not go on forever, and when the potential gets high enough the sodium channels close, shutting off the flow of sodium ions. In cardiac cells the depolarized transmembrane potential is maintained for about 300 msec, during and after which other membrane processes return it to equilibrium and allow it to recover gradually its excitability. This excursion of the transmembrane potential is called an action potential and the stimulus that initiated it is said to be a superthreshold stimulus.

The crucial ingredient for the propagation of an action potential is that while one section of membrane is depolarized during an action potential, there is a flow of current to other sections of polarized membrane that is sufficient to raise the potential there above the local threshold for an action potential response, which in turn depolarizes more tissue.

To describe this process in terms of mathematical expressions, we must keep track of all the currents. The currents are of two types, namely transmembrane and parallel to the membrane. To model these we make a simplifying assumption, namely, that the voltage potential varies only in one spatial direction, so that both in the intracellular and extracellular space gradients of voltage occur in only the direction of propagation. The assumption that there are no gradients of voltage across the cross-section of a cell is usually called the core conductor assumption. Second, it is convenient, but not necessary (nor true), to assume that the extracellular space is isopotential (equivalently the extracellular resistance is zero). If we denote the transmembrane potential by the variable v, then at any point in the cell's interior the current in the direction of propagation is given by

$$i = -\frac{A}{R_c}\frac{\partial v}{\partial x},\tag{17.1}$$

where A is the cross-sectional area of the cell and R_c is the resistivity of the cytoplasm in the cell's interior, measured in units of Ωcm (this is nothing more than a statement of Ohm's law). The transmembrane (outward) current has two ingredients, the ionic currents and the capacitive current, because the passive membrane acts like a capacitor. These we express as

$$i_T = Cp\frac{\partial v}{\partial t} - pf,\tag{17.2}$$

where C is the capacitance per unit area of the cell membrane, p is the perimeter of the cross-section of the cell orthogonal to the direction of propagation, and f is the total inward ionic current per unit area of membrane. Thus, at any point the rate of current loss to the extracellular space is given by

$$\frac{\partial i}{\partial x} = -i_T,\tag{17.3}$$

or, expressing this current balance in terms of the potential v,

$$Cp\frac{\partial v}{\partial t} = \frac{\partial}{\partial x}\left(\frac{A}{R_c}\frac{\partial v}{\partial x}\right) + pf.\tag{17.4}$$

It is a simple calculation to show that the inclusion of nontrivial extracellular resistance is equivalent to taking a larger intracellular resistivity in Equation (17.4).

The full description of the action potential requires knowledge of the behavior of the ionic currents, and for this there are many models from which to choose. The Hodgkin-Huxley model [13] describes the behavior of transmembrane currents for squid axon and establishes the format in which all other ionic models are constructed. For cardiac tissue there is the Beeler-Reuter model [3] for myocardium, the McAllister-Noble-Tsien model [22] and DiFrancesco-Noble model [9] for Purkinje fiber, and the

Yanagihara-Noma-Irisawa model [32] for sinoatrial node. There are as well a number of modifications of these to account better for certain features such as the fast sodium current [10,11].

There is one feature of all of the ionic models that we will exploit here, and that is that the inward sodium current is always very fast compared to the other time-dependent currents of the model. For the study of action potential propagation, it is adequate to keep track of only the fast sodium current and any instantaneous currents, such as the rectifying potassium current i_{K_1}, and their interactions with the transmembrane potential. All of the membrane models have the common feature that in the upstroke of the action potential the fast ionic currents can be modeled quite well as a function of transmembrane potential, say $f(v)$, having three zeros, $V_0 < V_1 < V_2$, where V_0 is the stable polarized state of the membrane. The slope of the function f at V_0 is $-1/R_m$, where R_m is the membrane resistivity, measured in units of Ωcm^2. For potentials in the range $V_0 < v < V_1$, $f(v)$ is negative, while in the range $V_1 < v < V_2$, $f(v)$ is positive. The value V_1 is the threshold above which there is an autocatalytic production of voltage because of the activated inward sodium current. The state V_2, while being a rest state for the upstroke current, is only a pseudo-steady state of the full membrane dynamics, because when an action potential is near V_2, other currents whose dynamics are slow compared to the evolution of the upstroke currents are turned on.

To find the function f for a specific ionic model, it is sufficient to assume that the sodium activation is instantaneous by taking m, the sodium activation, to be equal to $m_0(v)$, and to hold all other time-dependent activations and inactivations at their polarized steady-state levels.

With this consideration in mind, throughout this chapter whenever it is necessary to know something about the structure of the function f, we will assume that $f = f(v)$ has three zeros, $V_0 < V_1 < V_2$, V_0 is the resting (polarized) state of the membrane, the slope of f at V_0 is $-1/R_m$, and $f(v)$ is negative for $V_0 < v < V_1$ and positive for $V_1 < v < V_2$. It will also turn out to be important that

$$\int_{V_0}^{V_2} f(v)\,dv > 0.$$

Notice that by simply rescaling the potential, the smallest and largest of the three zeros of $f(v)$ can always be taken to be 0 and 1, respectively. From time to time, we shall illustrate results using a specific example of the function f, either the cubic function

$$f(v) = \frac{1}{\alpha R_m} v(v - \alpha)(1 - v) \qquad (17.5)$$

or the piecewise linear function

$$f(v) = \frac{H(v - \alpha) - v}{R_m} \qquad (17.6)$$

where $H(x)$ is the Heaviside step function and $0 < \alpha < \frac{1}{2}$. Neither of these functions are especially good as quantitative approximations to the actual fast ionic currents, but they have the correct qualitative behavior.

We end this section with a summary of the parameters involved. In Table 17.1 are listed the dimensional physical parameters with their definitions and physical dimensions as well as a list of the dimensionless parameters used in this and subsequent sections.

TABLE 17.1. Summary of physical and dimensionless parameters.

Parameter	Definition	Dim. Units
A	Cross-sectional area of cell	cm^2
c	Propagation speed	cm/sec
C	Membrane capacitance	$\mu Farad/cm^2$
L	Length of cell	cm
Λ	Space constant	cm
p	Perimeter of cellular cross-section	cm
r_g	Effective intercellular resistance	Ω
R_c	Cytoplasmic resistance	Ω-cm
R_e	Effective cable resistance	Ω-cm
R_m	Membrane resistance	Ω-cm^2
S	Surface area of cell active membrane	cm^2
V	Volume of cytoplasm per cell	cm^3

Dimensionless parameters:

$$r = \frac{Sr_g}{R_m}, \quad q = \frac{SL^2 R_c}{V R_m}$$

17.3   Traveling Wave Solutions

We begin our discussion of propagation in myocardium by a summary of the features of Equation (17.4). We will subsequently need to modify this equation to account for some important features of myocardium that this model simply cannot explain. Everything in this section is well known in the mathematical literature on diffusion-reaction equations (see, e.g., Fife [12]), but is included here to give a basic understanding of mathematical features of the equation.

The first step is to see under what conditions Equation (17.4) can support propagation. By propagation we mean there is a solution that provides a transition between two rest states of the equation. By introducing dimensionless space, time, and potential variables, we can rewrite Equation (17.4) in the dimensionless form

$$u_t = u_{xx} + g(u) \tag{17.7}$$

for an infinitely long cable with $-\infty < x < \infty$, where $g = R_m f$ is the dimensionless ionic current, $g(0) = g(\alpha) = g(1) = 0$, $g'(0) < 0$, $g'(1) < 0$, and u represents the dimensionless potential. This equation is often denoted as the Nagumo equation or the bistable equation in the mathematical literature.

This equation is a nonlinear **parabolic equation**. Historically, wavelike behavior was thought to occur only in hyperbolic equations, so the discovery that parabolic equations could support nonlinear waves was of major importance.

A reasonable first attempt is to see if there are translation invariant solutions, that is, solutions of the form $u(x,t) = U(x+ct)$ for some value of c. Any such solution must satisfy the differential equation $U'' - cU' + g(U) = 0$, and this, being an ordinary differential equation, is easier to study than the original partial differential equation. The statement that a traveling wave solution exists can be summarized formally as follows:

Theorem 17.1: *Suppose $g(u)$ is a continuously differentiable function with $g(0) = g(\alpha) = g(1) = 0$, $g'(0) < 0$, $g'(1) < 0$, and with no other zeros on the interval $0 < u < 1$. If $\int_0^1 g(u)du > 0$, then there is a unique speed $c = c_0 > 0$ for which a unique traveling wave solution of the Nagumo equation (17.7) exists of the form $u(x,t) = U(x+c_0 t)$, with $U' > 0$ providing a transition between $u = 0$ and $u = 1$ and satisfying $0 \leq U \leq 1$.*

To prove the existence of a traveling wave solution, we look for solutions of the equation $U'' - cU' + g(U) = 0$ that connect the rest points $U = 0$ and $U = 1$ by plotting trajectories of the equation in the UU' phase plane. The rest points at $U = 0$ and $U = 1$ with $U' = 0$ are both saddle points in the UU' phase plane, while the point $U = \alpha$ is either a node or a spiral point. Since the rest points at $U = 0$ and $U = 1$ are saddle points, the goal is to see if the parameter c_0 can be chosen so that the trajectory that leaves $U = 0$ at $x = -\infty$ can be made to connect with the trajectory that approaches $U = 1$ at $x = +\infty$.

Before we look at trajectories, we can determine the sign of c_0. If a connecting trajectory $U(x)$ between $U = 0$ and $U = 1$ exists, we can multiply the equation $U'' - c_0 U' + g(U) = 0$ by $U'(x)$ and integrate from $x = -\infty$ to $x = +\infty$ with the result that

$$c_0 \int_{-\infty}^{\infty} U'^2 dx = \int_0^1 g(u)du. \tag{17.8}$$

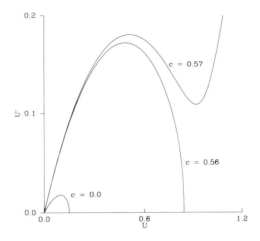

FIGURE 17.1. Trajectories in the UU' phase plane for the equation $U'' - cU + U(U - 0.1)(1.0 - U) = 0$, leaving the rest point $U = 0, U' = 0$, with $c = 0.0$, 0.56 and 0.57.

In other words, if a traveling wave solution exists, the sign of c_0 is the same as the sign of the area under the curve $g(u)$ between $u = 0$ and $u = 1$. If the area under the curve $g(u)$ between 0 and α is smaller in absolute magnitude than the area under $g(u)$ between α and 1, then the traveling solutions move the state variable u from $u = 0$ to $u = 1$ corresponding to the leading edge of a polarizing action potential. If the area under the curve $g(u)$ between 0 and 1 is negative then the traveling wave still exists, but it corresponds to a repolarizing wave, that is, the wave moves backward rather than forward, repolarizing rather than depolarizing the medium. If this is the case, the medium is not sufficiently excitable to support action potential spread. When $g(u)$ is the cubic polynomial $g(u) = \alpha u(u - \alpha)(1 - u)$, the speed c is positive for $0 < \alpha < 1/2$.

We now suppose that $\int_0^1 g(u)du > 0$, so that $c_0 > 0$. We want to see what happens to the unique trajectory in the UU' plane that leaves the saddle point at $U = 0$, $U' = 0$ in the direction $U' > 0$. The idea is to pick some value of c and to integrate along this trajectory to determine if it eventually approaches the rest point at $U = 1$, $U' = 0$, and if it does not, to pick another value of c for which the trajectory might come closer. In fact, this is exactly the method that Hodgkin and Huxley used to integrate their equations numerically to find traveling wave solutions [13]. A few such attempts are depicted in Figure 17.1.

For a few values of c we can determine explicitly how the trajectory behaves. First, if $c = 0$, we can find an explicit equation for this trajectory by multiplying the governing equation by U' and integrating to find $U'^2 + G(U) = 0$, $G(U) = \int_0^U g(u)du$. This trajectory cannot exist near $U = 1$, so the trajectory leaves the rest point $U = 0$ with $U' > 0$ and intersects the $U' = 0$ axis at some value of $U < 1$. It cannot be the connecting trajectory.

Next, suppose c is a very large number. In the UU' phase plane the slope of the trajectory leaving the rest point at $U = 0$ is the positive root of $\lambda^2 - c\lambda + g'(0) = 0$, which is certainly bigger than c. Define K as the smallest positive number for which $g(u)/u \leq K$ for all u on the interval $0 \leq u \leq 1$, and let b be any fixed positive number smaller than c. On the line $U = bU'$ in the UU' phase plane the slope of trajectories satisfies

$$\frac{dU'}{dU} = c - \frac{g(U)}{U'} = c - b\frac{g(U)}{U} \geq c - bK.$$

By picking c large enough, we are sure that $c - bK > b$, so that once trajectories are above the curve $U = bU'$ they stay above it and, of course, we know that the trajectory leaving the rest point at $U = 0$ starts out above this curve. Thus, this trajectory always stays above the line $U = bU'$ and therefore passes above the rest point at $U = 1$.

Now we have two trajectories, one with $c = 0$, which misses the rest point at $U = 1$ by crossing the $U' = 0$ axis at some point $U < 1$, and one with c very large, which misses this rest point by staying above the rest point at $U = 1$. Since trajectories depend continuously on the parameters of the problem, there is a continuous family of trajectories depending on the parameter c that fill in between these two special trajectories, and therefore there is at least one trajectory that hits the point $U = 1$, $U' = 0$.

In fact, the value of $c = c_0$ for which this heteroclinic connection occurs is unique. Notice that slopes of trajectories in the UU' plane are monotonically decreasing functions of the parameter c. Suppose at some value of $c = c_0$ there is a known connecting trajectory. For any value of c that is larger, the trajectory leaving the saddle point at $U = 0$ must lie above the connecting curve, but the trajectory approaching the saddle point at $U = 1$ as x approaches $+\infty$ must lie below the connection, so there cannot be another connecting trajectory for a larger value of c. Similarly, there cannot be a connecting trajectory for a smaller value of c, so that the value of $c = c_0$ and hence the connecting trajectory is unique.

For quantitatively realistic functions $g(u)$, it is usually necessary to calculate numerically the speed of propagation of the traveling wave solution. However, in the two special cases of Equations (17.5) and (17.6) the speed of propagation can be calculated explicitly. Suppose that $g(u)$ is a cubic polynomial, expressed in the form

$$g(u) = -A^2(u - u_0)(u - u_1)(u - u_2) \tag{17.9}$$

where the zeros of the cubic are ordered with $u_0 < u_1 < u_2$. To find the connecting trajectory between the smallest zero u_0 and the largest zero u_2, we make the guess that

$$U' = -a(U - u_0)(U - u_2). \tag{17.10}$$

We substitute this guess into the governing equation $U'' - cU' + g(U) = 0$, and find that we must have

$$a^2(2U - u_0 - u_2) - ca - A^2(U - u_1) = 0. \qquad (17.11)$$

This is a linear function of U that can be made identically zero only if we choose

$$a = \frac{A}{\sqrt{2}}, \quad c = \frac{A}{\sqrt{2}}(u_0 - 2u_1 + u_2). \qquad (17.12)$$

In the special case that $u_0 = 0$, $u_1 = \alpha$, and $u_2 = 1$, this reduces to

$$c = \frac{A}{\sqrt{2}}(1 - 2\alpha) \qquad (17.13)$$

showing that the direction of propagation changes at $\alpha = 1/2$.

The traveling wave solutions were relatively easy to find because the tool used was phase-plane analysis. To understand more about these traveling wave solutions, such as how they are initiated or if they are stable, requires substantially more sophisticated mathematical tools.

The traveling wave solution for the Nagumo equation is stable. This means that starting from a large class of initial data, the solution of the Nagumo equation approaches the traveling wave solution as time becomes large. The precise statement of the stability result is as follows:

Theorem 17.2: *Suppose $g(0) = g(\alpha) = g(1) = 0$, $g'(0) < 0$, $g'(1) < 0$, $g(u) < 0$ for $0 < u < \alpha$, and $g(u) > 0$ for $\alpha < u < 1$, and suppose $U(x)$ satisfies $U'' - cU' + g(U) = 0$, with $U(-\infty) = 0$ and $U(+\infty) = 1$. Suppose further that the initial function $u(x, 0) = \phi(x)$ satisfies*

$$\limsup_{x \to -\infty} \phi(x) < \alpha, \quad \liminf_{x \to \infty} \phi(x) > \alpha \qquad (17.14)$$

then there are constants $z, K > 0$, and $b > 0$, so that $u(x,t)$, the solution of the Nagumo equation, satisfies

$$|u(x,t) - U(x + ct - z)| < Ke^{-bt}. \qquad (17.15)$$

The proof of this theorem can be found in Fife [12]. In words, this theorem says that for any initial configuration of the function $u(x,0)$ that is greater than α far to the right and is less than α far to the left, the solution approaches some translation of the traveling wave solution at an exponential rate.

The Nagumo equation has a threshold response. This means that it is possible to initiate a propagating wavefront if the initial stimulus is sufficiently large, but if the initial stimulus is too small the medium returns directly to its rest state. Just how large is *large enough*?

To give specific meaning to the idea of *large enough* we need to construct appropriate comparison functions. If we define the function $G(u)$ by

$$G(u) = \int_0^u g(u')du',$$

then, since $\int_0^1 g(u)du > 0$, there is a unique number $u_0 > \alpha$ for which $G(u_0) = 0$. Furthermore, $G(u) > 0$ and $G'(u) = g(u) > 0$ for $u_0 < u < 1$. For any number $\beta > u_0, \beta < 1$, define the length b_β as

$$b_\beta = 2 \int_0^\beta \frac{du}{\sqrt{2(G(\beta) - G(u))}},$$

and the function $q_\beta(x)$ as the solution of $q'' + f(q) = 0$ whose first integral is $(q'^2/2) + G(q) = G(\beta)$ and satisfies $q(0) = 0$, $q'(0) = \sqrt{2G(\beta)}$. Then $q_\beta(x) > 0$ on the interval $0 < x < \beta, q_\beta(0) = q_\beta(b_\beta) = 0$, and the maximum of q_β is attained at $x = b_\beta/2$ where $q_\beta(b_\beta/2) = \beta$. The function $q_\beta(x)$ is easily sketched in the phase plane. We can now state a threshold result.

Theorem 17.3: *Suppose $u(x,t)$ is the solution of the Nagumo equation with $0 \le u(x,0) \le 1$ and suppose there are numbers β and x_0 so that $u(x,0) \ge q_\beta(x-x_0)$ on the interval $x_0 < x < x_0+b_\beta$. Then $\lim_{t\to\infty} u(x,t) = 1$.*

The proof of this statement is given in Aronson and Weinberger [1] and is a direct application of comparison arguments (see, e.g., Protter and Weinberger [23]) that are very important in the mathematical study of nonlinear parabolic equations.

According to this result, if the initial disturbance is large enough, the solution of the Nagumo equation eventually approaches 1, that is, an action potential propagates throughout the medium. However, if the initial disturbance is not large enough, even though it may exceed α in places, the solution will die to zero. An example of a precise statement to this effect is as follows:

Theorem 17.4: *Suppose $u(x,t)$ is a solution of the Nagumo equation, and for some number ρ, $0 \le \rho < \alpha$,*

$$\int_{-\infty}^\infty \max(u(x,0) - \rho, 0)dx < \sqrt{\frac{2\pi}{s(\rho)e}}(\alpha - \rho) \qquad (17.16)$$

where $s(\rho) = \sup\{(g(u)/u) - \rho, \ \alpha < u < 1\}$. Then $\lim_{t\to\infty} u(x,t) = 0$.

The proof of this statement follows from comparison arguments and can be found in [1].

In words, the theorem says that even though the initial function $u(x, 0)$ may be very large in places, if its integral is not large enough the solution will decay to rest. To get some idea of the behavior of the lower bound on the integral, if $g(u)$ is the cubic polynomial $g(u) = au(1-u)(\alpha-u)$, a sketch of the function $\sqrt{(2\pi/s(\rho)e)}(\alpha - \rho)$ shows it to be very well approximated by the straight line $2a\sqrt{2\pi/e}\,(\alpha - \rho)/(1 - \alpha)$, as a function of ρ.

17.4   Continuous Cable Theory

Continuous cable theory can be loosely defined as the collection of facts that can be derived from Equation (17.4) without specific knowledge of the function f [16]. Because the rules that result are rather easy to derive, remember, and use, much of the theory invoked by physiologists is based on continuous cable theory. In this section we will use scaling arguments and dimensional analysis to find simple expressions for the space constant of tissue, the speed of wavefront propagation, and the stimulus threshold as functions of the physical parameters of the cells. We will emphasize dimensionless quantities so that the dependencies on physical parameters is made explicit.

First, the space constant for a cable is defined to be that number Λ so that the profile of the potential resulting from a constant subthreshold stimulus applied at a single point is an exponential proportional to $e^{-x/\Lambda}$. To find the space constant analytically, we suppose that the voltage is time independent and deviates only slightly from the resting potential, so we can linearize the governing equation to obtain

$$0 = \frac{1}{p}\frac{\partial}{\partial x}\left(\frac{A}{R_m}\frac{\partial v}{\partial x}\right) - \frac{v}{R_m}. \tag{17.17}$$

It follows that exponential solutions of the form $v = e^{-x/\Lambda}$ must have

$$\Lambda = \sqrt{\frac{AR_m}{pR_c}}. \tag{17.18}$$

If L is a typical length scale, for example the length of a cardiac cell, this expression for space constant can be represented in terms of nondimensional variables as

$$\frac{\Lambda}{L} = \frac{1}{\sqrt{q}}, \quad q = \frac{L^2 pR_c}{AR_m}. \tag{17.19}$$

Next, by simple scaling arguments, it is easy to see that if there is a translation invariant solution of Equation (17.4), it is of the form

$$v(x,t) = V\left(\sqrt{\frac{pR_c}{AR_m}}x + \frac{c_0 t}{CR_m}\right) \tag{17.20}$$

where the function V is a function of one dimensionless independent variable and satisfies the equation $V'' - c_0 V' + R_m f(V) = 0$, which is independent of the parameters of cable structure. In fact, V is linearly related to the dimensionless function U found in the last section. The dimensionless number c_0 is uniquely determined by properties of $g = R_m f$ using the shooting arguments of the last section, but is also independent of all other physical parameters of the cable structure. It follows that the speed of propagation is given by

$$c = \frac{c_0}{C} \sqrt{\frac{A}{p R_m R_c}} \tag{17.21}$$

or, in dimensionless quantities

$$\frac{c C R_m}{L c_0} = \frac{1}{\sqrt{q}}, \quad q = \frac{L^2 p R_c}{A R_m}. \tag{17.22}$$

Because the resting potential is a stable state there is a minimal current stimulus strength required for an action potential to be initiated. We suppose that a very short impulse of current with current density I is added to the cable at a single point (or a very small spatial area), and we want to determine if it will be successful at initiating an action potential. The governing equation we take as

$$C \frac{\partial v}{\partial t} = \frac{1}{p} \frac{\partial}{\partial x} \left(\frac{A}{R_c} \frac{\partial v}{\partial x} \right) + f + \frac{I}{p} \delta(x) \delta(t) \tag{17.23}$$

where $\delta(x)$ is the Dirac delta function and I is the current stimulus strength-duration. Recall that the Dirac delta function carries with it the dimensions of the inverse of its argument. We rescale space and time to eliminate dimensional parameters and find the nondimensional equation

$$\frac{\partial v}{\partial t'} = \frac{\partial^2 v}{\partial x'^2} + R_m f + \frac{I}{C} \sqrt{\frac{R_c}{p A R_m}} \delta(x') \delta(t') \tag{17.24}$$

where x' and t' are the dimensionless space and time variables, respectively. It follows that if V^* is the minimal amplitude of stimulus required to initiate an action potential in the dimensionless equation

$$\frac{\partial v}{\partial t'} = \frac{\partial^2 v}{\partial x'^2} + R_m f + V^* \delta(x') \delta(t') \tag{17.25}$$

then

$$I = V^* C \sqrt{\frac{A p R_m}{R_c}} \tag{17.26}$$

is the current stimulus threshold for the problem with dimensional parameters. The notation for the constant V^* was chosen as a reminder that this constant has units of volts but is independent of all other physical parameters of the problem. In dimensionless quantities, the stimulus threshold is expressed as

$$\frac{I}{V^*CLp} = \frac{1}{\sqrt{q}}, \quad q = \frac{L^2pR_c}{AR_m}. \tag{17.27}$$

Cardiac cells are rather short (on the order of 100 μm) and are coupled through low-resistance gap junctions that are located in restricted locations of cell-to-cell contact called intercalated discs. The simple relationships from cable theory for space constant, speed of propagation, and stimulus threshold are probably adequate to describe propagation in a single cardiac cell, but for longer distances with many cells they are certain to be incorrect because they do not incorporate the effects of gap junctions. A simple attempt to include the effect of gap junctions is to continue to view a long chain of cells as a continuous cable, but to replace the cytoplasmic resistance in these formulae with the effective resistance R_e, where

$$R_e = R_c + \frac{Ar_g}{L} \tag{17.28}$$

where r_g is the intercellular coupling resistance between two cells and L is the length of a cell. The effect of this substitution is to replace the cells and their gap junctions with a uniform cable whose average resistance per unit length is the same as the original cell-junction configuration. For future reference, it is useful to write these quantities in terms of dimensionless quantities, as

$$\begin{aligned}
\frac{\Lambda}{L} &= \frac{1}{\sqrt{r+q}}, \\
\frac{cCR_m}{c_0L} &= \frac{1}{\sqrt{r+q}}, \\
\frac{I}{V^*CS} &= \frac{1}{\sqrt{r+q}}
\end{aligned} \tag{17.29}$$

where $r = Sr_g/R_m$, $q = SL^2R_c/VR_m$, $S = pL$ is the surface area of active membrane, and $V = AL$ is the volume of cytoplasm for each cell. Of course, L is the length of the cell in the direction of propagation. In the sequel, we shall refer to the formulae (17.29) as continuous cable theory.

These approximate formulae for space constant, speed of propagation, and stimulus threshold work reasonably well for well-recovered normal myocardial tissue, but unfortunately they are significantly flawed when, for example, the tissue is not well recovered or gap junction resistance is abnormally high.

Even though these formulae are rather simple, they are quite often misused or misunderstood. As a simple example, it is an oft-quoted rule of thumb that $c^2 R_e = constant$ (in the cardiology literature speeds are usually denoted as θ rather than as c so that one often sees $\theta^2 R_e = constant$). Since $c^2 R_e$ is not dimensionless, it cannot possibly be constant, because it must depend on other physical parameters. In fact, from Equation (17.29) one can easily determine that the *constant* is actually $c_0 V/(CSL)$. Thus, an experiment in which only R_e is varied should, according to continuous cable theory, obtain speeds that relate to R_e through $c^2 R_e = constant$. However, if the excitability of the cable or the direction in which propagation is measured are changed, then the *constant* will change as well and cable theory will appear to fail. Of course, this is not a failure of cable theory at all, but rather an incorrect application of the predictions of cable theory. Unfortunately, such misapplications are abundant.

17.5   Discrete Cable Theory

To apply cable theory to cardiac cells, one must hope that the effect of the discrete intercellular connections is small enough that it can be ignored or smoothed out into an effective smooth cable. Unfortunately, there are situations where this approximation is very bad indeed, and something quite different must be done to obtain reliable quantitative information. The main disagreement comes when the intercellular resistance is much more important than the cytoplasmic resistance. This can be effectively seen in the laboratory by applying stimuli to the cells well before they are fully recovered or by adding heptanol, an alcohol whose primary effect is to increase the resistance of gap junctions.

Cable theory was derived from the assumption that the cytoplasmic resistance was more important than the intercellular coupling resistance. Suppose instead that we start with the exactly opposite assumption, that the most significant resistance is intercellular, and cytoplasmic resistance is negligible. Of course, this is again an approximation, but there is substantial experimental evidence for its validity, since propagation, especially in relatively unrecovered tissue, or in tissue where gap-junction function is known to be impaired, is observed to be *saltatory*, appearing to propagate rapidly through individual cells and slowly through gap junctions [27]. Numerical simulations have also shown this effect [8,25]. We therefore assume that myocardium consists of many equipotential excitable cells (i.e., there are no potential differences within individual cells), each of whose transmembrane potential v is governed by the equation

$$CS\frac{dv}{dt} = Sf(v) + I \qquad (17.30)$$

where C is the membrane capacitance of the cell per unit surface area, $f(v)$ represents the inward ionic currents per unit area of membrane, I is the total of all input currents from other cells, and S is the total surface area of active membrane for an individual cardiac cell.

Cardiac cells are coupled by gap junctions through which currents can flow directly between two cells. It is usual to assume that gap junctions satisfy Ohm's law, that is, the current flow across a gap junction is linearly proportional to the voltage difference across the junction [31]. For a one-dimensional chain of coupled cells, we represent the voltage at the nth cell by v_n, from which it follows that

$$C\frac{dv_n}{dt} = f(v_n) + \frac{1}{Sr_g}(v_{n+1} - 2v_n + v_{n-1}) \tag{17.31}$$

where r_g is the total intercellular resistance between adjacent cells. This equation we shall denote as the discrete cable equation.

There is a significant difference between the behavior of solutions of this equation and those of the continuous cable equation, as we shall shortly see. However, it is tempting to see what agreement there is between the two models. If the values v_n can be interpolated by a smooth function V, as in $v_n = V((n + 1/2)L)$, where L is the length of a cardiac cell, then we can approximate the second-order difference of v_n by

$$v_{n+1} - 2v_n + v_{n-1} = L^2\frac{\partial^2 V}{\partial x^2} \tag{17.32}$$

and we arrive at the cable equation

$$C\frac{dV}{dt} = f(V) + \frac{L^2}{Sr_g}\frac{\partial^2 V}{\partial x^2}. \tag{17.33}$$

Since L is the fixed size of the cardiac cells it is not legitimate to argue that this comparison is valid in the limit that L goes to zero, because L is not a parameter that can be varied. The correct comparison between the two model equations occurs when profiles are quite smooth, which happens when r_g is small, with all other parameters held fixed. In that limit, the rules of cable theory hold and we conclude (by identifying coefficients) that for small r_g, the space constant, speed of propagation, and stimulus threshold are given approximately by

$$\frac{\Lambda}{L} = \frac{1}{\sqrt{r}},$$
$$\frac{cCR_m}{c_0 L} = \frac{1}{\sqrt{r}}, \tag{17.34}$$
$$\frac{I}{V^*CS} = \frac{1}{\sqrt{r}}$$

where $r = (Sr_g/R_m)$. When r is large these relationships fail for the same reasons that cable theory fails, namely, profiles are not smooth for the nonuniform cell-gap junction cable. However, when the cytoplasmic resistance R_c is exactly zero, the results of cable theory agree exactly with these formulae. The main source of the error must therefore be when r is large.

The most important difference between the continuous cable and the discrete cable is with the existence and speed of propagation of traveling waves [20]. In a continuous medium, the existence of a traveling wave solution does not depend on the size of the coupling coefficient. With cable theory, if a traveling wave solution of the form $u(x,t) = U(x+ct)$ exists for an equation with diffusion coefficient equal to 1, then for a different value of the diffusion coefficient D, the traveling wave solution is $u(x,t) = U((x/\sqrt{D})+ct)$, so that the speed of propagation is $c\sqrt{D}$.

This is not the case for a discrete model (17.31). For the discrete equation (17.31), we expect there to be some type of traveling wave of the form $v_n(t) = U(n+ct)$, for some function U, and some speed c. We also expect that a traveling wave solution, if it exists, will be monotone in n, satisfying $v_n \leq v_{n+1}$. One can show [20] that if a solution is initially monotonically increasing in n, then it remains so for all time.

That no propagation can occur for small coupling (large coupling resistance r_g) is easy to see. If $f(V_0) = f(V_1) = f(V_2) = 0$, and $f(v) < 0$ for $V_0 < v < V_1$, $f(v) > 0$ for $V_1 < v < V_2$, we can restrict our attention to values of v in $V_0 \leq v \leq V_2$, since one can show using comparison arguments that if $V_0 \leq v_n \leq V_2$ at $t = 0$, then the same inequality holds for all $t > 0$. With $V_0 \leq v_k \leq V_2$ for all k, and $v_k \leq v_{k+1}$,

$$C\frac{dv_n}{dt} \leq f(v_n) + \frac{1}{Sr_g}(V_2 - v_n).$$

Now suppose r_g is large enough so that there is a number $v^* < V_1$ so that the function $(1/Sr_g)(V_2 - v) + f(v)$ is negative for some values of $v < v^*$. If v_n has its initial value below v^*, then v_n must stay below v^* for all time, no matter what the neighboring values are. In other words, there can be no propagating solution that takes the values of v_n from 0 to 1. For the cubic function $f(v) = (v - V_0)(v - V_1)(V_2 - v)/[R_m(V_0 - V_1)(V_0 - V_2)]$ (which has slope $-1/R_m$ at $v = V_0$), this occurs whenever $Sr_g/R_m > 4(V_2 - V_0)/(V_1 - V_0)$.

In the very special case that f is the piecewise linear function (17.6), we can calculate exactly those values of coupling resistance for which there is a standing solution of Equation (17.6). If there is a standing solution (i.e., time independent) with $\lim_{n\to\infty} v_n = 1$ and $\lim_{n\to-\infty} v_n = 0$, then initial data that are initially below the standing solution are blocked from propagating. We find that for

$$\frac{Sr_g}{R_m} > \frac{(1-2\alpha)^2}{\alpha(1-\alpha)} = r^* \tag{17.35}$$

there are two standing solutions, thereby precluding propagation. Notice that for α very close to zero (when the medium is very excitable), the critical value of coupling resistance is very large (propagation is harder to stop), but for α nearly $\frac{1}{2}$, the critical coupling resistance is small. In other words, if r_g is some fixed number, we expect propagation failure for relatively less excitable media. Numerical simulation show the same qualitative behavior for the cubic nonlinearity.

Traveling wave solutions to the discretized equation (17.31) exist [33], but only if the coupling resistance is sufficiently small. The proof of existence of these traveling wave solutions is mathematically quite sophisticated.

The speed of propagation for the discrete problem cannot be determined analytically. However, the speed of propagation is very well approximated by the formula

$$\left(\frac{cCR_m}{c_0L}\right)^2 = \frac{1}{r} - \frac{1}{r^*}, \qquad r = \frac{Sr_g}{R_m} \tag{17.36}$$

where r^* is a dimensionless constant that depends on excitability of the medium. Notice that this formula is only a slight modification of Equation (17.34) and has the correct behavior for small intercellular coupling where it is asymptotic to the result from continuous cable theory, but it also implies that propagation is impossible for r sufficiently large. It is also reasonable that r^* should be a monotonically increasing function of excitability. A more excitable cell requires less stimulating current to raise it to threshold and, therefore, propagation of action potentials is possible with larger values of r. For the piecewise linear function (17.6), the critical parameter r^* is given by Equation (17.35).

Since propogation is impossible for r larger than r^*, it follows that the stimulus threshold becomes unbounded as r approaches r^* from below. Using dimensional analysis, one can show that the stimulus threshold for the discrete problem in Equation (17.31) is of the form

$$\frac{I}{CS} = h(r), \qquad r = \frac{Sr_g}{R_m} \tag{17.37}$$

where the function h is a function of its one dimensionless argument r and the excitability of the tissue, and is in units of voltage. A reasonable approximation (based on numerical simulation) to the threshold stimulus is provided by the formula

$$\frac{I}{CS} = V^*\frac{1}{\sqrt{r}} + U^*\frac{1}{\frac{1}{r} - \frac{1}{r^*}}, \qquad \text{provided } r < r^* \tag{17.38}$$

where $r = (Sr_g/R_m)$. Here the parameters V^* and U^* depend only on tissue excitability and are in units of voltage. This formula again reflects the fact that for small intercellular resistance the collection of cells resembles a uniform cable, but for large gap-junction resistance propagation may be impossible.

The formulae (17.35) and (17.38) have very important implications for numerical simulations. To simulate numerically the continuous cable Equation (17.4), it is typical to replace the second partial derivative with respect to x with a second order finite difference yielding the equations

$$C\frac{dv_n}{dt} = \frac{A}{pR_e\Delta x^2}(v_{n+1} - 2v_n + v_{n-1}) + f(v_n) \qquad (17.39)$$

where $v_n = v(n\Delta x)$. The number Δx is the numerical discretization size and has nothing at all to do with cell length. However, Equations (17.31) and (17.39) are exactly the same if we make the identification $(A/pR_e\Delta x^2) = (1/Sr_g)$. Thus, the numerical simulation will produce results consistent with the discrete model even though it is intended to simulate the continuous cable model. For example, the simulation will exhibit propagation failure and have infinite stimulus thresholds whenever

$$\frac{pR_e\Delta x^2}{R_m} > r^* \qquad (17.40)$$

even though this is totally inconsistent with the continuous cable theory that is being approximated. Another way to say the same thing is that in a discretization of continuous cable theory, propagation failure and infinite stimulus thresholds are purely numerical artifacts, being consequences of the discretization. One rather subtle way that this can occur is if the excitability of the membrane is changed while all other parameters as well as the discretization size are kept fixed. Since this has the effect of changing r^*, the level of excitability at which the propagation failure occurs will be dependent on the discretization size, which is not a physical parameter of the model.

The way to test for numerical artifact directly in a simulation is to change the discretization size and see if the parameter values at which the important critical phenomena occur change as well. We claim that if the discretization size is changed, parameter values at which propagation block and infinite stimulus threshold are observed will change in accordance with Equation (17.39), that is, propagation block will disappear if Δx is sufficiently decreased. A numerical simulation that claims to show such propagation block or infinite stimulus thresholds without demonstrating conclusively that the reported phenomena are independent of discretization size is suspect. It is simply not adequate to present the results of a numerical simulation using only one spatial discretization size, no matter how small.

17.6   A Modified Cable Theory

So far we have considered two models of propagation in myocardial tissue. The first model assumes that all resistances can be uniformly distributed along an equivalent cable, and then the relatively easy formulae of cable theory result. The second model assumes that cytoplasmic resistance can be ignored and all resistances can be lumped into the discrete coupling resistances. The formulae that result are slightly more complicated, but have the added feature that they predict propagation failure for sufficiently large coupling resistance.

It would be nice to have a theory for propagation and its failure that makes neither of these simplifying assumptions, but yet provides quantitatively reliable and easy to use formulae. In this section we will discuss the derivation of a modified cable theory that includes effects of continuous cytoplasm and discrete gap junctions. We will find formulae for the space constant, the critical coupling strength for propagation failure, the speed of propagation, and the critical stimulus strength-duration curve.

To the extent that it is possible, we should solve the differential equation

$$C\frac{\partial v}{\partial t} = \frac{1}{p}\frac{\partial}{\partial x}\left(\frac{A}{R_c}\frac{\partial v}{\partial x}\right) + f \tag{17.41}$$

in the cell interior, subject to boundary conditions at the junction between cells

$$-\frac{A}{R_c}\frac{\partial v}{\partial x}(x = nL^-) = -A'f(v(x = nL^-)) - \frac{v]}{r_g}$$

$$-\frac{A}{R_c}\frac{\partial v}{\partial x}(x = nL^+) = A'f(v(x = nL^+)) - \frac{v]}{r_g} \tag{17.42}$$

$$v] = v(x = nL^+) - v(x = nL^-).$$

Here, A' is the area of active tissue in the region of intercellular contact. The inclusion of the A' term is necessitated by the fact that the region of cell wall-to-wall contact may actually contain some amount of active membrane across which a transmembrane potential is measured and across which ionic currents may flow. If cells are connected end to end it is likely that $A' = 0$, but for side-to-side connections much of the contact area consists of active tissue, and so $A' \neq 0$. Since our ultimate goal is to understand something about end-to-end and side-to-side propagation, the term A' must be included so that the surface area of a cell is the same for both end-to-end and side-to-side calculations.

To determine the space constant analytically for a long collection of cells coupled with gap junctions, we linearize the governing equations in a neighborhood of the resting potential, and find the equation

$$\frac{1}{p}\frac{\partial}{\partial x}\left(\frac{A}{R_c}\frac{\partial v}{\partial x}\right) - \frac{v}{R_m} = 0 \tag{17.43}$$

subject to the boundary conditions (17.42) given above, linearized in the neighborhood of the resting potential. Here v is the deviation of the transmembrane potential from its resting value. We seek a solution of these equations that is decaying geometrically, so that $v(x + L) = \mu v(x)$ for some constant $\mu < 1$. The constant μ relates to the usual space constant L through $\mu = e^{-L/\Lambda}$. Following a straightforward calculation, we find that

$$\frac{r_g \mu}{2\rho}(E(\beta + \rho)^2 - \frac{1}{E}(\rho - \beta)^2) = \mu^2 - \mu(E(1 + \frac{\beta}{\rho}) + \frac{1}{E}(1 - \frac{\beta}{\rho})) + 1 \quad (17.44)$$

where $E = e^{aL}, a^2 = (pR_c/AR_m), \rho = (aA/R_c)$, and $\beta = (A'/R_m)$. If the space constant is many cell lengths, this relationship can be simplified substantially by replacing E by the first few terms of its Taylor series expansion $E = 1 + aL + ((aL)^2/2) + \ldots$. We find to leading order that

$$r_g = L^2 \left(\frac{R_m}{S\Lambda^2} - \frac{R_c}{V} \right) \quad (17.45)$$

which is exactly the same as found from continuous cable theory [see Equation (17.29)]. This is not a good approximation if the space constant is on the order of one or two cell lengths, in which case the full formula (17.44) should be used, but for most physiological situations this approximation is adequate. We shall use this formula in what follows to calculate the effective intercellular coupling resistance for cardiac tissue.

The next problem is to determine the critical coupling strength that precludes propagation. To do this we determine under what conditions there exists a standing transition between the stable steady states of the function f, since a standing transition precludes the possibility of propagation. Thus, we seek a solution of the time-independent problem

$$0 = \frac{1}{p} \frac{\partial}{\partial x} \left(\frac{A}{R_c} \frac{\partial v}{\partial x} \right) + f(v) \quad (17.46)$$

subject to the conditions (17.42) at the interface between cells. We assume that the function $f(v)$ has three zeros, two of which correspond to the depolarized and polarized states of the tissue, and we seek a solution that is a monotonic function connecting the two stable states (actually only quasisteady states) of the membrane.

As formulated, this problem is still too difficult to solve because we have insufficient information about f, so we examine the greatly simplified model problem with $f(v)$ given by Equation (17.6). Despite the crudeness of this approximation, from it we can extract analytical information that is very useful.

Following a standard but very tedious calculation, we find that propagation fails whenever the effective intercellular coupling r_g exceeds r_g^*, where r_g^* is the positive root of the quadratic polynomial

$$\begin{aligned}
&r_g^{*2}\alpha(\alpha - 1)(E^2(\rho + \beta)^2 - (\rho - \beta)^2) \\
&+ r_g^*(2\alpha - 1)^2(E^2(\rho + \beta) + \rho - \beta) \\
&+ (2\alpha - 1)^2(E^2 - 1) = 0.
\end{aligned} \quad (17.47)$$

If the space constant in normal tissue is many cell lengths, $r^* = (Sr_g/R_m)^*$ is accurately approximated by

$$r^* = \frac{(2\alpha - 2)^2}{\alpha(1 - \alpha)} + q\left(1 + \frac{(2\alpha - 1)^2}{3\alpha(1 - \alpha)}\right), \quad q = \frac{SL^2 R_c}{VR_m}. \tag{17.48}$$

The critical value of (nondimensional) intercellular coupling r^* is a monotonically increasing function of the parameter q. Since q depends on L, a chain of elongated cells that is coupled side to side is more likely to experience propagation failure than a chain of the same cells coupled end to end with exactly the same intercellular coupling for the two chains. Also note that the effect of cytoplasmic resistance is to increase the critical coupling resistance r^* above that for isopotential cells. However, for physiologically realistic parameter values, the number q is a very small number so that these effects are quite small. We also see that the effect of increased excitability is to increase r^*. One shortcoming of this formula is its dependence on excitability through the parameter α, for which there is no good estimate for real tissue. However, it is likely that this formula gives reasonable qualitative information for real tissue.

Now that we have some idea of the behavior of r^*, we can estimate the speed of propagation and stimulus strength-duration threshold in a chain of coupled cells, taking into account both the intercellular resistance and the cytoplasmic resistance. The formulae should have the features that they agree with cable theory when coupling resistances are small or absent, they should predict propagation failure for large intercellular coupling, and they should agree with the discrete theory when cells are isopotential. These features are reflected in the expressions

$$\left(\frac{cCR_m}{c_0 L}\right)^2 = \frac{1}{r + q}\left(1 - \frac{r}{r^*(q)}\right) \tag{17.49}$$

for the speed of propagation, and

$$\frac{I}{CSV^*} = \frac{1}{\sqrt{r + q}} + \frac{U^*}{V^*}\left(\frac{rr^*(q)}{r^*(q) - r}\right) \tag{17.50}$$

provided $r < r^*$ for the stimulus strength-duration threshold curve, where $r = (Sr_g/R_m)$, $q = (SL^2 R_c/VR_m)$. In the discussion that follows we shall refer to these formulae as modified cable theory.

17.7 Comparison with Experiments

In this section we will show some of the consequences of the modified cable theory as they relate to some recent experiments on propagation in myocardium. We will show how the expressions for space constant, speed of propagation, and stimulus-threshold are in agreement with experimental evidence while continuous cable theory without these corrections is noticeably flawed.

In thin layers, the myocardium is a collection of elongated cells that are aligned in one direction, called the longitudinal, or axial, direction. We must restrict our attention to thin layers because through the thickness of the myocardial wall the orientation of the long axis changes. Myocardium is anisotropic because propagation in the longitudinal direction is typically faster than in the lateral, or transverse, direction. Attempts to model the two-dimensional myocardium as a continuum have met with difficulty because experiments have shown that there can be a preferred direction for propagation failure, that is, propagation may succeed in one of the directions and fail in the opposite (orthogonal) direction if the stimulus is applied very early during the recovery of a cell [26,27] or if the intercellular coupling resistance is artificially increased [2,6,7]. In a continuum there can be no preferred direction for propagation failure.

When a stimulus is applied to a piece of myocardium at a single point, the wave of activity spreads as a growing ellipse. However, to apply cable theory to a two-dimensional medium, one must assume that isopotential curves are straight lines orthogonal to one of the axis directions, that is, waves are planar. In this way, one can assume that all cells along an isopotential line behave exactly the same way and they have no influence on each other, since, ideally, no currents are flowing parallel to the isopotential lines. This can be effectively accomplished in the laboratory by applying current stimuli along a line (rather than at a single point) and measuring potentials in a region where isopotential lines are straight parallel lines. Alternately, one can use long, narrow strips of tissue cut with edges parallel or perpendicular to the cell axis [6,7].

Our first task is to calculate the effective intercellular coupling resistance for the two directions using measured space constants. It is reasonable to suppose that a typical space constant is many cell lengths, so that we can use the formula

$$r_g = L^2 \left(\frac{R_m}{S\Lambda^2} - \frac{R_c}{V} \right) \tag{17.51}$$

to represent the relationship between space constant and intercellular coupling resistance. This formula determines how large r_g must be to account for a particular space constant if cell dimensions and cytoplasmic resistance are known. From this formula, we find that the ratio of couplings in longitudinal and transverse directions is

$$\frac{r_L}{r_T} = \frac{L_L^2}{L_T^2} \frac{\Lambda_L^{-2} - \gamma}{\Lambda_T^{-2} - \gamma}, \quad \gamma = \frac{R_c S}{V R_m}. \tag{17.52}$$

For canine crista terminalis typical speeds of propagation in normal well recovered tissue are 0.51 meters/sec in the longitudinal direction and 0.17

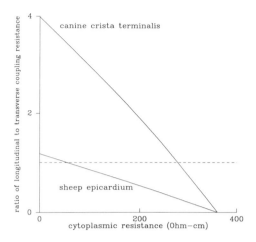

FIGURE 17.2. Ratio of longitudinal to transverse coupling resistance for canine crista terminalis and sheep epicardium plotted as a function of cytoplasmic resistance.

meters/sec in the transverse direction, a ratio of 3:1 [4,26,27,29]. In sheep epicardium, average speeds of propagation of 0.51 meters/sec in the longitudinal direction and 0.09 meters/sec in the transverse direction, a ratio of 5.5:1, were reported in [6,7]. It is reasonable to assume that for normal tissue the ratio of space constants is the same as the ratio of speeds. Thus, for cells six times longer than wide (for example, 100 μm long by 17 μm wide, a ratio of 6:1), and for space constants in the ratio of 3:1 for dog and 5.5:1 for sheep, the ratio of coupling resistances is

$$\frac{r_L}{r_T} = 4\frac{1-y}{1-0.11y} \qquad (17.53)$$

for canine crista terminalis, and

$$\frac{r_L}{r_T} = 1.2\frac{1-y}{1-0.031y} \qquad (17.54)$$

for sheep epicardium, where $y = (R_c S \Lambda_L^2 / V R_m)$. Plots of these functions are shown in Figure 17.2.

These formulae suggest a substantial and significant difference between tissue of dog and sheep, at least those from that the data were measured. If we completely ignore the effects of cytoplasmic resistance (set $y = 0$), these formulae imply that in canine crista terminalis, intercellular resistance is four times larger in the longitudinal direction than in the transverse direction, whereas in sheep epicardium the ratio of longitudinal to transverse intercellular resistance is 1:2, only slightly greater than 1. The ratio of longitudinal to transverse coupling is monotonically decreasing as a function of y and remains larger than 1 if y is smaller than 0.77 for dog and smaller than 0.17 for sheep. For typical cell sizes, membrane resistance of 7000 Ωcm^2, and a longitudinal space constant of 0.09 cm, this implies that longitudinal intercellular resistance is larger than transverse resistance if the cytoplasmic resistance is smaller than 282 Ωcm in dog or 62 Ωcm in sheep. For all of the calculations in the sequel, we take cells to be 0.01 cm long, 0.00167 cm wide, (S/V)= 2400 cm^{-1}, and $R_m = 7000$ Ωcm^2. If the longitudinal space constant is larger than 0.09 cm there is a reduction in the value of R_c at which this transition occurs.

From our calculation in the previous section, the ratio of critical longitudinal to transverse resistances for typical cell parameters calculated is nearly 1, independent of the excitability parameter α or the cytoplasmic resistance R_c. Thus, if the ratio (r_L/r_T) of longitudinal to transverse coupling resistance is greater than 1 we expect block to occur preferentially in the longitudinal direction. Alternatively, if the ratio of longitudinal to transverse coupling resistances is less than 1, we expect block to occur preferentially in the transverse direction. In other words, longitudinal block is preferred in dog crista terminalis if the cytoplasmic resistance is less than 282 Ωcm. For sheep epicardium, transverse block is preferred if cytoplasmic resistance is greater than 62 Ωcm.

The actual cytoplasmic resistance is on the order of 150 Ωcm, although single cell measurements in frog myocardium suggest values as low as 120 Ωcm [15]. In any case, for values in this range we come to the startling conclusion that canine crista terminalis is more strongly coupled in the transverse direction than in the longitudinal direction, but sheep epicardium is more strongly coupled in the longitudinal direction than in the transverse direction. This observation is consistent with the experimental results on the different tissue. Spach and collaborators [26,27] observed that block was more likely to occur in the longitudinal direction consistent with the hypothesis that longitudinal coupling is weaker than transverse coupling, whereas Jalife and collaborators [6,7] observed that block was more likely to occur in the transverse direction, consistent with the hypothesis that transverse coupling is weaker than longitudinal coupling [28].

A morphological explanation of this difference is unknown, but this result suggests that there can be significant differences in effective intercellular coupling between different preparations or tissues. The determining factor in any collection of cells is the ratio of cell dimensions compared to the ratio of speeds of propagation in those directions. In a medium where cells are nearly isopotential, if the ratio of cell dimensions is 6:1, say, then the speeds of propagation should also be in about the same ratio (since the spread through isopotential cells is nearly instantaneous compared to propagation between cells). Ratios of speeds of propagation that differ significantly from the ratio of cell dimensions implicate the intercellular resistance as important. For example, if the ratio of longituditunal to transverse speeds of propagation is significantly smaller than the ratio of cell dimensions, then longitudinal intercellular resistances is suggested as being larger than transverse intercellular resistance. On the other hand, if the ratio of speeds is of the same order or larger than the ratio of cell dimensions, then intercellular coupling in the transverse direction is suggested as larger than longitudinal intercellular coupling. Of course, Equation (17.51) contains a number of physical parameters that must be known accurately to obtain a correct estimate of the effective intercellular resistance for different tissue.

When speeds are measured it is typical and convenient to report them in units of length per unit time. Similarly, space constants are measured as length in units of centimeters. However, if speeds were reported in units of number of cells per unit of time and space constants in units of cell number, we would probably have a completely different intuition of which is the *fast* direction and which is the *slow* direction of propagation. With cells six times longer than wide, and longitudinal speeds three times larger than transverse, transverse propagation excites twice as many cells per unit time than longitudinal propagation. Clearly, in the economy of cells per unit time transverse propagation is the faster mode of propagation. An interesting exercise is to calculate how long it takes to excite two long and skinny patches of myocardium with exactly the same number of cells, one cut with the long side parallel to the long axis of the cells and the other with the long side transverse to the long axis of the cells with a stimulus applied at one of the skinny ends of the tissue. The revealing answer is that the transverse plane wave excites the patch twice as quickly as does the longitudinal plane wave, if cells have dimensions in the ratio of 6:1 and speeds in the ratio of 3:1.

According to cable theory, the only way to cause propagation failure in an excitable cable is to make either r_g or R_c infinite. The modified cable theory predicts that there are two ways to observe propagation failure in one coordinate direction. One way is to decrease r^* by decreasing the excitability of the tissue, and the second is to increase the intercellular resistance. Furthermore, if the intercellular resistances in the two coordinate directions are different, then the direction with the highest resistance should be the most likely to fail. Both Spach and collaborators and Jalife and collaborators have demonstrated that propagation in one direction can be successful while propagation in the perpendicular direction fails. Spach accomplished this by applying a premature stimulus, thereby decreasing the critical number r^*, and Jalife by changing the gap-junctional resistance by superfusion with heptanol. That the gap-junction resistance was not infinite in the heptanol experiment is consistent with the belief that all gap junctions in the cell should be affected by the addition of heptanol in the same way, keeping the ratio of the two intercellular resistances fixed, and since longitudinal propagation was successful, the resistance could not have been infinite in the transverse direction. In Figure 17.3 we show the ratio of speeds for the modified and the original cable theory as a function of

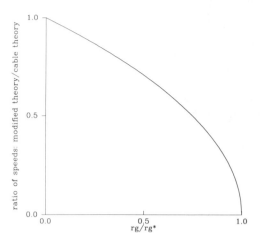

FIGURE 17.3. Ratio of the speeds of propagation from the modified theory to cable theory, plotted as a function of (r_g/r_g^*).

coupling resistance normalized by r_g^*. Observe that for small coupling resistance the two theories agree, but for large coupling resistance the modified theory predicts substantially decreased speeds and failure of propagation for r_g above r_g^*.

Jalife and collaborators [6,7] reported the results of providing a superthreshold stimulus to a well-recovered piece of sheep epicardium (stimuli were given at a timing interval of 1500 msec) and measuring the speed of longitudinal and transverse propagation. During the course of the experiment the piece of epicardium was perfused with heptanol so that as the experiment proceeded, the gap-junction resistance was presumed to increase. It was found that as the heptanol concentration increased, velocities in both directions decreased as expected, but the velocity in the direction transverse to the cell axis fell off more rapidly than in the longitudinal direction. They observed that the ratio of transverse to longitudinal speed of propagation was a monotonically decreasing function of heptanol concentration, and that transverse propagation failed before longitudinal propagation.

Balke and colleagues [2] performed a similar experiment with dog epicardium, although with smaller concentrations of heptanol, so that block did not occur. They also observed that the ratio of transverse to longitudinal velocity of propagation decreased as a function of heptanol concentration. In addition, they tried, using continuous cable theory, to make quantitative predictions of the speeds of propagation. They found that for modest concentrations of heptanol, continuous cable theory gave reasonable quantitative agreement with their experimental results. However, for larger concentrations of heptanol, cable theory predicted transverse speeds that were consistently higher than the experimentally observed speeds. Their explanation of this overestimate was that high levels of heptanol may cause deviations from uniform anisotropy, leading to a failure of the model.

Using the modified cable theory, we can explain the results of these experiments. We suppose that initial gap-junction resistance in the longitudinal direction is smaller than that in the transverse direction, and that the action of heptanol increases the gap-junction resistance, keeping them in the same ratio for both directions. In Figure 17.4 we show the ratio of transverse to longitudinal velocities of propagation plotted as a function of the longitudinal coupling resistance normalized by the longitudinal r_g^*. The dashed curve shows the results from continuous cable theory and the solid curve shows the results from the modified cable theory. The ratio of the transverse to longitudinal speeds of propagation is seen to decrease until transverse failure occurs at a value of longitudinal coupling resistance well below the critical r_g^*. Observe that for small coupling resistances there is agreement between the continuous cable and modified cable theories, showing a decrease in the ratio of the speeds, but for larger resistances, continuous cable theory predicts a ratio consistently larger than the modified theory (and the experimental results). For this plot, R_c was taken to be 150 Ωcm. At larger values of R_c, the agreement between the two curves at small values of r_g is improved, but it necessarily fails as r_g approaches r_g^*. With the inclusion of discrete effects in the modified theory, the predicted speeds are reduced compared to a cable model without such effects. In other words, the overestimates of predicted speeds using continuous cable theory can be accounted for by the fact that continuous cable theory includes no decrement of speeds due to discreteness of cells.

Finally, Jalife and collaborators [6,7] measured the stimulus threshold in thin strips of sheep epicardium, one cut in the longitudinal direction and one cut in the transverse direction. They found that the stimulus strength-duration threshold was a decreasing function of recovery time (longer recovery time required smaller stimuli), and that for stimulus times before a certain minimum recovery time, no propagation was possible. As the recovery time was reduced to this minimum time, the stimulus strength-duration threshold became unbounded. The unexpected observation was that for large recovery times, the stimulus strength-duration threshold was smaller for transverse propagation than for longitudinal propagation, but as the recovery time before application of the stimulus was decreased, this ordering was reversed, making the threshold for transverse propagation larger than for longitudinal propagation, and transverse propagation was the first to fail.

Continuous cable theory does not give an explanation of these observations. For continuous cable theory, the ratio of stimulus threshold values is exactly the same as the ratio of the space constants divided by the ratio of the lengths of a cell in the two coordinate directions, and none of these are affected by the recovery time. Thus, continuous cable theory predicts that

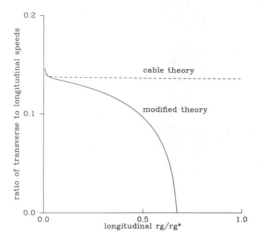

FIGURE 17.4. Ratio of transverse to longitudinal speeds of propagation for sheep epicardium as a function of longitudinal (r_g/r_g^*), assuming the ratio of transverse to longitudinal coupling resistance is fixed.

the ratio of stimulus strength-duration thresholds for transverse and longitudinal propagation should be a constant independent of recovery time before the stimulus is applied.

The predictions of the modified theory are substantially different from cable theory. In Figure 17.5 is shown the stimulus threshold curve plotted as a function of (r_g/r_g^*), normalized so that the threshold is 1 when $r_g = 0$. The result from continuous cable theory is shown as a dashed curve and the result from the modified theory is shown as a solid curve. The important observation is that continuous cable theory predicts a monotonically decreasing curve as a function of intercellular resistance, while the modified theory is biphasic. Thus, according to the modified theory, an experiment in which heptanol is gradually added to a tissue and the stimulus threshold curve measured will show first a decrease in threshold followed by a dramatic increase just before block of propagation. Notice that this will also exhibit directional dependence since the initial value of r_g (before any heptanol is added) is presumably different for the two fiber directions. In a bidirectional experiment, the stimulus threshold curves will cross as a function of heptanol concentration, since the direction with the largest initial r_g will at first have the smallest stimulus threshold, but as coupling resistances change (presumably keeping the ratio fixed) the direction with largest initial r_g will be the first to experience an increase in stimulus threshold leading to block. Continuous cable theory predicts a monotonic decrease in the stimulus threshold as a function of heptanol concentration. (As far as I know, this experiment has not yet been done.)

A similar argument explains the experimental results of Jalife and col-

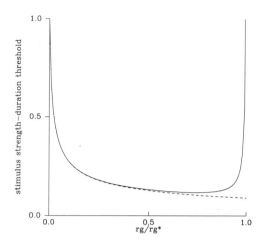

FIGURE 17.5. Stimulus strength-duration threshold plotted as a function of (r_g/r_g^*), normalized to be 1 at $r_g = 0$.

laborators [6,7]. The stimulus threshold curve depends on excitability, and therefore recovery time, in two ways. First, as the excitability is decreased, the constant V^* must increase, causing the entire curve (Figure 17.5) to increase, and second, as excitability decreases so also must r^*. For a bidirectional experiment, the direction with the largest intercellular coupling resistance will have the smallest stimulus threshold at large recovery times. As excitability decreases, the stimulus threshold must increase for both directions; however, the direction of highest intercellular coupling will be the first to experience the dramatic increase and subsequent block that occurs as r^* decreases. As a result, there must be a crossover of the two stimulus threshold curves as excitability decreases. Such a crossover has been observed [6,7].

17.8   Discussion

The study of propagation in anisotropic discrete myocardium has become big business. In the wake of the discovery that discreteness of coupling cannot be overlooked came a number of experimental and numerical studies of this effect [5,8,17,18,20,21,24,25]. The implications of this may in fact be extremely important to the understanding of the initiation and maintenance of life-threatening arrhythmias, such as ventricular tachycardia and fibrillation [19,30].

In Equations (17.49) and (17.50) are summarized the first attempts to describe these consequences with analytical formulae. Although they represent a vast improvement over previous formulae, there are many other realities that have been ignored whose importance is poorly understood. For example, myocardial cells are not perfect cylinders or rectangular boxes organized on a neat rectangular grid. Cells may have side-to-side connections, but the intercalated discs are orthogonal to the long axis of the cell [14], so that the walls of a cell look something like a staircase, and so on. How to incorporate these effects into analytical relationships is not known, nor is it known how important such effects are to the validity of the formulae.

Physiologists also place much importance in two other quantities, namely, the time constant of the rising action potential, denoted τ_{foot}, and the maximum rate of rise of the potential $(\partial v/\partial t)_{max}$. This chapter has not included a discussion of these quantities because, even though they are easily characterized in continuous cable theory (but often misunderstood), for real myocardium where propagation is not uniform along a cell or fiber (but saltatory), these quantities are space-dependent; that is, their values depend on where on the myocardium they are measured. In contrast, the space constant, the speed of propagation, and the threshold stimulus are averaged quantities and exact knowledge of where measurements are taken is less important.

REFERENCES

[1] D.G. Aronson and H.F. Weinberger. Nonlinear diffusion in population genetics, combustion and nerve pulse propagation. In J.A. Goldstein, editor, *Proceedings of the Tulane Program on Partial Differential Equations and Related Topics*, pages 5–49, Springer-Verlag, Berlin, 1975.

[2] C.W. Balke, M.D. Lesh, J.F. Spear, A. Kadish, J.H. Levine, and E.N. Moore. Effects of cellular uncoupling on conduction in anisotropic canine ventricular myocardium. *Circ. Res.*, 63:879–892, 1988.

[3] G.W. Beeler and H. Reuter. Reconstruction of the action potential of myocardial fibres. *J. Physiol. (Lond.)*, 268:177–210, 1977.

[4] L. Clerc. Directional differences of impulse spread in trabecular muscle from mammalian heart. *J. Physiol (Lond.)*, 255:335–346, 1976.

[5] W.C. Cole, J.B. Picone, and N. Sperelakis. Gap junction uncoupling and discontinuous propagation in the heart. *Biophys. J.*, 53:809–818, 1988.

[6] C. Delgado, B. Steinhaus, M. Delmar, D.R. Chialvo, and J. Jalife. Directional differences in excitability and margin of safety for propagation in sheep ventricular epicardial muscle. *Circ. Res.*, 67:97–110, 1990.

[7] M. Delmar, D.C. Michaels, T. Johnson, and J. Jalife. Effects of increasing intercellular resistance on transverse and longitudinal propagation in sheep epicardial muscle. *Circ. Res.*, 60:780–785, 1987.

[8] P.J. Diaz, Y. Rudy, and R. Plonsey. Intercalated discs as a cause for discontinuous propagation in cardiac muscle: A theoretical simulation. *Ann. Biomed. Eng.*, 11:177–189, 1983.

[9] D. DiFrancesco and D. Noble. A model of cardiac electrical activity incorporating ionic pumps and concentration changes. *Phil. Trans. Roy. Soc. Lond. B.*, 222:353–398, 1985.

[10] J.-P. Drouhard and F.A. Roberge. Revised formulation of the Hodgkin-Huxley representation of the sodium current in cardiac cells. *Comp. Biomed. Res.*, 20:333–350, 1987.

[11] L. Ebihara and E.A. Johnson. Fast sodium current in cardiac muscle, a quantitative description. *Biophys. J.*, 32:779–790, 1980.

[12] P.C. Fife. *Mathematical Aspects of Reacting and Diffusing Systems.* Lecture Notes in Biomathematics, vol. 28, Springer-Verlag, Berlin, 1979.

[13] A.L. Hodgkin and A.F. Huxley. A quantitative description of membrane current and its application to conduction and excitation in nerve. *J. Physiol.*, 177:500–544, 1952.

[14] R.H. Hoyt, M.L. Cohen, and J.E. Saffitz. Distribution and three-dimensional structure of intercellular junctions in canine myocardium. *Circ. Res.*, 64(3):563–574, 1989.

[15] J.R. Hume and W. Giles. Active and passive electrical properties of single atrial cells. *J. Gen. Physiol.*, 78:19–42, 1981.

[16] P.F. Hunter, P.A. McNaughton, and D. Noble. Analytical models of propagation in excitable cells. *Progr. Biophys. Mol. Biol.*, 30:99–144, 1975.

[17] R.W. Joyner. Effects of the discrete pattern of electrical coupling on propagation through an electrical syncytium. *Circ. Res.*, 50:192–200, 1982.

[18] M. Kawato, A. Yamanaka, S. Urushiba, O. Nagata, H. Irisawa, and R. Suzuki. Simulation analysis of excitation conduction in the heart: Propagation of excitation in different tissues. *J. Theor. Biol.*, 120:389–409, 1986.

[19] J.P. Keener. On the formation of circulating patterns of excitation in anisotropic excitable media. *J. Math. Biol.*, 26:41–56, 1988.

[20] J.P. Keener. Propagation and its failure in coupled systems of discrete excitable cells. *SIAM J. Appl. Math.*, 47:556–572, 1987.

[21] J.M. Kootsey and J. Wu. Anisotropic 2-D model of electrical propagation in cardiac muscle. *IEEE/EMBS*: 1988.

[22] R.E. McAllister, D. Noble, and R.W. Tsien. Reconstruction of the electrical activity of cardiac Purkinje fibers. *J. Physiol.*, 251:1–59, 1975.

[23] M.H. Protter and H.F. Weinberger. *Maximum Principles in Differential Equations*. Prentice-Hall, Englewood Cliffs, N.J., 1967.

[24] F.A. Roberge, A. Vinet, and B. Victorri. Reconstruction of propagated electrical activity with a two-dimensional model of anisotropic heart muscle. *Circ. Res.*, 58:461–475, 1986.

[25] Y. Rudy and W.-L. Quan. A model study of the effects of the discrete cellular structure on electrical propagation on cardiac tissue. *Circ. Res.*, 61:815–823, 1987.

[26] M.S. Spach and J.M. Kootsey. The nature of electrical propagation in cardiac muscle. *Am. J. Physiol.*, 244:H3–H22, 1983.

[27] M.S. Spach, W.T. Miller, D.B. Geselowitz, R.C. Barr, J.M. Kootsey, and E.A. Johnson. The discontinuous nature of propagation in normal canine cardiac muscle. *Circ. Res.*, 48:39–54, 1981.

[28] M.S. Spach and P.C. Dolber. Relating extracellular potentials and their derivatives to anisotropic propagation at the microscopic level in human cardiac muscle: Evidence for uncoupling of side-side fiber connections with increasing age. *Circ. Res.*, 58:356–371, 1986.

[29] S. Wiedmann. Electrical constants of trabecular muscle on mammalian heart. *J. Physiol. (Lond.)*, 118:348–360, 1970.

[30] A.L. Wit, S. Dillon, and P.C. Ursell. Influences of anisotropic tissue structure on reentrant ventricular tachycardia. In P. Brugada and H.J. Wellens, editors, *Cardiac Arrhythmias: Where to go from here?*, pages 27–37, Futura Publishing Co, Mount Kisco, NY, 1987.

[31] J.W. Woodbury and W.E. Crill. The potential in the gap between two abutting cardiac cells. *Biophys. J.*, 10:1076–1085, 1970.

[32] K. Yanagihara, A. Noma, and H. Irisawa. Reconstruction of sino-atrial node pacemaker potential based on the voltage clamp experiments. *Jap. J. Physiol.*, 30:841–857, 1980.

[33] B. Zinner. *Traveling Wavefront Solutions for the Discrete Nagumo Equation*. Ph.D. thesis, University of Utah, 1988.

18

Cellular Automata Models of Cardiac Conduction

Bo E.H. Saxberg[1]
Richard J. Cohen[1]

ABSTRACT The application of cellular automata models of cardiac conduction (CAMCC) to cardiac arrhythmias is examined. We define the relation between simple two- and three-state CAMCC and contrast them with models of continuous variables. Of interest for the analysis of reentrant arrhythmias is the presence of stable vortex or spiral patterns of activity in CAMCC with uniform parameters of conduction. Such organizing patterns are also observed in other nonlinear, nonequilibrium systems; the cellular automata models preserve the large-scale features of these patterns while collapsing the core of the vortex to a line discontinuity whose length is a reflection of an intrinsic length constant, the conduction velocity multiplied by the refractory period. A discussion is made of the interpretation of the discrete spatial lattice used in CAMCC; the proper interpretation for scaling is that of spatially sampling activity at points, not averaging over local volumes. We also consider the interpretation of the assignment of spatial (dispersion vs. gradients) and temporal inhomogeneities in the parameters of conduction of a CAMCC.

18.1 Introduction

In this chapter we discuss the application of finite-state models to understanding a variety of features of the electrical patterns of activity of the heart. These models consist of a spatially extended lattice of points; each point is allowed to take on any of a set of discrete states, and there is a dynamical rule that is applied to the lattice to iterate the evolution forward in time, so that the state at each point on the lattice is allowed to change based on the present and past values of the lattice states.

Such models, often termed cellular automata, have become of special interest with the direct applicability to computer simulation, which represents any system by a finite set of discrete states. In fact, any sort of a finite

[1] Harvard-M.I.T. Division of Health Sciences and Technology, M.I.T., Cambridge, Massachusetts 02139

element model realized on a computer is technically a cellular automata. The general study of cellular automata rules and behavior was introduced by von Neumann [84], and has recently been pioneered by Wolfram and others [59,90–92], with application to a wide variety of physical processes as well as being of interest in their own right. We will here concentrate on cellular automata models that have been applied to the understanding of abnormalities in the patterns of electrical activity in the heart.

We will distinguish the cellular automata models of cardiac conduction (CAMCC) from computer simulations of more detailed continuum phenomena by the arbitrary determination that the dynamical transition in CAMCC occurs between a "few" states. Continuum models represent continuum electrical properties of the myocardial syncytium (capacitance, resistance) and features of transmembrane ion transport, analogous to models of nonlinear coupled differential equations similar to the Hodgkin-Huxley equations for nerve cells, but with extensions appropriate for myocardial cells [7,20,50]. A simplified continuum model for cardiac activity is the Fitzhugh-Nagumo equation [27], studied by Winfree, Keener, and others [44,51,60,81,88]. In these models, the dynamical transitions are between states that are close to each other in some metric space, and the discrete representation on a digital computer is only a manifestation of the spatial and temporal resolution in approximation to a continuous analytical process. We will refer to these state variables and the corresponding models as limit-continuous, with the understanding that in the limit of infinite resolution (in space and/or time) their behavior is continuous. By contrast, the CAMCC attempt to simplify the dynamical description of the behavior of the system by making an operational definition of discrete transitions between a few states.

In Figure 18.1 is shown a two-dimensional square lattice of sampling length a, where each lattice point is spatially coupled to its four nearest neighbors, that is, the evolution of states at a given lattice point at any given discrete time step is a function of the current and previous states at that local point and those of its four nearest neighbors. The lattice spacing, a, generally represents a distance on the order of or larger than the electrotonic coupling length, that is, approximately 1 mm, much larger than cellular length scales. Other topologies can be chosen; for examples with figures in a hexagonal topology, see Kaplan and colleagues [43]. Topologies with nonnearest neighbor interactions can also be used, as will be discussed later with regard to electrotonic effects.

A major advantage of such CAMCC is their computational simplicity (by virtue of discrete transition rules among few states) and hence their speed. This is particularly useful in the analysis of spatially extended, complex patterns of electrical activity, such as those occurring in reentrant arrhythmias, where the two-dimensional or three-dimensional extent of the tissue must be represented at some minimal level of spatiotemporal resolution to describe the propagated motion of self-sustained activity. Since a propaga-

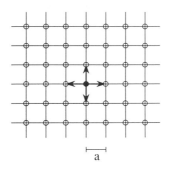

FIGURE 18.1. Two-dimensional Square Lattice for CAMCC, with four nearest neighbors and lattice spacing, a.

tion rule with some maximal physiological conduction velocity over some minimum point-to-point distance defines some minimal time step (i.e., temporal resolution for a change in state), the resolution in space and time are proportional. The computational time for a three-dimensional lattice will therefore increase as the fourth power of the spatiotemporal resolution. Efficient discrete CAMCC algorithms are a useful way to overcome this scaling problem.

An early CAMCC was that of Moe and coworkers [54], who were interested in reentry and fibrillation, and examined the relation of these arrhythmias to the dispersion of refractoriness hypothesis. Subsequently, Smith and Cohen [69] examined a similar model in relation to reentrant fibrillation, as well as abnormalities in atrioventricular (AV) nodal conduction [70], as was also done recently by Chee and colleagues [12]. Extensions of CAMCC to include features such as anatomical constraints, anisotropy due to fiber orientation (e.g., in conduction velocity), and dynamical changes in conduction parameters have been made [6,25,49,58,65,78,79].

There has also long been an interest in understanding the relation between the cellular electrical activity in the myocardium and the surface potentials recorded clinically by electrocardiogram (ECG). CAMCC models with varying amounts of anatomical and physiological detail were used by Selvester and coworkers [67,71] and others [1,16] to construct simulated ECG or body surface maps under physiological conditions or abnormal conditions such as regional ischemia.

18.2 CAMCC

18.2.1 DESCRIPTION

The simplest form of CAMCC to represent the features of cardiac electrical behavior is diagrammed in Figure 18.2, where the continuum form of the transmembrane voltage during an action potential at a cell in time is shown. For the details of the nonlinear behavior making up the electrical behavior of active membranes, the reader is referred to Jack and coworkers [39] and Plonsey [61]. Grossly a given region of membrane can be considered to remain at the resting transmembrane potential (approximately −90 mV) until depolarization (i.e., a shift to more positive transmembrane potential) in a neighboring region occurs. By virtue of passive electrical coupling (e.g., a cable model) in space, this causes a local depolarization that, once past some threshold potential, induces a rapid further depolarization in transmembrane potential. It is this last effect, the rapid rise in transmembrane potential once it is raised above some threshold, that reflects the intrinsically nonlinear, metastable behavior of the electrically active membrane. The transmembrane potential is then held approximately level during the plateau (depolarized to approximately +10 mV) until recovery begins, and the transmembrane potential is subsequently restored to its resting, repolarized value of −90 mV. A key feature of this activity is the existence of a refractory period. This is a time duration after the initiation of an action potential at a point that must pass before another action potential can be initiated. An experimental distinction is made between the absolute refractory period and the relative refractory period. The absolute refractory period refers to a time before which no depolarizing stimulus applied to the membrane can successfully generate an action potential, regardless of the strength of that stimulus (usually measured as the current sourced by an applied electrode). The relative refractory period refers to the assessment of the refractory period relative to the stimulus (current) strength; that is, a region that is refractory at low stimulus strength may respond and sustain an action potential at a higher stimulus strength. Our interest is in patterns of propagation, and the refractory period we refer to will be the largest duration after one action potential that will create a functional block to a second propagating action potential—a *functional refractory period.*[2] Later we will discuss how the CAMCC can be modified to account for relative refractory periods in response to variable stimulus strengths (whether due to electrotonic effects in propagation or to variable exogenous stimulus strength), as well as the possibility of slowed conduction with premature depolarization.

---

[2] The stimulus strengths can therefore be inferred as those that exist by virtue of the spatial coupling between depolarized and repolarized neighboring regions in the myocardium.

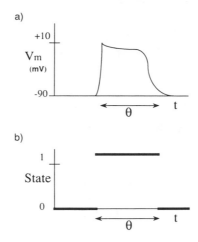

FIGURE 18.2. (A) Action potential: transmembrane voltage, V_m, vs. time, t. (B) Discrete two-state cellular automata: State 0, resting; State 1, depolarized and refractory; θ = refractory period.

The CAMCC approximation is a discrete transition between the resting or repolarized state and the depolarized, refractory state. In this statement we have made the simplification of identifying the duration of the action potential with the refractory period. These are not necessarily the same. Although they are normally comparable in magnitude and they do follow each other in their normal dynamical behavior [10,37,55,85], variations, as in postrepolarization refractoriness, can occur with ischemia [19,24,40,47]. As we will discuss below, it is the value of the refractory period that is important for determining the pattern of propagation; the duration of the action potential is important for calculation of the epicardial voltage or ECG expected from a particular pattern of model behavior. The lattice rule for the dynamical evolution of such a two-state CAMCC is as follows: (1) a lattice point moves from the resting to the depolarized, refractory state if one of its neighbors did so on the previous time step, and (2) a lattice point remains depolarized and refractory for a period of time, the refractory period, after which it returns to the resting state.

In Figure 18.3 a time history of propagation of a wave according to these rules is shown on a one-dimensional lattice; the velocity of propagation is one lattice spacing in one time step. In this "edge-triggered" model, the transition between states at one lattice point is triggered by the transition between states of a neighboring lattice point. We therefore need to know the direction of the state transition of neighboring states to define the dynamical rule for the two-state model that will propagate the states in time. To make a model that depends on only the values of the states themselves we introduce three states: (1) a resting state as before, (2) a depolarized

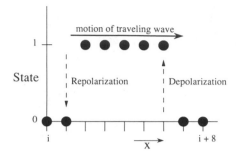

FIGURE 18.3. Two-state model traveling wave on one-dimensional spatial lattice. The orientation of the state transition defines the leading and trailing edge of the solitary wave, in this case moving to the right.

and activating state during which a given point will depolarize any of its resting nearest neighbors, and (3) a refractory state during which a given point is unable to depolarize neighboring points but has not itself recovered to the resting state.

The transitions in this three-state model are represented in Figure 18.4 with reference to the time course of the action potential. Because the activation from neighbor to neighbor occurs not by a state transition (or "edge") but by the occupation of an "activating" state, we will refer to this as a "state-triggered" model. The activating state represents the duration of time that a given region of membrane can source sufficient current to depolarize nearby membrane beyond the threshold voltage to start an action potential. The three-state model allows the explicit manipulation of the duration of this activating state as a physiological variable, in addition to the duration of the refractory period. The two- and three-state models are equivalent in discrete time (e.g., a cellular automata model) when the duration of the activating state in the three-state model is just one time step. This can be interpreted as implying the duration of the current source is less than or equal to one time step (e.g., the strong sodium channel defining the leading edge of the action potential for a discrete time resolution greater than 1 msec in normal propagation).

At this point, we make a careful distinction between the discrete nature of *functional* states, as described above, and the limit-continuous (in space and time) transmembrane voltage states. The points representing a limit-continuous function like the transmembrane voltage can be arbitrarily close together depending on the spatial and temporal sampling resolution of the model. This is in contrast to the few discrete functional states defining the temporal evolution of the CAMCC. We can couple the discrete state model to a limit-continuous action potential by assigning each lattice point a specific dynamical parameter to represent separately the local transmembrane

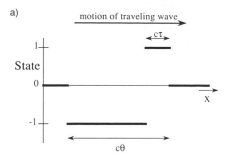

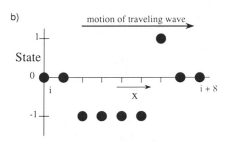

FIGURE 18.4. (A) Discrete three-state cellular automata. c = conduction velocity of the solitary wave; θ = refractory period; τ = activating time. (B) Three-state model traveling wave on one-dimensional spatial lattice.

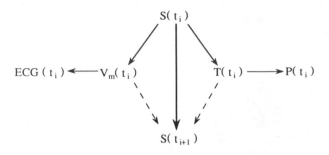

FIGURE 18.5. Variables in CAMCC (at discrete times t_i). Discrete: S = discrete functional state of CAMCC. Limit-continuous: V_m = transmembrane voltage; ECG = surface electrocardiogram (forward map); T = local myocardial tension (stress); P = pressure (by contraction on volume). Dashed lines indicate that an intermediate model can use continuous time variables to affect discrete functional state transitions.

voltage. Thus, when a given lattice point undergoes the transition from a resting functional state to the active functional state (i.e., that point has been activated, which also corresponds to the initiation of depolarization), we simultaneously represent the onset (at threshold) of the time course of the local action potential in the transmembrane voltage variable. The criterion for conduction from point to point remains only a function of the discrete functional states. The dynamical behavior of the cellular automata rules on the lattice would then permit the calculation of simulated epicardial voltages or torso surface potentials (ECG) by the forward map, for example, similar to the map used in the nonpropagating model by Miller and Geselowitz [52,53], and as mentioned earlier, several CAMCC models have taken a similar approach for this purpose. This is represented in Figure 18.5 by the heavy arrows connecting the evaluation at time t_i of the discrete functional state of the lattice, $S(t_i)$, and the consequent values of the transmembrane voltage and the ECG. A similar means can be used to couple the dynamical history of a CAMCC to mechanical activity by invoking local contraction around a lattice point as a function of the changes in functional states [13]. The temporal history of mechanical activity could then be represented by a limit-continuous parameter (the distribution of tension, T, or the blood pressure, P), as was the action potential waveform in the forward ECG map (see Figure 18.5).

Models intermediate between CAMCC and the continuum models mentioned earlier [7] can be formed by permitting limit-continuous variables, triggered by discrete state transitions, for example, to affect the occurrence of the discrete state transitions themselves. If we have a limit-continuous representation of local cellular transmembrane voltage coupled to the discrete functional states, we might include the values of the transmembrane voltage in the determination of electrotonic summation effects in marginal propagation. The criterion for the transition from the resting state to the activating state, a discrete state transition, would then become a function of the dynamical behavior in neighboring voltage, a limit-continuous parameter.

We represent this in Figure 18.5 by a dashed arrow indicating information flow from the limit-continuous parameter to the determination of the next functional discrete state. Intermediate models can describe features of cardiac conduction that the CAMCC will not describe. For example, the existence of a relative refractory period, supernormal excitability, and a variable safety factor for propagation all require the specification of a changing (in response to previous stimulus activity) current threshold for stimulation to the depolarized state. This current threshold can in turn be represented in a propagation algorithm of an intermediate model by including electrotonic summation as a function of neighboring voltage states (representing the ability to source current to depolarize nearby membrane). Such an intermediate model could also account for, for example, electrotonic effects during repolarization whereby a significant amount of neighboring repolarized tissue could shorten the local action potential.

An approximation to the electrotonic coupling effect in depolarization can be made for a nearest neighbor CAMCC by requiring a variable number of neighboring points to be recently stimulated to the depolarized, activating state ($S = 1$ in the three-state model) before a given point will itself be depolarized past threshold. For example, a lattice point with a large current requirement for stimulation would move from the resting, repolarized state, $S = 0$, to $S = 1$ only when at least three nearest neighbors were in the activating state $S = 1$, as opposed to a normal current requirement, which would require only one of the nearest neighbors to be in the state $S = 1$. A similar approximation for electrotonic coupling during repolarization could be made by shortening the value of θ, the time for recovery to $S = 0$ (identified with the refractory period in the three-state model), if, for example, all of its nearest neighbors were in the repolarized state, $S = 0$. Formally then, a CAMCC involves only information flow along the solid arrows, so that the dynamical evolution of the discrete states is determined by the discrete states alone. Intermediate models include information flow along the dashed arrows.

In fact, even the CAMCC implicitly involves an almost everywhere limit-continuous parameter. This will be termed the *autochrone*,[3] $\alpha(x)$, which is defined at any point in space, x, to represent the time since the last local activation. The dynamical rules for the simplest state triggered model can

---

[3] The word is made of two parts, "chrone" to indicate that it is a measure of time, and "auto" to indicate that each point maintains its own local clock, "self-timing", which is reset to zero when the propagating wavefront passes through that point.

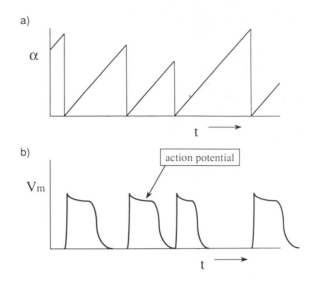

FIGURE 18.6. (A) Autochrone α corresponding to (B), series of local action potentials.

be restated in terms of this autochrone field:

1. A given lattice point can activate its neighbors for some time τ after the most recent activation, that is, for $\alpha \in [0, \tau]$

2. A given lattice point remains refractory for some time θ after the most recent activation, that is, for $\alpha \in [0, \theta]$.

The autochrone itself increments uniformly in time, except at the time of a local activation when it is discontinuously reset to 0 (see Figure 18.6). The determination of the loci of points where the autochrone equals 0 defines the spatial locus of the leading edge of the action potential. Note that the level sets of the autochrone are equivalent to the isochrones traditionally defined in cardiac electrical mapping. The loci of points where the autochrone equals 1 msec is equivalent to the corresponding isochrone, which represents the location of the propagating wavefront 1 msec previous to the current time. If we were able to determine the temporal history of the spatial pattern of the autochrone field amplitude, we would then know the temporal history of the electrical activity defined by action potential propagation. The discrete state functional rules determining the behavior of the CAMCC can then be interpreted as specifying the nontrivial behavior of the autochrone history, that is, the determination of where and when the autochrone is discontinuously reset to 0.

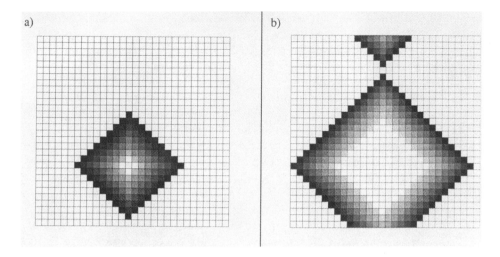

FIGURE 18.7. Propagation of depolarization wavefront in a two-dimensional cylindrical lattice. (A) Pattern resulting from a single initially depolarized point: Each point in the activating state (here one time step in duration) can activate (depolarize) any of its four nearest neighbors that are in the resting state (not activating or refractory). (B) Continued propagation of wavefront from a single-point stimulus in a resting cylindrical lattice (top row adjacent to the bottom row). The shading reflects the value of the autochrone, the time since the last activation measured in model time steps: 0 = black; 1–6 = proportionally lighter gray scale; 7 or more = white.

18.2.2   EXAMPLE OF BEHAVIOR

In Figures 18.7 to 18.10 we show a variety of classes of CAMCC behavior. All these figures are from a simple three-state CAMCC on a square lattice with a cylindrical topology wrapping the top and bottom edges, and with each cell being potentially activated by any one of its four nearest neighbors. Figure 18.7 illustrates propagation on the two-dimensional lattice from a point stimulation source. The black squares define the leading edge of the wavefront (autochrone equal to 0), and the shading reflects the increase in the value of the autochrone at unit rate with time. Figure 18.8A shows a unidirectional plane wave propagating from the lower edge that has just been blocked along half the plane wavefront. This half-plane block will initiate a reentrant vortex as shown in the succeeding Figures 18.8B–D. Figure 18.8D shows the stable, rotating vortex, which exists under conditions of homogeneous, isotropic refractory period, which is set to 6.5 time steps in this simulation. Figure 18.9A shows the form of the reentrant vortex one time step after the introduction of some finite, small dispersion in refractory periods to the pattern of Figure 18.8D. The dispersion in refractory periods is given by the approximately gaussian statistical distribution from which the refractory periods are independently assigned to each lattice point. This distribution in Figure 18.9A has a mean of 6.5 time steps and a standard deviation of 0.5 time steps; between Figure 18.8D and Figure 18.9A, we have not changed the mean of the refractory period distribution, but have increased the standard deviation from 0 to 0.5 time steps to introduce a small amount of dispersion. Figure 18.9B shows the structure of the reentrant vortex 15 time steps later with this dispersion of refractory periods; characteristically, for small dispersion of refractory periods the structure of the reentrant vortex persists. Figure 18.9C shows the form of the reentrant vortex one time step after the pattern of Figure 18.8D with the introduction of a relatively large dispersion of refractoriness: the standard deviation is increased from 0 to 2 time steps, while the mean is left unchanged at 6.5 time steps. Figure 18.9D, at 15 time steps later, shows the loss of the reentrant vortex structure, and this irregular appearance remains at 50 time steps (and more), as shown in Figure 18.10. In this last figure we see an example of the "fibrillatory" type of behavior seen with a large dispersion in refractory periods. In this type of behavior, wandering wavelets (i.e., the small connected lines of black dots defining a short segment of propagating wavefront) crawl around the lattice, annihilating or splitting on collision with each other.

18.3   Relation to Other Processes

The general features of the behavior of the CAMCC are not unique to cardiac conduction. The generation of rotating spiral waves under constrained initial conditions is also observed in chemical reactions, such as the Belousov-Zhabotinsky reaction [48,56,86–88,93], and in biological systems such as the *Dictyostelium discoideum* slime mold [21,80,88] and networks of neurons, as in the retina [9,30,88]. Such organizing patterns formed in nonequilibrium states can be perturbed by introducing a variety of inhomogeneities into the systems. One would conjecture that the generic features of the CAMCC might apply to these other problems in some cellular automaton limit, that is, that the cellular automata discrete functional state transition rules are similar for different processes whose underlying continuum dynamics may be quite different. Such unifying cellular automata features include the presence of an excitable, metastable field (the lattice points in the resting state of the CAMCC, for example), which is stimulated to a decay transition (e.g., depolarization) by spatial coupling (in the cardiac case, electrotonic coupling sourcing depolarizing current), and then is reexcited by some local process (e.g., repolarization) after some critical recovery time (refractory period). The decay process, occurring at the leading edge of the propagating wave, is permitted by the initial existence of the metastable state. Such a general view has been presented earlier [46], and the solitary propagating waves that result in such systems have

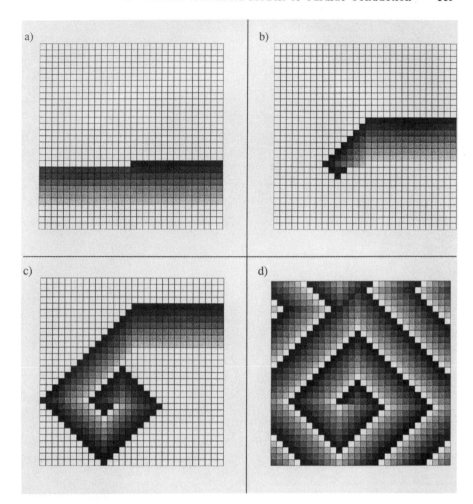

FIGURE 18.8. Development of vortex reentry. (A) Half-plane block to propagation to begin vortex reentry. A linear wavefront (propagating from below) is blocked from propagation along part of its length by forcing those lattice points to be in the refractory state (as if the wavefront encountered a transient region of prolonged refractoriness). The refractory period after this block will be made uniform and isotropic, equal to 6.5 time steps. (B), (C), and (D) Developing vortex reentry after half-plane block.

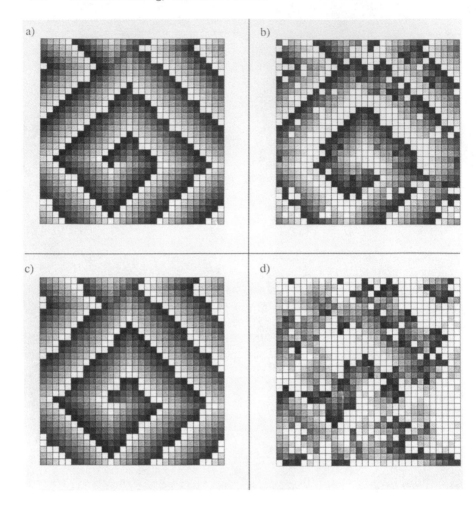

FIGURE 18.9. Effect of dispersion on refractory period. (A) One time step after the introduction of a small dispersion of refractoriness to the pattern of Figure 18.8D. The mean refractory period is still 6.5, but refractory periods now are independently assigned to the lattice from an approximately gaussian distribution with standard deviation 0.5 time steps. (B) Conduction pattern with small dispersion of refractoriness, 15 time steps after the pattern in (A). (C) One time step after the introduction of a large dispersion of refractoriness to the pattern of Figure 18.8D. Mean refractory period is still 6.5, but the standard deviation was increased from 0 to 2 time steps. (D) Conduction pattern with large dispersion of refractoriness, 15 time steps after the pattern in (C).

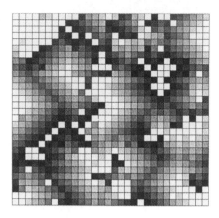

FIGURE 18.10. Conduction pattern with large dispersion of refractoriness, fifty time steps after Figure 18.9D (typical of pattern seen during this stable, self-sustained reentry for following time steps).

been termed "autowaves" [46,83]. There has been an attempt to generate a unifying continuum differential equation description for these systems, using the "reaction-diffusion" formalism as a continuous model for such chemical and biological systems [15,26,31,36,44,45,80,81]. With regard to cellular automata models displaying similar behavior to reaction-diffusion systems, Greenberg and colleagues [31] made an early mathematical analysis, constructing a winding number measure to demonstrate the stability of the spiral wave patterns seen in the case of uniform conduction properties. More recently, Gerhardt and coworkers [29] have used the phase plane portrait of excitable media as modeled by reaction-diffusion equations to construct a simplified cellular automata model for excitable media to study these spiral wave patterns. These endeavors have been one part of a general effort at understanding pattern formation in nonlinear, nonequilibrium systems [32,33,57].

With respect to the nonequilibrium nature of these systems, note that in the cardiac conduction problem the consumption of energy occurs in the recovery to the metastable, resting state. In the electrically active membrane, this process of recovery represents the restoration of the transmembrane ion gradients as a result of the action of active, energy-consuming ion transport. It is this process that represents the intrinsic nonequilibrium nature of such a system; its activity is permitted only by the existence of an energy flux passing through the system. In this view, the "resting state" of the myocardium is a misnomer, as this actually represents a high energy, metastable state of the system!

18.4   Representing Conduction Physiology

18.4.1   A DISCRETE REPRESENTATION

The advantage of the CAMCC is in its simplification of very complex non-linear behavior to a set of discrete state transition rules. The parameters governing these transitions, such as the refractory period or the local conduction velocity, can be determined by electrophysiological measurement on bulk tissue. The representation of the CAMCC is then one of a mesoscopic view [68], wherein the details of the behavior of the ion channels are not explicit, but instead are implicitly represented in the parameters governing the state transition rules. This can have certain advantages in the face of uncertainty about microscopic details. For example, by inserting a delay in the passage of information from point to point on the lattice, one can regulate the effective local conduction velocity for propagation of the active state (leading edge of the action potential). In experimental measurements, the local conduction velocity depends on myocardial fiber orientation, being two to three times as fast along the fiber as transverse [14,63,64,66]. By representing conduction delays on the lattice by a discrete tensor form (depending on the spatial orientation between coupled lattice points relative to local fiber orientation), the CAMCC conduction velocity can be made anisotropic based on an assignment of myocardial fiber orientation at each point. The actual microscopic rationale for this difference in conduction velocity appears to lie in the elongated structure of the fibers and the orientation and location of the gap junctions that electrically couple the cells [72–76]. A continuum approximation (to what is, at a microscopic level, actually a phenomenon of the discrete cellular structure itself) would involve assigning effective parameters of intracellular resistivity, and so forth, in an anisotropic fashion in an attempt to simulate the effect of the discrete cellular structure [11,18,42,62,82]. The CAMCC avoids this intermediate step and simply assigns the measured conduction velocity to the lattice. Changes in the conduction velocity as a function of previous patterns of stimulation (e.g., slowed conduction with premature stimulation) can be represented by making the conduction velocity depend on the autochrone history. Afterdepolarizations, which relate the transmembrane voltage history to previous patterns of stimulation (autochrone history), can also be incorporated in intermediate models, where it is now the local voltage at a point that may itself rise above threshold (independent of neighboring states) to trigger a spontaneous depolarization.

18.4.2   ELECTROTONIC INTERACTIONS

Several features of cardiac electrical activity can be included by adding electrotonic interactions to the CAMCC as discussed above (possibly including extension to intermediate models). Since the conduction velocity is represented in the CAMCC by some temporal delay in propagation of depolarization from one lattice point to the next, the existence of a latency is implicit. By including electrotonic effects representing a variable current strength of stimulation, the latency in response to exogenous stimuli of varying strengths can be made explicit. Similarly, the refractory period, both in its spatial variability as well as its dynamical behavior as a function of previous stimulation history, can simply be assigned to points on the CAMCC lattice from electrophysiological measurements. A relative refractory period can be introduced with electrotonic interactions. A variable neighboring volume (number of lattice points) of depolarized tissue would be required to stimulate a given point during its relative refractory period, representing the variable stimulus (current) strength required to cause a local depolarization.

Earlier we discussed a simple approximation of electrotonic interactions for a nearest neighbor CAMCC. Note that the existence of the electrotonic decay length in myocardial conduction places a lower limit on the spatial resolution of the CAMCC using a nearest neighbor conduction rule. For a depolarization at some given point, the electrotonic decay length, η, defines the length scale over which the transmembrane potential will decrease exponentially (due to passive cable properties of the membrane) as one moves away from that point. If several points of the lattice lie within the electrotonic decay length (Figure 18.11), one point, a, depolarizing to the active state next to a refractory nearest neighbor, b, could still activate a resting third point, c (i.e., raise it beyond the threshold transmembrane voltage), which was a second-nearest neighbor beyond the refractory nearest neighbor, b, provided that that second-nearest neighbor, c, was still within the electrotonic decay length of the original point, a. A CAMCC conduction rule on this lattice would have to consider second-nearest neighbor interactions. For a nearest neighbor CAMCC rule, the spatial sampling defined by the spatial lattice must therefore be interpreted as representing a length scale greater than the electrotonic decay length, so that a refractory nearest neighbor will block conduction from passing through.

In addition, the existence of electrotonic conduction causes the electrical activity of the myocardium to be smoothed over length scales smaller than the electrotonic length scale. As a limiting example, suppose the medium were to be divided into two alternating lattices, such that every other point were dead (permanently refractory), but would still support the passive electrotonic coupling. If the length scale of this lattice sampling were less than half the electrotonic decay length, the distance between alternating lattice points that support normal conduction would be less than the electrotonic decay length, and electrical activity would proceed unaffected (i.e., not blocked by the dead or refractory intervening region). Hence, conduction disturbances on length scales smaller than the electrotonic decay length are not significant in relation to the pattern of propagation of electrical activity, as they will not result in conduction block.

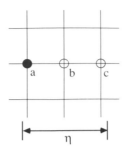

FIGURE 18.11. Three points lying within the electrotonic decay length, η. (A) Activating state. (B) Refractory state. (C) Resting, being depolarized by (A).

If, however, a model representing electrotonic summation is considered, a lattice sampling length smaller than the electrotonic decay length may be desired, as the conduction rule will involve summing over neighboring states in an area around a point (as a measure of the amount of depolarizing current sourced to that point). That area of summation has a characteristic radius determined by the electrotonic decay length; a lattice resolution smaller than the electrotonic decay length will give increased resolution and dynamic range in the specification of the different amounts of current sourced by different configurations of depolarized and repolarized tissue neighboring to a given point. Otherwise, as we indicated earlier in our simple nearest neighbor CAMCC (and hence for a lattice spacing greater than the electrotonic coupling length) one has a choice of one to four nearest neighbors being depolarized in a two-dimensional model, that is, just four different levels of non-zero current (or six in a hexagonal model, or eight in a square lattice where activation is by the eight nearest neighbors). In Figure 18.12 we show the appearance of a two-dimensional square lattice model with uniform conduction velocity, c, and lattice spacing, a, which is smaller than the electrotonic decay length, η, at a time η/c after depolarization at the central point. The small black-filled circles indicate those lattice points that are stimulated from discrete functional state $S = 0$ to $S = 1$ in this time interval; they represent the area of tissue depolarized above the threshold voltage by the central current source stimulus. The open circles indicate the lattice points that would be represented in a lattice model with spacing $a' = \eta$, and the four nearest neighbors at spacing η are shaded indicating their transition to the depolarized, activating state ($S = 1$) after this time interval, η/c, as would be expected in such a nearest neighbor CAMCC.

In the representation of nonnearest neighbor conduction, with ($a < \eta$), we can choose to use the time step interval, δt, for evaluating the lattice states as being either a/c or η/c. In the first case, the computational time increases for two reasons. First, the time steps are smaller, so more are

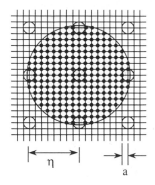

FIGURE 18.12. Electrotonic propagation: a = Lattice sampling length; η = electrotonic decay length. See text for description.

needed to cover the same amount of physiological propagation time. Second, the representation of nonnearest neighbor conduction is an extension of that for nearest neighbor conduction, where the conduction velocity (possibly anisotropic and inhomogeneous) between two points, at positions x_1 and x_2 such that $|x_1 - x_2| \leq \eta$, defines the time delay (latency) for activation of one lattice point by the other. Because all points within the electrotonic decay length of point 1 could eventually be stimulated past threshold by current sourced at point 1, the model must store for any given point the latent delays of activation from all neighbors within η, which can be large (in Figure 18.12 it is the number of black-filled circles). This increases both computational time and storage (as others have found with similar algorithms [78]).

The second case, in which the model time step interval is chosen as $\delta t = \eta/c$, leads to a Huygens wavefront propagation model, where at each time step a given point in the depolarized, activating state, $S = 1$, can cause the neighbors within a disc of radius η (or an ellipse for anisotropic conduction velocity due to fiber orientation, for example) to depolarize beyond threshold and so move from $S = 0$ to $S = 1$. Every point on an advancing wavefront will then stimulate resting tissue within a disc of radius $\eta = c\delta t$, which is similar to Huygens construction of a propagating wave. Several models have used a similar construction for a propagation algorithm, although not focused on electrotonic effects [6,67,71]. Because the temporal resolution is still sufficient to resolve conduction blocks (on the order of η/c), this time step interval provides a viable option for evaluating electrotonic effects without incurring enormous increases in computational time and memory, aside from the increased lattice spatial resolution (i.e., holding the time step interval fixed makes the computational time increase depend only on the spatial resolution to the power of the lattice dimension, as opposed to the lattice dimension plus 1 as discussed earlier).

Note that the response to a point stimulus in a resting lattice after a time,

η/c (Figure 18.12), will be identical with either choice of δt. In the choice of the larger value, $\delta t = \eta/c$, one will, however, sacrifice smoothness in the temporal representation of parameters (as would naturally be expected). This would be of more concern in an intermediate model, which might couple the transition to the activating state, $S = 1$, with a limit-continuous model of current sourcing to nearby tissue that changes in strength depending on the autochrone value of the source point (e.g., decreasing current strength as the autochrone increases, i.e., as the time since the last depolarization increases). At this point, we will introduce the activation field, $A_{i,j}(\alpha_i, d_{i,j})$, which will model current sourcing from the point i with autochrone α_i, to depolarize the point j at a distance $d_{i,j} = |\mathbf{x}_i - \mathbf{x}_j|$ from i. This is not an explicit representation of the actual physiological ionic currents, which would require the representation of the details of intracellular and extracellular resistance, membrane capacitance, and so forth, which we wish to subsume in a specification of mesoscopic parameters such as the conduction velocity, refractory period, electrotonic decay length, and so on. Rather, the activation field will allow us to distinguish patterns of propagation with different current requirements, allowing us to simulate failure or success of local propagation owing to electrotonic interactions. In the following, we will take the liberty of referring to the activation field as representing the amount of "current" sourced from a given point to its surroundings. A general, detailed model might endow $A_{i,j}(\alpha_i, d_{i,j})$ with a spatial exponential decay with length constant η and with a temporal dynamics as a function of the autochrone, α_i, that would represent the change in current sourced over time.

For a CAMCC, the simplest approximation is to make the current source strength from point i to point j, $A_{i,j}$, a constant (A) for some interval in time, for example, the activation time, τ.

$$A_{i,j}(\alpha_i, d_{i,j}) = \begin{cases} AB_\eta(\mathbf{x}_i, \mathbf{x}_j) & \text{if } \alpha_i \leq \tau_i \\ 0 & \text{otherwise} \end{cases} \qquad (18.1)$$

where

$$B_\eta(\mathbf{x}_i, \mathbf{x}_j) = \begin{cases} 1 & \text{if } |\mathbf{x}_i - \mathbf{x}_j| \leq \eta \\ 0 & \text{otherwise} \end{cases} \qquad (18.2)$$

defines the disc of points that can be stimulated by the point i. To represent anisotropic conduction by an ellipse instead of a disc, because of faster propagation longitudinal to fiber orientation (conduction velocity $c_\parallel$) than transverse (conduction velocity $c_\perp$), the major and minor axes of the ellipse would simply be $c_\parallel \delta t$ and $c_\perp \delta t$, respectively.

The evaluation by any given point, j, as to whether it will receive sufficient current to depolarize beyond threshold can be done by summing over the $A_{i,j}$, sourced from its neighbors at i, and comparing the sum to a threshold parameter, σ_j. If $\sum_i A_{i,j} \geq \sigma_j$, then a resting point j will be stimulated to the depolarized, activating state. Again, by making σ a function of the autochrone, $\sigma(\alpha)$, one can represent different requirements for current to stimulate past threshold as a function of time (e.g., in representing relative refractory periods). Larger values of σ correspond to a greater current requirement for stimulation. Smaller values of σ correspond to lower current requirements for stimulation (as might be of interest in representing supernormal excitability [77]).

One of the interesting suggestions of studies of fiber orientation induced anisotropy in conduction is that the effective safety factor (which reflects the balance between current source strength and current requirements for stimulation) differs between longitudinal and transverse propagation [73–76]. This can be represented by making the value of $A_{i,j}$ depend on the fiber orientation, $\hat{\mathbf{fo}}_i$ and $\hat{\mathbf{fo}}_j$ (assumed normalized vectors), at points i and j. In particular, a key point of interest is to be able to simulate the possibility of unidirectional block at the boundary between orthogonal fibers, where propagation from the longitudinal to the transverse mode is successful (i.e., a wavefront moving parallel to the fiber orientation at one point tries to cross to motion orthogonal to the fiber orientation at another point; see Figure 18.13A), while propagation from the transverse to the longitudinal mode (Figure 18.13B) will fail. Because the value of $A_{i,j}$ is defined for the point i sourcing current to the point j, we are at liberty to make $A_{i,j}$ nonsymmetric, so that $A_{i,j} \neq A_{j,i}$. In particular, if we define

$$\widehat{\Delta \mathbf{x}}_{i,j} = \frac{\mathbf{x}_i - \mathbf{x}_j}{|\mathbf{x}_i - \mathbf{x}_j|} \tag{18.3}$$

as the normalized vector pointing between the spatial locations of i and j, then we can introduce an asymmetry in the current source coupling by the introduction of a function, $\epsilon(z)$, whose argument, z, might be an antisymmetric function of the fiber orientation at the two points, i and j:

$$A_{i,j} = A(\alpha) B_{\eta, c_{\parallel}, c_{\perp}}(\mathbf{x}_i, \mathbf{x}_j)(1 + \epsilon(|\hat{\mathbf{fo}}_i \cdot \widehat{\Delta \mathbf{x}}_{i,j}| - |\hat{\mathbf{fo}}_j \cdot \widehat{\Delta \mathbf{x}}_{i,j}|)). \tag{18.4}$$

In the above we could take $\epsilon(z) = \epsilon z$ as the simplest form of the function ϵ. Here we have allowed the intensity of A to be time-dependent and have indicated that B should be ellipsoidal as a function of anisotropic conduction velocities. The effect of the third term is to yield the value $1 + \epsilon$ if propagation at i is longitudinal and at j will be transverse (Figure 18.13A), and the value $1 - \epsilon$ if moving from transverse to longitudinal propagation (Figure 18.13B). If propagation is between fibers parallel at i and j, then there is no contribution from ϵ. By decreasing the effective current sourced in moving from transverse to longitudinal propagation, blocked propagation is more likely in the face of a given current requirement (value of σ). If we wish to decrease the current sourced from both longitudinal to transverse and transverse to longitudinal propagation, we could alter the form of the function, $\epsilon(z)$, to $\epsilon(z) = -\epsilon|z|$. The effect of this would be to decrease the safety factor (increase the current requirement) for both the longitudinal to transverse and the transverse to longitudinal transition in propagation,

a)

b)

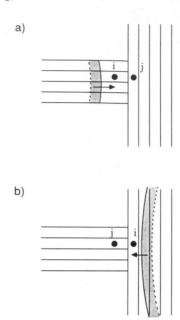

FIGURE 18.13. Conduction across fiber orientation anisotropy. (A) Wavefront (shaded) passing from a region of propagation longitudinal to myocardial fiber orientation to a region where propagation would immediately be transverse to fiber orientation. The points i and j are indicated as representing lattice points in a model of conduction (see text). (B) Wavefront passing from a region of transverse propagation to a region of longitudinal propagation.

making it possible for conduction to block in either direction at the boundary of orthogonal fibers. By changing the functional form of ϵ, we can alter the magnitude and the angular sensitivity of the effective asymmetry in current source (and hence likelihood of blocked conduction) as a function of fiber orientation at the two points i and j.

Another feature of interest is the variable effect of wavefront curvature and its interaction with fiber orientation through, for example, anisotropic safety factors or current requirements for propagation. (Interactions between velocity of wavefront propagation and wavefront curvature have been examined in analytical reaction-diffusion type models [81,94–96].) The topology of connection, that is, the number of neighbors included within the electrotonic decay distance of a given point on the lattice, limits the allowed resolution of the local curvature. This angular resolution is approximately a/η (for small a/η); in the simplest CAMCC with only coupling to the four next nearest neighbors, the angular resolution is only 90°, as seen in the earlier Figures 18.7 and 18.8. Compare this with Figure 18.12, where the lattice spacing $a < \eta$ yields a smoother approximation to the advanc-

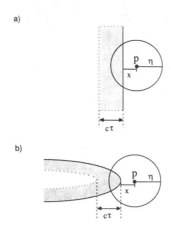

FIGURE 18.14. Curvature effect in electrotonic propagation: p = resting point, summing current within radius η; x = distance between p and leading edge of wavefront; τ = activation time; c = conduction velocity. (A) Planar wavefront. (B) Curved wavefront. See text for description.

ing wavefront than the nearest neighbor model. Phenomena that might be a function of this curvature will not be well represented in a lattice with coupling to only a few nearest neighbors (e.g., for $a > \eta$). For example, the ability to source current ahead of a sharply curved region versus line (or plane in three-dimensions) may be different in a case representing marginal propagation, and this in turn could advance or retard the propagation of pieces of different curvature. By including electrotonic summation effects, such an effect can be accounted for, as shown in Figure 18.14. In this figure, in both case A, planar wavefront, and case B, curved wavefront, the advancing wavefronts source current to stimulate a point, p, a distance x from the leading edge of the wavefront. The area behind the wavefront that can source current will be of thickness $c\tau$, since τ is assumed to represent the duration of the activating state. The point, p, however, will sum over current sources within the electrotonic decay length η and the total current sourced will therefore be proportional to the area of overlap between a disc of radius η centered at p and the area of activating points of thickness $c\tau$ behind the wavefront. In marginal propagation (through a region with a high current requirement, or σ) the difference in this area of overlap between the planar and curved wavefront may cause propagation to fail as the curvature increases, hence blocking or retarding conduction of the large curvature wavefront while still permitting normal propagation of the low curvature or planar wavefront.

The CAMCC and its extension to intermediate models thus have the virtue of providing the means to examine directly the implications of electrophysiological measurements for global behavior under a variety of conditions. Depending on the nature of the physiological conduction phenomena of interest, a variety of algorithms for the CAMCC can be constructed, from the simplest, nearest neighbor discrete state model to the introduction of electrotonic interactions and nonnearest neighbor models. The extension to intermediate models, coupling the state transitions to a variety of time-dependent limit-continuous variables (representing action potentials, current source strengths, requirements for depolarization past threshold, etc.) permit the analysis of the effects of a wide variety of features of conduction physiology on the global patterns of electrical activity.

18.5   Representing the Reentrant Core

By its simplification of the nonlinear features of the generation of the action potential, the CAMCC does lose information. The details of the nonlinear behavior of the active membrane are reflected in the structure of the core of reentrant vortices. In epicardial recordings of functional reentry, the vortices formed have a certain core size (approximately 0.5–1 cm) around which they circulate [2–5,17,35,38,41,89]. The nature of this core for spiral waves in analytical and continuum computer models has been of some interest [88]. In the CAMCC, the "core" around which the vortex evolves is a line discontinuity in the autochrone history. This line lies implicitly between pairs of lattice points and there is no explicit representation of the core by any one lattice point (Figure 18.8). The collapse of the core to a line discontinuity in the CAMCC reflects the fact that the nonlinear dynamics that cause the electrically active membrane to be metastable (local depolarization once transmembrane voltage threshold is reached) have been simplified in the CAMCC to a discrete map among a few discrete functional states, and phenomena that involve the details of this nonlinear behavior (e.g., the interior of the reentrant core) will not be represented. Each lattice point in the CAMCC has a discrete functional state space in addition to the intermittently discontinuous autochrone, as in Figure 18.15, where the cycle of a periodically stimulated lattice point (period T) is represented by the discrete functional state and the almost everywhere limit-continuous autochrone, α. There is no explicit representation of core states; the only states that are available are those that occur in the process of wavefront propagation (e.g., by a periodic wavetrain or vortex circulating with period T). As every lattice point must be assigned some discrete functional state and some autochrone value, the "core" must collapse to an implicit location between lattice points. By simplifying the nonlinear dynamics at the activating edge of the wavefront into a discrete state transition, we have in a sense "sharpened" the spatial extent of the dynamical representation so that the core singularity (where the autochrone would not be well defined) is collapsed into a line discontinuity. The analysis of the internal behavior of the core and the requirements that it might place on the conditions for

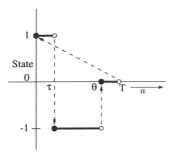

FIGURE 18.15. Loop in space consisting of the discrete functional state and the autochrone (α), for periodic stimulation at period T. The transitions among the discrete states are shown by the dashed arrows.

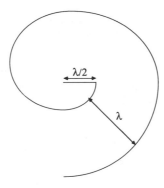

FIGURE 18.16. Periodic vortex motion around the core with velocity c creates periodic wavefronts at spacing, $\lambda = c\theta$, stimulating at the refractory period, θ.

the onset of reentry can only be done in the continuum models. As has been noted (pp. 200–201 Winfree, [88]), replacing the core of a vortex by inactive medium still permits the circulation of a spiral wave, now around this obstacle to propagation (and perhaps with altered period of circulation). The very existence of stable propagating reentrant vortices in CAMCC models as well as continuum models argues that the existence and stability of these structures is predominantly a function of the spatially extended autochrone field (which represents the history of wavefront passage), and not so much a function of the detailed representation of the core.

What the CAMCC can represent of the core is a length scale (by the spatial extent of the line singularity in the autochrone), λ, which is related to the requirement of a minimal spatial extent for reentry to be successful, given the conduction velocity, c, and refractory period, θ (Figure 18.16). Since circulation around the core occurs in time equal to the refractory period (in the case of homogeneous parameters), the length of the core will be $\lambda/2$, where $\lambda = c\theta$. A question for future study is whether the CAMCC representation of the core is sufficient for the analysis of global stability of behavior against reentrant arrhythmias (including reentrant fibrillation), as opposed to the microscopic local study of the details of the inner reentrant core. In addition, another point of interest is what features (perhaps electrotonic interactions with curvature effects, as discussed earlier) can be added to the simple CAMCC to mimic features of slowed conduction velocity near the core and wandering motion of the core itself, observed in cardiac mapping experiments [41] and seen in continuum models [88,95].

18.6   Rationale for Interpretation of Lattice Structure

The CAMCC represents the spatial extent of the behavior of the system on a discrete lattice. A question arises regarding the interpretation of the implicit limitation on spatial resolution imposed by the discrete lattice, when one seeks to interpret the behavior of the CAMCC in relation to a spatially continuous physical system (e.g., the syncytium model of cardiac conduction, spatially continuous chemical media, etc.). More precisely, in the case of cardiac conduction, the existence of cellular structures (the anisotropic location and density of gap junctions defining low-resistance connections between the cells) make the assumption of continuous electrical behavior problematic for cellular length scales. Because the sharpest dynamical transition in the propagating cardiac action potential, the leading edge defined by the strong sodium current, has a time course on the order of 1 msec, and hence an extent over many cells (the conduction velocity being on the order of 0.1–1 mm/msec [14,63,64,66]), the approximation of propagation of electrical activity through continuous media is based on smoothing spatially over length scales larger than those characterizing cellular sizes ($100\mu \times 20\mu$) [18,72].

Figure 18.17 illustrates the hierarchy of spatial representation. The CAMCC takes the approximately smooth spatial behavior of the continuum syncytial model and introduces a potentially coarse spatial representation. Two possible interpretations can be made.

1. Each point on the lattice represents the behavior averaged over a corresponding small spatial volume (e.g., cube for a cubic lattice in three-dimensions) centered on that point. In this view, the properties of model conduction (conduction velocity, refractory period, etc.) represent averaged properties over this small volume.

2. Each point on the lattice represents the behavior sampled at that point. The values of conduction parameters are those at that point.

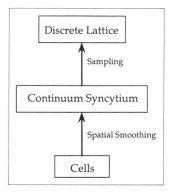

FIGURE 18.17. Hierarchy of spatial representation in models of cardiac conduction.

We will argue that the proper interpretation for scaling purposes is the second. For when one interprets the CAMCC as a spatially sampled model, the conduction rule that defines the temporal evolution of the CAMCC is unaffected by the value of the spatial sampling length. No matter what the sampling interval is, the temporal sharpness of the transition from the resting to the activating, depolarized state is the same, as one is examining the state transition for a point in the medium that remains dynamically unaffected by the sampling interval. Hence, the representation of the dynamics by discrete state transitions, as is required by the CAMCC, is well defined at any length scale of sampling.

By contrast, under the averaging interpretation the dynamical rule for the evolution of the behavior at a point on the lattice must change as the lattice spacing increases. Each cube of volume will contain some depolarized and some resting tissue, so that the dynamics between the discrete physiological states will become smoothed. In the limit of very large lattice spacing, over length scales larger than the reentry criterion, each cube may internally support self-sustained activity, and a dynamical rule involving transitions between three discrete activating, depolarized, and resting states will no longer be appropriate. In fact, assuming that the parameters of conduction are statistically identical between the cubes and that the pattern of self-sustained activity is homogeneously distributed (as in fibrillation), each such cube supporting self-sustained reentry must be equivalent to any other cube in averaged behavior and hence should be assigned the same averaged state!

The interpretation of the spatial lattice of the CAMCC as a spatial sampling over the physical system preserves the discrete dynamics between a few states independently of the length scale of the spatial sampling. This allows one to interpret the effects of lattice scaling by increasing or decreasing the spatial sampling length while the dynamical evolution rule of the

CAMCC remains unchanged.

18.7   Spatial Variation in Conduction Physiology

The parameters of conduction physiology used in the CAMCC can be assigned with spatial variation on the discrete lattice, reflecting inhomogeneities in the medium. The spatial variation in a given parameter can occur over different length scales, much as a temporally varying signal can have variation on different time (frequency) scales. In the physiology of cardiac electrical activity, there is an intrinsic length scale, λ, defined by the reentry criterion length, $\lambda = c\theta$, where as above c is the conduction velocity and θ the refractory period. As mentioned before, epicardial mapping data for reentrant patterns of activity indicate λ is on the order of 0.5 to 1.0 cm. It will be useful in discussions of the significance of spatial variations in conduction physiology to classify these variations relative to the intrinsic length scale λ: (1) spatial variations on length scales larger than λ will be termed spatial "gradients" in parameters, while (2) variations on length scales smaller than λ will be termed "dispersion" in parameters. Spatial variations, or inhomogeneities, will refer to fluctuations in physiological parameters on any length scale. Note that the length scale (or the spatial frequency) over which fluctuations occur is different from the magnitude of these fluctuations, just as the frequency of a sound is different from its intensity. One characteristic measure of such a length scale for fluctuations is the correlation length, ξ. Here we are assuming that the spectral character of spatial fluctuations is roughly that of low-pass filtered noise, so that at long length scales relative to the correlation length (low spatial frequencies), the fluctuations are uncorrelated because of biological variability, while at short length scales relative to ξ (high spatial frequencies) there is little fluctuation, that is, the physiological variables are approximately constant over length scales smaller than ξ. In Figure 18.18 we show the hierarchy of spatial length scales we have introduced; of these, the lattice spacing, a, is the one that is arbitrarily chosen for a particular simulation, but we have indicated a choice appropriate for the simplest CAMCC.

The distinction between dispersion and gradient preserves the historical view that spatial inhomogeneities, as in the dispersion of refractoriness hypothesis [34,54], lead to fractionated wavefronts and an increased likelihood of generating reentrant arrhythmias. It is spatial inhomogenities on length scales smaller than the reentry criterion length that will generate disorganized patterns of reentry as seen in Section 18.2.2; a spatially slow variation (relative to the reentry criterion length), or gradient, in either conduction velocity or refractory period will not cause wavefront fractionation but rather lead to smooth changes in behavior. In this case, with appropriate initial conditions an organized, structured pattern of reentry such as the vortices seen in Section 18.2.2 can be supported. It is the dispersion in

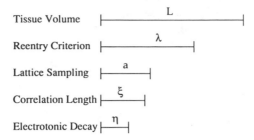

FIGURE 18.18. Hierarchy of length scales in models of cardiac conduction. Choice of lattice sampling size, a, is arbitrary, and defines the nature and resolution of the model. Here the choice is appropriate for the simplest CAMCC, which has only four nearest neighbors potentially activated by a given point, and for which the correlation length in physiological parameters (which may be different for different parameters) is assumed less than the lattice spacing. The relation between the correlation length and the electrotonic decay length is determined by physiology; we have simply indicated one possibility.

parameters of conduction that cause the patterns of electrical activity to become disorganized and less coherent. Note, however, that even the generation of organized, coherent vortex reentry requires a block to propagation, that is, some critical inhomogeneity.

In simple CAMCC, one often assigns parameters (such as the refractory period) independently from a statistical distribution to the lattice points to create spatial inhomogeneities. If an approximately gaussian statistical distribution is used to define the statistical variation in a parameter, the magnitude of the fluctuations is defined by the standard deviation. Because the values from lattice point to lattice point are assigned independently, the correlation length (or length scale of variation) for these parameters must be interpreted as being less than or equal to the lattice sampling length. Conversely, such a lattice can be said to represent a medium with spatial variation in conduction parameters on the length scale of the lattice sampling. Whether or not this represents dispersion in the parameters depends on the relation between the reentry criterion length and the lattice sampling length. However, a lattice sampling length larger than λ would not represent reentrant arrhythmias very well because the length scale of reentry would be smaller than the sampling length. In most CAMCC the lattice sampling length is less than λ, and the independent assignment of parameters results in a dispersion. This can be superimposed on a gradient. For example, endocardial to epicardial conduction can be assigned both a gradient in repolarization (shorter at the epicardium, longer at the endocardium, with a linear variation over the intervening length of approximately 1 cm, which means the length scale of variation, corresponding to the dominant spatial frequency, is approximately 4 cm), as well as a spatial dispersion in repolarization (reflecting, e.g., variable ischemia on length scales of a lattice spacing defined to represent 1 mm, i.e., approximately the electrotonic decay length).

18.8 Dynamical Inhomogeneities in Conduction Physiology

In CAMCC with static (time-independent) parameters of conduction physiology, the only variations possible are spatial fluctuations in these parameters. However, physiologically the conduction velocity and refractory period will change as a function of the history of local electrical activity, which can affect the initiation and stability of arrhythmias [28]. For example, the refractory period will shorten under rapid sequences of stimulation while it will lengthen under slow stimulation. As stated earlier, we can couple the dynamical behavior of the refractory period to that of the action potential duration. This dynamical behavior in turn can be described by a set of parameters that reflect the ability of the myocardial cells to repolarize in response to different stimulus sequences. One such model, generalized from electrical restitution experiments to measure dynamical changes in action potential duration, APD [8,22,23], takes the form:

$$APD(\alpha_0, T) = G(T, a, b, c) \frac{E(\alpha_0, \tau_1, \tau_2, \ldots, \tau_n, T)}{E(T, \tau_1, \tau_2, \ldots, \tau_n, T)}$$

where α_0 represents the value of the autochrone at the point at which the current action potential was stimulated (i.e., the test stimulus interval in an electrical restitution experiment). In this model the memory of the previous patterns of activation is represented by T, which could itself represent a complex function of the previous patterns of local activation, for example, a weighted convolution over previous stimulus intervals or action potential durations. (In electrical restitution experiments the cells are conditioned at some constant stimulus interval, T, until they achieve a steady state before the varying test stimulus interval, α_0, is applied.) Here the τ_i represent exponential decay constants for the function, E, while a, b, and c define the shape of the steady state curve G.[4] The τ_i and a, b, and c will be referred to as *characteristic parameters*. Characteristic parameters define the dynamical behavior of physiological conduction parameters such as the refractory period, action potential duration, or conduction velocity in response to previous patterns of electrical activity. We can introduce spatial variations in these characteristic parameters (reflecting physiological inhomogeneities), and so induce spatial inhomogeneities in the dynamical behavior of the

---

[4] G defines the action potential duration in the case of constant stimulus interval, $\alpha_0 = T$, in electrical restitution experiments.

conduction parameters. It is possible to have the actual values of the refractory period (or action potential duration) be the same at two points even though the corresponding characteristic parameters are quite different, either by variations in the history of activity at the two points or by, for example, the fact that both points might share the same limiting behavior at long time intervals between stimulations. This latter case might result in uniform conduction properties measured at long stimulus cycle lengths while uncovering dispersion (or gradients) at short stimulus cycle lengths.

Conversely, two points (or an entire lattice) sharing identical characteristic parameters may actually have different values for the refractory period, based on differences in the history of electrical activity that might occur by virtue of the spatial pattern of the depolarizing wavefront. In fact, there can be a feedback relation between the patterns of electrical activity and the parameters of conduction; incoherence and spatial disorganization in the pattern of electrical activity may induce spatial variation (dispersion) in conduction parameters, which in turn may increase the disorganization of the electrical activity. We are currently pursuing the significance of this coupling in the stability and occurrence of reentrant arrhythmias in CAMCC models.

Current work suggests that as the dispersion of conduction parameters increases, the coherence of reentrant vortex patterns decreases, and one ultimately approaches a fibrillatory pattern of self-sustained activity. In the analysis of the effects of dispersion in conduction parameters, CAMCC models with static parameters represent intrinsic inhomogeneities that are unchanging. Characteristic parameters, defining the dynamical behavior of conduction parameters in response to previous patterns of electrical activity, permit variable, dynamic inhomogeneities in conduction parameters representing both spatial fluctuations in the characteristic parameters and spatial fluctuations in the patterns of electrical activity.

This suggests that new experimental measurements are required to analyze the dynamical behavior of conduction velocity and refractory period as functions of:

1. Space: the distinction of gradients and dispersion, and the measurement of changes in parameters over different length scales.

2. Time: as functions of the history of previous electrical activity.

Electrical restitution curves are the first step in determining the effect of previous patterns of activity on the action potential duration (and hence by association the refractory period). It would be valuable to analyze carefully the dependence of conduction parameters, such as the refractory period, not just on the preceding stimulus interval, as has been customary, but on the previous m stimulus intervals and refractory periods, where m is increased until the system loses memory of the mth preceding stimulus. This would then permit the specification of the dynamical behavior of conduction physiology in response to arbitrary patterns of electrical activity (by a Markov vector of the $m - 1$ previous stimulus intervals and values of the conduction parameters), and would be important in examining by CAMCC modeling and theory the possible feedback relation between dynamical conduction parameter inhomogeneity and patterns of electrical activity.

18.9   Conclusion

CAMCC have been useful for the analysis of dynamical properties of the patterns of cardiac electrical behavior. In particular, they represent well the relation of inhomogeneities in parameters of conduction physiology to the global structure of patterns of cardiac electrical activity, from the organized, coherent wavefronts or reentrant vortices (spiral waves) also seen in a variety of other active media (chemical, Belousov-Zhabotinsky reactions, and cellular, *Dictyostelium discoideum* slime molds or neural fields of the retina), to the disorganized, incoherent structure of fibrillatory patterns.

In the representations of the complex, nonlinear dynamics of these active media, the simplification by discrete state transitions used in the CAMCC results in algorithms that are computationally simple and fast. This can be quite significant when studying questions that involve examining large-scale structural features of behavior where large lattices must be used, such as is the case for reentrant arrhythmias. However, this simplification of a discrete lattice results in limitations of the dynamical representation. In particular, details about the core of reentrant vortices are not well represented. The results of CAMCC must be qualified by this limitation, which may have significance, for example, in determining the exact threshold for change in patterns of behavior.

We will argue, however, that the global features of activity are insensitive to these microscopic details. The transition between classes of dynamical states (e.g., fibrillation vs. normal patterns of depolarization) may be quantitatively affected by the detailed representation of the microscopic membrane dynamics, but should not be qualitatively affected. The existence of organized spiral waves both in CAMCC and in cardiac or other continuous physical systems argues for the essential qualitative accuracy of the CAMCC representation. We have attempted in this chapter to clarify the interpretation of this relation by the specification of the general dynamical algorithm and the interpretation of the discrete spatial lattice. For the analysis of reentrant arrhythmias, the simplest CAMCC with specification of conduction velocity and refractory period is sufficient to generate both organized and disorganized self-sustaining patterns of reentry. The CAMCC and its extensions to intermediate models provide the ability to model many other forms of electrophysiological behavior thought to underlie dysrhythmogenesis and the variable response to exogenous stimulation. Ultimately the relation of the CAMCC to cardiac conduction physiology and arrhythmogenesis must be further examined in the future by experimental analysis, comparison of predictions from modeling by CAMCC versus continuous representations, and theoretical analysis.

Acknowledgements: This work was supported by NASA grant #NAGW-988 and a grant from the Whitaker Foundation. B.E.H. Saxberg is grateful for support from NIH Postdoctoral Fellowship #1-F32-HL08014.

References

[1] D. Adam, P. Levy, A. van Oosteran, and S. Sideman. On the generation of surface potential maps by a distributed conduction velocities model. In *Computers in Cardiology*, pages 483–486. IEEE Computer Society Press, Washington, DC, 1988.

[2] M.A. Allessie, F.I.M. Bonke, and F.J.G. Schopman. Circus movement in rabbit atrial muscle as a mechanism of tachycardia. *Circ. Res.*, 33:54–62, 1973.

[3] M.A. Allessie, F.I.M. Bonke, and F.J.G. Schopman. Circus movement in rabbit atrial muscle as a mechanism of tachycardia II. The role of nonuniform recovery of excitability in the occurrence of unidirectional block, as studied with multiple microelectrodes. *Circ. Res.*, 39:168–177, 1976.

[4] M.A. Allessie, F.I.M. Bonke, and F.J.G. Schopman. Circus movement in rabbit atrial muscle as a mechanism of tachycardia III. The "leading circle" concept. *Circ. Res.*, 41:9–18, 1977.

[5] M.A. Allessie, W.J.E.P. Lammers, F.I.M. Bonke, and J. Hollen. Intraatrial reentry as a mechanism for atrial flutter induced by acetylcholine and rapid pacing in the dog. *Circulation*, 70:123–135, 1984.

[6] P.M. Auger, A.L. Bardou, A. Coulombe, and J.M. Chesuais. Computer simulation of ventricular fibrillation and of defibrillating electric shocks. In *Computers in Cardiology*, pages 387–390. IEEE Computer Society Press, Washington, DC, 1989.

[7] G.W. Beeler and H. Reuter. Reconstruction of the action potential of ventricular myocardial fibers. *J. Physiol.*, 268:177–210, 1977.

[8] M.R. Boyett and B.R. Jewell. A study of the factors responsible for rate dependent shortening of the action potential in mammalian ventricular muscle. *J. Physiol.*, 285:359–380, 1978.

[9] J. Bures, V.I. Koroleva, and N.A. Gorelova. Leao's spreading depression, an example of diffusion-mediated propagation of excitation in

the central nervous sytem. In V.I. Krinsky, editor, *Self-Organization, Autowaves and Structures Far from Equilibrium*, pages 180–188. Springer-Verlag, New York, 1984.

[10] M.J. Burgess. Relation of ventricular repolarization to electrocardiographic T wave-form and arrhythmia vulnerability. *Am. J. Physiol.*, 236:H391–H402, 1979.

[11] M.J. Burgess, B.M. Steinhaus, K.W. Spitzer, and P.R. Ershler. Nonuniform epicardial activation and repolarization properties of in vivo canine pulmonary conus. *Circ. Res.*, 62:233–246, 1988.

[12] M-N Chee, S.G. Whittington, and R. Kapral. A model of wave propagation in an inhomogeneous excitable medium. *Physica D.*, 32:437–450, 1988.

[13] E.A. Clancey, J.M. Smith, and R.J. Cohen. A simple electrical-mechanical model of the heart applied to the study of electrical-mechanical alternans. (submitted.)

[14] L. Clerc. Directional differences of impulse spread in trabecular muscle from mammalian heart. *J. Physiol.*, 255:335–346, 1976.

[15] D.S. Cohen, J.C. Neu, and R.R. Rosales. Rotating spiral wave solutions of reaction diffusion equations. *SIAM Appl. Math.*, 35:536–547, 1978.

[16] R.L. Cohn, S. Rush, and E. Lepeschkin. Theoretical analyses and computer simulation of ECG ventricular gradient and recovery waveforms. *Trans. Biomed. Eng.*, BME-29:413–422, 1982.

[17] J.M.T. de Bakker, B. Henning, and W. Merx. Circus movement in canine right ventricle. *Circ. Res.*, 45:374–378, 1979.

[18] P.J. Diaz, Y. Rudy, and R. Plonsey. Intercalated discs as a cause for discontinuous propagation in cardiac muscle: A theoretical simulation. *Ann. Biomed. Eng.*, 2:177–189, 1983.

[19] E. Downar, M.J. Janse, and D. Durrer. The effect of acute coronary artery occlusion on subepicardial transmembrane potentials in the intact porcine heart. *Circulation*, 56:217–224, 1977.

[20] J.P. Drouhard and F.A. Roberge. A simulation study of the ventricular myocardial action potential. *IEEE Trans. Biomed. Eng.*, BME-29:494–502, 1982.

[21] A.J. Durston. Dictyostelium discoideum aggregation fields as excitable media. *J. Theor. Biol.*, 42:483–504, 1973.

[22] V. Elharrar, H. Atarashi, and B. Surawicz. Cycle length-dependent action potential duration in canine cardiac Purkinje fibers. *Am. J. Physiol.*, 247:H936–H945, 1984.

[23] V. Elharrar and B. Surawicz. Cycle length effect on restitution of action potential duration in dog cardiac fibers. *Am. J. Physiol.*, 244: H782–H792, 1983.

[24] N. El-Sherif, B.J. Scherlag, R. Lazzara, and P. Samet. Pathophysiology of tachycardia- and bradycardia-dependent block in the canine proximal His-Purkinje system after acute myocardial ischemia. *Am. J. Cardiol.*, 33:529–540, 1974.

[25] D. Eylon, D. Sadeh, Y. Kantor, and S. Abboud. A model of the hearts conduction system using a self-similar fractal structure. In *Computers in Cardiology*, pages 399–402. IEEE Computer Society Press, Washington, DC, 1989.

[26] P.C. Fife. Nonequilibrium cooperative phenomena in physics and related fields. In M.G. Velarde, editor, *Current Topics in Reaction-Diffusion Systems*, pages 371–411. Plenum Press, New York, 1984.

[27] R. Fitzhugh. Impulse and physiological states in models of nerve membrane. *Biophysics J.*, 1:445–466, 1961.

[28] L.H. Frame and M.B. Simson. Oscillations of conduction, action potential duration, and refractoriness. *Circulation*, 78:1277–1287, 1988.

[29] M. Gerhardt, H. Schuster, and J.J. Tyson. A cellular automaton model of excitable media including curvature and dispersion. *Science*, 247:1563–1566, 1990.

[30] N.A. Gorelova and J. Bures. Spiral waves of spreading depression in the isolated chicken retina. *J. Neurobiol.*, 14:353–363, 1983.

[31] J.M. Greenberg, B.D. Hassard, and S.P. Hastings. Pattern formation and periodic structures in systems modeled by reaction-diffusion equations. *Bull. Am. Math. Soc.*, 84(6):1296–1327, 1978.

[32] H. Haken. *Advanced Synergetics.* Springer-Verlag, Berlin, 1983.

[33] H. Haken. *Synergetics: An Introduction.* Springer-Verlag, Berlin, 1983.

[34] J. Han and G.K. Moe. Nonuniform recovery of excitability in ventricular muscle. *Circ. Res.*, 14:44–60, 1964.

[35] K. Harumi, C.R. Smith, J.A. Abildskov, M.J. Burgess, R.L. Lux, and R.F. Wyatt. Detailed activation sequence in the region of electrically induced ventricular fibrillation in dogs. *Jap. Heart J.*, 21:533–544, 1980.

[36] S. Hastings. Persistent spatial patterns for semi-discrete models of excitable media. *J. Math. Biol.*, 11:105–117, 1981.

[37] B.F. Hoffman, C.Y. Kao, and E.E. Suckling. Refractoriness in cardiac muscle. *Am. J. Physiol.*, 190:473–482, 1957.

[38] R.E. Ideker, G.J. Klein, L. Harrison, et al. The transition to ventricular fibrillation induced by reperfusion following acute ischemia in the dog: A period of organized epicardial activation. *Circulation*, 63:1371–1379, 1981.

[39] J.J.B. Jack, D. Noble, and R.W. Tsien. *Electrical Current Flow in Excitable Cells.* Oxford University Press, Oxford, 1975.

[40] M.J. Janse and A.G. Kleber. Electrophysiological changes and ventricular arrhythmias in the early phase of regional myocardial ischemia. *Circ. Res.*, 49:1069–1081, 1981.

[41] M.J. Janse, F.J. van Capelle, H. Morsink, et al. Flow of injury currents and patterns of excitation during early ventricular arrhythmias in acute regional myocardial ischemia in isolated porcine and canine hearts. *Circ. Res.*, 47:151–165, 1980.

[42] R.W. Joyner, F. Ramon, and J.W. Moore. Simulation of action potential propagation in an inhomogeneous sheet of coupled excitable cells. *Circ. Res.*, 36:654–661, 1975.

[43] D.T. Kaplan, J.M. Smith, B.E.H. Saxberg, and R.J. Cohen. Nonlinear dynamics in cardiac conduction. *Math. Biosci.*, 90:19–48, 1988.

[44] J.P. Keener. Nonlinear oscillations in biology and chemistry. In H.G. Othmer, editor, *Spiral Waves in Excitable Media*, pages 115–127. Springer-Verlag, New York, 1986.

[45] N. Kopell and L.N. Howard. Target pattern and spiral solutions to reaction-diffusion equations with more than one space dimension. *Adv. Appl. Math.*, 2:417–449, 1981.

[46] V.I. Krinsky. Self-organization, autowaves and structures far from equilibrium. In V.I. Krinsky, editor, *Autowaves: Results, Problems, and Outlooks*, pages 9–19. Springer-Verlag, New York, 1984.

[47] R. Lazzara, N. El-Sherif, and B.J. Scherlag. Disorders of cellular electrophysiology produced by ischemia of the canine his bundle. *Circ. Res.*, 36:444–454, 1975.

[48] B.F. Madore and W.L. Freedman. Computer simulations of the Belousov-Zhabotinsky reaction. *Science*, 222:615–616, 1983.

[49] M. Malik and T. Cochrane. Shell computer model of cardiac electropotential changes. *J. Biomed. Eng.*, 7:266–274, 1985.

[50] R.E. McAllister, D. Noble, and R.W. Tsien. Reconstruction of the electrical activity of cardiac Purkinje fibers. *J. Physiol.*, 251:1–59, 1975.

[51] A.B. Medvinsky, A.V. Panfilov, and A.M. Pertsov. Properties of rotating waves in three dimensions. Scroll rings in myocardium. In V.I. Krinsky, editor, *Self-Organization, Autowaves and Structures Far from Equilibrium*, pages 195–199. Springer-Verlag, New York, 1984.

[52] W.T. Miller and D.B. Geselowitz. Simulation studies of the electrocardiogram I. The normal heart. *Circ. Res.*, 43:301–315, 1978.

[53] W.T. Miller and D.B. Geselowitz. Simulation studies of the electrocardiogram II. Ischemia and infarction. *Circ. Res.*, 43:315–323, 1978.

[54] G.K. Moe, W.C. Rheinboldt, and J.A. Abildskov. A computer model of atrial fibrillation. *Am. Heart J.*, 67:200–220, 1964.

[55] E.N. Moore, J.B. Preston, and G.K. Moe. Durations of transmembrane action potentials and functional refractory periods of canine false tendon and ventricular myocardium: Comparisons in single fibers. *Circ. Res.*, 17:259–273, 1965.

[56] S.C. Muller, T. Plener, and B. Hess. The structure of the core of the spiral wave in the Belousov-Zhabotinskii reaction. *Science*, 230:661–663, 1985.

[57] G. Nicolis and I. Prigogine. *Self-Organization in Non-Equilibrium Systems*. John Wiley & Sons, New York, 1977.

[58] M. Okajima, T. Fujino, T. Kobayashi, and K. Yamada. Computer simulation of the propagation process in excitation of the ventricles. *Circ. Res.*, 23:203–211, 1968.

[59] N.H. Packard and S. Wolfram. Two dimensional cellular automata. *J. Statistical Phys.*, 38:901–946, 1985.

[60] A.V. Panfilov and A.T. Winfree. Dynamical simulations of twisted scroll rings in three dimensional excitable media. *Physica D.*, 17:323–330, 1985.

[61] R. Plonsey. *Bioelectric Phenomena*. McGraw-Hill, New York, 1969.

[62] F.A. Roberge, A. Vinet, and B. Victorri. Reconstruction of propagated electrical activity with a two-dimensional model of anisotropic heart muscle. *Circ. Res.*, 58:461–475, 1986.

[63] D.E. Roberts, L.T. Hersh, and A.M. Scher. Influence of cardiac fiber orientation on wavefront voltage, conduction velocity, and tissue resistivity in the dog. *Circ. Res.*, 44:701–712, 1979.

[64] T. Sano, N. Takayama, and T. Shimamoto. Directional difference of conduction velocity in the cardiac ventricular syncytium studied by microelectrodes. *Circ. Res.*, 7:262–267, 1959.

[65] B.E.H. Saxberg, M.P. Grumbach, and R.J. Cohen. A time dependent anatomically detailed model of cardiac conduction. In *Computers in Cardiology*, pages 401–404. IEEE Computer Society Press, Washington, DC, 1985.

[66] R.H. Selvester, W.L. Kirk, and R.B. Pearson. Propagation velocities and voltage magnitudes in local segments of dog myocardium. *Circ. Res.*, 27:619–629, 1970.

[67] R.H. Selvester, J. Solomon, and D. Sapoznikov. Computer simulation of the electrocardiogram. In L.D. Cady, editor, *Computer Techniques in Cardiology*, pages 417–453. Marcel Dekker, New York, 1979.

[68] R. Serra, M. Andretta, M. Compiani, and M. Zanarini. *Introduction to the Physics of Complex Systems*. Pergamon Press, New York, 1986.

[69] J.M. Smith and R.J. Cohen. Simple computer model predicts a wide range of ventricular dysrhythmias. *PNAS*, 81:233–237, 1984.

[70] J.M. Smith, R.L. Ritzenberg, and R.J. Cohen. Percolation theory and cardiac conduction. In *Computers in Cardiology*, pages 201–204. IEEE Computer Society Press, Washington, DC, 1984.

[71] J. Solomon and R.H. Selvester. Simulation of measured activation sequence in the human heart. *Am. Heart J.*, 85:518–523, 1973.

[72] J.R. Sommer and P.C. Dolber. Cardiac muscle: Ultrastructure of its cells and bundles. In A. Paes de Carvalho, B.F. Hoffman, and M. Lieberman, editors, *Normal and Abnormal Conduction in the Heart.*, pages 1–28. Futura Publishing Co, Mount Kisco, New York, 1982.

[73] M.S. Spach. The discontinuous nature of electrical propagation in cardiac muscle. *Ann. Biomed. Eng.*, 2:209–261, 1983.

[74] M.S. Spach. The electrical representation of cardiac muscle based on discontinuities of axial resistivity at a microscopic and macroscopic level. In A. Paes de Carvalho, B.F. Hoffman, and M. Lieberman, editors, *Normal and Abnormal Conduction in the Heart*, pages 145–180. Futura Publishing Co, Mount Kisco, New York, 1982.

[75] M.S. Spach and J.M. Kootsey. The nature of electrical propagation in cardiac muscle. *Am. J. Physiol.*, 244:H3–H22, 1983.

[76] M.S. Spach, J.M. Kootsey, and J. Sloan. Active modulation of electrical coupling between cardiac cells of the dog. *Circ. Res.*, 51:347–362, 1982.

[77] J.F. Spear and E.N. Moore. The effect of changes in rate and rhythm on supernormal excitability in the isolated Purkinje system of the dog. *Circulation*, 50:1144–1149, 1974.

[78] C.A. Swenne, H.A. Bosker, and N.M. vanHemel. Computer simulation of compund reentry. In *Computers in Cardiology*, pages 445–448. IEEE Computer Society Press, Washington, DC, 1987.

[79] N.V. Thakor and L.M. Eisenman. Three dimensional computer model of the heart: Fibrillation induced by extrastimulation. *Comp. Biomed. Res.*, 22:532–545, 1989.

[80] J.J. Tyson, K.A. Alexander, V.S. Manoranjan, and J.D. Murray. Spiral waves of cyclic AMP in a model of slime mold aggregation. *Physica D.*, 34:193–207, 1989.

[81] J.J. Tyson and J.P. Keener. Singular perturbation theory of traveling waves in excitable media (A review). *Physica D.*, 32:327–361, 1988.

[82] F.J.L. van Capelle and D. Durrer. Computer simulation of arrhythmias in a network of coupled excitable elements. *Circ. Res.*, 47:454–466, 1980.

[83] V.A. Vasiliev, Y.M. Romanovskii, D.S. Chernavskii, and V.G. Yakhno. *Autowave Processes in Kinetic Systems*. D. Reidel Publishing Co, Boston, 1987.

[84] J. von Neumann. *Theory of Self-Reproducing Automata*. University of Illinois Press, Urbana, 1966.

[85] S. Weidmann. The effect of the cardiac membrane potential on the rapid availability of the sodium carrying system. *J. Physiol.*, 127:213–224, 1955.

[86] A.T. Winfree. Rotating chemical reactions. *Sci. Am.*, June:82–95, 1974.

[87] A.T. Winfree. Spiral waves of chemical activity. *Science*, 175:634–636, 1972.

[88] A.T. Winfree. *When Time Breaks Down: The Three-Dimensional Dynamics of Electrochemical Waves and Cardiac Arrhythmias*. Princeton University Press, Princeton, NJ, 1987.

[89] A.L. Wit, M.A. Allessie, F.I.M. Bonke, W. Lammers, J. Smeets, and J.J. Fenoglio Jr. Electrophysiologic mapping to determine the mechanism of experimental ventricular tachycardia initiated by premature impulses: Experimental approach and initial results demonstrating reentrant excitation. *Am. J. Cardiol.*, 49:166–185, 1982.

[90] S. Wolfram. Statistical mechanics of cellular automata. *Rev. Mod. Phys.*, 55:601–644, 1983.

[91] S. Wolfram, editor. *Theory and Application of Cellular Automata.* World Scientific, Singapore, 1986.

[92] S. Wolfram. Universality and complexity in cellular automata. *Physica D.*, 10:1–35, 1984.

[93] A.N. Zaikin and A.M. Zhabotinsky. Concentration wave propagation in two-dimensional liquid-phase self-oscillating systems. *Nature*, 225:535–537, 1970.

[94] V.S. Zykov. Analytical evaluation of the dependence of the speed of an excitation wave in a two-dimensional excitable medium on the curvature of its front. *Biophysics*, 25:906–911, 1980.

[95] V.S. Zykov. Kinematics of the non-steady circulation of helical waves in an excitable medium. *Biophysics*, 32:365–369, 1987.

[96] V.S. Zykov. Kinematics of the steady state circulation in an excitable medium. *Biophysics*, 25:329–333, 1980.

19

Estimating the Ventricular Fibrillation Threshold

A.T. Winfree[1]

ABSTRACT Implicit in the basic principles of reasonably uniform excitable media is a vortexlike mode of self-excitation. In heart muscle this would be a rotating action potential and it has in fact been found in both two- and three-dimensional settings. It rotates in 120 msec and has a core diameter of $\frac{2}{3}$ cm or less, conforming to rough estimates based on oversimplified physics. Similar estimates indicate that the point-stimulus threshold should be about 4 mA/cm^2, based on observed thresholds of total current and a theoretical estimate of the maximum wavefront curvature compatible with sustained propagation. The vortex diameter together with the stimulation threshold can be used to derive the electrical threshold for nucleation by a stable vortex pair: this 16 mA estimate also compares favorably with observations of the electrical threshold for instigation of fibrillation. Electrical defibrillation should require local potential gradients of about 6 V/cm or current densities near 20 mA/cm^2, also roughly as observed. Quantitative derivation of these thresholds became feasible only after the pertinent electrophysiology was simplified beyond the comfort level of competent theorists, but this is sometimes how a new starting point is secured for eventual refinement to a believable theory.

19.1 Introduction

This chapter is intended for a beginning graduate student in physical sciences who is intrigued by some of the unsolved (and even undefined) puzzles to be found in cardiology. I focus attention on puzzles involving the reentrant electrical arrhythmias that mediate the mysterious transition from orderly beating to the turbulence called fibrillation. My objective is to find out whether elementary considerations suffice to estimate the magnitudes of quantities closely associated with this transition. This gathering of numbers is assisted by limiting the subject matter to physiologically *normal* ventricular myocardium in the *dog*. Since many ideas have been circulating in nonquantitative form for many years, and most ideas change their

---

[1]Department of Ecology and Evolutionary Biology, University of Arizona, Tucson, Arizona 85721

appearance radically when one insists upon quantitative consistency, application of this stress may be expected to refine our understanding of the onset of fibrillation. The first thing discovered in this effort is that not all the essential numbers are yet available. Anyone looking for all the answers had better read another chapter. Here I mostly present questions and experimentally testable inferences from postulates, quantified in a light-hearted, back-of-the-envelope spirit intended to provoke students to improve upon these rough estimates.

This section provides a minimal background on "rotors" in excitable media. The rotor is a periodic solution of Equation (19.1) below, a stable two-dimensional rotating pattern of excitation typical of excitable media. In any uniform medium it has a characteristic size and rotation period (although there are sometimes more than one stable alternative [146,147,150]), it may rotate in either direction and it may be located anywhere, depending only on the stimulus or initial conditions that provoked it. A periodic wavetrain radiates away from it in every direction tangential to its perimeter; all together they constitute a spiral wave (either right- or left-handed) with pitch equal to the rotor's perimeter. For book-length discussions see Winfree [149,159] and Zykov [168]. Section 19.2 presents the essential form of the wave equation of electrophysiological media and discusses its scaling laws. Note that it is fundamentally different from "The Wave Equation" familiar in acoustics, electromagnetic theory, mechanics, and quantum mechanics. The literature of electrophysiology contains abundant documentation of the first term in this wave equation, representing local membrane currents and their time and voltage dependences. Less explicit attention has been given to the second term, quantified by the diffusion coefficient D, so Section 19.3 attempts a gathering of estimates of D. Section 19.4 tests the quantitative framework thus provided by attempting to estimate the stimulus needed to elicit a propagating action potential, then comparing with (inference from) old experimental data. This requires of us a realization that propagation speed depends on wavefront curvature; the estimated dependency still needs testing in the laboratory. One testable implication from our approach is presented in Section 19.5: by optimizing electrode shape it should be possible to pace myocardium with much less energy than currently used. Carrying on as though this experiment had been done and had confirmed our inferences up to this point, we are finally in a position to address the question advertised in this chapter's title: "How much current is needed to start fibrillation?" The answer has been obtained experimentally in scores of published papers but so far as I am aware no one has ever asked or answered "Why does it take so much (or so little)? Why not 1000 times less (or more)?" Section 19.6 (the longest) bases a proposed answer on the foregoing plus the idea that the onset of fibrillation in normal tissue consists of creating rotors. Conclusions compare plausibly with published experimental results, but those determinations vary enormously, indicating a need to control geometric factors that were ignored when experiments were conducted under a different conceptual framework. The prediction is ventured that the results of appropriately repeated measurements will be a lot more reproducible, allowing a more severe test of these notions.

Section 19.7 anticipates the measurable dependence of vulnerable period duration on stimulus magnitude. Section 19.8 says that the "upper limit of vulnerability" predicted from this theory, and since established experimentally, is compatible in magnitude with the foregoing estimates. Section 19.9 asks why defibrillation requires so much energy and suggests ways to markedly reduce the requirement. Section 19.10 addresses the "excitable gap" in myocardium, suggesting that its presence has more to do with the general principles of rotor solutions to Equation (19.1) below than with anisotropy and cellular graininess. It suggests that the tachycardia established in Section 19.6 develops into fibrillation when the excitable gap is "wrung out" of a vortex filament by a kind of progressive twisting that seems inevitable in three-dimensional tissue. No experiments have yet addressed this issue. Conclusions are detailed in Section 19.11. This exercise largely turned into a library search for consistent numbers; the fruit of this labor is preserved in the form of a bibliography of 168 mostly experimental papers that proved useful for such purposes; all are cited in the text.

19.2   The Continuum Approximation

In one approximation, useful at least in two-dimensions, myocardium resembles a uniformly anisotropic continuous medium. In Cartesian coordinates, the cable equation of electrophysiology in a uniformly anisotropic continuous medium has the form:

$$\frac{\partial V_m(x,y,z,t)}{\partial t} = -\frac{J_m}{C_m} + D_x\frac{\partial^2 V_m}{\partial x^2} + D_y\frac{\partial^2 V_m}{\partial y^2} + D_z\frac{\partial^2 V_m}{\partial z^2} \qquad (19.1)$$

where the units must be watched with care: t is time (msec), V_m is local membrane potential (mV), and J_m is a sum of (voltage and time dependent) local ion-channel current densities (μA/cm^2) of the form $\sum(V_m - V_i)g_i$. The sum is taken over every kind of ion for which there are channel proteins embedded in the cell membrane. V_i is the constant and uniform "Nernst potential" for that ionic species, proportional to the logarithm of its inside/outside concentration ratio. g_i is the membrane's conductivity toward that ion, usually curve-fitted by complicated differential expressions indicating the time and voltage dependences of many components of that conductivity. C_m is a constant membrane-specific capacitance (μF/cm^2), D_i (cm^2/msec) is a constant diffusion coefficient in this reaction-diffusion equation: $D_i = 1/\{C_m(1\mu F/cm^2)$ times fiber surface/volume ratio S_v (some thousands/cm: see below) times a directionally anisotropic resistivity ρ. that includes interstitial fluid, cytoplasm and gap junctions but not the membrane (some hundreds of Ωcm along the fiber axis: see below)}. See Joyner and coworkers [63] for a derivation.

In a fibrous medium like myocardium, taking x y as the epicardial plane and x as the fiber axis, $D_x > D_y = D_z$: Resistivity ρ is an order of magnitude smaller and wave speed consequently several-fold greater along axis x. Substitute rescaled distances $y' = y\sqrt{(D_x/D_y)}$ and $z' = z\sqrt{(D_x/D_z)}$ to recover the isotropic case with coefficients $D'_y = D'_z = D_x$. We call it D henceforth:

$$\frac{\partial V_m}{\partial t} = -\frac{J_m}{C_m} + D\nabla^2 V_m \tag{19.2}$$

In other words, solutions for the uniformly anisotropic continuum are exactly those of the isotropic case, appropriately expanded (including boundary conditions) in both directions transverse to the fast axis. No temporal aspect of the solution is affected at all. This theorem is strictly true of continua but in discrete media such as myocardium it is an approximation whose domain of adequacy needs explicit investigation (see van Capelle [132], Kawato and colleagues [68], Rudy and Quan [102], Keener [69,70,72,73], and Chapter 17 in this volume.)

The behavior of this dynamical system requires study in three distinct contexts:

a) In uniformly anisotropic media during application of a briefly constant direct-current stimulus through extracellular electrodes. While these constant-current boundary conditions are in effect we deal with potential gradients $\underline{\nabla U}$ (V/cm, not mV/cm as for transmembrane potentials above) and current densities S(mA/cm^2, not μA/cm^2 as for transmembrane currents above). In principle, potential distributions are not affected by resistivity in a uniform medium, so rescaling only complicates their derivation (e.g., consider a block of uniform material between ± 1 volt electrodes: Isopotentials are evenly spaced from $+1$ to -1 volt regardless of the magnitude of resistivity though the current flowing down that gradient will be inversely proportional to the resistivity). Conservation of current (divergence of current vector $= 0$) requires that the steady-state isopotential contours around any electrode configuration must satisfy

$$\left[\frac{\partial}{\partial x}\left(\frac{1}{\rho_x(x,y,z)}\frac{\partial}{\partial x}\right) + \frac{\partial}{\partial y}\left(\frac{1}{\rho_y(x,y,z)}\frac{\partial}{\partial y}\right) + \frac{\partial}{\partial z}\left(\frac{1}{\rho_z(x,y,z)}\frac{\partial}{\partial z}\right)\right]U = 0, \tag{19.3}$$

where ρ_x indicates resistivity in the x direction, and so forth. If each of these anisotropic resistivities is uniform (independent of x, y, and z) then in coordinates rescaled as above this simplifies to

$$\frac{\partial^2 U}{\partial x'^2} + \frac{\partial^2 U}{\partial y'^2} + \frac{\partial^2 U}{\partial z'^2} = 0, \ i.e., \nabla^2 U = 0 \qquad (19.4)$$

independent of resistivity and of its anisotropy. Thus, it proves most convenient in this context to forego the rescaling suggested above. This is essentially the situation in an isolated two-dimensional thin layer of myocardium (e.g., Tsuboi and colleagues [127], Schalij [107], Dillon and colleagues [24], Kadish and colleagues [65], Allessie and colleagues [3,4], Zuanetti and colleagues [167]). It may also be the case in three-dimensional thick myocardium if fiber direction remains substantially the same from layer to layer through a centimeter or more in depth, as it might perhaps in whale myocardium; but see case (c) regarding hearts on a smaller scale, as in the dog.

b) In uniformly anisotropic media during the absence of external stimuli. While such no-flux boundary conditions are in effect, propagation patterns in a uniformly anisotropic medium would be most conveniently addressed in coordinates rescaled as suggested above, distances at right angles to the fiber fast axis having been expanded two- to threefold. This procedure should be useful in context of thin-layer myocardial preparations and in the thick myocardium of a heart larger than the dog's. In point of fact, the asymmetric isochronal contours on epicardial maps (e.g., those to be found in Schalij [107], Zuanetti and coworkers [167], or van Capelle and Allessie [133]) are not entirely symmetrized by photographically expanding the map transversely in proportion to the observed ratio of longitudinal to transverse propagation velocity, perhaps indicating imperfections of the continuum approximation.

c) In dog myocardium, neither (a) nor (b) quite applies, for the interesting reason that fiber direction smoothly *turns* as much as 180° with increasing depth, z, from epicardium to endocardium. This distance being only about 1 cm, the rotation is substantial even over distances as short as the one-dimensional electrophysiological space constant (about 1 mm), thus, however, uniformly anisotropic the medium may be in each two-dimensional plane, it cannot be so regarded in three-dimensions. One effect of this rotation is a spatial averaging of x, y directional properties: The anisotropy observed in thin layers is attenuated in bulk. Frazier and colleagues ([36], Appendix) show experimentally that extracellular potential $U(x, y, z)$ is nearly hemispherically distributed, falling off as 1/radius from a point source in actual three-dimensional myocardium much as in the uniformly anisotropic idealization (a). And turning to propagation, isochronal rings turn out to be more nearly circular [36] than expected from thin-layer preparations in which no such averaging occurs [4,24,107,167]. It also turns out that a given current density has about the same effect regardless of direction [37].

Since (c) spoils the applicability of devices (a) and (b) for renormalizing away the anisotropy, but since it does so by largely *removing* the x, y anisotropy, we resort to the still simpler device of proceeding as though three-dimensional dog myocardium were isotropic. Whatever conclusions we reach will eventually need refinement by a factor as large as about two within the context of a more sophisticated theory that incorporates specific anatomy.

19.3   The Diffusion Coefficient of Electric Potential

As noted above, all spatial aspects of solutions to the propagation equation scale in proportion to $\sqrt{D}$; for example, if D is made fourfold larger (e.g., by making resistivity ρ fourfold smaller) then the activation front moves twice as quickly and the critical nucleus diameter (Section 19.4) becomes twice as large.

Fast-axis (fiber-axis) coefficient D is commonly represented as λ^2/τ, where λ is the one-dimensional space constant (cm) and τ is the time constant (msec) of the membrane near equilibrium. In myocardium λ is typically about 0.08 cm and τ is typically about 4 msec according to van Capelle [132]. D is probably about 0.001 cm^2/msec = 1 cm^2/sec. Estimates vary widely, mostly because of lingering uncertainty about S_v and ρ for ventricular myocardium.

Averaged over orientations, ρ is about 400 Ωcm according to van Oosterom and colleagues [135], Rush and colleagues [103], Roberts and Scher [101], and Rudy and Quan [102]. However Starmer and Whalen [122] and Frazier and colleagues [37] find 263 and 215 Ωcm, respectively, for living dog ventricle; even though it seems the lowest on record, we will use the estimate of Frazier and colleagues [37] below, where necessary for consistency with their—the only—measurement of certain threshold current densities. The ratio of ρ transversely to ρ longitudinally is variously estimated as 2 (Steinhaus and colleagues [123]: two-dimensional computer model), 2.2 (Rush and colleagues [103]: three-dimensional left ventricular wall), 3.0 (Frazier and colleagues [37]: three-dimensional right ventricular wall), 3.3 (Roberts and Scher [101]: three-dimensional left ventricular wall), 4 (Lesh and colleagues [80]: two-dimensional computer model), 14 (Roberge and colleagues [100]: two-dimensional computer model), or about 9, figuring from the observed roughly threefold speed ratio (Roberts and Scher [101]: three-dimensional left ventricular wall), (Spach and Dolber [118]: endocardial lining of right ventricular septum), (Tsuboi and colleagues [127]: two-dimensional right ventricular epicardium), (Kadish and colleagues [65]: two-dimensional ventricular epicardium) and the supposition that speed varies inversely as the square root of resistivity [as in Equation (19.1) for uniformly anisotropic continua]. The average is 5.7. Based on their supposition that ρ transversely is three-fold greater than longitudinally, Frazier and colleagues [37] obtain $\rho = 132$ Ωcm longitudinally where Roberts and Scher [101] obtain 213 Ωcm. This value plays a role in their evaluation of critical field strengths below.

S_v, at least, is necessarily a bit of a fudge factor, introduced to make a model's propagation speed agree with observation despite the many-fold ambiguity of time constants associated with the sodium gate, that dominate propagation speed into well-recovered membrane. Use of the continuum model [cable equation alias reaction-diffusion equation (19.1)] is also not obviously appropriate in a felt-like medium consisting of microscopic cylindrical fibers, each of which is properly so described; certainly their bulk behavior must depend on details of the way in which they are connected, for example, how much extracellular fluid remains between them and whether it provides an isotropically conductive interstitial fluid or is barricaded by adjacent fibers in certain directions. And one does not immediately see how propagation velocity in the bulk medium will be related to propagation velocity along individual fibers.

The end-product of estimation in various laboratories is that S_v and longitudinal ρ for normal myocardium are taken to be respectively 8000/cm and 200 Ωcm in Sharp and Joyner [110] (so $D = 0.60$ cm^2/sec), 5000/cm and 200 Ωcm in Steinhaus and coworkers [123] and Joyner and coworkers [62] (so $D = 1.0$ cm^2/sec), or 2000/cm and 488 Ωcm in Roberge and coworkers [100] who suggest that this 2000/cm may be an underestimate, possibly by almost a factor of 2 (so $D = 0.5$ to 1.0 cm^2/sec). Estimating D instead from observed one-dimensional space constants and time constants without explicit involvement of ρ and S_v, Zykov [169, p. 114] gives the lowest estimate (0.15–0.25 cm^2/sec), van Capelle [132] gives a middling value (1.7 cm^2/sec) for both ventricular and atrial trabeculae, and Kawato and colleagues [68] imply the highest (2.5 cm^2/sec).

By way of comparison, here are estimates of D in some other media:
340 cm^2/sec in squid axon (Showalter and colleagues [112]),
1.2 cm^2/sec in dog Purkinje fibers (as λ^2/τ from data compiled in van Capelle [132]),
2 to 3 cm^2/sec in frog muscle (Hodgkin and Nakajima [52]) or
0.25 cm^2/sec in frog muscle (Chapman and Fry [13], 6 years later without citing the above),
0.2 cm^2/sec in the sinoatrial and atrioventricular nodes (as λ^2/τ from data compiled in van Capelle [132]),
0.005 cm^2/sec in transverse tubules of frog muscle (Adrian and colleagues [1]),
1.5×10^{-5} cm^2/sec in reaction-diffusion schemes for liquid chemical excitable media,
4×10^{-6} cm^2/sec in reaction-diffusion schemes for c-AMP waves in *Dictyostelium discoideum* (Monk and Othmer [91], Tyson and colleagues [128, 131]),

8×10^{-7} cm^2/sec in reaction-diffusion schemes for intracellular morphogenesis (Crick [20]).

The basic equations of electrophysiology, well established and fully exploited in other contexts, have yet to be checked and exploited in cardiology, largely because experimentally determined parameters are only now beginning to acquire reliable values. Moreover, the propagation equation (19.1) must be regarded as an approximation more suitable in certain situations than in others; in particular, it fails badly if ρ is increased (e.g., by heptanol or ischemia) so much (perhaps about 20-fold) that D becomes so small (perhaps about 20-fold) that λ decreases (four- to fivefold) to only a few times cellular dimensions (0.01 cm per cell in the long direction). Even without artificial increase, myocardial cells seem sufficiently disjoint to spoil the accuracy of scaling according to Equation (19.1), as noted above. Landmarks are currently being placed around the domain of validity within which it will be appropriate to use the continuum approximation. Spach and colleagues [121] may have been the first to draw attention to this issue, remarking that the continuum approximation should suffice for analysis of behavior on the scale of more than one activation front thickness ($\frac{1}{2}$ to 1 mm). Roberge and colleagues [100] suggest that so long as the one-dimensional space constant exceeds 10 cell lengths (a marginal proposition in dog myocardium), Equation (19.1) should provide an adequate approximation. Rudy and Quan [102] find that over the normal physiological range of gap junction resistances, the electrical behavior of myocardium is close to that implicit in Equation (19.1). However, Keener [69,70,72,73] explores the question from a new perspective, for which I refer the reader to his chapter (Chapter 17) in this volume. Given this unresolved question, plus many-fold uncorrelated uncertainties about the values of D and of ρ parallel and transverse to fibers, and comparable uncertainties about the major time-constants of sodium and calcium activations in the ionic currents constituting J_m, it seems to me that quantitative inferences from the equations of electrophysiology still cannot be taken very seriously. But their time is coming. My purpose in this chapter is to draw attention to a few of the quantitative ambiguities that seem close to resolution and to some of the underemphasized values of resolving them consistently. The lingering difficulty does not necessarily impugn the reliability of carefully chosen, robustly qualitative inferences, to which I allude throughout this chapter while trying to assemble consistent estimates of critical quantities.

19.4   The Stimulation Threshold

In any excitable medium there is a threshold for uniform (zero-dimensional) excitation, a V_m known to be about 24 mV from equilibrium in fully repolarized and fully recovered (diastolic) myocardial membrane. In one-dimensional settings, a new feature complicates the idea of "threshold":

to instigate a propagating action potential one must also bring a "liminal length" of medium above threshold [105]. In two-dimensional and three-dimensional experiments with myocardium, it is impractical to either manipulate or record V_m intracellularly; instead one records extracellular potentials merely to detect the presence or absence of a propagating front as a function of position and time, having manipulated extracellular fields $U(x, y, z)$ and corresponding current densities $S(x, y, z)$ to cause activation somewhere else. Unfortunately it is terribly difficult to deduce from the pre-1980 literature of cardiac stimulation any quantitative estimate even of bulk extracellular fields and currents; there are too many ambiguities. Rush and colleagues [104], Starmer and Whalen [122], and Lepeshkin and colleagues [79] published the first major efforts, establishing the concept that threshold in a given direction should be measured in terms of potential gradient magnitude or current density (the two being proportional through a uniform and constant conductivity). They estimated the threshold of steady extracellular potential gradient as 1.2 V/cm, equivalent to current density about 4 mA/cm^2 if resistivity ρ is 300 Ωcm. Frazier and colleagues [37] confirmed and considerably refined this estimate (to 0.64 V/cm parallel, or 1.9 V/cm transverse to fiber grain on their supposition that transverse resistivity $\rho_y = \rho_z$ is three times longitudinal resistivity ρ_x), additionally determining its dependence on pulse duration.

The required bulk current has quite heterogeneous consequences at the cellular level, crossing cell membranes and flowing through gap junctions in all directions. If 1.2 V/cm were uniformly divided across the 100 cells thus 200 membranes spanning 1 cm in the longitudinal direction (or across the 500 cells thus 1000 membranes spanning 1 cm in the transverse direction), each membrane would be subjected to a 6 mV (1.2 mV) drop in either direction, fourfold (20-fold) less than the intracellular threshold of 24 mV: evidently the current is focused into "hot-spots." Closer estimates can be made from the resolution of the threshold potential gradient into longitudinal (0.64 V/cm) and transverse (1.9 V/cm) components by Frazier and colleagues [37]. Again allowing 100 cells thus 200 membranes per cm in the longitudinal direction and 500 cells thus 1000 membranes per cm in the transverse direction, each membrane would be subjected to 3.2 mV or 1.9 mV, suggesting 10-fold hot-spots. These may be caused by gap junctions, ignored in the foregoing rough estimates, but I know of no direct experimental evidence for their existence. The remaining inequality (3.2/1.9) indicates that measurements still need refinement, or that the ratio of orthogonal resistivities is actually closer to 5.7 than to 3.0, or that a deficiency of the continuum approximation is showing here.

In two-dimensional or three-dimensional context yet another new feature complicates the idea of threshold: A wavefront will not propagate if it is so sharply curved that excitation fails because of electrotonic loading from the greater volume of unexcited membrane ahead. Thus, in order to nucleate a propagating action potential, a point source must not only transgress the appropriate local threshold, but more, the volume thus fired up must be bounded by a surface of more than some critical radius of curvature[2] (in units rescaled as above if the medium is a uniformly anisotropic continuum). Estimates of that radius have been made before, but they need refinement. Pertsov and colleagues [98] determined the critical radius in a simplified electrophysiological model, but there is no way to convert this model's space units to centimeters while simultaneously scaling D and propagation speed c to realistic values [18].[3] Tyson and Keener [130] determined it from the analytically estimated dependence $c(r) = c(0) - D/r$ of propagation speed on activation front curvature,[4] $1/r$, in the limit of small curvature, by linear extrapolation to speed 0 at $r_{critical} = D/c(0)$. In electrophysiological models propagation speed departs radically from this linear relation, propagation failing completely at some larger curvature [25,169], so this estimate for two-dimensional experiments should be regarded as a lower bound: Tyson and Keener [130] use an estimate of the longitudinal D_x (from Joyner and colleagues [64], probably two- to fourfold too low) together with the transverse propagation speed and the implicit supposition of cylindrical symmetry around the electrode, to obtain $r_{critical} = 0.6$ cm^2/sec $\div$ 30 cm/sec $= 0.02$ cm. For three-dimensional experiments (using a point source on or in the epicardium) the correct formula in the large-radius limit is $c(r) = c(0) - 2D/r$ [8,25,130] so with revised estimates for D and c (in the longitudinal direction) we obtain $2D/c(0) = 2 \times 1$ cm^2/sec $\div$ 75 cm/sec $= 0.027$ cm as critical radius of curvature (radius evaluated after rescaling). As noted above, the D ratio should be the square of the c ratio according to the scaling of Equation (19.1).

---

[2]Brooks et al. [9] (page 56) *incorrectly* conclude that initiation of a propagating action potential requires exceeding threshold current density simultaneously over a sufficient area, so that the threshold current requirement is the same, independent of electrode shape or size (so long as it has less than that critical area). Hoffman and Cranefield [53] correct this by observing that further optimization of electrode shape and size can further lower stimulus thresholds by two orders of magnitude. So far as I am aware the matter has not been pursued, nor has wavefront *curvature* been introduced as the necessary criterion supplementary to local current density. For point sources, Rushton's [105] notion of *liminal* length in one-dimensional context has effect similar to the two-dimensional and three-dimensional curvature criterion, but as far as I understand it, the underlying mechanism is quite different and for other electrode geometries the effect is quite different.

[3]This was also done incorrectly in Winfree [148]: Parameter ε in the FitzHugh-Nagumo model should have been 10-fold smaller (about 0.02 rather than 0.20). This makes no difference for the behaviors extracted in that paper (it just makes the medium activate somewhat sluggishly, as though triggered from a partly depolarized state), but it might were other behaviors featured.

[4]Take care to distinguish front curvature from curvature of the propagation path (curvature of the integral curve defined by the normal to the front), and from curvature of the vortex filament (see below), and to rescale all pictures *before* evaluating any curvature.

If the value is correct, its derivation still may have been inadmissible, because the critical radius thus estimated is less than a reasonable estimate of activation front thickness (1 msec risetime × 0.05 cm/msec = 0.05 cm). The familiar linear dependence of front speed on radius of curvature is derivable only in the limit of such large radius that there is only a small effect on speed. Extrapolating the two-dimensional dependence to speed zero at critical radius $D/c(0)$, one necessarily approaches a radius close to the thickness of the front. This follows from two simple principles. First, front thickness $\approx c(0)$ × risetime of the excitation process in fully recovered tissue. Second, in excitable media asymptotic front speed $c(0)$ = $\sqrt{(D/\text{risetime})}$ × a factor f that depends on the "recovery" variables (e.g., potassium channel activity, as distinguished from the "excitation" variables U and sodium channel activity) just ahead of the front. This can be appreciated from Equation (19.1): Magnifying the rate term J/C_m k-fold to shorten the risetime of the propagator species k-fold would be exactly compensated by writing t as τ/k and x as $X/\sqrt{k}$ (and similarly for y and z); in τ, X units the speed is unaffected, so in the original units it is increased $\sqrt{k}$-fold. The $\sqrt{D}$−dependence was derived above. Derivations of front speed for various specific models reveal $c(0) = f\sqrt{(D/\text{risetime},}$ appropriately defined) [7,11,31,34,41,71,78,86,95,108,109,129,164]; this was appreciated as long ago as 1906 as cited by Showalter and Tyson [112]. Coefficient f seldom strays far outside the range ±2. Its maximum range over all possible cubic rate laws for J/C in Equation (19.1) is ±$\sqrt{6}$, taking as risetime the reciprocal of the average J/C in the voltage range spanned by the action potential (otherwise the range is narrower), and freezing recovery variables at equilibrium. Thus (front thickness)/(linearly extrapolated two-dimensional estimate of critical radius) = f^2. If $|f| > 1$ then as the radius of curvature approaches critical, this radius becomes less than the thickness of the front, and so the isoconcentration contours constituting the front then necessarily have quite different curvatures, and the concept of "front curvature" loses clear meaning.

This same situation arose in seeking a critical curvature for propagation in the Belousov–Zhabotinsky chemically excitable medium. The front thickness of its propagator species ($HBrO_2$, analogous to V_m in electrophysiology, to be distinguished from the visual indicator and "controller" species, ferroin) has not been determined experimentally, but using accepted kinetic parameters, theory indicates 50 to 100 microns [55]. For curvatures no tighter than that, the role of differential geometry seems conceptually clear and experimentally corroborated, but the experimental determination of speed as a function of front curvature [33] relies on measurement of much tighter front curvatures, raising many questions about the meaning of front curvature and even of front speed in this context. So we need at least one other back-of-the-envelope corroborating estimate, and then a comparison with appropriate experiment.

Another way to estimate $D/c(0)$ comes from the definition of D as λ^2/τ, and the approximation of $c(0)$ as λ/τ times a "safety factor" SF typically on the order of 3 to 5 (Benedek and Villars, p. 196 [7]; van Capelle [132] tabulates values between 2 and 3 for myocardium): $D/c(0) = \lambda/SF =$ about 0.02 to 0.03 cm.

These estimates can be checked against measurements of diastolic threshold using a unipolar electrode of tip radius smaller than the estimate, on the supposition that boundary curvature is what limits the expansion of a tiny nucleus of depolarized tissue. In an isotropic medium extracellular current density is uniform on any surrounding hemispherical surface, so the product of threshold current density by total area perpendicular to that flow, $2\pi r^2_{critical}$, must be the measured total current, $I_{threshold}$. This of course supposes sufficient duration of applied current. Frazier and colleagues ([37] Figure 6B) shows that the required current density is essentially constant at 4 to 5 mA/cm^2 independent of direction for DC pulses of durations exceeding 4 msec. For very short durations, the product of duration times current density approaches a constant, 4 mCoulombs/cm^2. In general, for a pulse of duration P msec, threshold current density $= (4/P + 3)$ mA/cm$^2 = 3$ mA/cm^2 $(1 + 1.3$ msec/P msec) (Frazier and colleagues [37] Figure 6B).[5,6] We can use this curve to normalize experimentally determined thresholds for comparison.

Almost anything you want can be found somewhere in the 50-year literature of careful experiments in cardiological electrophysiology (mostly in *American Heart Journal, Circulation, Circulation Research, Journal of the American College of Cardiology,* and *American Journal of Physiology*), but usually divorced from appropriate context of quantitative theory. There are many apparent contradictions: almost anything and its converse can be found by sufficiently patient search, and errors are seldom debunked. So it is essential to use *Science Citations Index* for later papers citing the given

---

[5] Jones and Geddes [61] also report a stimulus strength vs. duration curve for pacing dog myocardium in diastole, in this case with a rather complex and incompletely specified electrode geometry and in terms of mA vs. msec: $I_{threshold}$ $= 0.45$ mA $(1 + 0.17$ msec/P msec). This should be proportional to the corresponding measurement in terms of local current density, but is not: The time constant is about seven times shorter (shorter even than the 1-msec risetime of the impulse), astonishingly so considering that the time constant of diastolic membrane is 4 to 5 msec [132]. Such contradictions abound and typically pass without comment in the cardiological literature. The job of theorists begins with at least pointing to them.

[6] Overlooking the twist of fiber direction along the z axis, and working in unscaled units, we should deal not with a hemisphere but with a prolate hemispheroid, perhaps three times longer in the x direction than in y and z directions. The interested student could refine the estimate made here using hemispheres. But the result might not be an improvement. What is really needed is a more gradually twisted preparation, perhaps from a much larger heart than the dog's, or a way of dealing mathematically with the twist of fiber direction from one xy plane to the next.

experiment, and to read and re-read the "Materials and Methods" sections for crucial circumstances that often were not appreciated as relevant at the time. For example, electrode shape is not always reported quantitatively, sometimes deliberately with explicit argument (which is false) that they are unimportant (see footnote number 2). Measured thresholds correspondingly vary widely, up to about 1 mA. Since there are many possible reasons for ostensibly high "threshold" currents (e.g., rapid prior pacing, poor electrode contact, electrode polarization, electrode tip radius exceeding $r_{critical}$, tissue damaged to radius exceeding $r_{critical}$ by prior current pulses) we seek the smallest in terms of some combination of current and duration:

1. The smallest current seems to be about 0.03 mA given for 2 msec in Michelson and colleagues [87] (Site 1, Table 1). At this duration, threshold current density would be 5 mA/cm^2 thus $r_{critical} = \sqrt{\{0.03}$ mA$/(2\pi\ 5$ mA/cm$^2)\} = 0.030$ cm, slightly higher than the low estimates made above. If current could spread almost spherically, rather than just hemispherically, from the exposed end of the fine wire electrode used (radius 0.005 cm), then "2π" ought to be replaced by more nearly "4π," revising the estimate to $r_{critical} = 0.021$ cm.

2. The next higher threshold current found is 0.04 mA at 1 msec, using an embedded source with apparently hemispherical surround [58] (a disk or hemisphere of radius 0.005 cm). At this short duration about 7 mA/cm^2 is required to stimulate, thus $r_{critical} = \sqrt{(0.04}$ mA$/(2\pi 7$ mA/cm$^2) = 0.030$ cm.

3. The shortest duration seems to be 0.5 msec at 0.10 mA using a similar electrode [134] (with electrode radius < 0.015 cm). At this duration a current density of 11 mA/cm^2 is required, so $r_{critical} = \sqrt{\{0.10}$ mA$/(2\pi 11$ mA/cm$^2)\} = 0.038$ cm.

In almost all other reports, electrode dimensions were not clearly reported [9,36,37,53,96,97]. Presumably the electrodes were bigger than $r_{critical}$ because higher threshold measurements, in the order of 0.2 to 0.5 mA, were obtained. Confidence in the simple estimation here ventured ($r_{critical} = 0.02$ to 0.04 cm) would benefit from a contemporary replication of the essential experiment with explicit attention to the circumstances here supposed to be pertinent. It is important for a practical application (see Section 19.5 on Minimum-Energy Pacing).

This calculation indicates that activation front propagation in ventricular myocardium cannot be expected to show much dependence on curvature in the range practically measurable (radii > 0.1 cm). Nor would the "wandering wavelets" model of atrial fibrillation [90] be incompatible with quantitative electrophysiology in the ventricle (although it might be if my estimate is incorrectly too small). The inconspicuousness of this effect may explain the curious absence of measurements in the published literature.

The new principle, if it is correct, that point sources will radiate pulses only as often as they *both* transgress a threshold condition *and* exceed critical curvature indicates that a single aberrant cell cannot serve as an ectopic focus in otherwise normal myocardium: a compact ball of roughly 100 would be required. I believe this argument only provides a different way to quantify the familiar idea that excitation will not spread from a tiny source exposed to too much "current loading."

19.5   Minimum-Energy Ventricular Pacing

The critical radius may have an important practical application as the spherical electrode size that minimizes energy demand on an implanted cardiac pacemaker. Filling the critical-radius sphere with metal eliminates uselessly hyperstimulated (and perhaps iatrogenically damaged) tissue close to a smaller source and confining the electrode surface inside that sphere spares power dissipation over superfluous area. An additional optimization comes in choosing pulse duration: total energy is proportional to pulse duration P times (current density, S)^2, which, at $S =$ threshold, is $P(4/P + 3)^2 = 16/P + 24 + 9P$, minimized at $P = 4/3$ msec thus $S = 6$ mA/cm^2. I have not seen these principles mentioned or used.

Are existing electrodes *empirically* optimized anyway? The total energy per pulse is P times the volume integral of $S^2\rho$, S being current density and ρ being specific bulk resistivity averaged over all directions, estimated in dog ventricle as 220 Ωcm [37], 263 Ωcm [122], 376 Ωcm [135], 387 Ωcm [101], 400 Ωcm [102], and 410 Ωcm [103]—342 Ωcm on average:

$$W = 342\Omega\text{cm}\,\frac{4}{3}\,\text{msec}\int\left[\frac{(6\frac{\text{mA}}{\text{cm}^2}4\pi r^2_{critical})}{4\pi r^2}\right]^2 4\pi r^2 dr$$

$$= 342\Omega\text{cm}\,\frac{4}{3}\,\text{msec}(6\frac{\text{mA}}{\text{cm}^2}r^2_{critical})^2\left[\frac{-4\pi}{r}\right]^{r\,=\,r_{critical}}_{r\,=\,\infty}$$

$$= 200000(0.03)^3 \text{ nanojoules} = 5.4 \text{ nanojoules.} \tag{19.5}$$

The integral is taken only to $r_{critical}$ on the assumption that we are using a spherical electrode of that (presumably optimal) radius. If our estimate of minimum current were four times too low (i.e., if the much greater abundance of threshold determinations around $\frac{1}{4}$ to $\frac{1}{2}$ mA were more correct), then our estimate of $r_{critical}$ would be two times too small and so our estimate of W would be eight times too small. An upper estimate of W is about 50 nanojoules. Lown and colleagues [82] and Widman and colleagues [143] report that 1000 nanojoules is usual practice; Karpawich and colleagues

[67] report a record minimum, reliably pacing the hearts of children with only 200 nanojoules. The estimate obtained here for the minimum-energy ventricular pacing threshold suggests that further improvements may still be feasible, even allowing for the large safety factor necessary in practice under diverse and changing conditions. A note of caution: at the scale of our estimated $r_{critical}$, myocardium is *not* a continuum, so Equation (19.5) should be regarded as a rough estimate.

19.6   The Ventricular Fibrillation Threshold

Sudden cardiac death in human beings is usually the result of ventricular tachyarrhythmia [39]. Tachyarrhythmia (fast abnormal beating) is commonly attributed to abnormal action potential initiation (connected with the threshold criteria above) or more commonly attributed to abnormal, specifically reentrant, propagation. Reentry can be essentially one-dimensional (e.g., an impulse cycling along an accessory pathway as in Wolffe-Parkinson-White syndrome) or it can be two- or three-dimensional, constituting a vortex-like or vortex-linelike action potential. These have characteristic period around 100 to 120 msec in normal dog myocardium [17,38,111], so short that many parts of the tissue cannot follow at such short intervals. Wavefronts then break up and the tachyarrhythmia becomes fibrillation [92].

Fibrillation can be instigated by diverse procedures, especially in physiologically abnormal tissue. No quantitative understanding of fibrillation, nor even a quantitative theory of its instigation, has yet commanded widespread acknowledgment, even in the case of physiologically normal tissue. It seems a timely challenge, given a century of experimental literature since fibrillation was recognized and named, three quarters of a century since the invention of standardized procedures for electrical measurement of cardiac activity and recognition that fibrillation is an electrical instability, half a century since the vulnerable phase for electrical instigation of fibrillation was studied well enough to define standard procedures for assaying a "ventricular fibrillation threshold" (VFT), and just a few years since two- and three-dimensional mapping of cardiac activation reached adequate resolution to reveal the fine structure of fibrillation and the processes of its onset.

It would not be hopelessly difficult to construct a few alternative working hypotheses about the electrical mechanisms of fibrillation. There is the standard cellular automaton "wandering wavelets" model (originally invoked for *atrial* fibrillation by Moe and coworkers [90] and Han and Moe [47], and since generalized by Smith and Cohen [117] and Kaplan and coworkers [66]), emphasizing heterogeneity in the local properties of badly damaged myocardium. By default this is implicitly applied also to normal myocardium although I am not aware that any decisive tests have been implemented. Herbschleb and coworkers [49,50] proposed an alternative intended to evade the dual embarrassments that both electrocardiograms and epicardial electrode voltage traces exhibit strikingly periodic behavior in the early stages of fibrillation (whereas far less organization is expected under standard assumptions), and that the period (< 120 msec [17,38,11]) is far shorter than both the minimum recovery time of any contemporary electrophysiological equations (of which we might still have none appropriate for ventricular myocardium) and the refractory period of isolated myocardial cells [51]. The motivated student can look up this excellent literature for clear statements of these two distinct interpretations and perhaps elaborate quantitative, even decisive tests to exclude one or the other or both. In my view this is the central problem obstructing the construction of a theory of fibrillation today: we do not really know to what extent the cells collectively viewed through extracellular electrodes or optical detectors are all simultaneously responding in essentially the same way.

Rather than restate what has already been presented so clearly by others elsewhere, I use this chapter to encourage the elaboration of multiple working hypotheses [12] by showing how easy it is to construct yet another quite distinct alternative and to make it sufficiently quantitative to be testable and potentially excludable. This one focuses on the VFT, a practically important concept since electrical *de*fibrillating devices cannot possibly be optimized before obtaining a clear understanding of the conditions under which fibrillation is initiated (or *re*-initiated) by electrical stimuli.

At this point a choice must be made: In addressing fibrillation and its onset, shall we examine the real subject of medical concern, namely, diseased and damaged myocardium, or shall we first address the same topics in physiologically normal myocardium? Preoccupation with diseased and damaged myocardium seems very natural and has been the nearly universal choice for obvious practical reasons. But the absence of any quantitative, predictive model in the half-century history of travelling this road suggests some insurmountable obstacle. Accordingly, it might be wise to try the other path for a while, partly because it remains still underexploited and partly because it allows us to use the understanding accumulated by membrane electrophysiologists who have come close to defining the mechanisms of excitation and recovery in uniform, physiologically normal tissue. The danger, of course, is that it may turn out that the word "fibrillation" covers two unrelated phenomena, the mechanism in heterogeneously abnormal myocardium having no relation to normal electrophysiology. But that cannot be known until both are deciphered. Since neither has been deciphered and the problem looks very hard, having defied generations of cardiologists, my choice is to start where some well-developed understanding might be tapped for new and untried approaches, namely, with uniform healthy tissue.

Suppose as a working hallucination that fibrillation in physiologically normal myocardium is commonly instigated by creating rotors [149, 156,159]. Rotors are vortexlike action potentials, the generic two- and three-dimensional mode of spontaneous activity in excitable media. In an isotropic medium (or rescaled uniformly anisotropic medium) the dynamics and diameter of the rotor are related through diffusion. As a rule of thumb, the size of the rotor is the diffusion distance during one rotation. If the nominal perimeter of the rotor is $c\tau$, c being an estimate of wave speed reduced by convexity at period τ, then its diameter $d = c\tau/\pi \approx 2\sqrt{(2D\tau)}$, so $d \approx 8\pi D/c$ [55], many-fold larger than the critical radius of curvature estimated above as $2D/c(0)$.[7] So in this context we need not be concerned about curvature effects. In normal myocardium rotors have rotation period around 100 msec [17,38,111]. So the expected diameter is roughly $8\pi D/c =$ 25 cm^2/sec $\div$ 40 cm/sec $\cong$ 2/3 cm (Winfree [158] cites experimental data indicating diameters ranging 0.6 to 2.0 cm. The estimate used here is near

---

[7]Before applying this rule of thumb in cardiology it is of interest to check it against reality outside cardiology. The dimensionless ratio $\sqrt{(c\pi d/D)} = \sqrt{(c\lambda/D)} = \sqrt{(c^2\tau/D)} = \sqrt{(\lambda^2/D\tau)} = \sqrt{(8\pi^2)} \approx 9$ by this rule of thumb, or ranges from about 4 to 10 in the mathematical model derived in Keener [71] and Tyson and Keener [130] (depending on an unknown fitting parameter). This twofold range corresponds to a twofold range in the estimate of d. Computed rotors in biochemical models of *Dictyostelium* give ratio 4 [91] or 7 [128], straddling the observed value 5 [126]. Rotors computed using the FitzHugh-Nagumo electrophysiological model with diverse parameters give ratios ranging from 4.3 to 12 or more [115]; with Zykov's model instead, the range is 4.4 to 8.4 [83]. Rotors computed using modified versions of the Beeler-Reuter electrophysiological model give 12 to 20 [19,158]. Rotors computed from the Oregonator model of the Belousov-Zhabotinsky reagent give 8 to 11, depending on chemical parameters [55]. Taking D to be 1.5×10^{-5} cm^2/sec, experiments using the Belousov-Zhabotinsky reagent give 4 to 9 as the temperature rises from 6°C to 45°C [42], using the standard recipe of Winfree [153]. Keener and Tyson [74] tabulate experiments with various recipes from which we find ratios ranging from 6 to 10. The spiral observed by Muller et al. [93,94] gave 7.5. Thus my impression is that there is a practical minimum ≈ 4 in the measured dependence of the ratio on kinetic parameters, and 9 is typical of nearby contemporary experiments, but at parameter extremes much larger values are encountered: for example, λ becomes arbitrarily large as the time-constant ratio of excitation/recovery processes becomes too large for stable propagation (while c and τ remain little affected). The asymptotic values of c and λ are variably underestimated by measurements taken close to the rotor where wavefronts are curved.

It is also of interest in this connection that, contrary to universal assumption, the rotor in a given uniform, continuous medium is *not necessarily* unique. Depending on the parameters of excitability, and in particular, in media exhibiting "supernormality," there may exist alternative stable types of rotor, each with its own core diameter and wavelength λ, radiating waves at its own characteristic period T. These compete for space since the collision boundary moves toward the longer period rotor at speed $(1/T_1 - 1/T_2)/(1/\lambda_1 + 1/\lambda_2)$; but until the boundary reaches the longer period rotor, both behave as though isolated. For example, FitzHugh-Nagumo kinetics with parameters as in Skaggs et al. [114] and Courtemanche et al. [19] supports both the rotor of period 11 time units discussed there, and another of period 17 time units and correspondingly longer wavelength. The dispersion curve is nonmonotonic in such cases [146,147].

the low end of that range and might need correction when more data become available.) What might happen next will be considered in Section 19.10 below, but here we ask only what is necessary to nucleate persistent rotors, thus identifying the VFT with the criteria for creating stable rotors in normal tissue. One potentially quantitative criterion was suggested in Winfree [159] and used to design an experiment, the results of which substantially matched predictions [17,38,111,148,152,158]. Thus encouraged, we elaborate the same notions here as a quantitative interpretation of the ventricular fibrillation threshold. Since this term has occasionally been used in different senses, I restrict its meaning for this chapter as follows:

An electrode pair of whatever geometry is used to drive constant direct electric current through physiologically normal ventricular myocardium for 1 to 10 msec; this stimulus is applied at the most favorable moment for instigating fibrillation (the "vulnerable phase" of duration addressed in Section 19.7 below); the VFT is the minimum current (orders of magnitude exceeding diastolic threshold) such that fibrillation results—usually promptly, although sometimes only after as many as 20 cycles of 5 to 10 Hz reentry.

It should be noted that there are two other commonly used definitions of "VFT," both of which we avoid here in order to focus on the simplest experimental situation:

a) In the "gated train" technique [10,30,43,54,56,92,124,163] the single brief critically timed pulse is replaced by a train of many such pulses, typically applied at 100 Hz for 100 msec or longer, overlapping the vulnerable period. Under these conditions the threshold current is lower (by roughly half) and the interpretation is complicated by overlapping the "protective zone" [124] and by electrical activation of heterogeneous catecholamine release from nearby nerve endings [29,124].

b) Another definition involves application of one to three successive stimuli to ischemic or infarcted tissue [5,23,39,60,76]. In this case the threshold stimulus can be orders of magnitude smaller (comparable to the diastolic threshold) but only if the stimulus is applied at exactly the right site, which must be found by exploration. This procedure is believed to enhance the preexisting heterogeneity of ischemic or infarcted tissue, making wavefronts vulnerable to immediate fragmentation, from which fibrillation follows [92].

As in this multiple-pulse protocol, the VFT as defined for this chapter is also lower in ischemic tissue, but not by two orders of magnitude, rather, more nearly by about half [10,43,145], probably for reasons involving increased intercellular resistance in damaged tissue, which rescales space in Equation (19.1).

The VFT depends on electrode geometry, being much higher for large electrodes applied to the chest surface, for example, than for a point-like electrode touching the exposed epicardium. So any quantitative theory must attend to electrode geometry, or, more to the point, the three-dimensional distribution of current density. To simplify, the following discussion is limited to paired point electrodes, that is, finite bipoles.

Starting from the conclusion of Winfree [148,152,158,159], a rotor will arise in a two-dimensional (or three-dimensional) continuum along the intersection of two critical contours in two-dimensions (or surfaces in three-dimensions). One is an isostimulus contour, a locus of uniform current density $S = S^*$, believed to be several-fold greater than the stimulus threshold current density.[8] The other is an isophase contour along which, at the moment the stimulus is applied, cells are all at a certain stage $T = T^*$ of excitation/recovery. [Therefore Bruce Hill (personal communication) suggests calling it R^* for "recovery" or "repolarization", but I persevere in the old usage just for continuity with two decades of literature.] T^* is an old activation isochron if and only if the medium is uniform. If there be geographical gradients of refractory period duration then the iso-T contours differ from the isoactivation contours (isochrons), for example, as demonstrated experimentally in Salama and colleagues [106] (in guinea pig ventricle, not dog). In Frazier and colleagues [38] it was also experimentally demonstrated that normal dog ventricular myocardium is sufficiently uniform in the part of the right ventricle studied, so that no such effects intrude: repolarization contours were smooth and parallel to prior isochrons. But in general this shortcut cannot be taken: iso-T contours must be measured.

It is to be understood that each of these critical contours provides only a landmark in a smooth field: there must be a substantial gradient of S through level S^* and of T through time T^*. The criterion for creating a rotor is that a certain range of S straddling S^* should crisscross through a certain range of T straddling T^*. Since current density S doubles within the space of several millimeters around typical small electrodes, and T strays from T^* by more than the critical few milliseconds within several millimeters because of the 0.5 mm/msec propagation speed, intersection (T^*, S^*) suffices to locate the rotor within a square centimeter area. Rotors do not possess any more exact location, since the vaguely defined "core" is no more than 1 centimeter across and probably "meanders" over a comparable range (it does in Beeler-Reuter model computations: Skaggs and Winfree, unpublished lab books [116]; Winfree [148,158]; Courtemanche and Winfree [19]). Application of *uniform* current density S^* does not create rotors, because

---

[8] Were stimulation provided by intracellular electrodes, S^* would be close to half of the maximum depolarization during a normal action potential, placing the membrane's state-space image roughly in the middle of the excitation-recovery loop. Since the 24 mV threshold is about 1/4 maximum depolarization, S^* would be twice the stimulus threshold. Using extracellular electrodes, estimation becomes more complicated. Stimulation requires only that enough hot-spots in the membrane exceed 24 mV depolarization, but S^* may require that the membrane on average is moved to midcycle. Thus, S^* is twice the stimulus threshold times some factor characterizing the heterogeneity of transmembrane currents, thus "several times" stimulus threshold. Measured, it turns out to be 20 mA/cm^2 ÷ 4 mA/cm^2 = 5 times.

a gradient transverse to the T *gradient*, straddling S^*, is required.

The mechanisms involved in provoking fibrillation by multiple impulses (under alternative VFT definitions mentioned above) may be similar to those mediating the spontaneous transition from tachycardia to fibrillation in normal tissue driven by short-period pulses from a rotor rather than from a VFT-testing electrode. But those mechanisms bear little relation to the mechanisms postulated here for *creating* the rotor. In this "method of successive adjacent stimuli" [151] the first stimulus (normally originating where Purkinje fibers trigger the endocardium) establishes a timing gradient containing the T^* contour, and the second stimulus must be given while the T^* contour lies within the geographical scope (at least a rotor diameter) of the S^* contour.

Point electrodes are necessarily surrounded by closed-ring iso-S contours, and iso-T contours are smooth curves that cut any such ring twice (an even number of times, in principle, but twice if typically flat). Thus, we are dealing with paired intersections and the vortices are created in mirrorimage counter-rotating pairs. Unless they are at least a core diameter apart (estimated above as $\frac{2}{3}$ cm) they overlap substantially and promptly recombine, having emitted only one or a few of the pulses characteristic of rotors in myocardium. (These pulses are initially around 150 msec apart, shortening pulse after pulse as described below, through the 110 msec of ventricular flutter, toward the 85 msec of ventricular fibrillation.) Thus, instigation of fibrillation must involve creation of intersections at least $\frac{2}{3}$ apart (or less in ischemic tissue with smaller D), and the VFT will correspond to that minimum diameter.

The critical values T^* and S^* were determined experimentally [38]. T^* equals the phase at which tissue has recovered excitability at twice the diastolic threshold for that kind of stimulus, plus or minus a few milliseconds (which is typically 170 msec after local activation). For present purposes we are not concerned about T^*: we only seek the minimum current, $I = I^*$, required to instigate fibrillation when the time is optimized. According to the present hypothesis, this will occur when the ring $S = S^*$ is large enough (the current I having been increased enough) to intersect an iso-T contour at points $\frac{2}{3}$ cm apart in normal tissue (Figure 19.1). The following calculations will need revision when this critical distance is refined; for the sake of proceeding with a definite value I use the $\frac{2}{3}$ cm obtained above, which seems consistent with the limited available data [148]. The critical field strength is 5 to 6 V/cm. This was determined with 3 msec pulses. It is presumably but one point on a strength-duration curve; less |potential gradient| is needed with longer duration, and more with lesser duration, but that curve has not been measured as far as I know.

Another ambiguity: This potential gradient is an average over many fiber directions. Were it resolved in to longitudinal and transverse components differing by the same factor as found for the stimulus threshold potential gradient (about 3) then they could be multiplied by the corresponding

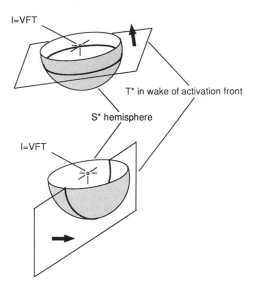

FIGURE 19.1. The surface along which local stimulus magnitude, S, passes through a critical value, $S^* = 20$ mA/cm^2, is shown as a shaded hemisphere around an epicardial point source of current, I. Behind the activation front there is a moving surface shown as an unshaded rectangle along which the phase of recovery, T, is passing through a critical value, T^*. The pivot of a reentrant vortex arises everywhere along their intersection, the heavy circle or semicircle depending on the direction of T^*. If $I \geq VFT$ the circle is wide enough ($\geq \frac{2}{3}$ cm) for antipodal vortex cores to survive without mutual interference.

resistivity components to obtain longitudinal and transverse measures of S^* in terms of current density. Adhering for the sake of consistency to the figures of Frazier and coworkers [37,38] we obtain $S^* = 20$ mA/cm^2 in every direction. This is a guess that remains to be tested; it may need revision. But the arguments to follow can easily be repeated with a modestly revised value for S^*.

Before proceeding it is useful to check whether 5 V/cm and 20 mA/cm^2 are physiologically reasonable field strength and current density. Might such stimuli be damaging membranes or frying cytoplasm? If so, then we must abandon pretense of dealing with physiologically normal tissue with known electrophysiological properties. By way of comparison, diastolic threshold field strength is about 0.6 V/cm [37]. Thus, S^* must be achieving 24 mV × 5/0.6 = 200 mV displacements near the "hot-spots" mentioned above. This is only a little above the normal range of membrane potential excursions. In fact Lepeshkin and coworkers [79] find *little or no* lasting effect on membrane responsiveness following exposure to extracellular fields as intense as 20 V/cm, but *much* at 50 to 100 V/cm (with duration about 2 msec). Yabe and colleagues [162] show that above 60 V/cm pacemaking and conduction are temporarily disabled. Even admitting the possibility of twofold underestimate of the damaging intensity, fields near the S^* contour (where rotors are expected to arise at the onset of fibrillation) apparently do not upset normal physiological behavior. Next checking power dissipation (see Rush and coworkers [104]), we multiply critical field strength 5 V/cm by current density 20 mA/cm^2 to get 0.1 watt/cm$^3 = \frac{1}{40}$ cal/sec/cm^3 = only $\frac{1}{40}$ °C/sec in water. This presents no problem, especially with mere 5-msec stimuli, so it is sensible to proceed, remembering only that tissue 10 times closer to the electrodes is exposed to 100 times the field found near S^*, thus 10,000 times the power dissipation, so its condition must be looked on askance.

Around a unipolar electrode resting on the exposed surface of an isotropic medium, with nonconducting air in the other hemisphere, a hemisphere of diameter $\frac{2}{3}$ cm (radius $r^* = \frac{1}{3}$ cm, safely less than the 1 cm thickness of the left ventricular wall) would be maintained at current density S^* by a current $I = S^* 2\pi r^{*2} = 16$ mA. Myocardium is not isotropic but the peculiar three-dimensional nonuniformity of its anisotropy makes the hemisphere approximation better than it would be were the two-dimensional anisotropy uniform throughout. Using epicardial point electrodes 2.5 to 3.0 cm apart, which function practically as two unipolar electrodes for purposes of isostimulus hemispheres no more than 1 cm in diameter, van Tyn and MacLean [136] report VFT = 12 to 22 mA , overlapping our estimate. Turning it around, to find the radius of the S^* hemisphere we find r^* such that $S^* 2\pi r^{*2} = 12$ to 22 mA: $r^* = \sqrt{(12 \text{ to } 22)/(20\ 2\pi)} = 0.30$ to 0.41 cm, thus diameter 0.6 to 0.8 cm, straddling our estimate of rotor diameter as $\frac{2}{3}$ cm.

The critical current density hemisphere in myocardium thicker than r*

(left ventricle) passes current $I_{3-D} = 20$ mA/cm^2 $2\pi r^{*2}$. A critical-current-density *cylinder* of the same radius in *thinner* myocardium (right ventricle) passes current $I_{2-D} = 20$ mA/cm^2 $2\pi h r^*$, h being the thickness. The ratio of three- to two-dimensional critical currents is thus r^*/h. If we take h in the right ventricular free wall to be about $\frac{1}{4}$ of the 1 cm thickness of the left side, then this ratio exceeds 1, so we might expect to see higher VFTs on the left wherever left and right VFTs are distinguished in an experiment using identical procedures on both sides. This is indeed confirmed below [10,29,54,113,124,163].

This measurement ostensibly depends on the electrodes being far enough apart so that the surrounding fields are spherically symmetric. But most VFT measurements are done with more compact bipolar electrodes. The conceptual motivation was that one wants to measure VFT locally (since it may vary regionally), with geographical precision unobtainable with more widely spaced electrodes. Such determinations of the VFT might be expected to be higher, since in a tighter bipole more of the current is short-circuited between poles and more total current is required to generate the same current density at any given distance: current density falls off as 1/distance^3 rather than as 1/distance^2. Moreover, the high fields between nearby electrodes probably fry the intervening tissue, more effectively short-circuiting the bipole. Quantitatively, over what range of bipole spacings could we expect this effect to become conspicuous? Bipole behavior may be expected to depart conspicuously from unipole behavior for purposes of nucleating rotors at least one rotor diameter apart when the bipole spacing falls below about one rotor radius, namely, about $\frac{1}{3}$ cm. Some published experiments have spacing in this range.

So we need pictures of iso-S contours, $S(x, y) =$ constant, generated by a bipole, as a function of bipole spacing and current strength. On each such picture we must locate the S^* contour and its maximum diameter to be compared to our $\frac{2}{3}$ cm minimum. Such pictures are readily obtained, but we must first address a preliminary question of interpretation. As noted in Section 19.2 (case a), steady-state isopotential contours are independent of uniformly anisotropic resistivity. We proceed even though it is not quite clear that we want the steady-state solution for stimuli of duration shorter than a few milliseconds. But the isopotential contours are not the isocurrent-density contours: those *are* affected by anisotropic resistivity. In two dimensions we know how to handle that by rescaling space, at least in the continuum approximation of *uniform* anisotropy. But in three-dimensional myocardium, anisotropy is *not* uniform: fiber direction rotates systematically with depth below the epicardium. Perhaps on this account, the un-rescaled activation maps of Frazier and coworkers [36] show more isotropic wavefronts than would be expected in any two-dimensional layer, as though the effects of fiber direction were averaged over a wide range of angles. Moreover, published reports of VFT using bipoles (see Table 19.1 below) never recorded bipole orientation relative to local epicardial fiber direction, so the available results are in that sense averaged (or scattered) over all orientations. It would therefore not be rewarding to solve this problem now in full generality. As noted in Section 19.2 (case c), we instead make the expedient approximation that the isopotential contours coincide with (averaged) isocurrent-density contours, as though the medium were, on average, isotropic. When and if VFT measurements are conducted in two dimensions using fully anisotropic thin layers of myocardium (over an endocardial infarct as in Dillon and colleagues [24], or dermatomed as in Tsuboi and colleagues [127], Kadish and colleagues [65] and Zuanetti and colleagues [167], or spared after endocardial freezing as in Allessie and colleagues [3,4]), then the current distributions can usefully be solved more exactly.

The potential around a finite point-pair bipole is a standard exercise in undergraduate electrostatics. In the angle-averaged or isotropic approximation, current flows along potential gradient vectors at density proportional to the total current through the two electrodes. One of these isogradient contours is our S^* contour. The contour lines around a bipole of any spacing, $2a$, look the same as those around any other when the diagrams are rescaled to superimpose the electrodes, and the stimulus labels, L, represent current density per unit total current, multiplied by a^2. Figure 19.2 such a universal diagram for:

$$2\pi S_x[x,y] \;=\; \frac{I(x-a)}{(y^2+(x-a)^2)^{3/2}} - \frac{I(x+a)}{(y^2+(x+a)^2)^{3/2}}$$

$$2\pi S_y[x,y] \;=\; \frac{Iy}{(y^2+(x-a)^2)^{3/2}} - \frac{Iy}{(y^2+(x+a)^2)^{3/2}}$$

$$S[x,y] \;=\; \sqrt{(S_x[x,y]^2 + S_y[x,y]^2)}. \tag{19.6}$$

The contour lines are easily labeled where they cross the bipole axis, at $x =$ any x_o: The current density anywhere is just the linear superposition of current densities associated with all the electrodes, in this case a radial current density from one solitary electrode plus a radial current density toward the other. Each current density, integrated over a hemisphere of area $2\pi r^2$ (or sphere, if the electrode were embedded in the myocardium) totals I, the total current through the electrodes. Along the bipole axis, the two vectors are parallel so their sum is simply $|1/(2\pi(x_o-a)^2)\pm 1/(2\pi(x_o+a)^2)|$, using $+$ between poles and $-$ outside. Thus, $L = a^2 S/I = a^2/\pi(x_o - a)^2 \pm a^2/2\pi(x_o + a)^2$. Along the figure-$\infty$ contour through the center $L = 1/\pi$, i.e., $S = I/\pi a^2$. Critical contour S^* might be inside this (stronger) or outside (weaker) depending on I and a. As a rule of thumb, it would suffice to use the simpler unipole estimate inside ($L > 1/\pi$, i.e., $a^2 > I/\pi S^*$) but we proceed with the general bipole expression to secure an additional insight.

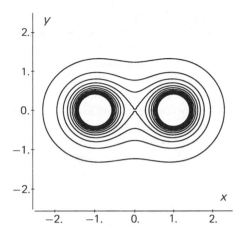

FIGURE 19.2. Contours of uniform current density, normalized by I/a^2, around point sources of opposite electrical charge spaced $2a$ apart in a uniformly resistive three-dimensionally isotropic conductor. This is a Mathematica^{TM} ContourPlot of $S(x, y)$ from Equation (19.6) with $I = 1$ mA and $a = 1$ cm. The outermost contour label is $L = Sa^2/I = \frac{1}{4\pi}$; labels increase by increments $\frac{1}{4\pi}$ through $1/\pi$ on the figure-∞ contour to $\frac{12}{4\pi}$ along the innermost. The smaller, higher current contours are omitted for clarity only; they are essentially concentric circles. For other bipole spacings and total currents, multiply L by I (in mA)/a (in cm)^2 and you have local current density S in mA/cm^2. Column L in Table 19.1 tells which contour shape represents the S^* locus in each published experiment. The sources, at the centers of the two white disks, may be imagined as finite metal electrodes (near which current density would otherwise be unbounded) of the same shape as isopotential contours.

With each choice of bipole half-spacing, a, we seek the least current, I^*, such that intersections with an appropriately placed T^* isochronal contour can be at least $\frac{2}{3}$ cm apart. It would be more accurate to reckon rotor dimensions and spacing anisotropically, or at least to contrast the extreme cases in which contours are elongated around a bipole axis parallel or perpendicular to the grain. Also, the possible intersections depend on the direction from which the activation front crosses the bipole field. Neither is ever attended to in experimental determinations of VFT, which may be a major reason for the roughly twofold spread of all such determinations.

We consider separately the two main cases: front parallel to the bipole axis (probably the usual situation, with electrodes on the epicardium and activation rising from endocardium to epicardium) or perpendicular (possible in parts of the myocardium that are activated laterally, or with electrodes mounted in plunge needles). We do not distinguish orientation of the bipole axis relative to fiber grain on the epicardium; experimental results for $L \approx 1/\pi$ (only, since S^* contours are symmetric far inside and outside the figure-∞ contour) might be less scattered if the bipole had always been aligned to the superficial grain or its orientation at least noted.

With wavefronts perpendicular to the bipole axis, two intersections occur, nucleating rotors of opposite handedness. If $L > 1/\pi$ this can happen when the T^* contour cuts the S^* rings around either electrode. If electrodes are 1 cm apart, propagation at 25 cm/sec would then give us two apparent vulnerable phases 40 to 80 msec apart—as is indeed mysteriously observed in a number of VFT measurements [9].

With wavefronts parallel to the bipole axis, either vortex rings (if excitation comes from the endocardium) or two or four intersections occur (if the ventricle is paced from one edge), nucleating rotors of alternating handedness: There are two if $L > 1/\pi$ or four if $L < 1/\pi$. In the latter case if the internal pair are too close together ($< \frac{2}{3}$ cm) and more remote from the outer intersections, then they can be expected to annihilate one another and so we overlook them, transferring attention to the larger distance between the outermost intersections. In the case of vortex rings, they might fuse, but once again we care only about the long axis of the fused ring.

Systematizing these remarks, the (I, a) plane can be covered by two kinds of contour map. Contours of uniform I/a^2 show the shape of the $S = S^*$ contour: contours $L = a^2 S^*/I$ spanning the range plotted in Figure 19.2 are traced in Figure 19.3. Contours of uniform y such that $S(a, y) = S^*$ reveal the maximum vertical range of the critical contour; the $\frac{1}{3}$ cm contour is superimposed on Figure 19.3. This shows how the expected VFT varies with a, in the case of T^* contours perpendicular to the bipole axis, namely, not much until $a < 0.2$ cm. The VFT stays close to 15 mA for larger bipole spacings.

Figure 19.3 is repeated as Figure 19.4, with modifications for the case of T^* contours parallel to the bipole axis. In this case we are more inter-

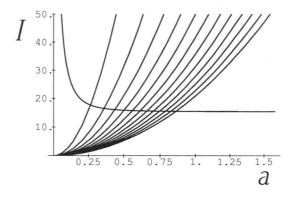

FIGURE 19.3. On the experimental parameter plane (I, a), twelve parabolic contours $\frac{1}{4\pi} < L = S^* a^2 / I < \frac{12}{4\pi}$ (left to right) represent the twelve current density contours of Figure 19.2 (outer to inner). The decreasing curve indicates the current I (mA) at each half-spacing a (cm), such that the S^* contour's vertical range spans $\frac{2}{3}$ cm (i.e., it passes very nearly $\frac{1}{3}$ cm above each electrode and to the outside of each electrode). This serves as a first approximation to the VFT for activation fronts arriving along the bipole axis (so the axis punctures iso-T surfaces).

ested in contours of uniform x such that $S(x, 0) = S^*$. Consider the region $L > 1/\pi$ (shaded range of parabolic contours), in which there are two separate S^* rings, thus four T^*, S^* intersections. They will mutually annihilate in pairs if the radius of the ring [(given very nearly by the $S(a, y) = S^*$ curve in Figure 19.3)] is less than $\frac{1}{3}$ cm, so just as before, the VFT must be above that curve—unless the middle pair are too close together and so annihilate one another. That happens along and above the dotted cup-shaped contour, $I \geq S^* a^2 / S(\frac{1}{3}/a, 0)$. In such circumstances, a smaller ring would suffice—how much smaller is hard to say, and it probably depends delicately on the position and angle of the T^* contour at the moment the stimulus transiently creates the S^* contour. One bound comes from recognizing that this situation lies largely within the unipole domain according to the rule of thumb proffered above that $L = S^* a^2 / I > 1/\pi$. Again denoting by r^* the radius of the hemisphere on which current density $= S^*$, we have $I = (2\pi r^{*2}) S^*$, and we require both $a + r^* > \frac{1}{3}$ cm, and $a - r^* < \frac{1}{3}$ cm. (As a check, these conditions give parabolas that almost exactly match the dotted cup-shaped contour found from the bipole formula.) Further, we require $r^* > a - r^*$ so that the middle two rotors, not the outer two, shall fall together: this implies $L < 2/\pi$. This iso-L parabola provides a "drop-net" under the VFT inside the dotted cup-shaped contour in Figure 19.4.

Putting together our expectations for activation fronts parallel and perpendicular to the bipole axis, we find a shaded zone of ambiguity between the two curves: there is no unique ventricular fibrillation threshold under

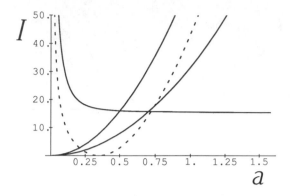

FIGURE 19.4. Figure 19.3 is modified for activation fronts parallel to the bipole axis (see text). To the right of the higher parabola $S^*a^2/I > 1/\pi$ and rotors can arise and persist above whichever is lower of the lower parabola $S^*a^2/I = 2/\pi$ or the descending curve replotted from Figure 19.3. Inside the cup-shaped contour the two interior T^*, S^* intersections are too close to each other and not too close to the outer two intersections, thus leaving only the outer pair and allowing us to space vortices $\frac{2}{3}$ cm apart with less current than required in Figure 19.3. As in Figure 19.3, I and a are indicated in mA and cm, respectively.

the circumstances described (bipolar electrodes randomly oriented to the activation front). It would be no surprise if published measurements with bipolar electrodes turn out to be scattered throughout this region.

A quick search for measurements of the VFT (in the original sense, viz., a single DC pulse during the vulnerable phase) reported in terms of mA total current turns up 16 papers using normal canine myocardium (Table 19.1). There are probably many more, especially papers on other subjects in which the VFT is assayed as part of the control experiments. In Figure 19.5 the tabulated minima are superimposed on an abstract of the prior figures.

The most conspicuous feature of Table 19.1 is the large scatter of VFT assays. This is partly because stimulus durations differ, so threshold currents are not directly comparable. In the different physiological circumstances of diastolic threshold measurement Jones and Geddes [61] and Frazier and colleagues [37] show rough constancy of the current × duration product, at least at shorter durations. But those stimulus-duration curves are not appropriate for normalizing these measurements, since both incorporate the time constant of membrane near equilibrium, whereas the dynamics of inducing fibrillation most likely involves active membrane with quite different time constants. Not only durations but also pulse profiles differ; Jones and Geddes [61], for example, report peak current in a truncated exponential whereas most others use a square pulse.

The observed scatter might have something to do with the electrical

TABLE 19.1. Single-Pulse VFT Measurements (in normal dog ventricular myocardium).

SOURCE	a[cm]	D[ms]	I[mA]	$L = S^* a^2/I$
Right Ventricle				
Han et al. [44]	0.05	10	23–31	<0.01
Han et al. [45]	0.05	10	11–17	<0.01
Tamargo et al. [124]	0.25	5	15±3	0.10–0.07
Chen et al. [17]	0.25	≥10	20–80	0.06–0.02
Euler and Moore [30]	0.35	10	>13	<0.19
Wiggers and Wegria [144]	0.40	20–30	10–29	0.32–0.11
Wegria and Wiggers [139]	0.40	10–40	12	0.27
Shumway et al. [113]	0.50	10	10–17	2.00–0.29
Gang et al. [39]*	0.50	2	9–43	0.56–0.12
Euler [29]	0.50	10	15±2	0.38–0.29
Kowey et al. [75]	0.75	10	16–38	0.70–0.30
Matta et al. [84]*	0.75	2	12–67	0.28
Left Ventricle				
Tamargo et al. [124]	0.25	5	33±3	0.04
Wegria et al. [138]	0.40	20	20–30	0.16–0.11
Wiggers and Wegria [144]	0.40	20–30	10–29	0.32–0.11
Wegria and Wiggers [139]	0.40	10–40	12	0.27
Shumway et al. [113]	0.50	10	14–31	0.35–0.16
Shumway et al. [113]	0.50	10	12–21	0.42–0.24
Jones and Geddes [61]	0.50	3→5	50→25	0.10→0.20
Euler [29]	0.50	10	24±2	0.23±0.020
Damiano et al. [21]	0.75	5	14–35	0.80–0.32
van Tyn and MacLean [136]	>1.25	10	11–24	2.84–1.30

* using endocardial catheter bipoles, not epicardial.

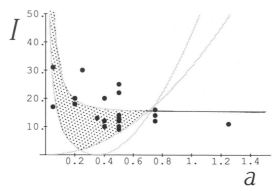

FIGURE 19.5. The data of Table 19.1 are superimposed on an abstract of Figure 19.4, shading the zone of ambiguous threshold level whose boundary is defined by arcs of the several curves in Figure 19.4. Since each threshold determination is tabulated as a range in Table 19.1, the upper end of which is often unavailable or unaccountably high, I plot only the observed minimum currents (in mA). In cases where only mean and standard deviation were published, I plot mean minus 1 standard deviation. The paucity of dots in the lower left shaded zone, where L is large, may be related to shorting of such narrow bipoles through fried tissue, thus spuriously large readings. Half-spacing a is indicated in cm.

anisotropy of myocardium, which is substantially less resistive along the fiber grain than in the two transverse directions. This effect is present in the thick left-ventricular myocardium, although as noted above it is minimized by twist in dog myocardium, averaging the directional preferences of all layers. Reports of VFT measurements never mention the bipole's alignment relative to epicardial fiber grain or the orientation of T isochrons. On our hypothesis that fibrillation will or will not arise, depending on the distance (anisotropically measured) between intersections of a relatively flat T^* contour with an oblong S^* contour of unknown orientation, substantial variation should be expected. Controlling these factors might improve the precision of future measurements. That is one testable implication of this hypothesis.

Scatter is also not surprising when a catheter electrode is used (as in Matta and coworkers [84] and Gang and coworkers [39]) since its contact with the endocardium is impossible to reproduce exactly. Some of the scatter is also due to real regional differences: Much as anticipated above, threshold is at least 50% higher in the thicker left ventricle than in the thinner right [10,29,54,113,124,163]. Shumway and colleagues [113] also note significant differences between anterior and posterior left ventricle. It should also be noted that spuriously low VFT estimates can be obtained in damaged myocardium, and that damage does occur through repeated application of these strong currents from unmoved electrodes during a long series of repetitious tests at high current. It was also not always appreciated that current strength is physiologically excessive between electrodes < 1 cm apart or very close to a very small electrode tip, resulting in coagulation and short-circuiting, thus introducing some fictitiously high readings [136].

The upshot of this analysis seems to be that it is not hard to create a quantitative alternative to standard ideas about the onset of fibrillation, and the particular alternative here considered—the theory of rotors—remains unexcluded after confrontation with VFT measurements. More than that, we seem to have encountered the viable seed of a quantitative theory of VF induction in normal tissue. New experiments are suggested by which orienting the bipole parallel or perpendicular to the isochronal contours (and, ideally, to epicardial fiber grain) much of the data scatter should be removed. Then a VFT curve (rather than distribution) should be revealed that might or might not agree more closely with the quantitative implications of this simple model.

It should be noted that according to this interpretation, at currents close to threshold the nascent counter-rotating vortices are created in precariously close proximity as seen on the epicardial surface. Their lifetimes are thought to be limited to the extent that they overlap. They might be additionally limited for another reason. Neither vortex is strictly two-dimensional; in fact they are either (depending on T^* surface orientation) intramural vortex rings or parts of a single vortex line diving from the clockwise rotor on the epicardial surface intramurally back to the surface at the anticlockwise rotor like paired sunspots in the solar photosphere. This vortex filament is the intersection of the T^* *surface* (the ghost of a past activation front, unless during the interval since activation a spatial gradient of recovery rate has tilted it from that position; see above) with the S^* *surface* (a surface of revolution of Figure 19.2 about the bipole axis). If the myocardium is thicker than the vortex core ($\frac{1}{3}$ cm may suffice, since z-propagation is about twofold slower than the directional average of x and y propagation speeds) then a filament exposing vortices on the epicardium might never encounter the endocardium. Adjacent paired vortices *have* been observed on one surface of the human ventricle while not observed on the other [27]. Thus, the new experiment suggested in Winfree [158] may already have been done (without observing the results) during VFT measurements! According to theory, the curvature of a vortex filament induces a radial motion (inward or outward) at a rate proportional to D and to local curvature. If that motion is inward then the half-ring tends to collapse, so the separation of vortices necessary for persistent arrhythmia may be somewhat larger than estimated above (the rotor core diameter.) Such ring shrinkage is guaranteed mathematically only if all state variables diffuse at equal rates. The actual direction in myocardium awaits determination in the laboratory or using a reliable membrane equation. If the motion is instead outward (as in some simple models of myocardial excitability) or if the myocardium is so thin that the S^* surface touches the endocardium, then the filament will soon break against the endocardial surface, becoming two transmural vortex filaments like those observed by Chen and colleagues [17] and Frazier and colleagues [38].

Experiments suggest that the range of stimulus magnitude (I) between creating vortices (I°) and creating *stable* vortices (I^*) is not large. If the threshold for "repetitive extrasystole" is I° and the ventricular fibrillation threshold is I^*, then $I^* = $ roughly $I^\circ + 11$ mA or roughly $1.5I^\circ$ (Matta and coworkers [84], Fig. 3) or $I^* = 1.2\ I^\circ + 4.3$ mA (Kowey and coworkers [75], Fig. 2) or (with "gated train" stimuli) $I^* = 0.54\ I^\circ + 7.5$ mA [56]. Survival durations of adjacent computed vortex action potentials could be compared using exactly the protocol of Winfree [148].

In terms of the model we are trying to test here, in physiologically normal myocardium the *ratio* (> 1) of the ventricular fibrillation threshold (Section 19.6) to the stimulation (pacing) threshold (Section 19.4) should be independent of pulse duration but should depend markedly on electrode geometry. In the extreme case of a pointlike source on the epicardial surface we have seen that the ratio is about 16 mA/0.05 mA = 320, a figure that could appear several-fold lower if a larger diastolic threshold were obtained because of electrode size exceeding $r_{critical}$. In fact, ratios in the hundreds are typically reported in the earlier literature [89,144]. In the opposite extreme case of uniform field between large flat electrodes the ratio of thresholds should be (threshold current density S^* at time T^*) / (threshold current density during diastole): 20 mA/cm^2 $\div$ 4 mA/cm^2 = 5, a lower limit that would never quite be reached in practice. In the survey of Jones and Geddes [61] we find ratios as large as 38 but never lower than 5. These ratios seemed to depend on duration, but Voorhees and coworkers [137] corrected and updated them, finding ratios as large as 21 but never lower than 7 independent of duration, using external paddle electrodes.

19.7   Vulnerable Period Duration

Vulnerability to a single stimulus varies with time during the normal cycle of excitation and recovery. At each time there is a threshold stimulus magnitude required to elicit fibrillation. The least of these, at the moment of greatest vulnerability, is called the VFT. The shape of the curve, however, seems to have received little quantitative attention. It is usually depicted as a vee or a parabola about 30 msec wide not far above the minimum. The theory ventured here makes a definite prediction that could be checked quantitatively. At any stimulus magnitude (current) I, a point source is surrounded by a hemisphere of radius $r^*(I)$ along which current density = S^*: $S^* 2\pi r^*(I)^2 = I$. If $r^*(I) = \frac{1}{3}$ cm (or whatever is the nominal rotor radius) then the hemisphere can intersect a T^* surface along an arc of diameter adequate to accommodate a rotor pair—but only if the stimulus is applied exactly when that surface is passing through the hemisphere's diameter. Were the hemisphere larger (were I greater), there would be a finite range

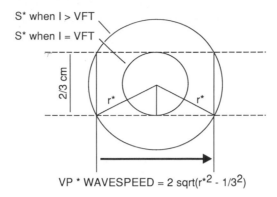

$$VP * WAVESPEED = 2\ sqrt(r^{*2} - 1/3^2)$$

FIGURE 19.6. The circles are diameters of spheres along which $S = S^*$ at electrode current I=VFT (just sufficient to achieve diameter $\frac{2}{3}$ cm) or greater. An isorecovery contour moving left to right can be cut by such circles at sites $\frac{2}{3}$ cm apart (along the dashed lines) during the indicated range of times, VP. Figure 19.7 plots VP against I. Were the left side endocardium and the electrode not intramural but resting on the epicardium, then only the left side of this diagram would be realized (an isocurrent hemisphere rather than sphere) and VP would be half the indicated duration.

of vulnerable phases. Figure 19.6 shows that this range is VP = 1 to 2 times (depending on wave direction) $\sqrt{(r^{*2} - \frac{1}{3}^2)}$ /speed $\cong 1.5\sqrt{(I/2\pi S^* - \frac{1}{3}^2)}$ cm $\div$ (0.025 cm/msec) from endocardium toward epicardium transversely to fibers $\cong \sqrt{(25I(\text{mA}) - 400)}$ msec, as sketched in Figure 19.7. This might account for vulnerable phase durations typically 20 to 40 msec obtained with stimuli of 30 to 80 mA typically used in such experiments. Because of the complex actual activation sequence in a human or canine heart, the vulnerable period does not become much larger even when, in a large heart, the upper limit of vulnerability does become much larger; it can never exceed the ventricular activation time of 70 to 80 msec in a canine or human heart. But the lower parts of the curve could usefully be measured with attention to anisotropy, activation direction, and the distinction between spherical and hemispherical stimuli.

19.8 The Upper Limit of Vulnerability

An *upper limit* of vulnerability (a maximum stimulus beyond which fibrillation is *not* evoked) is implicit in the supposition that rotors underlie the onset of ventricular fibrillation in normal myocardium [156,159]. Once again in context of electrically induced fibrillation, the upper limit can be estimated quantitatively by estimating the current at which the S^* surface

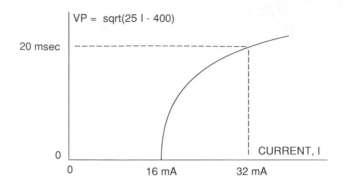

FIGURE 19.7. The vulnerable phase duration, VP, is zero at critical current VFT, and grows parabolically as current further increases. Converting principles to numbers via the known constants of myocardium (see text), VP is found to be $\sqrt{400}$ msec = 20 msec at twice VFT.

would be so far from the electrode site that it is everywhere near or would be beyond the tissue's borders. This distance presumably varies radically from one preparation to the next, and is in any case never reported, nor is an upper limit to vulnerability commonly reported. For back-of-the-envelope estimation, the heart resembles a solid 200-cm^3 conducting ball. Suppose the base of the ventricles to be a disk of radius about 2.5 cm (somewhat less than the radius of a 200 g sphere of water). The outermost viable S^* surface, near the base, would have area somewhat less than $\pi 2.5^2 \text{cm}^2$, about 26-fold greater than the area of the S^* hemisphere of radius $\frac{1}{3}$ cm required to initiate rotors. Thus, the current at the upper limit of vulnerability should be about 26 times the VFT. Two experimental estimates of this ratio were found:

Tables 1 and 2 of Chen and colleagues [15] using healthy dog myocardium reveal energy ratios ranging from 181 to 337. Taking 261 as mean energy ratio, $\sqrt{261} = 16$ is the corresponding ratio of voltages or currents. Lesigne and colleagues [81] found the ventricular fibrillation threshold in healthy dog myocardium at 0.04 joules and the upper limit of vulnerability at 38 joules. The ratio is thus 950 in terms of energy or 31 in terms of voltage or current. These two results (16 and 31) straddle our back-of-the-envelope estimate of 26.

19.9   Minimum-Energy Defibrillation

This upper limit of vulnerability may be significant in relation to power or energy requirements for defibrillation by a single, strong direct-current shock. If fibrillation in normal ventricular myocardium, at early stages when it can still be extinguished without tissue damage, consists of many vortex filaments activating tissue asynchronously, but locally according to normal physiological processes, then the fibrillating tissue is pervaded by T^* surfaces. If the shock strength does not everywhere exceed this upper limit of vulnerability, then the hazard remains that the shock will itself create rotors by creating T^*, S^* intersections as suggested above in a simpler context. If this were the main determinant of success of failure then the current needed for reliable defibrillation would be near this same value. It is [14,15,81]. (However, the interested student should check Witkowski and colleagues [162] for a similar experiment differently interpreted: defibrillation fails only by leaving some portion of the myocardium still fibrillating, but there seems no basis for distinguishing persistent "old" fibrillation from freshly created "new" fibrillation.)

The energy required is of course at least proportional to the mass of the heart. A more significant measure would be the local current *density* required. If fibrillation is initially a tangle of moving, breaking vortex lines analogous to the more orderly vortex lines postulated [156,159] then observed [17,38] to underlie ventricular tachycardia, then fibrillation could be resolved by the same techniques employed in any excitable medium: the problem is that image of the medium in its state space is stably stretched across the normal excitation-recovery loop, and needs only to be lifted clear to enable a return to synchronous uniformity [152,155,158]. This requires a depolarization nearly equal to that required (S^*) to move relatively refractory tissue *into* the center of the loop during vulnerable-phase induction. Thus, uniform application[9] of $S^* = 20$ mA/cm^2 should suffice. Note that this concept does not involve the direction or timing of the voltage gradient, nor the orientation of the rotor with respect to that gradient or the fiber direction at the moment of stimulation.

The medium's image in state space as visualized by extracellular recordings does seem to fill the excitation-recovery loop as postulated and each point exhibits strikingly regular 85 msec electrical periodicity [160], as though the myocardium were indeed pervaded by moving vortices. But nothing like the putative vortices are visible in the first high-resolution movie of a fibrillating epicardium [160]. This gives us pause to notice that the idea that defibrillation would be accomplished by lifting the state-space image actually does not depend on vortices: it depends only on the medium's image during fibrillation being confined to the interior of the

---

[9]The notion of uniformly applied current S presents a conceptual problem because the microscopically local transmembrane current necessarily ranges at least from $-S$ to $+S$, and probably 10-fold wider if current is focused into "hotspots." This problem has not been explicitly resolved, but I imagine that it will be resolved by quantifying the prompt relaxation of micron-size disparities of membrane potential in the first few milliseconds after the stimulus. Something interesting will be discovered in attending to this explicitly.

excitation-recovery loop, whatever may be the geographical organization of that mapping. So we still guess that current density $S^* = 20$ mA/cm^2, applied uniformly, would resolve fibrillation in normal myocardium. The energy delivered in 3 msec at 20 mA/cm^2 throughout a volume of about 1 cm^3 with resistivity $\rho \cong 350$ Ωcm is ≈ 0.003 sec $\times$ (0.02 A/cm^2)^2 $\times$ 1 cm^3 $\times$ 350 Ωcm $\cong 0.0004$ joule/cm^3, or $\frac{1}{12}$ joule for a 200-g ventricle. In theory this would suffice. In practice it does not because present-day defibrillating devices still create extremely nonuniform fields. Rush and colleagues [104] find fourfold variations in local field intensity within the ventricles, using large external electrodes. Witkowski and colleagues [161] find six- to eight-fold variations using a large epicardial patch electrode. Chen and colleagues [16] find 15- to 20-fold variations using (smaller, nearer) epicardial electrodes. With epicardial electrodes, in order to make sure that no cm^3 falls short of the required energy density, at least 15 times the required field (thus 225 times the required energy) must be applied, thus 0.09 joule/cm^3 on average (18 joules for a 200-g ventricle); in the best cases as little as 0.020 joule/cm^3 sufficed [14–16]. With the more uniform field of Witkowski and coworkers [161], 5 joules sufficed. Echt and coworkers [28], Mirowski and coworkers [88], and Tchou and coworkers [125] use 10 to 25 joules for humans.

Quantitative ambitions are frustrated in this area largely because, despite many measurements, the widely used concept of a "defibrillation threshold" of total applied energy does not seem to exist [22,85,99]. The problem seems to be that the effective agent of defibrillation is current density, but contemporary electrode arrangements create such diversely heterogeneous fields in the myocardium [14–16,104,161] that every application of such a stimulus encounters the spatial heterogeneity of fibrillation in a different way, with different results. I imagine that a sharp threshold will be determined just as soon as uniform fields can be produced, and it will be orders of magnitude lower than present-day ill-defined energy "thresholds."

Attempts have been made to establish a local threshold for electrical defibrillation in terms of a minimum *local field strength* or *current density* that must be everywhere exceeded. Wharton and colleagues [141,142] and Zhou and colleagues [165] show that it needs about 6 V/cm throughout most of the ventricles. Witkowski and colleagues [161] defibrillated successfully only when field strength everywhere exceeded 5 V/cm. With improved technique this has recently been reduced to 4 V/cm [166]. These estimates straddle S^*, measured in terms of field strength as 5 V/cm [37].

These numerical coincidences do not validate the vortex vision of fibrillation: they merely fail to exclude it on empirical quantitative grounds. But if it were correct, might it suggest any further improvement in defibrillation technique beyond the requirement [14–16,104,161] that the field intensity must be made uniform at least to minimize energy and possibly also to avoid creation of new rotors? What might be the ultimate lower limit of required energy density? The answer may be "almost zero"! In principle, elimination of vortex filaments in a three-dimensional excitable medium requires only that conduction be blocked for the duration of one vortex period (in myocardium, 85–120 msec) along a two-dimensional surface whose topological boundary is the vortex filament. In the simplest case—a planar vortex ring—this surface is a disk spanning the ring: the reentrant excitation is extinguished when it fails to pass through the ring. In a vortex tangle the corresponding (Seifert) surface is topologically intricate, but it need be no thicker than about 1 mm and so involves only a tiny fraction of the entire myocardial mass. If a modest energy density, focused in such a surface, could momentarily block propagation, then it would eliminate reentry with proportionally less energy than required to infuse the entire volume.

The striking contrast between the electric energy required in a dog heart to initiate fibrillation and that required to defibrillate—on the order of 10^4-fold using contemporary nonuniform fields—invites explanation. It seems that to initiate fibrillation only a modest volume need be affected, from which short-period waves then unstably invade an arbitrarily large mass of adjacent myocardium, creating additional vortices there when they break up; but using contemporary procedures, the entire volume must be massively countershocked to restore order [160]. In smaller hearts the ratio is smaller or even zero in hearts so small that they defibrillate spontaneously. The ratio is presumably larger in larger hearts, in direct proportion to mass. But reduced to terms of local current density, it may be 1:1 if the notions advanced in this chapter prove viable.

19.10   Transition from Reentrant Tachycardia to Turbulence

The fate of transmural vortex lines may depend on a three-dimensional feature of myocardium not mentioned above, namely, the electrophysiological difference between endocardium to epicardium. If the Purkinje fiber web penetrating the subendocardium becomes involved in the vortex, it probably affects the local vortex period. With different rotation rates at its extremities, the intramural filament accumulates twist, which is dissipated only at free ends of the filament. In a thick medium this may have either or both of two effects [48]. The filament may spring into a helical shape, expanding intramurally and presumably distorting until it fragments against surfaces elsewhere. Or it may become so twisted that wavefronts are forced too close together for continued stable propagation.

This brings us to the unresolved question of "the excitable gap," a digression from which we will shortly return to examine the way in which twist may mediate the transition from stable rotation (ventricular tachycardia or flutter) to turbulence (ventricular fibrillation) in fully three-dimensional preparations (only). Rotors in excitable media usually have a rotation period *longer* than the ostensible refractory period of the medium. The interval between full recovery of excitability and arrival of the next excitation from a rotor is called the "excitable gap." Without it, the entire wavetrain would be unstable toward even the slightest local temporary prolongation of refractoriness. There is a generic reason for its existence, and there is a supplementary special reason in ventricular myocardium.

Generically, the refractory period is not measured correctly when the stimulus electrode is nearly a point source. In that case, as noted in Section 19.4, curvature effects confound the measurement. Since propagation speed is diminished both by prematurity (short interval since prior activation) and by curvature, critical prematurity (the refractory period) assayed by a sharply curved front will appear longer than it otherwise would. The refractory period should therefore be measured with a large flat electrode.

A beginning has been made toward measurement of propagation speed in dog and rabbit myocardium as a function of period or wavelength but a definitive curve has not yet emerged. It is important to use normal myocardium that is *uniformly* anisotropic (a condition that does not obtain over adequately large regions in small hearts), to distinguish longitudinal from transverse propagation, to fire action potentials periodically for a long enough time to achieve steady properties, and to use an electrode shape capable of driving the medium to its shortest stable period—which may be shorter in zero-dimensional (uniform) context than in one-dimensional propagation, which in turn may be shorter than in two- or three-dimensions because higher dimensional wavefronts have more possibilities for instability when confronted with relatively refractory tissue. This experiment has not yet been properly executed even in the much easier case of chemically excitable media. In myocardium it is not clear even in principle that the experimental difficulties *can* be overcome.

Resorting to an example from physical chemistry, the dependence of wave speed on prematurity has been twice measured and twice computed for the Belousov-Zhabotinsky excitable medium (continuous, isotropic). Computations from the "Oregonator" model [26,55,74] show the minimum period for repeated excitation to be substantially less than the rotor's period. But in both laboratory measurements they were indistinguishable [32,55]. Those measurements were both done with pointlike sources of radius only a few times $D/c = 2 \times 10^{-5}$ cm^2/sec $\div$ 0.01 cm/sec $= 0.002$ cm. In contrast, Krinsky and Agladze [77] overwhelmed a rotor in this medium with volleys of 20% shorter period waves emitted from a flat electrode, proving the existence of at least that much excitable gap. Turning to cellular biochemistry, in the slime mold *Dictyostelium discoideum*, two distinct biochemical models of the c-AMP rotor show about 20% excitable gap, as does the real thing [91,128,131].

Rotors in ventricular myocardium also have a period about 20% (30 msec) longer than the ostensible refractory period [3,4]. Allessie and colleagues interpret this excitable gap as an effect of anisotropy, arguing that its electrophysiological mechanism depends on propagation being markedly faster along the fiber grain. In contrast, introduction of uniform anisotropy into the electrophysiologist's continuum Equation (19.1) has no effect on the temporal aspect of its solutions. The logical inference then must be either that the gap is a consequence of features in which real myocardium departs from uniformly anisotropic continuity, or that it is in fact present even in isotropic situations. One version of the Beeler-Reuter [6] model of myocardial excitability suggests the latter, although its excitable gap occupies only a small and fluctuating fraction of the period [19]. It may come and go, here and there, during the spontaneous meander typical of rotors even in uniform, isotropic, continuous excitable media [55,83].

The supplementary special reason for an excitable gap in ventricular myocardium is that, whatever the period of excitation may be, the refractory period shortens to stay below it, the gap vanishing (if it ever does) only at intervals less than 100 msec [see [140], Fig. 5; [58], Fig. 1; [35], Fig. 3; [107], Fig. 2.2 (rabbit, not dog)]. During fibrillation, extracellular electrodes monitoring a region of diameter in the order of one one-dimensional space constant (less than 1 mm) show startlingly regular repetitive activation at intervals of about 85 msec [160]. This brevity might be arrived at pulse by pulse, largely during the first several repeats of reentrant activation [43,46,58]. Using rabbit ventricular myocardium Schalij [107] obtained 160 to 170 msec refractory periods while pacing at 350-msec intervals. The next refractory period shortened to 130 to 140 msec after a shock delivered 170 msec after a pacing pulse. Repeating this treatment, the refractory period keeps shortening, down to 90 msec. This resembles experience during creation of rotors in dog ventricular myocardium [38]: the refractory period is 170 msec, but after the S^* shock at time T^* close to the end of refractoriness, the first cycle of reentrant tachycardia has period 130 to 140, then 120 msec, shortening further to about 100 msec.

The refractory period generally stays less than the excitation interval, but there is a limit, and although two-dimensional rotors do not normally approach that limit, they may in a twisted vortex line, for two reasons. First, twist shortens the rotor's period in every case examined thus far (all computationally: see Winfree [154]). Second, even without that effect, the geometry of twist forces wavefronts closer together as shown in Figure 19.8. As twist accumulates due to a persistent difference of rotation period between ends of the vortex line, the wavefront spacing approaches the minimum compatible with stable propagation: the excitable gap is squeezed out. Twist cannot accumulate in this way along a very short vortex filament because it spans a proportionally slight range of membrane properties and because twist comes unravelled at the ends as quickly as it accumulates intramurally; but it must accumulate intramurally along a filament spanning thick myocardium with different refractory periods near the two bounding surfaces.

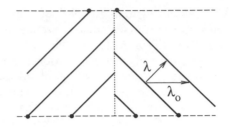

FIGURE 19.8. A vertical vortex filament (imagine epicardium and endocardium as parallel horizontals as dotted) is surrounded by a spiral wave on every horizontal plane. Here we examine a vertical plane: The spiral punctures it at points indicated by dots on the bounding horizontal planes. The horizontal spacing of the dots, λ_o, is the vortex period × wave speed, both of which decrease with increasing twist. These dots would lie in vertical columns were all the spirals identical, but the columns tilt if the spirals rotate with changing altitude, i.e., if the vortex filament has twist. The greater the twist the more these diagonals (wavefronts) approach the horizontal and the shorter is the perpendicular distance, λ, between them compared with the horizontal distance, λ_o. With enough twist, λ contracts to less than the minimum compatible with stable propagation.

How much twist is needed to wring out the excitable gap? A simple estimate comes from contemplation of a straight vortex filament showing a standard two-dimensional rotor in every perpendicular cross-section. If the rotor turns from slice to slice, turning fully 360° in distance H, then the three-dimensional wavelength λ is shorter than the "two-dimensional" (horizontal) wavelength λ_o ; specifically $1/\lambda^2 = 1/\lambda_o^2 + 1/H^2$ (see Figure 19.8 and Henze and coworkers [48]). Suppose $\lambda = (1 - \varepsilon)\lambda_o$ represents the critical minimum wavelength and $\varepsilon \ll 1$ (the excitable gap is already small, as in Beeler-Reuter "myocardium"). Then $(1/(1 - \varepsilon)^2 - 1)/\lambda_o^2 = 1/H^2$, i.e., $H = \lambda_o/\sqrt{(2\varepsilon)}$. How much twist is that? $\sqrt{(2\varepsilon)}$ full turn per wavelength, and the wavelength is about 100 msec × 0.03 cm/msec = 3 cm. If ε were 10% this would be 54°/cm. Considering that myocardial fiber orientation itself turns about three times that fast across the 1-cm thickness of the left ventricular wall, this does not seem implausible. No experiment has yet sought this effect. Frazier and colleagues [38] mapped three slices across a transmural vortex line spanning about 1 cm and showed no twist at all, but the vortex was only a few hundred milliseconds old: it should be remapped after 1 to 2 sec (about 15 turns), near the time when the transition from flutter to fibrillation may occur.

In this situation wavefronts should be unstable toward patchy failure of propagation—*conduction block*, as it is called in the cardiological literature. Every regional inhomogeneity that affects refractory period or conduction speed, no matter how slightly, then becomes a snag on which the precariously propagating front may stumble, creating a hole in the two-dimensional surface of the activation front. Since internal edges of wave-fronts become vortex lines themselves, vortex turbulence is the immediate result.

This process may provide one among many ways to account for the transition to fibrillation from flutter in normal three-dimensional ventricular myocardium. If the myocardium were particularly inhomogeneous as a result of past ischemia or other disease, it would only be more vulnerable to the same process. It is interesting to note that the transition from flutter to fibrillation is suppressed in experimental preparations from which the endocardium has been removed to such an extent that the surviving epicardium is essentially two-dimensional (Dillon and colleagues [24] using an infarct, Schalij [107] and Allessie and colleagues [3,4] using liquid nitrogen). Removing the subendocardial lining of Purkinje fibers without seriously compromising the three-dimensionality of propagation merely doubled the VFT (Damiano and colleagues [21] using iodine); in ischemic tissue similar treatment completely eliminated the otherwise spontaneous transition from tachycardia to fibrillation (Janse and colleagues [57,59] using phenol).

VF has long been recognized as a spatially heterogeneous process, so it has been easy to suppose that its indispensable cause may be some kind of heterogeneity. One need not look far for candidates, particularly in diseased myocardium. So it comes as a surprise to discover that in the presence of such heterogeneity as is normally supposed to be responsible for transition from tachycardia to fibrillation, no transition occurs unless the tissue is functionally three-dimensional; and that in three-dimensions, fibrillation results from the same stimulus even in the absence of more than an infinitesimal degree of microscopic heterogeneity. It seems necessary to ask now whether such heterogeneity per se is the cause, or only an incidental catalyst, or even an irrelevancy, in normal myocardium. The problem goes away if the traditional sense of "heterogeneity" (microscopic discontinuity of cellular refractory period duration) is expanded to include larger scale gradients (possibly smooth) of rotor period in the third dimension. It goes away even more convincingly if our purview, artificially restricted to physiologically normal tissue, is reexpanded to include ischemic and infarcted tissue: then both kinds of "heterogeneity" are plentifully present and doubtless both contribute to disruption of short-period waves. But their distinctive roles should be sorted out.

This section cannot end without mention that from the viewpoint adopted here it does seem peculiar that rotors in physiologically normal *atrial* myocardium (essentially two-dimensional) have no excitable gap, and make the transition to fibrillation without delay [2]. It is possible that atrial myocardium is much less uniform than a comparable sheet of ventricular myocardium; certainly its irregularities are conspicuous [119,120]. It may also be noteworthy that in some excitable media rotors spontaneously fragment and multiply, even if the medium is perfectly two-dimensional and perfectly uniform: for example, Beeler-Reuter membrane ([19], Figure 26-5 in Winfree [157]) and the abstract cellular excitable medium of Gerhardt and colleagues (Figure 5 in [40]). Thus there exist at least a few plausible mechanisms for this transition, which need to be distinguished experimentally.

19.11   Conclusion

This preliminary effort may be the first to provide quantitative derivations of stimulus threshold, fibrillation threshold, vulnerable period duration, upper limit of vulnerability, and mechanism of onset of fibrillation from elementary principles. All numerical values are rough but I make no apologies for that. My purpose is to show that it is possible to make use of contemporary electrophysiology to contrive quantitatively testable hypotheses about the mechanisms of fibrillation in physiologically normal heart muscle. Whether they are individually correct or not is of less concern than that they be informed by the excellent literature of propagation in myocardium and other excitable media, be generally consistent with accepted experiments, be potentially quantifiable, be sufficiently abundant that the method of multiple working hypotheses can be played out to advantage in the laboratory, and be specific in pointing out quantities that ought to be controlled with care or measured in feasible experiments. This is a game that any graduate student can play during the next few years.

Acknowledgements: This project was funded by the National Science Foundation. A critical query from Rahul Mehra triggered the line of thought in Section 19.4, which led to the rest. I thank J.J. Tyson for discussions of D and the excitable gap, and F.X Witkowski, L. Glass, and especially B. Hill for critical comments on the "finished" manuscript, which resulted in many changes.

REFERENCES

[1] R.H. Adrian, W.K. Chandler, and A.L. Hodgkin. The kinetics of mechanical activation in frog muscle. *J. Physiol.*, 204:207–230, 1969.

[2] M.A. Allessie, F.I.M. Bonke, and F.J.G. Schopman. Circus movement in rabbit atrial muscle as a mechanism of tachycardia. III. The "leading circle" concept: A new model of circus movement in cardiac tissue without the involvement of an anatomical obstacle. *Circ. Res.*, 41:9–18, 1977.

[3] M.A. Allessie, M.J. Schalij, C.J. Kirchoff, L. Boersma, M. Huybers, and J. Hollen. Cell to cell signalling: From experiments to theoretical

models. In A. Goldbeter, editor, *The Role of Anisotropic Impulse Propagation in Ventricular Tachycardia*, pages 565–575. Academic Press, New York, 1989.

[4] M.A. Allessie, M.J. Schalij, C.J.H.J. Kirchoff, L. Boersma, M. Huyberts, and J. Hollen. Electrophysiology of spiral waves in two dimensions: The role of anisotropy. In J. Jalife, editor, *Mathematical Approaches to Cardiac Arrhythmias.* Ann. NY Acad. Sci., 591:247–256 (1990).

[5] P.J. Axelrod, R.L. Verrier, and B.L. Lown. Vulnerability to ventricular fibrillation during acute coronary arterial occlusion and release. *Am. J. Cardiol.*, 36:776–781, 1975.

[6] G.W. Beeler and H. Reuter. Reconstruction of the action potential of ventricular myocardial fibers. *J. Physiol.*, 268:177–210, 1977.

[7] G.B. Benedek and F.M.H. Villars. *Physics with Illustrations from Biology and Medicine.* Addison-Wesley, Reading, MA, 1979.

[8] P.K. Brazhnik, V.A. Davydov, V.S. Zykov, and A.S. Mikhailov. Vortex rings in excitable media. *Zhurnal. Eksper. Teoret. Fiziki*, 93:1725–1736, 1987.

[9] C.M. Brooks, B.B. Hoffman, E.E. Suckling, and O. Orias. *Excitability of the Heart.* Grune & Stratton, New York, 1955.

[10] M.J. Burgess, D. Williams, and P. Ershler. Influence of test site on ventricular fibrillation threshold. *Am. Heart J.*, 94:55–61, 1977.

[11] R.G. Casten, H. Cohen, and P.A. Lagerstrom. Perturbation analysis of and approximation to the Hodgkin–Huxley theory. *Q. Appl. Math.*, 32(4):365–402, 1975.

[12] T.C. Chamberlin. The method of multiple working hypotheses. *Science*, 15:92–97, 1890. Reprinted in *Science*, 148:754-759, 1965.

[13] R.A. Chapman and C.H. Fry. An analysis of the cable properties of frog ventricular myocardium. *J. Physiol.*, 283:263–282, 1978.

[14] P-S Chen, N. Shibata, E.G. Dixon, R.O. Martin, and R.E. Ideker. Comparison of the defibrillation threshold and the upper limit of ventricular vulnerability. *Circulation*, 73:1022–1028, 1986.

[15] P-S Chen, N. Shibata, P.D. Wolf, et al. Epicardial activation during successful and unsuccessful ventricular defibrillation in open chest dogs. *Cardiovasc. Rev. Rep.*, 7:625–648, 1986.

[16] P-S Chen, P.D. Wolf, F.J. Claydon, et al. The potential gradient field created by epicardial defibrillation electrodes in dogs. *Circulation*, 74(3):626–636, 1986.

[17] P-S Chen, P.D. Wolf, E.G. Dixon, et al. Mechanism of ventricular vulnerability to single premature stimuli in open chest dogs. *Circ. Res.*, 62:1191–1209, 1988.

[18] M. Courtemanche, W.E. Skaggs, and A.T. Winfree. Stable three-dimensional action potential circulation in the FitzHugh-Nagumo model. *Physica D.*, 41:173–182, 1990.

[19] M. Courtemanche and A.T. Winfree. A two-dimensional model of electrical waves in the heart. *Pixel*, 1(3):24–31, 1990.

[20] F. Crick. Diffusion in embryogenesis. *Nature*, 225:420–422, 1970.

[21] R.J. Damiano, P.K. Smith, H. Tripp, et al. The effect of chemical ablation of the endocardium on ventricular fibrillation threshold. *Circulation*, 74:645–652, 1986.

[22] J.M. Davy, E.S. Fain, P. Dorian, and R.A. Winkle. The relationship between successful defibrillation and delivered energy in open-chest dogs: Reappraisal of the "defibrillation threshold" concept. *Am. Heart J.*, 113:77–84, 1987.

[23] J.M.Y. de Bakker, M.J. Janse, F.J.L. van Capelle, and D. Durrer. Endocardial mapping by simultaneous recording of endocardial electrograms during cardiac surgery for ventricular aneurysm. *J. Am. Coll. Cardiol.*, 2:947–953, 1983.

[24] S. Dillon, M.A. Allessie, P.C. Ursell, and A.L. Wit. Influence of anisotropic tissue structure on reentrant circuits in the epicardial border zone of subacute canine infarcts. *Circ. Res.*, 63:182–206, 1988.

[25] D.-F. Ding. A plausible mechanism for the motion of untwisted scroll rings in excitable media. *Physica D.*, 32:471–487, 1988.

[26] J.D. Dockery, J.P. Keener, and J.J. Tyson. Dispersion of traveling waves in the Belousov-Zhabotinsky reaction. *Physica D.*, 30:177–191, 1988.

[27] E. Downar, L. Harris, L.L. Mickelborough, N. Shaikh, and I.D. Parson. Endocardial mapping of ventricular tachycardia in the intact human ventricle: Evidence for reentrant mechanisms. *J. Am. Coll. Cardiol.*, 11:783–791, 1988.

[28] D.S. Echt, K. Armstrong, P. Schmidt, P.E. Oyer, E.B. Stinson, and R.A. Winkle. Clinical experience, complications, and survival in 70 patients with the automatic implantable cadioverter/defibrillator. *Circulation*, 71:289–296, 1985.

[29] D.E. Euler. Norepinephrine release by ventricular stimulation: Effect on fibrillation thresholds. *Am. J. Physiol.*, 238:H406–H413, 1980.

[30] D.E. Euler and E.N. Moore. Continuous fractionated electrical activity after stimulation of the ventricles during the vulnerable period: Evidence for local reentry. *Am. J. Cardiol.*, 46:783–791, 1980.

[31] R. FitzHugh. Mathematical models of excitation and propagation in nerve. In H.P. Schwann, editor, *Biological Engineering*, pages 1–85. McGraw-Hill, New York, 1969.

[32] P. Foerster, S.C. Muller, and B. Hess. Critical size and curvature of wave formation in an excitable chemical medium. *Proc. Natl. Acad. Sci. USA*, 86:6831–6834, 1989.

[33] P. Foerster, S.C. Muller, and B. Hess. Curvature and propagation velocity of chemical waves. *Science*, 241:685–687, 1988.

[34] D.A. Frank-Kamenetsky. *Diffusion and Heat Exchange in Chemical Kinetics*. Nauka, Moscow, 1955.

[35] M.R. Franz and A. Costard. Frequency-dependent effects of quinidine on the relationship between action potential duration and refractoriness in the canine heart in situ. *Circulation*, 77:1177–1184, 1988.

[36] D.W. Frazier, W. Krassowska, P-S Chen, et al. Transmural activations and stimulus potentials in three-dimensional anisotropic canine myocardium. *Circ. Res.*, 63:135–146, 1988.

[37] D.W. Frazier, W. Krassowska, P-S Chen, et al. Extracellular field required for excitation in three-dimensional anisotropic canine myocardium. *Circ. Res.*, 63:147–164, 1988.

[38] D.W. Frazier, P.D. Wolf, J.M. Wharton, A.S.L. Tang, W.M. Smith, and R.E. Ideker. Stimulus-induced critical point: Mechanism for the electrical initiation of reentry in normal canine myocardium. *J. Clin. Invest.*, 83:1039–1052, 1989.

[39] E.S. Gang, J.T. Bigger, and F.D. Livelli. A model of chronic ischemic arrhythmias: The relation between electrically inducible ventricular tachycardia, VFT, and myocardial infarct size. *Am. J. Cardiol.*, 50:469–477, 1982.

[40] M. Gerhardt, H. Schuster, and J.J. Tyson. A cellular automaton model of excitable media including the effects of curvature and dispersion. *Science*, 247:1563–1566, 1990.

[41] P. Gray, K. Showalter, and S.K. Scott. Propagating reaction-diffusion fronts with cubic autocatalysis: The effects of reversibility. *J. Chim. Phys.*, 84:1329–1333, 1987.

[42] R.L. DeHaan and A.T. Winfree. Unpublished laboratory books, 1973.

[43] J. Han. Ventricular vulnerability during acute coronary occlusion. *Am. J. Cardiol.*, 24:857–864, 1969.

[44] J. Han, G.P. deJalon, and G.K. Moe. Fibrillation threshold of premature ventricular responses. *Circ. Res.*, 18:18–25, 1965.

[45] J. Han, D. Millet, B. Chizzonitti, and G.K. Moe. Temporal dispersion of recovery of excitability in atrium and ventricle as a function of heart rate. *Am. Heart J.*, 71:481–487, 1966.

[46] J. Han and G.K. Moe. Cumulative effects of cycle length on refractory periods of cardiac tissues. *Am. J. Physiol.*, 217:106–109, 1969.

[47] J. Han and G.K. Moe. Nonuniform recovery of excitability in ventricular muscle. *Circ. Res.*, 14:44–60, 1964.

[48] C. Henze, E. Lugosi, and A.T. Winfree. Stable helical organizing centers in excitable media. *Can. J. Phys.*, 68:683–710, 1990.

[49] J. Herbschleb, I. van derTweel, and F. Meijler. The apparent repetition frequency of ventricular fibrillation. *Comput. Cardiol.*, 249–252, 1982.

[50] J. Herbschleb, I. van derTweel, and F. Meijler. The illusion of travelling wavefronts during ventricular fibrillation. *Circulation*, 68 (Supp. III):343, 1983.

[51] J. Heschler and R. Speicher. Regular and chaotic behavior of cardiac cells stimulated at frequencies between 2 and 20 hz. *Eur. Biophys.*, 17:273–280, 1989.

[52] A.L. Hodgkin and S. Nakajima. The effect of diameter on the electrical constants of frog skeletal muscle fibers. *J. Physiol.*, 221:105–120, 1972.

[53] B.F. Hoffman and P.F. Cranefield. *Electrophysiology of the Heart*. McGraw-Hill, New York, 1960.

[54] L.N. Horowitz, J.F. Spear, and E.N. Moore. Relation of endocardial and epicardial ventricular fibrillation thresholds of the right and left ventricles. *Am. J. Cardiol.*, 48:698–701, 1981.

[55] W. Jahnke, W.E. Skaggs, and A.T. Winfree. Chemical vortex dynamics in the Belousov-Zhabotinsky reaction and in the 2-variable Oregonator model. *J. Phys. Chem.*, 93:740–749, 1989.

[56] P. Jaillon, I. Schnittger, J.C. Griffin, and R.A. Winkle. The relationship between the repetitive extrasystole threshold and the ventricular fibrillation threshold in the dog. *Circ. Res.*, 46:599–605, 1980.

[57] M.J. Janse, A.G. Kleber, A. Capucci, R. Coronel, and F. Wilms-Schopman. Electrophysiological basis for arrhythmias caused by acute ischemia. *J. Mol. Cell. Cardiol.*, 18:339–355, 1986.

[58] M.J. Janse, A.B.M. van der Steen, R.T. van Dam, and D. Durrer. Refractory period of the dog's ventricular myocardium following sudden changes in frequency. *Circ. Res.*, 24:251–262, 1969.

[59] M.J. Janse, F. Wilms-Schopman, R.J. Wilensky, and J. Tranum-Jensen. Role of the subendocardium in arrhythmogenesis during acute ischemia. In D.P. Zipes and J. Jalife, editors, *Cardiac Electrophysiology and Arrhythmias*, pages 353–362. Grune & Stratton, Orlando, 1985.

[60] M.J. Janse and A.L. Wit. Electrophysiological mechanisms of ventricular arrhythmias resulting from myocardial ischemia and infarction. *Physiol. Rev.*, 69:1049–1169, 1989.

[61] M. Jones and L.A. Geddes. Strength-duration curves for cardiac pacemaking and ventricular fibrillation. *Cardiovasc. Res. Ctr. Bull.*, 15: 101–112, 1977.

[62] R.W. Joyner, B.M. Ramza, R.C. Tan, J. Matsuda, and T.T. Do. Effects of tissue geometry on initiation of a cardiac action potential. *Am. J. Physiol.*, 256:H391–H403, 1989.

[63] R.W. Joyner, R. Veenstra, D. Rawling, and A. Chorro. Propagation through electrically coupled cells: Effects of a resistive barrier. *Biophys. J.*, 45:1017–1025, 1984.

[64] R.W. Joyner, M. Westerfield, and J.W. Moore. Effects of cellular geometry on current flow during a propagated action potential. *Biophys. J.*, 31:183–194, 1980.

[65] A. Kadish, M. Shinnar, E.N. Moore, J.H. Levine, C.W. Balke, and J.F. Spear. Interaction of fiber orientation and direction of impulse propagation with anatomic barriers in anisotropic canine myocardium. *Circulation*, 78:1478–1494, 1988.

[66] D.T. Kaplan, J.M. Smith, B.E.H. Saxberg, and R.J. Cohen. Nonlinear dynamics in cardiac conduction. *Math. Biosci.*, 90:19–48, 1988.

[67] P.P. Karpawich, M. Hakimi, D.L. Cavitt, and R. Schallhorn. Clinical comparison of low threshold platinized and new steroid-eluting platinized transvenous pacing leads in children. *Am. J. Cardiol.*, 64:423, 1989.

[68] M. Kawato, A. Yamanaka, S. Urushibara, O. Nagata, H. Irisawa, and R. Suzuki. Simulation analysis of excitation conduction in the heart: propagation of excitation in different tissues. *J. Theor. Biol.*, 120:389–409, 1986.

[69] J.P. Keener. Causes of propagation failure in excitable cells. In R. Rensing, editor, *Temporal Disorder in Human Oscillatory Systems*, pages 134–140, Springer-Verlag, Berlin, 1987.

[70] J.P. Keener. The effects of gap junctions in propagation in myocardium: A modified cable theory. In J. Jalife, editor, *Mathematical Approaches to Cardiac Arrhythmias.* Ann. NY Acad. Sci., 591:257–277, 1990.

[71] J.P. Keener. A geometrical theory for spiral waves in excitable media. *SIAM J. Appl. Math.*, 46:1039–1056, 1986.

[72] J.P. Keener. A mathematical model for the vulnerable phase in myocardium. *Math. Biosci.*, 90:3–18, 1988.

[73] J.P. Keener. Propagation and its failure in coupled systems of discrete excitable cells. *SIAM. J. Appl. Math.*, 47:556–572, 1987.

[74] J.P. Keener and J.J. Tyson. Spiral waves in the Belousov-Zhabotinsky reaction. *Physica D.*, 21:307–324, 1986.

[75] P.R. Kowey, R.L. Verrier, and B. Lown. The repetitive extrasystole as an index of vulnerability during myocardial ischemia in the canine heart. *Am. Heart J.*, 106:1321–1325, 1983.

[76] J.B. Kramer, J.E. Saffitz, F.X. Witkowski, and P.B. Corr. Intramural reentry as a mechanism of ventricular tachycardia during evolving canine myocardial infarction. *Circ. Res.*, 56:736–754, 1985.

[77] V.I. Krinsky and K.I. Agladze. Interaction of rotating waves in an active chemical medium. *Physica D.*, 8:50–56, 1983.

[78] L. Kuhnert, H.J. Krug, and L. Pohlmann. Velocity of trigger waves and temperature dependence of autowave processes in the Belousov-Zhabotinsky reaction. *J. Phys. Chem.*, 89:2022–2026, 1985.

[79] E. Lepeschkin, J.L. Jones, S. Rush, and R.E. Jones. Local potential gradients as a unifying measure for thresholds of stimulation, standstill, tachyarrhythmia and fibrillation appearing after strong capacitor discharges. *Adv. Cardiol.*, 21:268–278, 1978.

[80] M.D. Lesh, M. Pring, and J.F. Spear. Cellular uncoupling can unmask dispersion of action potential duration in ventricular myocardium. *Circ. Res.*, 65:1426–1440, 1989.

[81] C. Lesigne, B. Levy, R. Saumont, P. Birkui, A. Bardou, and B. Rubin. An energy-time analysis of ventricular fibrillation and defibrillation thresholds with internal electrodes. *Med. Biol. Eng.*, 14:617–622, 1976.

[82] B. Lown, M.D Klein, and P.I. Hershberg. Coronary and precoronary care. *Am. J. Medicine*, 46:705–724, 1969.

[83] E. Lugosi. Analysis of meander in the Zykov kinetics. *Physica D.*, 40:331–337, 1989.

[84] R.J. Matta, R.L. Verrier, and B. Lown. Repetitive extrasystole as an index of vulnerability to ventricular fibrillation. *Am. J. Physiol.*, 230:1469–1473, 1976.

[85] W.C. McDaniel and J.C. Schuder. The cardiac ventricular defibrillation threshold: Inherent limitations in its application and interpretation. *Med. Instrum.*, 21:170–176, 1987.

[86] H.P. McKean. Nagumo's equation. *Adv. Math*, 4:209–223, 1970.

[87] E.L. Michelson, J.F. Spear, and E.N. Moore. Initiation of sustained ventricular tachyarrhythmias in a canine model of chronic myocardial infarction: Importance of the site of stimulation. *Circulation*, 63:776–784, 1981.

[88] M. Mirowski. The automatic cardioverter-defibrillator: an overview. *J. Am. Coll. Card.*, 6:461–466, 1985.

[89] G.K. Moe, A.S. Harris, and C.J. Wiggers. Analysis of initiation of fibrillation by electrographic studies. *Am. J. Physiol.*, 134:473–492, 1941.

[90] G.K. Moe, W.C. Rheinboldt, and J.A. Abildskov. A computer model of atrial fibrillation. *Am. Heart J.*, 67:200–220, 1964.

[91] P.B. Monk and H.G. Othmer. Relay, oscillations and wave propagation in a model of Dictyostelium discoideum. *Lect. Math. Life Sci.*, 21:87–122, 1989.

[92] E.N. Moore and J.F. Spear. Ventricular fibrillation threshold. *Arch. Int. Medicine*, 135:446–453, 1975.

[93] S.C. Muller, T. Plesser, and B. Hess. Two-dimensional spectrophotometry of spiral wave propagation in the Belousov-Zhabotinsky reaction II. Geometric and kinematic patterns. *Physica D.*, 24:87–96, 1987.

[94] S.C. Muller, T. Plesser, and B. Hess. Two-dimensional spectrophotometry of spiral wave propagation in the Belousov-Zhabotinsky reaction I. Experiments and digital data representation. *Physica D.*, 24:71–86, 1987.

[95] A. Nitzan, P. Ortoleva, and J. Ross. Nucleation in systems with multiple stationary states. *Faraday Symp. Chem. Soc*, 9:241–253, 1974.

[96] O. Orias, C.M. Brooks, E.E. Suckling, J.L. Gilbert, and A.A. Siebens. Excitability of the mammalian ventricle throughout the cardiac cycle. *Am. J. Physiol.*, 163:272–282, 1950.

[97] D.G. Palmer. Interruption of T waves by premature QRS complexes and the relationship of this phenomenon to ventricular fibrillation. *Am. Heart J.*, 63:367–373, 1962.

[98] A.M. Pertsov, A.V. Panfilov, and F.U. Medvedeva. Instabilities of autowaves in excitable media associated with critical curvature phenomena. *Biofizika*, 28:100–102, 1983.

[99] M.F. Rattes, D.L. Jones, A.D. Sharma, and G.J. Klein. Defibrillation threshold: A simple and quantitative estimate of the ability to defibrillate. *PACE*, 10:70–77, 1987.

[100] F.A. Roberge, A. Vinet, and B. Victorri. Reconstruction of propagated electrical activity with a two-dimensional model of anisotropic heart muscle. *Circ. Res.*, 58:461–475, 1986.

[101] D.E. Roberts and A.M. Scher. Effect of tissue anisotropy on extracellular potential fields in canine myocardium in situ. *Circ. Res.*, 50:342–351, 1982.

[102] Y. Rudy and W.-L. Quan. A model study of the effects of the discrete cellular structure on electrical propagation on cardiac tissue. *Circ. Res.*, 61:815–823, 1987.

[103] S. Rush, J.A. Abildskov, and R. McFee. Resistivity of body tissues at low frequencies. *Circ. Res.*, 12:40–50, 1963.

[104] S. Rush, E. Lepeshkin, and A. Gregoritsch. Current distribution from defibrillation electrodes in a homogeneous torso model. *J. Electrophysiol.*, 2(4):331–342, 1969.

[105] W.A.H. Rushton. Initiation of the propagated disturbance. *Proc. Roy. Soc. Lond. B.*, 106:210–243, 1937.

[106] G. Salama, R. Lombardi, and J. Elson. Maps of optical action potentials and NADH fluorescence in intact working hearts. *Am. J. Physiol.*, 252:H384–394, 1987.

[107] M. Schalij. *Anisotropic Conduction and Ventricular Tachycardia.* Ph.D. thesis, University of Limburg, 1988.

[108] A.C. Scott. The electrophysics of a nerve fiber. *Rev. Mod. Physics,* 47:487–533, 1975.

[109] A.C. Scott. *Neurophysics.* John Wiley & Sons, New York, 1977.

[110] G.H. Sharp and R.W. Joyner. Simulated propagation of cardiac action potentials. *Biophys. J.,* 31:403–424, 1980.

[111] N. Shibata, P-S Chen, E.G. Dixon, et al. Influence of shock strength and timing on induction of ventricular arrhythmias in dogs. *Am. J. Physiol.,* 255:H891–H901, 1988.

[112] K. Showalter and J.J. Tyson. Luther's 1906 discovery and analysis of chemical waves. *J. Chem. Educ.,* 64:742–744, 1987.

[113] N.E. Shumway, J.A. Johnson, and R.J. Stish. The study of ventricular fibrillation by threshold determinations. *J. Thorac. Surg.,* 34:643–653, 1957.

[114] W.E. Skaggs, E. Lugosi, and A.T. Winfree. Stable vortex rings of excitation in neuroelectric media. *IEEE Trans. Circ. Sys.,* 35(7):784–787, 1988. There is a typographical in the equation: parameters 0.7 and 0.5 are interchanged.

[115] W.E. Skaggs and A.T. Winfree. Unpublished computations. 1987.

[116] W.E. Skaggs and A.T. Winfree. Unpublished labatory book. 1988.

[117] J.M. Smith and R.J. Cohen. Simple finite-element model accounts for wide range of cardiac dysrhythmias. *Proc. Natl. Acad. Sci. USA,* 81:233–237, 1984.

[118] M.S. Spach and P.C. Dolber. The relation between discontinuous propagation in anisotropic cardiac muscle and the vulnerable period of reentry, In D.P. Zipes and J. Jalife, editors, *Cardiac Electrophysiology and Arrhythmias,* pages 241–252. Grune & Stratton, Orlando, FL, 1985.

[119] M.S. Spach, P.C. Dolber, and P.A.W. Anderson. Multiple regional differences in cellular properties that regulate repolarization and contraction in the right atrium of adult and newborn dogs. *Circ. Res.,* 65:1594–1611, 1989.

[120] M.S. Spach, P.C. Dolber, and J.F. Heidlage. Interaction of inhomogeneities of repolarization with anisotropic propagation in dog atria. *Circ. Res.,* 65:1612–1631, 1989.

[121] M.S. Spach, W.T. Miller, D.B. Geselowitz, R.C. Barr, J.M. Kootsey, and E.A. Johnson. The discontinuous nature of propagation in normal canine cardiac muscle. *Circ. Res.*, 48:39–54, 1981.

[122] G.F. Starmer and R.E. Whalen. Current density and electrically induced ventricular fibrillation. *Med. Instrum.*, 7:3–7, 1973.

[123] B.M. Steinhaus, K.W. Spitzer, M.J. Burgess, and J.A. Abildskov. Electrotonic interactions in a model of anisotropic cardiac tissue. In *Proceedings of the 1986 Society for Computer Simulation*, pages 421–426, Reno, NV, 1986.

[124] J. Tamargo, B. Moe, and G.K. Moe. Interaction of sequential stimuli applied during the relative refractory period in relation to determination of fibrillation threshold in the canine ventricle. *Circ. Res.*, 37:534–541, 1975.

[125] P.J. Tchou, N. Kadri, J. Anderson, J.A. Caceres, M. Jazayeri, and M. Akhtar. Automatic implantable cardioconverter/defibrillators and survival of patients with left ventricular dysfunction and malignant ventricular arrhythmias. *Ann. Int. Med.*, 109:529–534, 1988.

[126] K.J. Tomchik and P.N. Devreotes. Adenosine 3',5'-monophosphate waves in Dictyostelium discoideum: A demonstration of isotope dilution-fluorography. *Science*, 212:443–446, 1981.

[127] N. Tsuboi, I. Kodama, J. Toyama, and K. Yamada. Anisotropic conduction properties of canine ventricular muscles. *Jap. Circ. J.*, 49:487–498, 1985.

[128] J.J. Tyson, K.A. Alexander, Manoranjan V.S., and J.D. Murray. Spiral waves of cyclic AMP in a model of slime mold aggregation. *Physica D.*, 34:193–207, 1989.

[129] J.J. Tyson and P.C. Fife. Target patterns in a realistic model of the Belousov-Zhabotinskii reaction. *J. Chem. Phys.*, 73:2224–2237, 1980.

[130] J.J. Tyson and J.P. Keener. Singular perturbation theory of traveling waves in excitable media (a review). *Physica D*, 32:327–361, 1988.

[131] J.J. Tyson and J.D. Murray. Cyclic-AMP waves during aggregation of dictyostelium amoebae. *Development*, 106:421–6, 1989.

[132] F.J.L. van Capelle. *Slow Conduction and Cardiac Arrhythmias*. Ph.D. thesis, University of Amsterdam, 1983.

[133] F.J.L. van Capelle and M.A. Allessie. Computer simulation of anisotropic impulse formation. In A. Goldbeter, editor, *Cell to Cell Signalling: From Experiments to Theoretical Models*, pages 577–588. Academic Press, London 1989.

[134] R.Th. van Dam, D. Durrer, J. Strackee, and H. van der Tweel. The excitability cycle of the dog's left ventricle determined by anodal, cathodal, and bipolar stimulation. *Circ. Res.*, 4:196–204, 1956.

[135] A. van Oosterom, R.W. de Boer, and R.Th. van Dam. Intramural resistivity of cardiac tissue. *Mol. Biol. Eng. Comput.*, 17:337–343, 1979.

[136] R.A. van Tyn and L.D. MacLean. Ventricular fibrillation threshold. *Am. J. Physiol.*, 201:457–461, 1961.

[137] W.D. Voorhees, K.S. Foster, L.A. Geddes, and C.F. Babbs. Safety factor for precordial pacing: Minimum current thresholds for pacing and for ventricular fibrillation by vulnerable-period stimulation. *PACE*, 7:356–360, 1984.

[138] R. Wegria, G.K. Moe, and C.J. Wiggers. Comparison of the vulnerable periods and fibrillation thresholds of normal and idioventricular beats. *Am. J. Physiol.*, 133:651–657, 1941.

[139] R. Wegria and C.J. Wiggers. Factors determining the production of ventricular fibrillation by direct currents. *Am. J. Physiol.*, 131,:104–118, 1940.

[140] T.C. West, E.L. Frederickson, and D.W. Amory. Single fiber recording of the ventricular response to induced hypothermia in the anaesthetized dog. *Circ. Res.*, 7:880–888, 1959.

[141] J.M. Wharton, P.D. Wolf, P-S Chen, et al. Is an absolute minimum potential gradient required for ventricular defibrillation? (abstract). *Circulation*, 74:II–342, 1986.

[142] J.M. Wharton, P.D. Wolf, N. Danieley, et al. Cardiac potential and potential gradient fields generated by single, combined, and sequential shocks during ventricular defibrillation. (submitted) 1990.

[143] W.D. Widman, L. Eisenberg, S. Levitsky, A. Mauro, and W.W. Glenn. Ventricular fibrillation complicating electrical pacemaking: A comparison of direct current and radio-frequency pacemaker stimulation. *Surg. For.*, 14:260–263, 1963.

[144] C.J. Wiggers and R. Wegria. Quantitative measurement of the fibrillation thresholds of the mammalian ventricles with observations on the effect of procaine. *Am. J. Physiol.*, 131:296–308, 1940.

[145] C.J. Wiggers, R. Wegria, and B. Pinera. The effects of myocardial ischemia on the fibrillation threshold-the mechanism of spontaneous ventricular fibrillation following coronary occlusion. *Am. J. Physiol.*, 131:309–316, 1940.

[146] A.T. Winfree. Alternative stable rotors in an excitable medium. *Proceedings of Puschino USSR Meeting on Autowaves,* Physica D., 1991. (in press.)

[147] A.T. Winfree. Discrete spectrum of rotor periods in an excitable medium. *Phys. Lett. A.,* 149:203–206, 1990.

[148] A.T. Winfree. Electrical instability in cardiac muscle: Phase singularities and rotors. *J. Theor. Biol.,* 138:353–405, 1989.

[149] A.T. Winfree. *The Geometry of Biological Time.* Springer-Verlag, New York, 1980.

[150] A.T. Winfree. Multiple stable solutions to the kinetic equations of an excitable medium. In: *Integral Methods in Science and Engineering.* Hemisphere, Washington, DC, 1991. (in press.)

[151] A.T. Winfree. Organizing centers for chemical waves in two and three dimensions. In R. Field and M. Burger, editors, *Oscillations and Traveling Waves in Chemical Systems,* pages 441–472. John Wiley & Sons, New York, 1985.

[152] A.T. Winfree. Rotors in normal ventricular myocardium: Einthoven lecture. 1990.

[153] A.T. Winfree. Spiral waves of chemical activity. *Science,* 175:634–636, 1972.

[154] A.T. Winfree. Stable particle-like solutions to the nonlinear wave equations of three-dimensional excitable media. *SIAM Review,* 32:1–53, 1990.

[155] A.T. Winfree. Stably rotating patterns of reaction and diffusion. *Prog. Theor. Chem.,* 4:1–51, 1978.

[156] A.T. Winfree. Sudden cardiac death. *Sci. Am.,* 248:144–161, 1983.

[157] A.T. Winfree. Ventricular reentry in three dimensions. In D.P. Zipes and J. Jalife, editors, *Cardiac Electrophysiology, from Cell to Bedside,* pages 224–234, WB Saunders Co, Philadelphia, 1990.

[158] A.T. Winfree. Vortex action potentials in normal ventricular muscle. In J. Jalife, editor, *Mathematical Approaches to Cardiac Arrhythmias.* Ann. NY Acad. Sci., 591:190–207, 1990.

[159] A.T. Winfree. *When Time Breaks Down: The Three-Dimensional Dynamics of Electrochemical Waves and Cardiac Arrhythmias.* Princeton University Press, Princeton, 1987.

[160] F.X. Witkowski and P.A. Penkoske. Activation patterns during ventricular fibrillation. In J. Jalife, editors, *Mathematical Approaches to Cardiac Arrhythmias.* Ann. NY Acad. Sci., 591:219–231, 1990.

[161] F.X. Witkowski, P.A. Penkoske, and R. Plonsey. Mechanism of cardiac defibrillation in open-chest dogs using unipolar DC-coupled simultaneous activation and shock potential recordings. *Circulation,* 82:244–60, 1990.

[162] S. Yabe, W.M. Smith, J.P. Daubert, P.D. Wolf, and R.E. Ideker. The strength of monophasic and biphasic shocks that cause conduction block. *J. Am. Coll. Cardiol.,* 13:67A, 1989.

[163] M.S. Yoon, J. Han, and R.A. Fabregas. Effect of ventricular aberrancy on fibrillation threshold. *Am. Heart J.,* 89:599–604, 1975.

[164] A.M. Zhabotinsky. A study of self-oscillatory chemical reaction. III. Space behavior. In B. Chance, E.K. Pye, A.K. Ghosh, and B. Hess, editors, *Biological and Biochemical Oscillators.* Academic Press, New York, 1978.

[165] X. Zhou, J.P. Daubert, P.D. Wolf, W.M. Smith, and R.E. Ideker. Importance of the shock electric field for defibrillation efficacy. *Circulation,* 78:II–219, 1986.

[166] X. Zhou, P.D. Wolf, D.L. Rollins, W.M. Smith, and R.E. Ideker. Potential gradient needed for defibrillation with monophasic and biphasic shocks. *PACE,* 12:651, 1989.

[167] G. Zuanetti, R.H. Hoyt, and P.B. Corr. β-adrenergic mediated influences on microscopic conduction in epicardial regions overlying infarcted myocardium. *Circ. Res.,* 67:284–302, 1990.

[168] V.S. Zykov. *Simulation of Wave Processes in Excitable Media.* Manchester University Press, Manchester, England, 1988.

20

Basic Mechanisms of Ventricular Defibrillation

Raymond E. Ideker et al.[1]

ABSTRACT Recordings were made simultaneously from many electrodes placed on and in the hearts of animals to study the basic principles of ventricular defibrillation. The findings are listed below. Earliest activations following a shock slightly lower in strength than needed to defibrillate (a subthreshold defibrillation shock) occur in those cardiac regions in which the potential gradients generated by the shock are weakest. Activation fronts after subthreshold shocks do not appear to be continuations of activation fronts present just before the shock. An upper limit exists to the strength of shocks that induce fibrillation when given during the "vulnerable period" of regular rhythm. This upper limit of vulnerability correlates with and is similar in strength to the defibrillation threshold. To defibrillate, a shock must halt the activation fronts of fibrillation without giving rise to new activation fronts that reinduce fibrillation. The response to shocks during regular rhythm just below the upper limit of vulnerability is similar to the response to subthreshold defibrillation shocks. Shocks during regular rhythm initiate rotors of reentrant activation leading to fibrillation when a critical point is formed, at which a certain critical value of shock potential gradient field strength intersects a certain critical degree of myocardial refractoriness. This critical point may explain the existence of the upper limit of vulnerability. The critical point may also partially explain the finding that the relationship between shock strength and the success of the shock in halting fibrillation is better represented by a probability function rather than by a discrete threshold value. Very high potential gradients, approximately an order of magnitude greater than needed for defibrillation, have detrimental effects on the heart, including conduction block, induction of arrhythmias, decreased wall motion, and tissue necrosis.

[1] Raymond E. Ideker, Anthony S.L. Tang, David W. Frazier, Nitaro Shibata, Peng-Sheng Chen, and J. Marcus Wharton, Department of Medicine and Pathology, Duke University Medical Center, Durham, North Carolina, 27710 and the Engineering Research Center in Emerging Cardiovascular Technologies, Duke University, Durham, North Carolina, 27706

20.1   Introduction

At least three factors have engendered a renewed interest in the basic mechanisms of defibrillation. One, the development of the automatic implantable cardioverter-defibrillator, has shown that it is possible to extend the lives of a significant number of people prone to sudden cardiac death [42]. Increasing the rate of success of shocks and decreasing the strength of shocks needed for defibrillation would increase the usefulness of the automatic implantable cardioverter-defibrillator. Two, the hardware and software have been developed to allow recording simultaneously from many sites in the heart to determine the activation sequences before and after defibrillation shocks as well as the potentials created in the heart by the shocks [69]. Three, recent advances in understanding of the nonlinear dynamics of excitable media appear to apply to ventricular fibrillation and defibrillation [65]. This chapter reviews some of the recent findings about how a shock succeeds or fails in halting ventricular fibrillation.

Early proposed mechanisms of defibrillation focused on the termination of activation fronts by the shock. In the 1930s and 1940s, the "total extinction" hypothesis was proposed which states that, to defibrillate, the shock must halt all of the activation fronts present throughout the ventricles during fibrillation [24,60]. In the 1970s, this explanation was supplanted by the "critical mass" hypothesis [45,74]. The critical mass mechanism posits that activation fronts must be halted only within a certain critical mass of the myocardium, thought to be about 75% of the ventricular mass in the dog. Activation fronts in the remaining 25% were thought to be insufficient to maintain fibrillation and thought soon to die out. As discussed below, more recent proposed mechanisms of defibrillation have centered on the ability of the shock (1) to initiate new activation fronts, (2) to alter refractoriness, and (3) to cause electrophysiological abnormalities and damage in myocardial regions where the shock field is strong.

The development of computer-assisted cardiac mapping techniques in which recordings are made simultaneously from up to hundreds of electrodes placed directly in the heart [55] has led to the acquisition of much new information about the initiation and maintenance of cardiac arrhythmias [16,51,66]. Computer-assisted mapping has recently been applied to the study of ventricular defibrillation by extending the technique so that recordings of the millivolt signals generated by cardiac activation are made, followed a few milliseconds later by recordings of defibrillation shock potentials of hundreds of volts, followed again a few milliseconds later by recordings of the millivolt signals generated by cardiac activation in response to the shock [68,69]. Much of the following information about the mechanism of defibrillation was learned using this mapping technique.

20.2   Location of Earliest Activation Following Failed Defibrillation Shocks

Recordings have been made in dogs of cardiac activation immediately before and after a defibrillation shock as well as of the potentials caused by the shock itself at 56 to 120 electrodes placed on or in the ventricles with several different configurations of epicardial or catheter defibrillation electrodes [10,56,58,67]. In Figure 20.1, an example of the potential distribution is shown for a 500V shock delivered via a catheter electrode in the right ventricular apex or cathode and a cutaneous patch electrode over the left thorax or anode, taken from the work of Tang and colleagues [56]. Shock potentials and cardiac signals just after the shock were recorded from 52 plunge needles inserted through the right and left ventricular free walls. Each plunge needle contained two electrodes, one to record from just beneath the epicardium (the subepicardium shown in the top panel), the other to record from the inner third of the ventricular wall adjacent to the cavity (the subendocardium shown in the bottom panel). Another 18 electrodes were placed to record potentials from the interventricular septum and from the atria [56]. All panels show the potentials superimposed on the epicardial surface. The two panels on the left represent the left anterolateral cardiac surface and the two panels on the right represent the right posterolateral surface.

The most negative potentials were recorded in the posterolateral right ventricle, adjacent to the catheter cathode electrode. The potentials changed rapidly in this region, as indicated by the closely spaced isopotential lines. This indicates a high potential gradient in the region. Distant from the catheter electrode, in the anterior left and right ventricles, the potentials changed more slowly indicating a lower potential gradient. The least negative potentials were in the left ventricular apex, consonant with the fact that this portion of the heart was closest to the cutaneous anode electrode. In all regions, the subendocardial potentials were more negative than the subepicardial potentials, indicating a transmural component to the potential gradient field caused by the current exiting the heart to travel from the intracardiac cathode to the extracardiac anode.

The change in transmembrane potential by an extracellular stimulus is thought to be directly proportional to the extracellular potential gradient [36,50]. Thus, the extracellular potential gradient field created throughout the ventricles by the shock is thought to determine the strength of the shock needed for defibrillation [38]. In the study by Tang and colleagues [56], the potential gradient field was estimated from the shock potentials and the locations of the recording electrodes [10]. At the end of the study, the animal heart was cut into slices approximately 2 mm thick, and, with a hand-held digitizer, the locations of the recording electrodes were entered into a computer from the individual slices [37,56]. In all studies to

POTENTIAL DISTRIBUTION

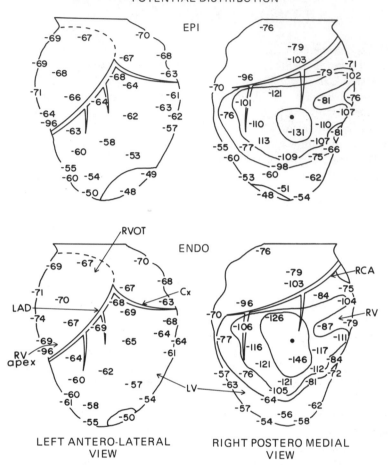

FIGURE 20.1. The potential field from an unsuccessful 500-V, 6-msec monophasic truncated exponential defibrillation shock. Numbers denote the location of the recording electrodes and the potential in volts recorded at each site referenced to the left leg. Isopotential lines are drawn 25 V apart. Solid circles represent unsatisfactory electrode recordings. The dashed line indicates the upper border of the right ventricular outflow tract. LV = left ventricle; RV = right ventricle; RVOT = right ventricular outflow tract; LAD = left anterior descending coronary artery; RCA right coronary artery; and Cx = left circumflex coronary artery. (Reproduced with permission from Tang et al. [56]. ©1988 IEEE.)

date [10,56], the potential gradient distributions have been very uneven for configurations in which at least one defibrillation electrode is in or on the heart, with high gradients near these electrodes and much weaker gradients in regions of the heart distant from these electrodes. For example, Figure 20.2 shows the magnitude of the potential gradient field for the same shock for which the potentials are shown in Figure 20.1. While not shown, the X, Y, and Z components of the gradient were also calculated. The highest gradients were in the posterolateral right ventricular apex around the catheter, as suggested by the closely spaced isopotential lines in Figure 20.1. The lowest gradients were in the anterobasal right and left ventricular free walls, distant from the catheter. The gradient field was very uneven, with the highest gradient (37 V/cm) more than 18 times larger than the lowest gradient (2 V/cm).

When shocks were much weaker than required to defibrillate, earliest sites of activation after the shock leading to the resumption of fibrillation were located throughout the ventricular myocardium [52]. As shock strength was increased, early sites of activation were no longer observed in the high gradient regions around the defibrillation electrodes. Following unsuccessful shocks just slightly weaker than needed to defibrillate, earliest activation leading to the resumption of fibrillation arose almost exclusively in regions of low potential gradient distant from the defibrillation electrodes. For example, Figure 20.3 illustrates the first three postshock cycles of activation leading to the resumption of fibrillation after the shock for which the gradient field is shown in Figure 20.2. The shock was administered 10 sec after the electrical induction of ventricular fibrillation. Earliest recorded activation occurred in the subepicardium of the anterobasal left ventricular free wall, within the low gradient region shown in Figure 20.2. The activation front arising in this region conducted only for a short distance before blocking, as indicated by the wide black line. The front was able to conduct to the subendocardium, however, as shown by the long arrow. From there activation spread caudally toward the apex but could not spread to the adjacent basal regions because of block, as shown by the wide black lines on the endocardium. Instead, the front curved around the line of block to activate the adjacent basal region. By this time, the tissue first activated after the shock is assumed to have had time to recover so that it was reexcited by the fronts curving around both lines of blocks (second postshock beat). Activation then spread caudally along the subendocardium to the anterior subepicardium. The fronts then formed two mirror image rotors, clockwise on the right ventricular side and counterclockwise on the left ventricular side, in both the subepicardium and subendocardium. This pair of rotors then continued (third postshock beat), forming what has been called a figure eight reentrant pattern in the cardiac electrophysiological literature [17].

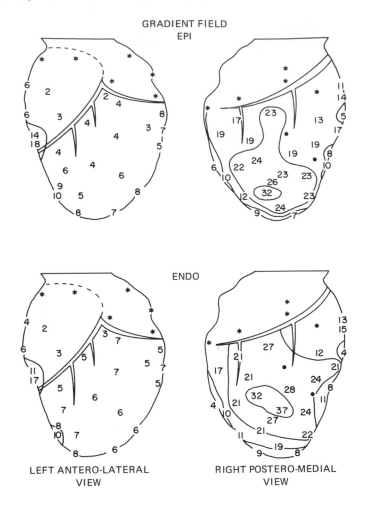

FIGURE 20.2. The extracellular potential gradient field calculated from the potentials shown in Figure 20.1 and from the locations of the recording sites. The numbers are the potential gradients at each recording site in V/cm. Asterisks represent the top row of electrodes on the atria and on the right ventricular outflow tract where potential gradients were not calculated because no recording sites are above them. Isogradient lines are drawn 10 V/cm apart. (Reproduced with permission from Tang et al. [56]. ©1988 IEEE.)

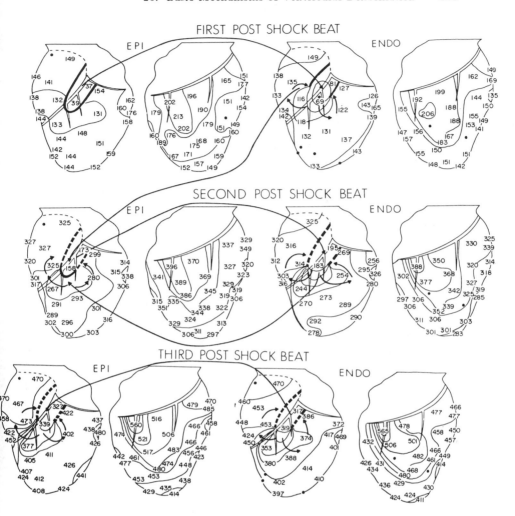

FIGURE 20.3. The first three cycles of activation leading to the resumption of fibrillation after an unsuccessful 500-V shock. The numbers are the times in msec of local activations at each recording site, timed from the beginning of the shock. Isochronal lines are drawn 20 msec apart. The solid bars signify conduction block, while the dashed bars signify a frame shift from one isochronal map to the next. Such frame lines are necessary whenever a continuous process such as reentrant activation is illustrated by a series of static maps. (Reproduced with permission from Tang et al. [56]. ©1988 IEEE.)

20.3   Origin of Activation after Unsuccessful Defibrillation Shocks

According to both the total extinction hypothesis [40,59] and the critical mass hypothesis [45,74], earliest postshock activation occurs in low gradient regions because the potential gradient field is too weak to halt the activation fronts of fibrillation present in these regions at the time of the shock. After the shock, these activation fronts spread away from the low gradient regions causing disorganized activation throughout the remainder of the ventricles so that fibrillation continues. Thus, to defibrillate, the shock must either halt all activation fronts (total extinction hypothesis) or halt those activation fronts within a certain critical mass (critical mass hypothesis).

Mapping results do not confirm these hypotheses though [7,9,54]. Following shocks slightly lower in strength than required for defibrillation, excitation appears to occur via new activation fronts rather than via continuation of the activation fronts present just before the shock [9]. Activation after the shock is first recorded following a long pause, called the isoelectric window by Chen and colleagues [7,8]. For example, earliest activation is recorded 37 msec after the shock in Figure 20.3. Such long pauses are not observed during ventricular fibrillation before the shock recorded with the same set of electrodes. As shock strength is decreased to levels further below the level needed for defibrillation, the postshock isoelectric window becomes shorter, and at fractions of a joule, no longer can be detected (Figure 20.4), suggesting that unsuccessful defibrillation shocks of very low strength do not halt the activation fronts of fibrillation [54]. For shocks just slightly weaker than needed to defibrillate, the isoelectric window shortens as the recording electrodes are placed more closely together [7,9], suggesting that very slow conduction is the cause of the isoelectric window, perhaps because of the latency associated with stimulation of highly refractory tissue [47]. These findings are consistent with the interpretation that shocks slightly weaker than necessary for defibrillation directly excite most tissue that is not absolutely refractory, with activation fronts arising after the shock in highly refractory tissue at the border of the directly excited regions.

20.4   The Upper Limit of Vulnerability

The possibility that the new activation fronts reinitiating fibrillation after a failed defibrillation shock are caused by stimulation of refractory tissue by the shock raises the question of how a shock ever succeeds in defibrillating. Strong electrical stimuli induce fibrillation if given during the repolarization interval of normal rhythm when the cells are relatively refractory, the so-called vulnerable period. Since activation occurs continuously dur-

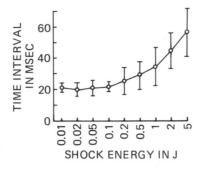

FIGURE 20.4. Effect of shock energy on the time of earliest recorded postshock activation. Mean values are shown for unsuccessful defibrillation shocks given via electrodes on the left ventricular apex and right atrium to seven dogs 10 sec after the electrical induction of fibrillation. Standard deviations are indicated by brackets. The shortest time interval following the beginning of the shock for which activations could be reliably detected is about 20 msec. The time until earliest recorded activation shortens as shock energy is decreased. (Reproduced with permission from Shibata et al. [52].)

ing fibrillation, repolarization should occur continuously also, so that at any time some portion of myocardium should be in the vulnerable period of refractoriness during which a stimulus of sufficient strength can induce fibrillation. The fact that it is possible to defibrillate with large shocks implies the existence of an upper limit of strength above which shocks will not induce fibrillation when given during the vulnerable period. Such a limit has been found to exist and is highly correlated with the strength of the shock needed to defibrillate (Figure 20.5) [6,18,39]. In Figure 20.5, the shock strength required for defibrillation was determined within 10% by a technique similar to one described by Bourland and coworkers [4]. The upper limit of vulnerability in Figure 20.5 was found by giving shocks at numerous times during the vulnerable period of regular paced rhythm and determining within 10% the largest shock strength that induced fibrillation [6].

20.5   The Upper Limit of Vulnerability Hypothesis for Defibrillation

The existence of an upper limit of vulnerability and the high correlation between this upper limit and the shock strength required for defibrillation are consistent with the hypothesis that successful defibrillation requires a shock strength that reaches or exceeds the upper limit of ventricular vulnerability [6]. This hypothesis implies that activation sequences following

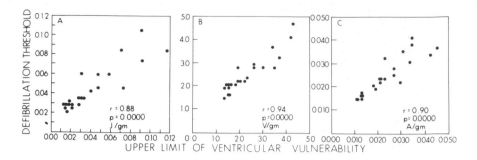

FIGURE 20.5. Correlation of the approximate strength of the shock required to defibrillate (the defibrillation threshold) and the upper limit of vulnerability for defibrillation electrodes on the right atrium (anode) and the left ventricular apex (cathode) in 22 dogs. Results are expressed in units of energy (Panel A), voltage (Panel B), and current (Panel C). All units are expressed per gram of heart weight. (Reproduced with permission from Chen et al. [6] by permission of the American Heart Association, Inc.)

unsuccessful shocks just slightly below the strength needed for defibrillation should be similar to activation sequences at the start of fibrillation induced by shocks of the same energy given via the same shocking electrodes during the vulnerable period of regular rhythm. Experimental results in animals have shown that the responses to both types of shocks are similar in many respects [52,54]. Excitation following both types of shocks usually followed a pause (the isoelectric window) of 30 to 50 msec, began from a few early sites in regions of low potential gradient, and spread over the remainder of the ventricles as large, coherent activation fronts (Figure 20.6). Shocks were delivered via electrodes on the left ventricular apex (cathode) and the right atrium (anode). A pause in activation followed both the shock during the vulnerable period (panel A) and the unsuccessful defibrillation shock (panel B). Earliest activation (arrows) was in the base of the ventricles following the shock during the vulnerable period (panel C) as well as following the unsuccessful defibrillation shock (panel D). In both cases, activation conducts toward the apex as large, coherent activation fronts. The isoelectric interval is 47 msec in panel C and 50 msec in panel D.

For all animals, no significant differences were present in the time from the shock until earliest activation, the total number of early sites, the location of the earliest site, or the total time for the first postshock activation fronts to traverse the ventricles [54].

The above results are consistent with the existence of two criteria for successful defibrillation. One, the potential gradient field of the shock must be strong enough to abolish the activation fronts of fibrillation throughout all or a critical mass of myocardium [40,45,59,60]. This criterion is met by shocks much weaker than required for defibrillation, but defibrillation is still not successful. Thus, this criterion is a necessary but not sufficient condition for defibrillation. Two, the shock must not give rise to new activation fronts that reinitiate fibrillation. This criterion requires a higher potential gradient than the first criterion and is responsible for most of the shock energy needed for defibrillation. The mechanism by which a shock slightly weaker than necessary for defibrillation fails to halt fibrillation is similar to that by which a shock of the same strength initiates fibrillation during the vulnerable period of regular rhythm, and the long pause following both types of shocks is caused by the latency of stimulation of relatively refractory tissue.

20.6 Electrical Initiation of Ventricular Fibrillation

Based on the phase-resetting properties of certain periodic excitable oscillators such as cardiac pacemakers [2,27,28,63–65], Winfree has proposed a topological model demonstrating the unavoidable existence of a critical point where the interaction of a critical stimulus level with a critical time point within the cycle of the oscillator causes an undefined response. He proposes that a rotor of excitation isochrones will be formed at the critical point (which he calls a phase singularity). Winfree has used these concepts to explain the behavior of chemical waves that propagate through an excitable medium, the Belousov-Zhabotinsky reagent [62], and the behavior of cardiac activation following a shock, employing the Fitzhugh-Nagumo model of excitable tissue [65] (see Chapter 19 for more details).

The experimental findings of Frazier and colleagues [19] correspond closely to the modeling results of Winfree. A rotor of activation is produced when refractoriness is dispersed uniformly through a myocardial region, and a shock is given that creates a range of potential gradients dispersed at an angle to the dispersion of refractoriness (Figure 20.7).

Panel A shows the distribution of activation times and refractoriness following the last of 10 beats of regular (S1) pacing in a dog. The small open circles represent the locations of 117 recording electrodes in a 9 × 13 grid on the epicardial surface of the right ventricle. The grid is approximately 3 × 3 cm. Regular pacing was performed simultaneously from a row of eight epicardial stimulating wires to the right of the recording electrodes (S1). The solid isochronal lines represent the spread of the activation front away from the S1 electrodes. The isochronal lines show the location of the activation front at 10-msec intervals after the S1 stimulus. Approximately parallel isochronal lines were created by the row of S1 electrodes. The recovery periods (dashed lines) were calculated at 32 sites evenly spaced across the array using local 2 mA cathodal stimuli. The refractory periods were similar at all electrode sites (166 ± 3 msec), indicating minimal inhomo-

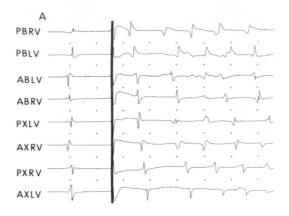

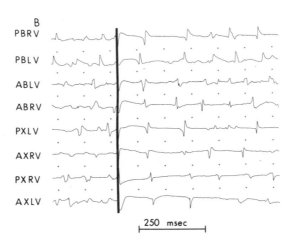

FIGURE 20.6. Examples of recordings and isochronal activation maps in the same dog following the largest shock (2 J) that induced fibrillation during the vulnerable period of normal rhythm (Panels A and C) and the largest shock that failed to defibrillate (2 J) 10 sec after the electrical induction of fibrillation (Panels B and D). Epicardial maps from 56 electrodes are shown in Panels C and D. Polar projections of the ventricles are shown with the stippled apical defibrillation electrode in the center and the atrioventricular groove at the periphery. Numbers represent the locations of electrodes with satisfactory recordings and give the time of activation for those locations in msec from onset of the shock.

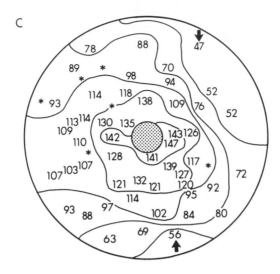

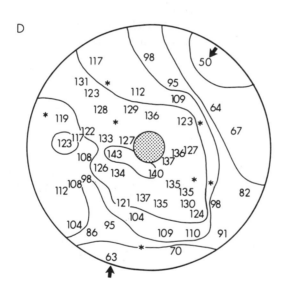

FIGURE 20.6. (cont.) Asterisks indicate electrode sites where adequate recordings were not obtained. The isochronal lines are 20 msec apart. A = anterior; P = posterior; B = base; X = apex; R = right; L = left; V = ventricle; and Lat = lateral. (Reproduced with permission from Shibata et al. [52].)

geneous refractoriness. Because local activation times correlated closely to local tissue refractoriness, uniform parallel isorefractory lines were created.

Panel B shows the potential gradients generated by a large premature (S2) shock of 3 msec and 150 volts delivered from a mesh electrode (4.5 × 1 cm) at the bottom of the recording electrodes, creating isogradient lines perpendicular to the isorefractory lines. Isogradient lines are spaced every 1 V/cm. Such shocks were given to scan the vulnerable period.

Panel C shows the initial activation pattern at the start of VF. Activation times, shown in msec, are measured from the start of the premature S2 stimulus. Isochrones are at 10-msec intervals. Initial conduction appeared as an activation front that ended blindly in the center of the plaque, arising distant from the S2 site and spreading toward areas of later S1 refractoriness, forming a rotor of reentry around an arc of temporary conduction block (hatched line). The hatched area represents the region directly excited by the S2 stimulus. The solid line at the border between the directly excited region and the region of earliest poststimulus activation represents the frame line transition between this activation map and the map for the next cycle of reentry. Counterclockwise reentry was generated. The initial site of reentry (the critical point) was at a potential gradient of approximately 5 V/cm and at a point where the tissue was just emerging from its refractory period from the last regular S1 stimulus at the time the large S2 stimulus was given.

In the cardiac electrophysiological literature, this activation pattern is called leading circle reentry [1]. The center of the reentry rotor is reliably formed where a critical degree of refractoriness intersects a critical level of potential gradient. For a 3-msec truncated exponential shock waveform of low tilt, this critical point is formed where a shock field of 5 V/cm intersects tissue just coming out of its refractory period as determined by a local 2-mA stimulus [19]. The critical values of stimulus potential gradient and tissue refractoriness will probably be different for other shock waveforms [12,73].

20.7   The Critical Point and the Upper Limit of Vulnerability

On a strength-interval plot, the vulnerable region of fibrillation includes all of the combinations of shock strengths and shock timings during the relative refractory period that induce fibrillation [22,34,61]. The existence of an upper limit of shock strength initiating ventricular fibrillation indicates that the vulnerable region on a strength-interval plot is bounded on all sides, including the top. The upper limit of vulnerability can be explained by the existence of the critical point [63,65] and by the fact that the potential gradient field decreases with distance from the shocking electrodes

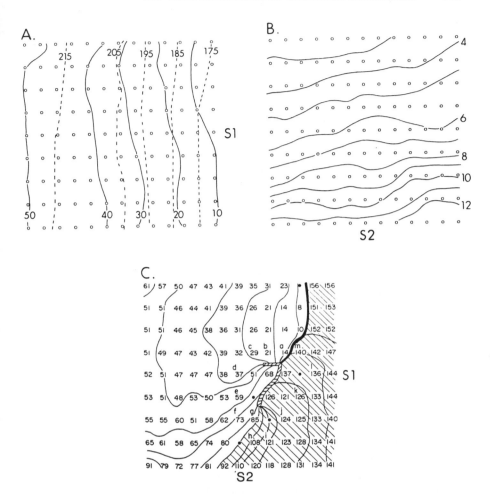

FIGURE 20.7. Initiation of reentry and ventricular fibrillation following orthogonal interaction of myocardial refractoriness and the potential gradient field created by a large stimulus. Panel A shows the distribution of activation time and refractoriness just before the stimulus, Panel B the potential gradient field of the stimulus, and panel C the initial activation pattern just after the stimulus. (Modified with permission from Frazier et al. [19] by copyright permission of the American Society of Clinical Investigation.)

delivering the large premature stimuli [10,58]. The lower limit of vulnerability, that is, the ventricular fibrillation threshold, is the smallest stimulus strength that will induce fibrillation. At the lower limit of vulnerability the stimulus is sufficiently strong that the critical level of potential gradient is just far enough away from the stimulating electrode to allow a reentry rotor to form and be perpetuated (Figure 20.8, isogradient line *a*). The formation of a critical point will occur when the critical degree of refractoriness intersects the critical isogradient line. As the stimulus strength is increased, the critical isogradient line moves farther away from the shocking electrode and so increases in length (Figure 20.8, isogradient line *f*, for example). Because reentry rotors will be induced at all places where the critical isorecovery and isogradient lines intersect, fibrillation can then be induced over a wider range of intervals at sites more distant from the shocking electrode, as previously shown [53]. In the idealized example shown in Figure 20.8, the critical isorecovery and isogradient lines intersect at two points, forming a pair of mirror image rotors, one centered at each critical point, as has been observed experimentally [6,53]. Stimulation very early or very late will fail to induce fibrillation because the critical isogradient and critical isorecovery lines will not intersect; rather, the critical isogradient line will fall entirely within excitable or refractory tissue. When the shock strength is further increased so that the critical isogradient line is moved off the ventricles, fibrillation will not occur at any interval because no critical point will be created (Figure 20.8, isogradient line *i*). Thus, an upper limit of vulnerability is present.

The induction of fibrillation as is normally done with stimulating wires or small catheter electrodes will not show an upper limit of vulnerability, since the stimulus currents are not usually increased so high that the potential gradients exceed the critical isogradient level over all of the ventricular myocardium. Such extremely high currents would probably induce severe myocardial damage close to the stimulating wires that serve as point sources for the current [20]. The vulnerable region therefore appears open at the top for point stimulation, which is the pattern traditionally reported [22,23].

The critical point hypothesis needs to be expanded to account for all of the experimental findings during the electrical induction of arrhythmias. First, large, premature electrical stimuli frequently induce only one or two cycles of activation, called repetitive responses, which then halt spontaneously. Some episodes of repetitive responses occur at shock strengths and timings that are just outside the vulnerable region on a strength-interval plot. Other episodes, however, occur with an identical shock strength and timing that in other trials in the same animal induces fibrillation. While the critical point hypothesis provides a mechanism for the initiation of the initial reentry rotor, it does not yet explain how this rotor either stops spontaneously or degenerates into fibrillation. Second, not all cardiac maps of the initiation of fibrillation by large, premature stimuli reveal reentry rotors; in some cases activation appears to emanate in all direction from

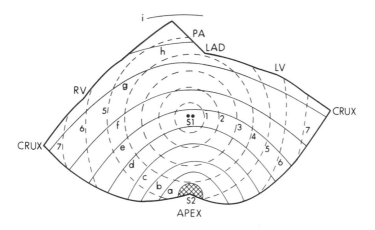

FIGURE 20.8. Hypothesized relationship between critical points, the vulnerable region, and defibrillation. The epicardial surface of the canine heart is depicted as if the ventricles were folded out after an imaginary cut was made from the crux to the apex. Isorecovery lines (dashed lines 1–7), representing different degrees of refractoriness, are concentric about the pacing site labeled S1. Large premature stimuli or defibrillation shocks are delivered from the apex of the heart through the electrode labeled S2 with the return electrode located elsewhere in the body away from the heart. Isogradient lines (solid lines a–i), representing different levels of extracellular potential gradient, are concentric about the S2 electrode with the smallest values in the ventricles occurring in the small region at the top of the ventricles representing the pulmonary outflow tract. RV = right ventricle; LV = left ventricle; PA = pulmonary artery; and LAD = left anterior descending coronary artery. (Modified with permission from R.E. Ideker et al. [26].)

a single point, as if arising from a focus of electrical activity [54]. It is possible, however, that this finding is an experimental artifact caused by not having recording electrodes sufficiently close together or in the right locations to detect the rotors.

20.8   The Critical Point and the Probability Function of Defibrillation

There is not a discrete defibrillation threshold shock strength, above which all shocks succeed and below which all shocks fail. Instead, a sigmoidal relationship is thought to exist between the shock strength and the probability of successful defibrillation [13,21] (Figure 20.9). Nonetheless, a single threshold value is frequently used in the study of defibrillation and, depending on chance and on the manner in which it is determined, it may differ from the average shock strength that caused defibrillation 50% of the time [41]. A potential gradient of approximately 6 V/cm has been reported to be the minimal field strength required for defibrillation with a 14-msec truncated exponential waveform [58]. Increasing the shock voltage so that the shock potential gradient exceeds this critical field strength throughout all of the ventricular muscle produces defibrillation in most cases. Decreasing the shock voltage below this critical field strength yields a greater percentage of unsuccessful defibrillation episodes, with earliest activation arising from areas of low potential gradient. As the shock strength is further decreased, the probability of successful defibrillation continues to decrease in an approximately sigmoidal relationship [13].

The concept of a critical point to date, although demonstrated only for the electrical induction of reentry, can be used to explain the probability of success for defibrillation in a manner similar to the explanation for the upper limit of ventricular vulnerability. Multiple activation fronts occur on the myocardium at all times during fibrillation [25], perhaps because of the presence of multiple, shifting pathways of reentry [43]. Since at any instant some portions of the ventricular myocardium are undergoing activation, other portions should be recovering. Thus, at any instant during fibrillation, the critical isorecovery line should also exist in several different sites throughout the ventricles. When the shock voltage is sufficiently high that the critical gradient is exceeded over the entire ventricular myocardium, a critical point should not occur, regardless of the locations of the critical isorecovery lines (Figure 20.8, isogradient line i). Because activation fronts do not conduct away from tissue directly excited by potential gradients exceeding the critical gradient level [19], all activation fronts of fibrillation present at the time of the shock will be halted, no new fronts will be created by the shock field, and defibrillation will occur. Thus, when the potential gradients over both ventricles exceed the critical gradient level,

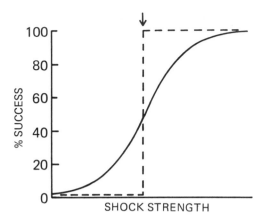

FIGURE 20.9. The probability curve for defibrillation. There is not a discrete defibrillation threshold (dashed line at arrow) above which all shocks succeed and below which all shocks fail. Rather, a dose-response type of curve exists (solid line), in which greater shock strengths are associated with greater percentages of success. (Modified with permission from Davy et al. [13].)

defibrillation shocks should succeed 100% of the time. Decreasing the shock voltage so that the critical isogradient line is present on the ventricles creates the possibility that, depending on the distribution of refractoriness at the time of the shock, the critical isogradient line can intersect a critical isorecovery line, creating a critical point. Thus, reentry rotors can be induced and lead to refibrillation, even though all fibrillatory activation fronts are extinguished by the shock. For shocks of slightly lower energy than the 100% successful shock, a high probability of success continues to exist. These shocks presumably produce the critical isogradient line in very small areas of myocardium where the probability that the critical degree of refractoriness will be present at the time of the shock is small (Figure 20.8, isogradient line h). As the shock voltage is further decreased, the critical isogradient line will cross progressively larger regions of myocardium, yielding a higher probability that it will intersect myocardium at the critical degree of refractoriness, creating a critical point and hence reentry. Low probabilities of successful defibrillation occur when the critical isogradient line crosses such a large volume of myocardium that the odds are high that it will intersect a critical isorecovery line at the instant of the shock (Figure 20.8, isogradient line g). Thus, successful defibrillation depends on avoiding the creation of critical points, which is best accomplished by increasing the shock voltage so that it exceeds the critical potential gradient level over the entire ventricular myocardium.

The critical point hypothesis may not be the total explanation for defibrillation in infarcted, ischemic, or diseased myocardium. With abnormal

myocardium, anatomic obstacles such as infarct scars or inherent disparities in refractoriness may exist prior to the application of the defibrillation
shock and create the conditions necessary for initiating reentry independent
of the interaction of critical isorecovery and isogradient lines. However, the
shock strength required for defibrillation appears to be similar for normal,
acutely ischemic, and chronically infarcted myocardium, although some
controversy exists [5,30,57]. Thus, the critical point concept may apply
also to abnormal myocardium. Further studies are necessary to determine
the role of critical points in the defibrillation of diseased hearts.

By itself, the critical point hypothesis does not totally account for the
probability of success curve for defibrillation because it does not explain
why one or two rapid cycles of activation frequently occur in the first few
100 msec following successful defibrillation shocks [7,45,74]. In the 1970s
these rapid activations were thought to support the critical mass hypothesis of fibrillation; they were thought to be continuations of activation fronts
present before the shock that arose from a ventricular volume smaller than
the critical mass so that they died out before engulfing the remainder of
the myocardium with the disorganized activation of fibrillation. As explained earlier in this chapter, cardiac mapping indicates that these activation fronts are not continuations of fronts present before the shock,
but are new activation fronts caused by the shock itself. The reason these
activation fronts sometimes stop spontaneously and other times reinduce
fibrillation may be similar to the reason activation fronts following shocks
during the vulnerable period of normal rhythm sometimes stop following a
few repetitive responses and other times lead to fibrillation.

A second question not answered by the critical point hypothesis is why
reentry rotors are not always seen after unsuccessful defibrillation shocks;
in more than half of unsuccessful defibrillation episodes with shocks slightly
weaker than required for defibrillation, activation appears to spread away in
all directions from a single point in a focal pattern [9,71]. This observation
may be similar to the observation discussed earlier that shocks during the
vulnerable period of regular rhythm can initiate activation patterns that
appear focal [54].

20.9   Very High Potential Gradients Have Detrimental Effects on the Heart

In addition to the beneficial effects of defibrillation, electrical fields can
have detrimental effects on the heart if the potential gradient is too high.
A very strong shock much greater than the upper limit of vulnerability can
induce ventricular fibrillation at any time during the cardiac cycle, not just
during the vulnerable period [18,39]. Strong shocks can also produce loss of
potassium from myocardial fibers, decreased conduction velocity, prolonged

depolarization, and neurostimulation of both cholinergic and adrenergic fibers [3,35,44,49]. At slightly higher levels of shock strength, decreased cardiac function, cessation of conduction, and inhibition of pacemaker cells can occur [31,35,48]. At still higher strengths, frank necrosis of myocardium can be produced [11].

The potential gradient or current density levels have been determined at which some of these detrimental effects are observed. For 10-msec truncated exponential waveforms, a shock field strength of approximately 200 mA/cm^2 causes a 50% decrease in left ventricular systolic pressure [46]. Inhibition for 4 sec of pacemaking activity in clusters of chick embryo cells occurs with square wave stimuli at approximately 80 V/cm [31]. Conduction block occurs at approximately 70 V/cm for a low tilt, truncated exponential waveform [70]. Ventricular tachycardia arises from myocardial regions exposed to potential gradients of 100 to 200 V/cm with low tilt, truncated exponential waveforms [58].

Strong defibrillation shocks can prolong depolarization and prevent activation for seconds to minutes. This finding caused Dudel to hypothesize that the paralysis of myocardium by electric shocks is the mechanism of defibrillation [15]. Such prolonged depolarization may be caused by alterations in intracellular ionic concentrations created by holes in the cell membrane, which have been reported to be created by potential gradients of 200 V/cm applied to in vitro chick embryo myocytes [32]. Since defibrillation requires a much smaller potential gradient than 200 V/cm [58,72] and since activation is normally observed within a fraction of a second following defibrillation shocks [7], electrical paralysis is not the usual mechanism of defibrillation. In fact, electrical paralysis is likely to be harmful, decreasing wall motion and probably inducing arrhythmias.

20.10   The Mechanism of Defibrillation

Most of the ideas discussed in this chapter are speculative. The mechanism or mechanisms of defibrillation are not yet known with certainty. Knowledge is advancing quickly about defibrillation, and the information in this chapter will probably soon be supplanted by additional findings. Adding to the excitement of this area is the steady improvement being made in the choice of shock strength required for defibrillation with implantable devices [5,14,29,33]. We hope that a better fundamental understanding of the mechanism of defibrillation will allow this rapid progress to continue.

Acknowledgements: Supported in part by the National Institutes of Health research grants HL-42760, HL-28429, HL-44066, HL-33637, HL-40092 and HL-17670, National Science Foundation Engineering Research Center Grant CDR-8622201, and by CPI Inc. and Physio-Control Corp. Dr. Tang is the recipient of a Canadian Heart Foundation Fellowship.

This chapter is an updated, revised version of a chapter written for the book *Cardiac Pacing* [26].

REFERENCES

[1] M.A. Allessie, F.I.M. Bonke, and F.J.G. Schopman. Circus movement in rabbit atrial muscle as a mechansim of tachycardia. III. The "leading circle" concept: A new model of circus movement in cardiac tissue without the involvment of an anatomical obstacle. *Circ. Res.*, 41:9–18, 1977.

[2] C. Antzelevitch and G.K. Moe. Electronic inhibition and summation of impulse conduction in mammalian sinoatrial node. *Am. J. Physiol.*, 245:H42–H53, 1983.

[3] M.F. Arnsdorf, D.A. Rothbaum, and R.W. Childers. Effect of direct current countershock on atrial and ventricular electrophysiological properties and myocardial potassium efflux in the thoracotomised dog. *Cardiovas. Res.*, 11:324–333, 1977.

[4] J.D. Bourland, W.A. Tacker Jr., and L.A. Geddes. Strength-duration curves for trapezoidal waveforms of various tilts for transchest defibrillation in animals. *Med. Instrum.*, 12:38–41, 1978.

[5] M.S. Chang, H. Inoue, M.J. Kallok, and D.P. Zipes. Double and triple sequential shocks reduce ventricular defibrillation threshold in dogs with and without myocardial infarction. *J. Am. Coll. Cardiol.*, 8:1393, 1986.

[6] P-S Chen, N. Shibata, E.G. Dixon, R.O. Martin, and R.E. Ideker. Comparison of the defibrillation threshold and the upper limit of ventricular vulnerability. *Circulation*, 73:1022–1028, 1986.

[7] P-S Chen, N. Shibata, E.G. Dixon, et al. Activation during ventricular defibrillation in open-chest dogs: Evidence of complete cessation and regeneration of ventricular fibrillation after unsuccessful shocks. *J. Clin. Invest.*, 77:810–823, 1986.

[8] P-S Chen, N. Shibata, P.D. Wolf, et al. Epicardial activiation during successful and unsuccessful ventricular defibrillation in open chest dogs. *Cardiovasc. Rev. Rep.*, 7:625–648, 1986.

[9] P-S Chen, P.D. Wolf, S.D. Melnick, N.D. Danieley, W.M. Smith, and R.E. Ideker. Comparison of activation during ventricular fibrillation and following unsuccessful defibrillation shocks in open chest dogs. *Circ. Res.*, 66:1544–1560, 1990.

[10] P-S Chen, P.D. Wolf, F.J. Claydon, et al. The potential gradient field created by epicardial defibrillation electrodes in dogs. *Circulation*, 74:626–636, 1986.

[11] C.F. Dahl, G.A. Ewy, E.D. Warner, and E.D. Thomas. Myocardial necrosis from direct current countershock, effect of paddle size and time interval between discharge. *Circulation*, 50:956–961, 1974.

[12] J.P. Daubert, D.W. Frazier, W. Krassowska, S. Yabe, W.M. Smith, and R.E. Ideker. Direct excitation of relatively refractory tissue by monophasic and biphasic shocks (abstract). *J. Am. Coll. Cardiol.*, 13:215A, 1989.

[13] J.M. Davy, E.S. Fain, P. Dorian, and R.A. Winkle. The relationship between successful defibrillation and delivered energy in open-chest dogs: Reappraisal of the "defibrillation threshold" concept. *Am. Heart J.*, 113:77–84, 1987.

[14] E.G. Dixon, A.S.L. Tang, et al. Improved defibrillation thresholds with large contoured epicardial electrodes and biphasic waveforms. *Circulation*, 76:1176–1184, 1987.

[15] J. Dudel. Elektrophysiologische grundlagen der defibrillation und künstlichen stimulation des herzens. *Med. Klin.*, 52:2089–2100, 1968.

[16] N. El-Sherif, W.B. Gough, and M. Restivo. Rentrant ventricular arrhythmias in the late myocardial infarction period: 14. Mechansims of resetting, entrainment, acceleration, or terminiation of reentrant tachycardia by programmed electrical stimulation. *PACE*, 10:341–371, 1987.

[17] N. El-Sherif, R.A. Smith, and K. Evans. Canine ventricular arrhythmias in the late myocardial infarction period: 8. Epicardial mapping of reentrant circuits. *Circ. Res.*, 49:255–265, 1981.

[18] A. Fabiato, P. Coumel, R. Gourgon, and R. Saumont. Le seuil de réponse synchrone des fibres myocardiques. application à la comparaison expérimentale de l'efficacité des différentes formes de chocs électriques de défibrillation. *Arch. Mal Coeur*, 60:527–544, 1967.

[19] D.W. Frazier, P.D. Wolf, J.M. Wharton, A.S.L. Tang, W.M. Smith, and R.E. Ideker. Stimulus-induced critical point: Mechanism for the electrical initiation of reentry in normal canine myocardium. *J. Clin. Invest.*, 83:1039–1052, 1989.

[20] W.E. Gaum, V. Elharrar, P.D. Walker, and D.P. Zipes. Influence of excitability on the ventricular fibrillation threshold in dogs. *Am. J. Cardiol.*, 40:929–939, 1977.

[21] J.H. Gold, J.C. Schuder, and H. Stoeckle. Contour graph for relating per cent success in achieving ventricualr defibrillation to duration, current, and energy content of shock. *Am. Heart J.*, 98:207–212, 1979.

[22] B.F. Hoffman, E.F. Gorin, F.S. Wax, A.A. Siebens, and C.M. Brooks. Vulnerability to fibrillation and the ventricular-excitability curve. *Am. J. Physiol.*, 167:88–94, 1951.

[23] B.F. Hoffman, E.E. Suckling, and C.M. Brooks. Vulnerability of the dog ventricle and effects of defibrillation. *Circ. Res.*, 3:147–151, 1955.

[24] D.R. Hooker, W.B. Kouwenhoven, and O.R. Langworthy. The effect of alternating currents on the heart. *Am. J. Physiol.*, 103:444–454, 1933.

[25] R.E. Ideker, G.J. Klein, L. Harrison, et al. The transition to ventricular fibrillation induced by reperfusion following acute ischemia in the dog: A period of organized epicardial activation. *Circulation*, 63:1371–1379, 1981.

[26] R.E. Ideker, A.S.L. Tang, D.W. Frazier, N. Shibata, P-S Chen, and J.M. Wharton. Ventricular defibrillation: Basic concepts. In N. El-Sherif and P. Samet, editors, *Cardiac Pacing*, pages 713–726. Saunders, Orlando, 1991.

[27] J. Jalife and C. Antzelevitch. Phase resetting and annihilation of pacemaker activity in cardiac tissue. *Science*, 206:696–697, 1979.

[28] J. Jalife, V.A.J. Slenter, J.J. Salata, and D.C. Michaels. Dynamic vagal control of pacemaker activity in the mammalian sinoatrial node. *Circ. Res.*, 52:642–656, 1983.

[29] D.L. Jones, G.J. Klein, G.M. Guiraudon, et al. Internal cardiac defibrillation in man: pronounced improvement with sequential pulse delivery to two different lead orientations. *Circulation*, 73:484–491, 1986.

[30] D.L. Jones, A. Sohla, and G.J. Klein. Internal cardiac defibrillation threshold: effects of acute ischemia. *PACE*, 9:322–331, 1986.

[31] J.L. Jones, E. Lepeschkin, R.E. Jones, and S. Rush. Response of cultured myocardial cells to countershock-type electric field stimulation. *Am. J. Physiol.*, 235:H214–H222, 1978.

[32] J.L. Jones, C.C. Proskauer, W.K. Paull, E. Lepeschkin, and R.E. Jones. Ultrastructural injury to chick myocardial cells in vitro following "electric countershock." *Circ. Res.*, 46:387–394, 1980.

[33] K.M. Kavanagh, A.S.L. Tang, D.L. Rollins, W.M. Smith, and R.E. Ideker. Comparison of the internal defibrillation thresholds for monophasic and double and single capacitor biphasic waveforms. *J. Am. Coll. Cardiol.*, 14:1343–1349, 1989.

[34] B.G. King. *The Effect of Electric Shock on Heart Action with Special Reference to Varying Susceptibility in Different Parts of the Cardiac Cycle.* Ph.D. thesis, Columbia University, 1934.

[35] G. Koning, A.H. Veefkind, and H. Schneider. Cardiac damage caused by direct application of defibrillation shock to isolated Langendorff-perfused rabbit heart. *Am. Heart J.*, 100:473–482, 1980.

[36] W. Krassowska, T.C. Pilkington, and R.E. Ideker. Periodic conductivity as a mechanism for cardiac stimulation and defibrillation. *IEEE Trans. Biomed. Eng.*, BME-34:555–560, 1987.

[37] C. Laxer, R.E. Ideker, W.M. Smith, L.D. German, L. Harrison, and T.C. Pilkington. Computer acquisition of a database for relating myocardial infarct geometry to cardiac electrical potentials. *Proc. Comp. Cardiol.*, 339–342, 1980.

[38] E. Lepeschkin, J.L. Jones, S. Rush, and R.E. Jones. Local potential gradients as a unifying measure for thresholds of stimulation, standstill, tachyarrhythmia and fibrillation appearing after strong capacitor discharges. *Adv. Cardiol.*, 21:268–278, 1978.

[39] C. Lesigne, B. Levy, R. Saumont, P. Birkui, A. Bardou, and B. Rubin. An energy-time analysis of ventricular fibrillation and defibrillation thresholds with internal electrodes. *Med. Biol. Eng.*, 14:617–622, 1976.

[40] B. Lown. Electrical reversion of cardiac arrhythmias. *Br. Heart J.*, 29:469–489, 1976.

[41] W.C. McDaniel and J.C. Schuder. The cardiac ventricular defibrillation threshold: Inherent limitations in its application and interpretation. *Med. Instrum.*, 21:170–176, 1987.

[42] M. Mirowski. The automatic implantable cardioverter-defibrillator: An overview. *J. Am. Coll. Cardiol.*, 6:461–466, 1985.

[43] G.K. Moe, W.C. Rheinboldt, and J.A. Abildskov. A computer model of atrial fibrillation. *Am. Heart J.*, 67:200–220, 1964.

[44] E.N. Moore and J.F. Spear. Electrophysiologic studies on the initiation prevention, and termination of ventricular fibrillation, In D.P. Zipes and J. Jalife, editors, *Cardiac Electrophysiology and Arrhythmias*, pages 315–322. Grune & Stratton, Orlando, 1985.

[45] M.M. Mower, M. Mirowski, J.F. Spear, and E.N. Moore. Patterns of ventricular activity during catheter defibrillation. *Circulation*, 49:858–861, 1974.

[46] M.J. Niebauer, C.F. Babbs, L.A. Geddes, and J.D. Bourland. Efficacy and safety of the reciprocal pulse defibrillator current waveform. *Med. Biol. Eng. Comp.*, 22:28–31, 1984.

[47] O. Orias, C.M. Brooks, E.E. Suckling, J.L. Gilbert, and A.A. Siebens. Excitability of the mammalian ventricle throughout the cardiac cycle. *Am. J. Physiol.*, 163:272–232, 1950.

[48] D.G. Pansegrau and F.M. Abboud. Hemodynamic effects of ventricular defibrillation. *J. Clin. Invest.*, 49:282–297, 1970.

[49] B. Peleska. Cardiac arrhythmias following condenser discharges and their dependence upon strength of current and phase of cardiac cycle. *Circ. Res.*, 13:21–32, 1963.

[50] R. Plonsey and R.C. Barr. Inclusion of junction elements in a linear cardiac model through secondary sources: Application to defibrillation. *Med. Biol. Eng. Comp.*, 24:137–144, 1986.

[51] S.M. Pogwizd and P.B. Corr. Reentrant and nonreentrant mechanisms contribute to arrhythmogenesis during early myocardial ischemia: Results using three-dimensional mapping. *Circ. Res.*, 61:352–371, 1987.

[52] N. Shibata, P-S Chen, E.G. Dixon, et al. Epicardial activation following unsuccessful defibrillation shocks in dogs. *Am. J. Physiol.*, 255:H902–H909, 1988.

[53] N. Shibata, P-S Chen, E.G. Dixon, et al. Epicardial mapping of the initiation of ventricular fibrillation by shocks during the vulnerable period (abstract). *J. Am. Coll. Cardiol.*, 7:183A, 1986.

[54] N. Shibata, P-S Chen, E.G. Dixon, et al. Influence of shock strength and timing on induction of ventricular arrhythmias in dogs. *Am. J. Physiol.*, 255:H891–H901, 1988.

[55] W.M. Smith and R.E. Ideker. Computer techniques for epicardial and endocardial mapping. *Prog. Cardiovasc. Dis.*, 26:15–32, 1983.

[56] A.S.L. Tang, P.D. Wolf, W.M. Claydon III, F.J. Smith, T.C. Pilkington, and R.E. Ideker. Measurement of defibrillation shock potential distributions and activation sequences of the heart in three dimensions. *Proc. IEEE*, 76:1176–1186, 1988.

[57] J.M. Wharton, V.J. Richard, C.E. Murry, D.L. Rollins, K.A. Reimer, and R.E. Ideker. Effect of chronic myocardial infarction on defibrillation (abstract). *Circulation*, 76:IV–108, 1987.

[58] J.M. Wharton, P.D. Wolf, P-S Chen, et al. Is an absolute minimum potential gradient required for ventricular defibrillation? (abstract). *Circulation*, 74:II–342, 1986.

[59] C.J. Wiggers. The mechanism and nature of ventricular defibrillation. *Am. Heart J.*, 20:399, 1940.

[60] C.J. Wiggers. The physiologic basis for cardiac resuscitation from ventricular fibrillation—Method for serial defibrillation. *Am. Heart J.*, 20:413, 1940.

[61] C.J. Wiggers and R. Wegria. Ventricular fibrillation due to single, localized induction and condenser shocks applied during the vulnerable phase of ventricular systole. *Am. J. Physiol.*, 128:500–505, 1940.

[62] A.T. Winfree. Spiral waves of chemical activity. *Science*, 175:634–636, 1972.

[63] A.T. Winfree. Sudden cardiac death. *Sci. Am.*, 248:144–161, 1983.

[64] A.T. Winfree. Unclocklike behavior of biological clocks. *Nature*, 253: 315–319, 1975.

[65] A.T. Winfree. *When Time Breaks Down: The Three-Dimensional Dynamics of Electrochemical Waves and Cardiac Arrhythmias*. Princeton University Press, Princeton, 1987.

[66] A.L. Wit, M.A. Allessie, F.I.M. Bonke, W. Lammers, J. Smeets, and J.J. Fenoglio Jr. Electrophysiologic mapping to determine the mechanism of experimental ventricular tachycardia initiated by premature impulses: experimental approach and initial results demonstrating reentrant excitation. *Am. J. Cardiol.*, 49:166–185, 1982.

[67] F.X. Witkowski and P.A. Penkoske. Relation of defibrillatory potential minima to post-defibrillatory activation. In *Proc. 10th Annual Conf. of the IEEE Engineering in Medicine and Biology Society*, pages 212–213, 1988.

[68] F.X. Witkowski and P.A. Penkoske. Simultaneous cardiac potential field and direct cardiac recordings during ventricular defibrillation (DF) (abstract). *Circulation*, 76:IV–108, 1987.

[69] P.D. Wolf, D.L. Rollins, W.M. Smith, and R.E. Ideker. A cardiac mapping system for the quantitative study of internal defibrillation. In *Proc. 10th Annual Conf. of the IEEE Engineering in Medicine and Biology Society*, pages 217–218, 1988.

[70] S. Yabe, W.M. Smith, J.P. Daubert, P.D. Wolf, D.L. Rollins, and R.E. Ideker. Conduction disturbances caused by high current density electric fields. *Circ. Res.*, 66:1190–1203, 1990.

[71] X. Zhou, P-S. Chen, P.D. Wolf, W.M. Smith, and R.E. Ideker. Activation patterns following unsuccessful defibrillation shocks (abstract). *Circulation*, 80:II–135, 1989. Presented at the 62nd Scientific Sessions of the American Heart Association, November, 1989.

[72] X. Zhou, J.P. Daubert, P.D. Wolf, W.M. Smith, and R.E. Ideker. The potential gradient for defibrillation (abstract). *Circulation*, 78:II–645, 1988.

[73] X. Zhou, P.D. Wolf, D.L. Rollins, W.M. Smith, and R.E. Ideker. Potential gradient needed for defibrillation with monophasic and biphasic shocks (abstract). *PACE*, 12:651, 1989.

[74] D.P. Zipes, J. Fischer, R.M. King, A. Nicoll, and W.W. Jolly. Termination of ventricular fibrillation in dogs by depolarizing a critical amount of myocardium. *Am. J. Cardiol.*, 36:37–44, 1975.

21

Mechanically Induced Changes in Electrophysiology: Implications for Arrhythmia and Theory

Max J. Lab[1]
Arun V. Holden[2]

ABSTRACT Despite the fact that heart disease together with arrhythmia is a potent cause of sudden death in the Western world the precise mechanisms remain unclear, and the treatment on the whole disappointing. The initiating cause of the first ectopic beat that precipitates lethal arrhythmia in the first hours of myocardial ischemia is not understood. The reasons for sustaining the arrhythmia are also not understood. Finally the origin of sudden death in myocardial failure of diverse etiology is not clear. We present evidence that mechanical changes, prevalent in ischemia and cardiac failure, can initiate electrophysiological changes by a process we call mechanoelectric feedback, which is the reverse of excitation-contraction-coupling. Pathological events could disturb this feedback system so producing a contributory mechanism for electrical instability leading to ventricular arrhythmia. We discuss the theoretical possibility that mechanoelectric feedback on a beat-to-beat basis provides an additional nonlinear dynamical feature which could produce chaotic responses leading to arrhythmia, and we describe an example or analog of how this could occur.

21.1 Introduction

Ventricular arrhythmia is common not only in patients with acute myocardial ischemia [36,37,55,68], but also in those with chronic congestive heart failure [23,58]. Despite the fact that heart disease together with arrhythmia is a potent cause of sudden death in the Western world (see also above references) the precise mechanisms remain unclear and the treatment

---

[1] Department of Physiology, Charing Cross and Westminster Medical School, London, England

[2] Department of Physiology and Centre for Non Linear Science, Leeds University, Leeds, England

inadequate. It is striking that these arrhythmias parallel the degree of mechanical myocardial dysfunction rather than electrophysiological changes and seem to be independent of etiology [6,73]. Since the electrical activity of the heart triggers its mechanical activity, most approaches to ventricular arrhythmia are based on abnormal electrical behavior: either irregularity in the ventricular action potential or reentrant conduction paths in the ventricle. However, evidence that changes in the stress/strain relationships within the myocardium may lead directly to arrhythmia is increasing (see [44,49] for reviews). Moreover, the concept that chaos is involved in arrhythmia generation is gaining support [29]. This chapter suggests that mechanical dysfunction associated with heart failure and regional acute ischemia plays an important role in the development of arrhythmia. This could be by a process known as mechanoelectric feedback or contraction-excitation feedback. We also speculatively examine the theoretical possibility that this feedback provides an additional nonlinear feature that results in chaotic responses leading to the arrhythmia generation.

Excitation-contraction coupling, briefly summarized in Figure 21.1, has been extensively investigated (see [27,39,80] for reviews). In this process the action potential initiates calcium release from the sarcoplasmic reticulum and calcium interacts with the contractile proteins to produce force and/or length changes. Relaxation involves both sequestration of calcium and its membrane extrusion. However, evidence is accumulating that this process is not strictly unidirectional and feedback paths exist from contractile activity to influence sarcolemmal membrane electrophysiology (as shown in Figure 21.1).

Excitation-contraction coupling may be followed by the heavy black arrows (mainly clockwise). Ion fluxes (1+) in Figure 21.1 determine the membrane potential, which can also provide a driving force for ion movements (1−). The changes in membrane potential are a function of the ionic equilibrium potentials and conductances, g. At rest g_{Na} is low and g_K is high. The latter mainly contributes to the negative resting potential which is maintained in the long term by an adenosine triphosphatase (ATPase)-dependent sodium/potassium pump. This keeps the internal sodium ion concentration, $[Na^+]_i$, low and $[K^+]_i$ high. Extracellular calcium ion concentration, $[Ca^{2+}]_o$, is relatively high while sarcoplasmic calcium concentration, $[Ca^{2+}]_s$, is very low. With the upstroke of the action potential g_{Na} is rapidly increased and the fast inward sodium current, i_{Na}, reverses the transmembrane potential. Despite the consequential increase in the outward driving forces for potassium, g_K decreases and the outward repolarizing current is less than expected. The cardiac action potential is thus prolonged. A slow inward current, carried mainly by calcium, i_{Ca}, also prolongs the action potential as does an electrogenic sodium/calcium exchange. The channels for i_{Ca} are influenced by cyclic adenosine monophosphate (c-AMP)-dependent protein kinase. Depolarization (2) causes a rise in sarcoplasmic calcium from the stores (3+) directly, and probably by calcium-

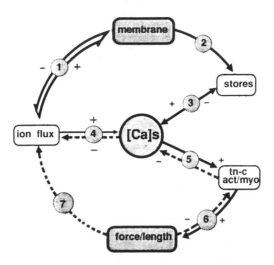

FIGURE 21.1. Diagram of some of the interactions between membrane potential and myocardial contraction.

induced calcium release. Normally in mammalian muscle i_{Ca} does not raise $[Ca^{2+}]_s$ immediately to any significant degree unless the action potential is long (4+). The calcium combines with Troponin-C (5+), which causes Troponin-I to allow actin and myosin (act/myo) interaction. The process, which needs ATP, results in force development (6+). As, or probably before, the membrane repolarizes during relaxation (6−), the sarcoplasmic reticulum sequesters calcium (5−; 3−). Calcium can also leave the sarcoplasm by a metabolically dependent calcium pump or by sodium/calcium exchange (4−). Greater binding to Troponin can also lower $[Ca^{2+}]_s$, but this is associated with increased force rather than faster relaxation. Length-dependent activation is incorporated in (6+/−). Force and length changes could link with membrane events (mechanoelectric feedback) by processes depicted by the dotted lines. For example, mechanical changes could change ionic flux by directly affecting permeability or diffusion gradients (7). Indirectly (6−;5−), force and length changes could influence the membrane by altering $[Ca^{2+}]_s$. This may influence ionic flux (4−), and hence membrane potential (1+), by modulation of electrochemical gradients for calcium, outward potassium currents, "leak" currents, and electrogenic sodium/calcium exchange.

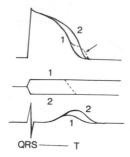

FIGURE 21.2. Diagram of mechanically induced changes in electrophysiology. Middle tracing depict the loads: 1 is a high and 2 is a low load, which can be associated with muscle shortening. The action potential (upper tracing) is shorter (1) with the high load. The ECG diagramed in the lower trace shows a shorter QT interval and a smaller T wave. The dashed line in the "load" trace indicates a mechanical change within a single beat. This is accompanied by an afterdepolarization in the action potential (arrow).

21.2   Experimental Evidence for Mechanoelectric Feedback

There is a common methodology in most of the experiments described here. They generally had simultaneous measurements of mechanical variables (force, length, pressure) and electrophysiological variables (intracellular action potentials, monophasic action potentials, electrocardiogram). During these recordings mechanical loading conditions were changed and the results were generally concordant in the different experimental situations. Mechanical maneuvers, comparable to those found in the failing heart and dyskinetic segments of the left ventricle during regional ischemia, were capable of influencing the cardiac action potential by changing its duration and excitability and/or inducing abnormal depolarizations, to produce ventricular arrhythmia.

21.2.1   MECHANICALLY INDUCED CHANGES IN ACTION POTENTIAL DURATION

As arrhythmias occur during beat-to-beat cardiac activity, mechanical changes during a cardiac cycle may be of paramount importance. However, continuous stretch of healthy heart muscle does shorten the action potential duration (equivalent to the QT interval of the ECG) and can generate spontaneous depolarizations [13,18,21,41,45,46,60,65].

The general changes in the action potential with intra or interbeat mechanical alterations may be as depicted in Figure 21.2. An increased load (beat 1) is accompanied with a shortened action potential by comparison with myocardium contracting against a reduced load (beat 2). The load reduction is, importantly, associated with muscle shortening to length (2) and "deactivation" of force (not shown in the diagram). That is, the instantaneous midcycle force now generated at the new shorter length (2) is smaller than the force it would have generated had the muscle started contracting at that new length [4]. This combination of reduced load and action potential prolongation has been observed experimentally in isolated papillary muscle [40,61], frog ventricular strip [61], intact perfused ventricles of frog [46] and rabbit [43], as well as intact ventricle in situ of dog [25,53] and pig [15]. During cardiac surgery in humans, reduced load also leads to prolonged action potential duration [54,76].

The dotted lines in Figure 21.2 represent the results of an intrabeat mechanical decrease of load. This leads to a change in action potential duration of the same beat that can be associated with an "afterdepolarization" (arrow in Figure 21.2) [40,44,46,60].

21.2.2   MECHANICALLY INDUCED CHANGES IN EXCITABILITY

The changes in action potential described above appear to be independent of external neurohumoral influences [50] and are reflected in changes in myocardial refractoriness and excitability [53]. Changes in the absolute refractory period of the in situ pig heart parallel those of the action potential duration during load manipulation [17]. Since alterations in cardiac action potential duration have consequences on ventricular excitability, changes in load via mechanoelectric feedback may thus affect the development of arrhythmia.

21.2.3   MECHANICALLY INDUCED CHANGES IN THE ELECTROCARDIOGRAM

The QT interval is prolonged with reduced load (Figure 12.2). This has been found under experimental conditions [44] and in man [22]. The T-wave of the electrocardiogram, which is some function of electrophysiological inhomogeneity of repolarization, also changes with altered mechanical loading (Figure 21.2). Repolarization in normal myocardium is heterogeneous and gives rise to the anomalous "upright" T wave of the ECG [64]. This mechanically induced change in the T wave implies an altered repolarization gradient.

21.3   Mechanism of Mechanoelectric Feedback

The mechanism almost certainly lies in the close interaction between membrane and mechanical events at the cellular level. In the intact heart in situ, however, there are confounding changes in baroreceptor reflexes when changing circulatory pressures, but one pilot study on a different aspect of mechanoelectric feedback [50] suggests that reflexes do not play a major role. Mechanisms for mechanically induced potentials have been presented [40,46] and discussed in reviews [15,44]. One possibility relates the mechanically induced electrical changes to changes in intracellular calcium [48]. These intracellular calcium changes would affect the action potential either by an electrogenic sodium/calcium exchange [63] or by a calcium-activated inward current [11] to prolong the action potential duration. If the mechanical influence is in the latter part of the cardiac cycle, the consequent changes in intracellular calcium could produce early afterdepolarizations. Mechanically activated ionic channels in the sarcolemma is a possibility that also needs entertaining.

21.4   Mechanoelectric Feedback and Arrhythmia: General Considerations

21.4.1   INITIATION OF ARRHYTHMIA

The first mechanically induced ventricular ectopic beat in pathological heart could be produced by one of several mechanisms: (1) an afterdepolarization reaching threshold—see Figure 21.3, (2) a diastolic depolarization reaching threshold—see above, (3) local intramyocardial current flow, where local asymmetry in contraction patterns could produce local differences in action potential duration, and (4) a reentrant beat where local dyskinesia could produce local changes in excitability and conduction.

21.4.2   SUSTAINING OF VENTRICULAR FIBRILLATION: VENTRICULAR GEOMETRY AND REENTRANT ARRHYTHMIAS

A perturbation that alters the normal dispersion of repolarization and excitability may potentiate the likelihood of reentrant arrhythmia, since reentry depends on nonuniform recovery of excitability and repolarization [14,69]. Regional differences exist in electrical restitution [51] such that changes in loading status may possibly exaggerate the electrical heterogeneity [43] that normally exists. This would be due to asymmetry of ventricular geometry and wall stress and thus predispose to reentrant arrhythmia. In this respect, ventricular dilatation of the isolated rabbit heart has

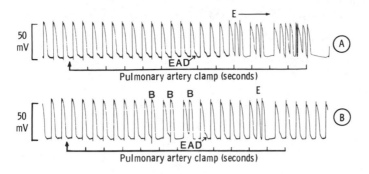

FIGURE 21.3. Monophasic action potentials recorded from the epicardium of the right ventricle during pulmonary artery occlusion in anesthetized dog. (A) Pulmonary occlusion (arrow) results in an early afterdepolarization (EAD) on the falling phase of the action potential. Irregular ectopic beats E appear a few seconds later. (B) Records from the same preparation where the first three ectopics produce a bigeminal B or paired pattern with the sinus beats.

been shown to lead to increased dispersion of refractoriness independent of coronary perfusion pressure and cycle length [67].

Given that a localized disturbance of heart tissue (e.g., produced by ischemia) could allow the formation of a broken wavefront and hence reentrant activity, a variant of mechanoelectric feedback may be visualized. Holden and colleagues [34] emphasized the importance of two factors allowing reentrant circulation. First, some form of anisotropy is necessary for the initiation of reentrant activity. Second, reentry requires a minimum ventricular size: there must be enough room for the ventricle to contain a wave of excitation rotating around an organizing filament. That is, any distension or hypertrophy of the heart will increase the likelihood of ventricular fibrillation, an observation known for many years [26].

21.5   Mechanoelectric Feedback and Arrhythmia: Clinical Considerations

It seems clear that *mechanical* changes can alter the action potential duration of the left ventricle on a regional basis and hence change repolarization indices such as the QT interval and the T wave of the ECG. Moreover, the action potential changes may be accompanied by a change in myocardial excitability. The excitability changes, together with the dispersion of repolarization, have direct clinical relevance since both of these factors are known to affect ventricular arrhythmogenesis.

21.5.1 INDIRECT INTRAMURAL LOAD CHANGES

Regional ischemia.

Primary electrophysiological disturbances are thought to cause arrhythmia in myocardial infarction, probably via reentry and abnormal current flow in and around the ischemic area. However, wall motion disturbances are striking in these regions and could conceivably result in mechanically induced arrhythmia. The contribution could be through any or all of the mechanisms mentioned above: mechanically induced early afterdepolarizations and ectopic impulse formation or inhomogeneity of mechanoelectric feedback, and thus dispersion of repolarization and arrhythmic current flow. A third possibility could be variations in refractory periods accompanying the inhomogeneous expression of mechanoelectric feedback, conduction velocities, and reentry paths. Interestingly, phase I arrhythmias occur in the first hour of ischemia and it is during this period that the myocardium is most compliant [66]. The compliance could enable the mechanical circumstances necessary for mechanically induced arrhythmia. In support of mechanoelectric feedback being a contributor to arrhythmia are the observations that the morbidity and mortality of coronary artery disease correlate less well with ventricular electrophysiological changes than when the degree of myocardial mechanical dysfunction is also incorporated in the correlations [3,6,42,73,74].

Congestive cardiac failure

Sudden death is probably arrhythmic in about half the patients in heart failure [2]. Heart failure is associated with structural [71], metabolic [35,75], and neurohumoral changes [24]. Recent studies using rabbit hearts with adriamycin-induced heart failure [20] suggest that an intrinsic mechanism, independent of any neural, humoral, or mechanical effect, may contribute to the electrophysiological changes seen in myopathic hearts. However, several observations as well as problems with conventional therapy suggest that we consider a mechanical bias in proposing mechanisms and treatment. The presence of unsustained ventricular tachycardia roughly doubles the mortality rate after adjusting for the mechanical index of left ventricular ejection fraction [10].

21.5.2 EXTRAMURAL (SYSTEMIC) LOAD CHANGES

Hypertension [57] and aortic stenosis [78] have a high incidence of ventricular arrhythmia that does not correlate well with etiology or coronary blood flow.

21.5.3 ARRHYTHMIA THERAPY

The above considerations raise some interesting possibilities relating to acute therapy through mechanical intervention. Can one stop an episode of ventricular fibrillation by rapidly and mechanically reducing the size of the heart? There have in fact been reports of cardioversion of ventricular fibrillation induced by cough [1,56,72], which may be regarded as a short sharp valsalva maneuver thus squeezing the heart, or a thump on the chest [5].

A feasible corollary would be the reduction in the incidence of ventricular fibrillation by internally reducing heart size; say by compartmentalizing more blood in the venous system by venodilation. The efficacy of conventional antiarrhythmic drugs in controlling arrhythmia in heart failure assessed by programmed stimulation [59] or continuous 24-hr ECG [77] is poor. A low ejection fraction is associated with a poor response. To date, no randomized clinical trial has shown improved survival following treatment of cardiac failure with conventional antiarrhythmic drugs. This may indicate that the arrhythmia is caused by mechanical dysfunction and that it thus cannot be controlled by antiarrhythmic drugs. Moreover, if the hypothesis that ventricular arrhythmias in heart failure are caused by abnormal stress/strain relationships within the myocardium is correct, then treatment with peripheral vasodilators, which would reduce wall stress indirectly by reducing preload, afterload, or both, should have a favorable effect on arrhythmia. Such an effect has been shown [8–10,79]. Heart size would also be reduced, and this may limit the possibility of sustaining ventricular fibrillation. Vasodilators have been shown experimentally to prolong the action potential and absolute refractory period [16], and these electrophysiological changes are generally antiarrhythmic. The reduced efficacy of conventional antiarrhythmic drugs and the apparent beneficial effects of vasodilator compounds support the hypothesis that mechanoelectric feedback is implicated in the generation of arrhythmia.

21.6 Theoretical Implications

The discussion above strongly suggests that mechanoelectric feedback may be involved in arrhythmia. There is, moreover, evidence that chaotic processes are concerned in some arrhythmias [29,30]. Although we do not present any definitive theoretical treatment of the mechanoelectric link here, we think it is relevant to suggest theoretical questions at this stage because, first, the concept itself may have clinical importance and could benefit by theoretical approaches, and second, the introduction of feedback pathways and time delays into a system can increase its range of dynamic behavior, and this may have bearing on the generation of chaos and/or cardiac arrhythmias.

21.6.1 MECHANICALLY INDUCED ECTOPICS

We choose our particular example of mechanoelectric feedback induced arrhythmia (Figure 21.3) because it is related to the parasystole and bigeminy recently modeled by Glass and coworkers using nonlinear system analysis [12,28]. In our example, occlusion of the pulmonary artery of dogs increased right ventricular pressure and led to the appearance of transient depolarizations that progressively increased in size until an apparent threshold was reached to produce right ventricular extrasystoles (Figure 21.3). These can appear as bigeminy (Figure 21.3B). The changes were not accompanied by ST segment shifts in the ECG and occurred within three to four beats. They are not caused by ischemia but are probably related to disturbances of wall segment motion [13]. The observations would be compatible with the mechanical intervention redistributing regional forces so that abnormal wall motion precipitates a discharging ventricular focus. The early beat has a small pressure/volume and the next sinus beat an exceptionally large one (postextrasystolic potentiation) and, once more, a differential wall motion distribution with an ectopic beat. Cyclic repetition of these mechanoelectric events produces the bigeminal pattern. Similar electromechanical disturbances have been seen in the left ventricular wall following aortic occlusion in the pig [44,60], and the electrophysiological changes that occur during load manipulation in this preparation correlate with the mechanical events [15]. Whether the analysis used by Glass and others [12,28] can be applied to the ectopy and bigeminy shown in Figure 21.3 is a matter for scrutiny. Moreover, precisely how one might establish the quantitative significance of mechanoelectric feedback as generating arrhythmia needs detailed exploration.

21.6.2 CHANGES IN RESTITUTION

One tentative approach in relating mechanoelectric feedback to chaotic dynamics using electrical restitution curves derives, purely by analogy, from the analyses of Chialvo and colleagues [7]. Briefly, they demonstrated a model of deterministic chaos in Purkinje fibers when the nonlinear system showed: (1) supernormality in the threshold strength/interval curve, (2) steepening of the electrical restitution curve, and (3) a discontinuity in the electrical restitution curve. The dynamics were modeled by empirical, experimentally determined curves. There are in fact experimental observations on mechanically induced electrophysiological changes that parallel the requirements of this analysis in an analogous way, and these are shown in Figure 21.4.

Increasing the afterload in intact pig heart in situ reduced absolute refractory period (Figure 21.4A). An increase in load in intact isolated rabbit ventricle shifted the electrical restitution curve down, but more importantly, steepened the time course of early recovery (Figure 21.4B). The

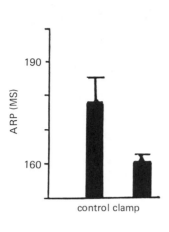
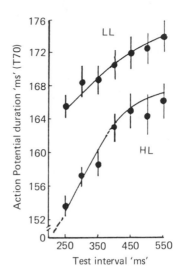

FIGURE 21.4. Experimental observations on mechanically induced electrophysiological changes. Left: Intact pig left ventricle in situ shows a significant reduction in absolute refractory period (ARP) when aorta is clamped so that blood pressure increases by 5-10%. β blockade did not influence the results. Right: Restitution curves from an intact Langendorf-perfused rabbit heart lightly loaded (LL) and heavily loaded (HL).

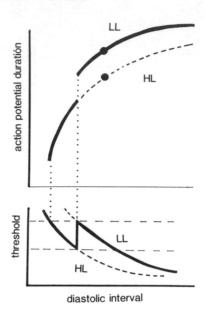

FIGURE 21.5. Diagram of restitution curve (top panel) and threshold (bottom panel) during light (LL) and heavy (HL) loads (see text). The threshold curves by analogy with the study of Chialvo et al. [7] can have three horizontal ranges with the middle range corresponding to the critical range predisposing to chaotic behavior and irregular rhythms.

reduced refractory period shown in Figure 21.4A, albeit in pig heart in situ, means that the restitution curve would also have shifted to the left, and this is indicated by the dashed line extending the curve. The effect these changes would have on an electrical restitution curve are diagrammatically indicated in the upper panel of Figure 21.5. The threshold strength/interval curves that would be associated with these restitution curves is indicated in the lower panel. The increase in load also shifts the strength/interval curve to the left. The vertical dotted lines mark inexcitability in the loaded and unloaded situations.

An ectopic beat would redistribute regional forces in the ventricle with a high mechanical load in a particular region. This beat therefore would have dynamical aspects different from a sinus beat in the same region by virtue of the electrophysiological events in effect jumping from a high load restitution curve to a normal load curve, and from one threshold curve to another. For successive beats there could be a type of "discontinuity" in the restitution curve, as described by Chialvo and colleagues [7], and a quasi-"supernormal" period in the strength/interval curve. The latter may be visualized if one joins the high load and low load curves for say two consecutive beats at diastolic intervals short enough to be commensurate with early ectopics. This strength/interval curve could be analogous to the curve described by Chialvo and colleagues [7]. This type of analogy is already very speculative, and further speculation is hazardous. Albeit only by analogy, the situation just described has much in common with the model proposed by the above authors for generating chaotic dynamics and arrhythmia. Whether or not mechanoelectrical feedback could change the rate-dependent process predicted solely on the basis of electrical parameters needs to be tested following the basic approach used in Chialvo and coworkers [7], but should be included measurements of developed force as a function of diastolic interval. In fact, a precise formalism may be constructed if the difference equation model by Chialvo and coworkers [7] is expanded to include additional terms for the recovery of force as a function of diastolic interval [D.R. Chivalvo, personal communication].

21.6.3 MECHANOELECTRICAL ALTERNANS

Restitution changes similar to those just described can lead to electrophysiological alternans. Electrophysiological alternans has been regarded as a period-doubling bifurcation that may be a route to chaos [30–32,70]. Alternans in action potential duration has been demonstrated to precede almost invariably ventricular fibrillation in experimental regional ischemia [19]. We think that this in itself does not constitute evidence that a period-doubling cascade to chaos produced the fibrillation [33]. There has in fact been argument that fibrillation may not be chaotic [38]. However, strikingly, the alternans in the experimental ischemia could be observed to be out of phase in adjacent areas providing a matrix for abnormal current flow. Although changes in electrical restitution per se can lead to alternans in action potential duration, it is becoming clear that steady-state electrical alternans as well as mechanical alternans may be secondary to alternans in intracellular calcium [62]. Indeed, calcium alternans has been reported in global "ischemia" in crystalloid solution perfused hearts [52]. Since we are interested in mechanoelectric feedback, most likely via calcium, it is worthy of note that mechanical force alternans is accompanied by calcium alternans [62]. If calcium were alternating in regional ischemia, regional mechanical alternans would be expected. This has been observed [47]. The mechanical alternans would result in an alternans of mechanoelectric feedback affecting action potential durations. The questions thus posed are: Can the types of analyses used for electrophysiological alternans [7,30–32,70] be applied to mechanical alternans, and to what extent could mechanoelectric feedback (mechanoelectric alternans) be incorporated in the analyses?

Notwithstanding the complexity of the possible theoretical considerations, attempts at doing the latter may prove fruitful. For if acute regional ischemia can generate combinations of period bifurcation, reentry, and abnormal electromechanical restitution, then the introduction of a mechano-

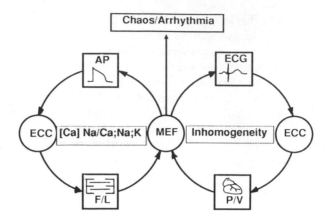

FIGURE 21.6. Diagram of interaction of excitation contraction coupling (ECC) and mechanoelectric feedback (MEF) at the cellular level (left-hand side) and gross level (right-hand side). AP=action potential; F/L=force/length; P/V=pressure/volume.

electric feedback via calcium, if not directly contributary, would enrich this electrophysiological milieu for potential chaotic behavior and arrhythmia.

21.7   Summary

We suggest that following an ectopic beat, which may be mechanically induced, interacting nonlinear time courses of recovery of restitution and excitability (ECC of left loop in Figure 21.6) are compounded by an instantaneous feedback (MEF in Figure 21.6) between mechanical conditions of the myocardium and these nonlinear recovery processes.

Intracellular calcium is most likely to be involved in mechanoelectric feedback via its influence on calcium activated currents, and electrogenic sodium/calcium exchange would facilitate these processes. It has been proposed that intracellular calcium is a linking factor that may be necessary for generating nonlinearities and chaos in excitable tissues [33]. Mechanoelectric feedback, via changes in intracellular calcium, could contribute to the nonlinearity of electrophysiological parameters in pathological heart so that small changes in initial loading or mechanical conditions could lead to arrhythmia. In the intact ventricle the situation would be compounded by both mechanical and electrical inhomogeneity (Figure 21.6, right-hand loop) thus enabling a milieu of altered excitability, arrhythmogenic current flow, and reentry. These proposals highlight the importance of rigorously testing the role of chaotic dynamics in mechanoelectric feedback.

It appears that even if mechanically induced changes are not central to the issue of arrhythmogenesis, they at least contribute to arrhythmia generation in the altered electrophysiological milieu of regional ischemia and also heart failure. Arrhythmia is a potent cause of sudden death in the Western world and therapeutic regimes in the past have concentrated on purely electrophysiological approaches. An interesting alternative would be to focus on changing hemodynamic and thus mechanical cardiac parameters.

Acknowledgements: Supported by grants from the British Heart Foundation and The Wellcome Trust.

REFERENCES

[1] C.E. Bartecchi. Emergency cardiac maneuvers–editorial. *South Med J.*, 82:1–2, 1989.

[2] J.T. Bigger. Prevalence and possible mechanisms of ventricular arrhythmias in congestive heart failure. In C. Wood, editor, *New Perspectives in Cardiovascular Medicine Arrhythmias in Heart Failure: New Frontiers*, pages 11–27. Oxford: Royal Society of Medical Services, 1988.

[3] S.H. Braat, C. deZwaan, P. Brugada, and H.J. Wellens. Value of left ventricular ejection fraction in extensive anterior infarction to predict development of ventricular tachycardia. *Am. J. Cardiol.*, 52:686–689, 1983.

[4] A. Brady. Time and displacement dependence of cardiac contractility: Problems in defining the active state and force-velocity relations. *Fed. Proc.*, 24:1410–1420, 1965.

[5] G. Caldwell, G. Millar, E. Quinn, R. Vincent, and D. Chamberlain. Simple mechanical methods for cardioversion: Defence of the precordial thump. *Br. Med. J.*, 291:627–630, 1985.

[6] R. Califf, J. Burks, V. Behar, J. Margolis, and G. Wagner. Relationship among ventricular arrhythmias, coronary artery disease and angiographic and electrocardiographic indicators of myocardial fibrosis. *Circulation*, 57:725–732, 1987.

[7] D.R. Chialvo, D.C. Michaels, and J. Jalife. Supernormal excitability as a mechanism of chaotic dynamics of activation in cardiac Purkinje fibers. *Circ. Res.*, 66:525–545, 1990.

[8] J.G. Cleland, H.J. Dargie, G.P. Hodsman, et al. Captopril in heart failure: A double blind controlled trial. *Br. Heart J.*, 52:530–535, 1984.

[9] J.G. Cleland, J.H. Dargie, S.G. Ball, et al. Effects of enalapril in heart failure: A double blind study of effects on exercise performance, renal function, hormones and metabolic state. *Br. Heart J.*, 54:305–312, 1985.

[10] J.N. Cohn, D.G. Archibald, S. Ziesche, et al. Effect of vasodilator therapy on mortality in chronic congestive heart failure. *N. Engl. J. Med.*, 314:1547–52, 1986. Results of a Veterans Administration Co-Operative Study.

[11] D. Colquhoun, E. Neher, H. Reuter, and C.F. Stevens. Inward current channels activated by intracellular Ca in cultured cardiac cells. *Nature*, 294:752–754, 1981.

[12] M. Courtemanche, L. Glass, M.D. Rosengarten, and A. Goldberger. Beyond pure parasystole: Promises and problems in modeling complex arrhythmias. *Am. J. Physiol.*, 257:H693–H706, 1989.

[13] J.W. Covell, M.J. Lab, and R. Pavalec. Mechanical induction of paired action potentials in intact heart in situ. *J. Physiol. (Lond.)*, 320:34P, 1981.

[14] P.F. Cranefield and A.L. Wit. Cardiac arrhythmias. *Ann. Rev. Physiol.*, 41:459–472, 1979.

[15] J. Dean and M.J. Lab. The effect of changes in load on the monophasic action potential of the in situ pig heart. *Cardiovasc. Res.*, 23:887–896, 1989.

[16] J.W. Dean and M.J. Lab. The effect of changes in afterload on the absolute refractory period of the pig heart. *P.A.C.E.*, 10:4 (Suppl ii):987, 1987.

[17] J.W. Dean and M.J. Lab. Regional changes in myocardial refractioness during load manipulation in the in-situ pig heart. *J. Physiol.*, 1990. (in press.)

[18] K.A. Deck. Anderungen des ruhepotentials und der kabeleigenschaflen von Purkinje-Baden bei der dehnung. *Pflügers Arch.*, 280:131–140, 1964.

[19] S. Dilly and M.J. Lab. Electrophysiological alternans and restitution during acute regional ischaemia in myocardium of anaesthetised pig. *J. Physiol.*, 402:315–333, 1988.

[20] J.D. Doherty, B.S. Manley, and S. Cobbe. Electrophysiological changes in an animal model of congestive cardiac failure. *Clin. Sci.*, 74 (Suppl 18):30P, 1988.

[21] J. Dudel and W. Trautwein. Das aktionspotential und mechanogram des herzmuskels unter dem einflus der dehnung. *Cardiologie*, 25:344–362, 1954.

[22] E.L. Ford and N.P. Campbell. Effect of myocardial shortening velocity on duration of electrical and mechanical systole. *Br. Heart J.*, 44:179–183, 1980.

[23] G.S. Francis. Development of arrhythmias in the patient with congestive heart failure: Pathophysiology, prevalence and prognosis. *Am. J. Cardiol*, 57:3B–7B, 1986.

[24] G.S. Francis. Neurohumoral mechanisms involved in congestive heart failure. *Am. J. Cardiol.*, 55A:15–21, 1985.

[25] M.R. Franz, D. Burkhoff, D.T. Yue, and K. Sagawa. Mechanically induced action potential changes and arrhythmia in isolated and in situ canine hearts. *Cardiovasc. Res.*, 23:213–223, 1989.

[26] W. Garry. The nature of fibrillating contraction of the heart. Its relation to tissue mass and form. *Am. J. Physiol.*, 31:397–414, 1914.

[27] W.R. Gibbons. Cellular control of cardiac contraction. In H.A. Fozzard, E. Haber, R.B. Jennings, Katz. A.M., and H.E. Morgan, editors, *The Heart and Cardiovascular System*, pages 747–778. Raven Press, New York, 1986.

[28] L. Glass, A.L. Goldberger, and J. Belair. Dynamics of pure parasystole. *Am. J. Physiol.*, 251:H841–H847, 1986.

[29] L. Glass, A.L. Goldberger, M. Courtemanche, and A. Shrier. Nonlinear dynamics, chaos and complex cardiac arrhythmias. *Proc. Roy. Soc. Lond. A.*, 413:9–16, 1987.

[30] L. Glass, M.R. Guevara, and A. Shrier. Universal bifurcations and the classification of cardiac arrhythmias. *Ann. NY Acad. Sci.*, 504:168–178, 1987.

[31] M.R. Guevara, L. Glass, and A. Shrier. Phase locking, period-doubling bifurcations, and irregular dynamics in periodically stimulated cardiac cells. *Science*, 214:1350–1353, 1981.

[32] M.R. Guevara, G. Ward, A. Shrier, and L. Glass. Electrical alternans and period doubling bifurcations. In *Computers in Cardiology*, pages 167–170. IEEE Computer Society, Long Beach, CA, 1984.

[33] A.V. Holden and M.J. Lab. Chaotic behavior in exitable tissues. *Ann. NY Acad. Sci.*, 591:303–315, 1990.

[34] A.V. Holden, M. Markus, and H. Othmer. *Nonlinear Wave Processes in Excitable Media*. Plenum, London, 1990.

[35] O.B. Holland, J.V. Nixon, and L. Kuhnert. Diuretic-induced ventricular ectopic activity. *Am. J. Med.*, 70:762–768, 1981.

[36] W.B. Kannel and D.L. McGee. Epidemiology of sudden death: Insights from the Framingham Study. In M.E. Josephson, editor, *Sudden Cardiac Death*, pages 93–105, FA Davis, Philadelphia, 1985.

[37] W.B. Kannel and A. Schatzkin. Sudden death: Lessons from subsets in population studies. *J. Am. Coll. Cardiol.*, 5(Suppl):141B9B, 1985.

[38] D.T. Kaplan and R.J. Cohen. Searching for chaos in fibrillation. *Ann. NY Acad. Sci.*, 591:367–374, 1990.

[39] A.M. Katz, H. Tabenaka, and J. Watras. The sacroplasmic reticulum. In H.A. Fozzard, E. Haber, R.B. Jennings, A.M. Katz, and H.E. Morgan, editors, *The Heart and Cardiovascular System*, pages 731–746. Raven Press, New York, 1986.

[40] R. Kaufmann, M.J. Lab, R. Hennekes, and H. Krause. Feedback interaction of mechanical and electrical events in the isolated ventricular myocardium (cat papillary muscle). *Pflügers Arch.*, 332:96–116, 1971.

[41] R. Kaufmann and U. Theophile. Automatie fordernde dehnungseffects am purkinje faden pappilarmuskeln und vorhoftrabekein von rhesusaffen. *Pflügers Arch.*, 291:174–89, 1967.

[42] M. Kelly, P. Thompson, and M. Quinlan. Prognostic significance of left ventricular ejection fraction after acute myocardial infarction. *Br. Heart J.*, 55:16–24, 1985.

[43] S.Y. Khatib and M.J. Lab. Differences in electrical activity in the apex and base of left ventricle produced by changes in mechanical conditions of contraction. *J. Physiol. (Lond.)*, 324:25–26, 1982.

[44] M.J. Lab. Contraction-excitation feedback in myocardium: Physiological basis and clinical relevence. *Circ. Res.*, 50:757–766, 1982.

[45] M.J. Lab. The effect on the left ventricular action potential of clamping the aorta. *J. Physiol. (Lond.)*, 202:73P–74P, 1969.

[46] M.J. Lab. Mechanically dependant changes in action potentials recorded from the intact frog ventricle. *Circ. Res.*, 42:519–528, 1978.

[47] M.J. Lab. Mechanoelectric coupling in myocardium and its possible role in ischaemic arrhythmia. In S. Sideman and R. Beyer, editors, *Activation, Metabolism, and Perfusion of the Heart Activity*, Martinus Nijhoff, Dordrecht, Boston, Lancaster, 1987.

[48] M.J. Lab, D. Allen, and C. Orchard. The effects of shortening on myoplasmic calcium concentration and on the action potential in mammalian ventricular muscle. *Circ. Res.*, 55:825–829, 1984.

[49] M.J. Lab and J.W. Dean. Analysis and simulation of the cardiac system. In S. Sideman and R. Bayer, editors, *Mechanoelectric Feedback: Prevalence and Relevance*, Martinus Nijhoff, 1989.

[50] M.J. Lab and M.I.M. Noble. Do extrinsic reflexes mediate the right ventricular arrhythmia produced by pulmonary artery occlusion in anaesthetised dogs? *Pflügers Arch.*, 1988.

[51] M.J. Lab and J. Yardley. Regional differences in electrical restitution of the pig left ventricle. *J. Physiol. (Lond.)*, 353:70P, 1984.

[52] H.C. Lee, R. Mohabir, N. Smith, and M.R. Franz. Clusin WT effect of ischemia on calcium-dependent fluorescence transients in rabbit hearts containing indo 1. Correlation with monophasic action potential. *Circulation*, 78:1047–1059, 1988.

[53] B.B. Lerman, D. Burkhoff, D.T. Yue, and K. Sagawa. Mechanoelectric feedback: Independent role of preload and contractility in modulation of canine ventricular excitability. *J. Clin. Invest.*, 76:1843–1850, 1985.

[54] J.H. Levine, T. Guarnieri, A.H. Kadish, R.I. White, H. Calkins, and J.S. Kan. Changes in myocardial repolarization in patients undergoing balloon valvuloplasty for congenital pulmonary stenosis: Evidence for contraction-excitation feedback in humans. *Circulation*, 77:70–77, 1988.

[55] B. Lown. Sudden cardiac death: The major challenge confronting contemporary cardiology. *Am. J. Cardiol.*, 43:313, 1979.

[56] I. Maroszan and L. Szatmari. Life-threatening arrhythmia, noted by coronarography, stopped by induced cough. *Orv. Hetil.*, 128:1555–1557, 1987.

[57] J.M. McLenachan, E. Henderson, K.I. Morris, C. Isles, and H.J. Dargic. Ventricular arhythmias in patients with hypertensive left ventricular hypertrophy. *N. Engl. J. Med.*, 317:787–792, 1987.

[58] T. Meinertz, T. Hoffmann, W. Kasper, et al. Significance of ventricular arrhythmias in idiopathic dilated cardiomyopathy. *Am. J. Cardiol.*, 53:902–907, 1984.

[59] M.D. Meissner, H.R. Kay, S.R. Spielman, A.M. Greenspan, S.P. Cutalek, and L.N. Horowitz. Acute antiarrhythmic drug efficacy is independantly related to left ventricular function. *J. Am. Coll. Cardiol.*, 7:130, 1986.

[60] M.J. Lab. Contribution of mechano-electric coupling to ventricular arrhythmias during reduced perfusion. *Int. J. Microcirc.*, 8:433–442, 1989.

[61] M.J. Lab. Transient depolarisation and action potential alteration following mechanical changes in isolated myocardium. *Cardiovasc. Res.*, 14:624–637, 1980.

[62] M.J. Lab and J. Lee. Changes in intracellular calcium during mechanical alternans in isolated ferret ventricular muscle. *Circ. Res.*, 66:585–595, 1990.

[63] L.J. Mullins. The generation of electric currents in cardiac fibers by Na/Ca exchange. *Am. J. Physiol.*, 263:C103–C110, 1979.

[64] D. Noble and I. Cohen. The interpretation of the T-wave of the electrocardium. *Cardiovasc. Res.*, 12:13–27, 1978.

[65] Z.J. Penefsky and B.F. Hofmann. Effects of stretch on mechanical and electrical properties of cardiac muscle. *Am. J. Physiol.*, 204:433–438, 1963.

[66] F.A. Pirzada, E.A. Ekong, P.A.S. Vokonas, C.A. Apstein, and W.B. Hood. Experimental infarction. XII sequential changes in left ventricular pressure-length relationships in the acute phase. *Circulation*, 53:970–975, 1976.

[67] M.J. Reiter, D.P. Synhorst, and D.E. Mann. Electrophysiological effects of acute ventricular dilatation in the isolated rabbit heart. *Circ. Res.*, 62:554–562, 1988.

[68] J. Roelandt and P.G. Hugenholtz. Sudden death: Prediction and prevention. *Eur. Heart J.*, 7 (Suppl. A):169–80, 1985.

[69] M.R. Rosen. The links between basic and clinical cardiac electrophysiology. *Circulation*, 77:251–263, 1988.

[70] G.V. Savino, L. Romanelli, D.L. Gonzalez, O. Piro, and M.E. Vallentinuzzi. Evidence for chaotic behaviour in driven ventricles. *Biophys. J.*, 56:273–280, 1989.

[71] J. Schaper and W. Schaper. Ultrastructural correlates of reduced cardiac function in human heart disease. *Eur. Heart J.*, 4(Suppl. A):137–142, 1983.

[72] D.D. Schultz and G.S. Olivas. The use of cough cardiopulmonary resuscitation in clinical practice. *Heart Lung*, 15:273–282, 1986.

[73] R.A. Schulze, J. Rouleau, P. Rigo, S. Bowers, W. Strauss, and B. Pitt. Ventricular arrhythmias in the late hospital phase of acute myocardial infarction-related to left ventricular function detected by gated cardiac blood pool scanning. *Circulation*, 52:1006–1011, 1975.

[74] R.A. Schulze, H.W. Strauss, and B. Pit. Sudden death in the year following myocardial infarction: Relation to ventricular premature contractions in the late hospital phase and left ventricular ejection fraction. *Am. J. Med.*, 62:192–199, 1977.

[75] D.E. Stewart, H. Ikram, E.A. Espiner, and M.G. Nicholls. Arrhythmogenic potential of diuretic induced hypokalaemia in patients with mild hypertension and ischaemic heart disease. *Br. Heart J.*, 54:290–297, 1985.

[76] P. Taggart, P.M.I. Sutton, T. Treasure, et al. Monophasic action potentials at discontinuation of cardiopulmonary bypass: Evidence for contraction-excitation feedback in man. *Circulation*, 77:1266–1275, 1988.

[77] The Cardiac Arrhythmia Pilot Study (CAPS) investigators. The effect of encainide, flecainide, imipramine, and moricizine on ventricular arrhythmias occurring 6-60 days after myocardial infraction. *Am. J. Cardiol.*, 61:501–509, 1988.

[78] K. Von Olshausen, F. Schwartz, J. Aptelbach, N. Rohrig, B. Kramer, and W. Kubler. Determinants of the incidence and severity of ventricular arrhythmias in aortic valve disease. *Am. J. Cardiol.*, 51:1103–1109, 1983.

[79] M.W.I. Webster, M.A. Fitzpatrick, M.G. Nicholls, H. Ikram, and J.E. Wells. Effect of enalapril on ventricular arrhythmias in congestive heart failure. *Am. J. Cardiol.*, 56:566–569, 1985.

[80] S. Winegrad. Membrane control of force generation. In H.A. Fozzard, E. Haber, R.B. Jennings, A.M. Katz, and H.E. Morgan, editors, *The Heart and Cardiovascular System*, pages 703–730. Raven Press, New York, 1986.

22

Nonlinear Dynamics at the Bedside

Ary L. Goldberger[1]
David R. Rigney[1]

ABSTRACT   Heart rhythms in health and disease display complex dynamics. The clinical data suggest that concepts developed in nonlinear mathematics, such as bifurcations and chaos, will be appropriate to describe some of these complex phenomena. Careful analysis will be needed to establish the presence of deterministic chaos in cardiac rhythms. Data that appear highly periodic such as normal sinus rhythm may in reality be quite variable. In contrast, chaotic-appearing rhythms such as ventricular fibrillation often contain strong periodicities.

22.1   Nonlinear Hearts

22.1.1   THE CASE OF THE BIFURCATING PATIENT

Clinical cardiology is a rich source of nonlinear phenomenology including abrupt changes (presumably *bifurcations*), self-sustained and complex *oscillations*, and erratic fluctuations suggesting *chaos* [11,12,14,17,21,24,25,35, 40] (Table 22.1). This section will focus primarily on cardiac bifurcations and periodic phenomena. Chaos is considered in a later section.

Bifurcationlike behavior is not uncommon in medicine, and some of the best examples are provided by patients on cardiology wards. The sudden appearance of an arrhythmia when some parameter of the system—say extracellular potassium or the concentration of a pharmacologic agent— is varied over a critical range is an example of bifurcationlike behavior.[2] Another is the abrupt onset (or offset) of a bundle branch block pattern when the heart rate crosses some critical threshold (rate-related bundle branch block as seen in Figure 22.1).

Both tachycardia- and bradycardia-dependent bundle branch blocks have been described. For instance, right or left bundle branch block may appear

---

[1]Cardiovascular Division, Beth Israel Hospital, Harvard Medical School, Boston, Massachusetts 02215

[2]In some nonlinear systems, abrupt changes may occur *intermittently* even when there are no changes of parameter values.

TABLE 22.1. Nonlinear dynamics in cardiology.

| Abrupt changes (bifurcations or intermittency) |
| Sustained and complex oscillations |
| Chaotic behavior |
| Hysteresis and bistability |

RATE-RELATED BUNDLE BRANCH BLOCK

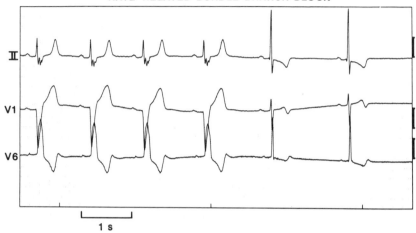

1 s

FIGURE 22.1. Rate-dependent bundle branch block is an example of an abrupt nonlinear change (bifurcation). In this example, left bundle branch block disappears when the heart rate slows below a critical threshold.

TABLE 22.2. Period-doublinglike behavior in clinical cardiology [12,14,18,21, 24,25,26,39,40,44].

Total electrical alternans in pericardial effusion/tamponade
ST-segment alternans
T-U wave alternans
Bidirectional tachycardia
QRS alternans in atrioventricular reentrant tachycardias
Pulsus alternans in heart failure
PR interval alternans

suddenly when the heart rate goes above 75 min^{-1}. When the heart rate slows again, normal conduction will resume. Often the onset and offset rates for these bundle branch blocks will not be the same. This type of *hysteresis* is another characteristic that suggests nonlinearity in the cardiac electrical system. (If, at a given heart rate, either normal conduction or a bundle branch block may exist then this would indicate *bistability*.)

An interesting class of bifurcation-type phenomena is referred to by clinicians under the rubric of *alternans* (Table 22.2). The term alternans applies to conditions characterized typically by a periodic beat-to-beat change in some aspect of cardiac electrical or mechanical behavior. These abrupt changes ($AAAA \rightarrow ABAB$) are reminiscent of subharmonic (period-doubling) patterns in perturbed nonlinear systems. Many different examples of alternans have been described by clinicians and a number of others have been reported in the laboratory [11,12,18,21,24–26,40]. Here, to be representative rather than exhaustive, we mention four examples of alternans behavior: *bidirectional tachycardia, ST segment alternans, pulsus alternans,* and *total electrical alternans* with pericardial effusion/tamponade. Other examples are listed in Table 22.2.

In *bidirectional tachycardia* (Figure 22.2), which may be supraventricular or ventricular, the electrical axis of the ventricular depolarization wave (QRS complex of the electrocardiogram) alternates in direction from one beat to the next. This potentially fatal arrhythmia (particularly bidirectional ventricular tachycardia) is usually seen with organic heart disease or digitalis intoxication. The actual mechanism, however, is uncertain. Does the alternans here represent a periodic shift in the locus of some ectopic pacemaker, a periodic shift in ventricular conduction in the presence of a unifocal pacemaker, or, as Jalife and Michaels [29] have proposed, a type of nonlinear entrainment of one ectopic pacemaker by another (modulated parasystole variant)?

Another type of electrophysiologic alternation is called *ST segment alternans* (Figure 22.3). In this condition, the ST segment (representing the beginning phase of ventricular repolarization in the electrocardiogram)

BIDIRECTIONAL TACHYCARDIA

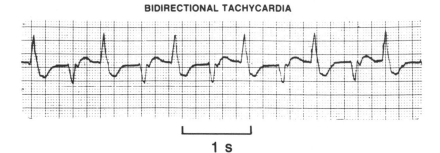

1 s

FIGURE 22.2. Bidirectional tachycardia is characterized by beat-to-beat alternation of the polarity of electrocardiographic waves. This instance of bidirectional ventricular tachycardia was related to digitalis intoxication.

ST SEGMENT ALTERNANS

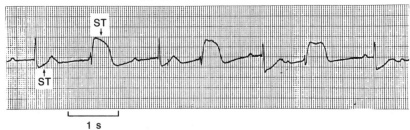

1 s

FIGURE 22.3. ST segment alternans is characterized by beat-to-beat variation in the amplitude of ST segment elevation in a single electrocardiographic lead. This pattern is a marker of electrical instability and may be a precursor of ventricular fibrillation.

changes amplitude on a beat-by-beat basis. ST segment alternans is of interest because it often precedes a lethal arrhythmia, ventricular fibrillation. Smith and Cohen [44] simulated this kind of repolarization shift in a discrete element model of the ventricular conduction system. They suggested further that this alternation was part of a Feigenbaum-type period-doubling sequence culminating in cardiac chaos (ventricular fibrillation). As discussed below this interpretation has been challenged [13].

Yet another example of alternans is so-called *pulsus alternans* (Figure 22.4) [1]. In contrast to the electrophysiological types of alternans just described (bidirectional tachycardia and ST alternans), pulsus alternans is purely mechanical. The diagnosis is made by observing a beat-to-beat alternation in systemic (or pulmonary artery) blood pressure. When this variation is of sufficient magnitude it may be detected by the alert clinician. Pulsus alternans in the systemic circulation is classically associated with

PULSUS ALTERNANS

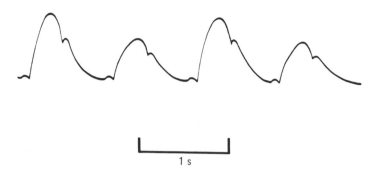

1 s

FIGURE 22.4. Pulsus alternans is characterized by a beat-to-beat alternation in the blood pressure. This pattern is usually a marker of severe ventricular dysfunction.

severe myocardial dysfunction and probably reflects periodic changes in the strength of the heartbeat. Do these oscillations, in turn, correspond to periodic changes in calcium fluxes at the cellular level?

The final example of alternans is referred to as *total electrical alternans* (Figure 22.5), which is seen with large pericardial effusions, usually with cardiac tamponade. Under normal conditions, the movement of the heart is constrained by the pericardium. But when excessive pericardial fluid accumulates, the heart is free to swing as a pendulum, suspended by the great vessels. In cases of uncomplicated pericardial effusion, the swinging usually has the same frequency as the heart rate (1:1 swinging). In other cases, the heart swings at half the heart rate. This 2:1 swinging is usually observed when the pericardial pressure reaches a critically high value such that diastolic filling is impaired and the heart becomes "choked off" (tamponade physiology). The 2:1 mechanical oscillation explains a characteristic electrocardiographic pattern in which the QRS-T vector alternates in direction on successive beats (total electrical alternans). This is similar to the QRS alternation seen with bidirectional tachycardia. However, with pericardial effusion and tamponade, the normal (sinus) heart rhythm may be maintained and the alternans is entirely mechanical in origin.

Recently, we [39] showed that the change from a 1:1 to a 2:1 swinging pattern in pericardial effusion may be explained by the nonlinearity of Newton's equation of motion as applied to the heart. Terms in the equation correspond to buoyancy and gravitational forces, forces due to ejection of blood into the great vessels, and damping forces. A transition between 1:1 and 2:1 swinging frequencies occurs when certain parameters are changed, notably when the heart rate increases above a critical value. The finding

TOTAL ELECTRICAL ALTERNANS

Lead II

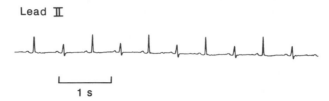

1 s

FIGURE 22.5. Patients with pericardial effusion, particularly when tamponade is present, may develop alternation of the electrocardiographic waveforms. This type of *total electrical alternans* is due to the periodic swinging motion of the heart in the pericardial fluid.

is consistent with the clinical observation of 2:1 swinging with electrical alternans in association with a reflex increase in heart rate during cardiac tamponade. This model may serve as a prototype for analyzing other types of alternans described above.

22.1.2   THE CHAOTIC HEARTBEAT—HEALTHY OR DISEASED?

A controversial aspect of nonlinear cardiology concerns the relevance of chaos theory. Many investigators have argued that the heart has chaotic dynamics. The debate is about *which* cardiac rhythms are chaotic in the technical sense. Classical cardiologists use the term chaos in a vernacular way to describe a number of apparently erratic, abnormal rhythms, including multifocal atrial tachycardia; complex, multiform bursts of ventricular ectopic beats and, of course, ventricular fibrillation (Figure 22.6). More recently, a number of dynamicists, probably representing the majority, have proposed that some of these pathologic rhythms represent deterministic chaos. For example, Ruelle [41], in a well-known article, cited ventricular fibrillation as a prime example of a physiological strange attractor.

Probably the most explicit formulation of the chaos theory of cardiac arrhythmia is that of Smith and Cohen [44] mentioned earlier. Their computer model extended an earlier analysis by Moe and colleagues [38] that described fibrillation as having turbulent dynamics. The Smith-Cohen model consists of a cylindrical array of coupled elements that explicitly incorporates a spatial dispersion of refractoriness. The model displays a variety of "rhythm" disturbances reminiscent of those seen clinically. In particular, "chaotic" fibrillatorylike activity is seen to follow alternanslike behavior. This prompted Smith and Cohen [44] to propose that cardiac chaos (i.e., fibrillation) is preceded by a period-doubling cascade "not dissimilar to that of other systems approaching a disorganized state." The validity of

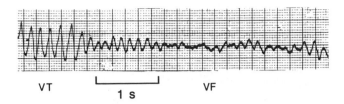

FIGURE 22.6. Ventricular tachycardia (VT) abruptly changes to ventricular fibrillation (VF) in the electrocardiographic recording made during a cardiac arrest, consistent with a nonlinear bifurcation.

this interpretation, however, has been challenged. Kaplan [31] exercised this finite element model and found that the "fibrillation" is a transient. Furthermore, his reanalysis of experimental data has failed thus far to confirm a period-doubling cascade actually leading to fibrillation. Based on nonlinear analysis of body surface electrocardiographic recordings during fibrillation, Kaplan [31] concluded that ventricular fibrillation is primarily a random process and not low-dimensional chaos.

However, the claim that fibrillation represents either nonlinear chaos or a strictly random process is inconsistent with other experimental studies. If fibrillation is chaotic, its frequency spectrum should be broad not narrow. Yet spectral analysis of fibrillatory electrocardiographic waveforms from animals and humans has consistently revealed relatively narrow, not broad spectra (Figure 22.7) [4,5,13,27,32]. Furthermore, Ideker and colleagues [28,47] have reported careful electrophysiologic recordings from fibrillating dogs. They observed surprisingly organized waves of epicardial excitation at the onset of fibrillation [28]. Even more striking was their subsequent observation of highly periodic activation pulses recorded at the endocardial surface minutes after the onset of ventricular fibrillation [47].

If fibrillation, initially the prime candidate for cardiac chaos, is not in fact chaotic, what is? Several years ago we proposed an alternative view of cardiovascular electrodynamics summarized by two propositions [15]:

1. The onset of ventricular fibrillation and related tachyarrhythmias causing sudden cardiac death is usually a periodic, not chaotic, process.

2. Normal sinus rhythm, contrary to conventional wisdom, may be chaotic.

What is the evidence to support the counterintuitive notion that normal sinus rhythm can be chaotic? Most clinicians and dynamicists accept the regularity of normal sinus rhythm as an article of faith. Casual inspection of an electrocardiographic rhythm strip from a normal subject—or palpation of one's own pulse—certainly suggest regularity. Yet, the normal heartbeat

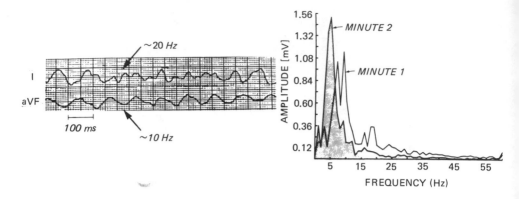

FIGURE 22.7. Ventricular fibrillation (canine experiment) looks "chaotic" at first glance (Figure 22.6). However, closer inspection reveals periodic oscillations (left panel), and the spectrum (right panel) of the epicardial or body surface electrocardiogram shows a relatively narrowband pattern during the first minute of fibrillation, with even further narrowing during the second minute. (Adapted with permission from Goldberger et al. [13].)

is, in fact, *not* metronomically regular (Figure 22.8). Even in resting subjects it fluctuates in a highly erratic manner, as revealed by time series analyses (although this physiologic irregularity is usually not subjectively perceptible). Further, the spectrum of the healthy heartbeat is often $1/f$-like, with superimposed spikes corresponding to physiological oscillations associated with breathing, baroreceptor reflexes, and other lower frequency control mechanisms [2,20]. This kind of broad spectrum with superimposed peaks is reminiscent of the "noisy periodicity" of strange attractors [34].

Time series plots showing erratic fluctuations, such as those observed with the heartbeat of healthy subjects, and their associated broad spectra, are consistent with, but not diagnostic of, chaos. Another test for chaos is to plot delay maps (phase space maps) for the time series in search of strange attractors (Table 22.3). Preliminary studies from our laboratory indicate that phase space trajectories of normal interbeat intervals follow a complicated, sometimes "spiderlike" pattern that may be strange attractors in the presence of noise (Figure 22.9). The trajectories typically converge toward a dense central "body," leave the center along irregular "arms," and then return to the central region [15].

The erratic fluctuations of normal sinus rhythm heart rate contrast with the dynamics observed in patients with heart disease and impending cardiac arrest. We [16] have reported two abnormal heart rate patterns in patients with severe left ventricular failure (a group at high risk of sudden death) and in patients who actually sustain a fatal or near-fatal tachyarrhythmia while wearing a portable electrocardiographic monitor (Figure 22.10). One pattern we have called the *oscillatory pattern* because it is characterized

TABLE 22.3. Testing for chaos in clinical data sets.

Time series with erratic fluctuations
Broadband spectrum
Strangelike attractor in phase space
Finite correlation dimension
Positive Lyapunov exponent

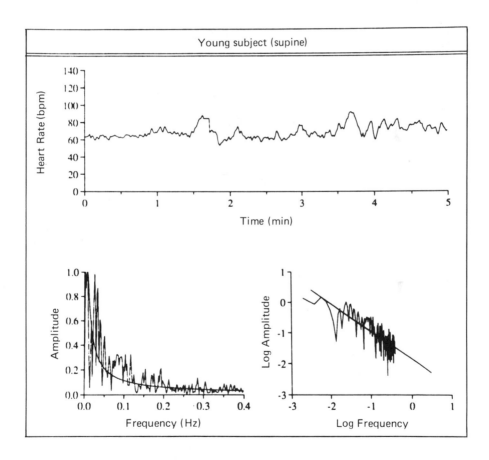

FIGURE 22.8. The heart rate of a healthy subject even at rest shows erratic fluctuations. The frequency spectrum is broad, with a 1/f-like (inverse power-law) distribution (Adapted with permission from Lipsitz et al. [33].)

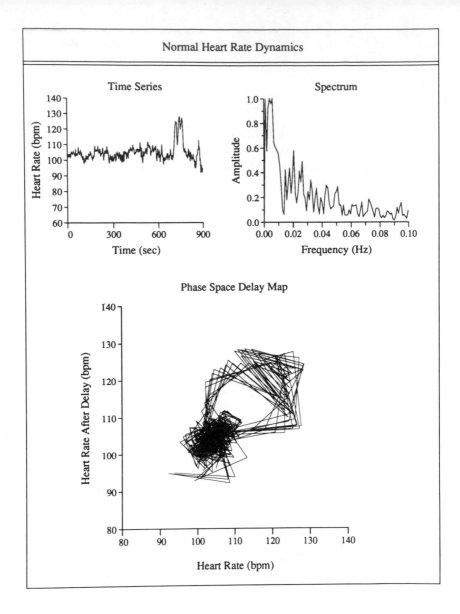

FIGURE 22.9. Time series, frequency spectrum, and two-dimensional delay map (phase space representation) of 900 sec of heart rate data from a normal individual. The delay map shows the trajectories of the heart rate vectors where the first variable is the current heart rate and the second variable is the heart rate after a fixed delay of 12 sec. Whether these trajectories actually represent a strange attractor remains to be established. There are multiple difficulties in attempting to compute a nonlinear dimension of such a physiologic time series (see text). Note, however, that the healthy sinus rhythm heart rate dynamics may be considerably more irregular than sinus rhythm dynamics in patients at high risk of sudden death. (Adapted with permission from Goldberger et al. [17]; cf. Figures 10–12.)

by low frequency (0.01–0.06 Hz) oscillations in sinus heart rate. The other dynamical pattern we have termed the *flat pattern* because it is characterized by a marked reduction in beat-to-beat variability. These pathologic patterns are reminiscent of the reduction in heart rate variability and the low frequency oscillations (sinusoidal pattern) reported previously in the fetal distress syndrome [10].

In contrast to the strange attractorlike phase space representations of normal sinus rhythm, these pathologic patterns are represented by more periodic type attractors. As anticipated, the oscillatory pattern is represented by more circular trajectories suggesting a noisy limit cycle (Figure 22.11). The flat pattern is represented by a phase portrait suggesting a fixed point attractor (Figure 22.12).

Additional tests for chaos include measurement of Lyapunov exponents and calculation of correlation dimensions [3,6,9,22,23,36,46] (Table 22.3). However, as emphasized by other investigators, the reliability of such measurements in biological data sets is currently uncertain. For example, with respect to heart rate data, many technical questions remain to be resolved. How many data points are needed? What is the shortest acceptable sampling interval? How do you test for stationarity? How much noise is acceptable? How should ectopic beats be handled? What are the differences between using interbeat intervals vs. heart rate (= interbeat interval^{-1}). Should the data be filtered (e.g., using singular value decomposition)?

We have previously discussed possible mechanisms underlying the normal, possibly chaotic variations in heart rate and the types of pathological periodicities seen in a variety of disease states [15,16,20]. The heart rate fluctuations both in health and disease are the result of a complex, neurohumoral feedback system involving the divergent influences of the sympathetic and parasympathetic branches of the autonomic nervous system. Efforts are currently underway in our laboratory to derive nonlinear equations of motion for these interactions that give rise to apparent chaos and periodicity when parameters are varied over physiologically realistic ranges. As we have previously emphasized, nonlinear dynamics of the heartbeat can be viewed an *epiphenomenon* since they are under neural control [16]. Therefore, chaos of the heartbeat—if it exists—reflects neurohumoral chaos. The use of the heart rate dynamics to assay subtle nonlinear changes in neuroautonomic control should be of considerable interest to cardiologists, given the growing volume of evidence implicating neuroautonomic dysfunction in heart failure [8] and in sudden cardiac death [30] syndromes.

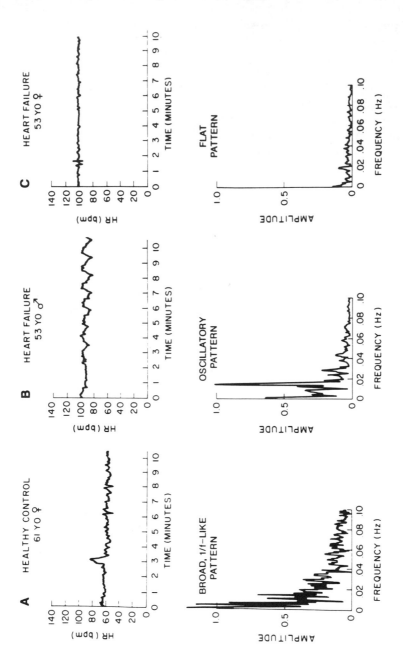

FIGURE 22.10. (A) Normal heart rate time series (top panel) shows marked variability and is represented by a broad $1/f$-like spectrum. Patients at high risk of sudden death may show alterations in normal heart rate regulation with either slow *oscillations* such as those shown in (B) or a marked overall loss of variability (*flat pattern*) as shown in (C).

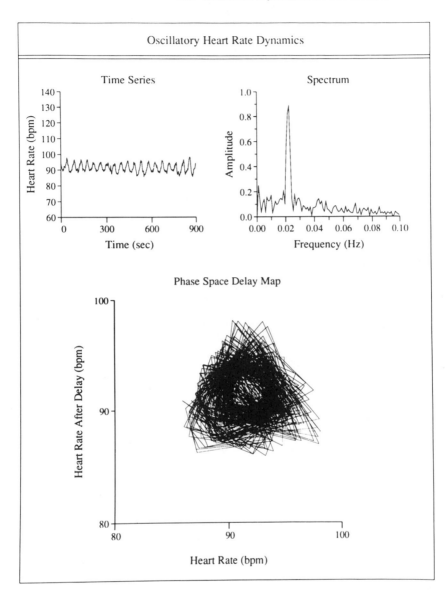

FIGURE 22.11. Two-dimensional phase space representation (delay map) of 900 sec of heart rate data from a patient who sustained a cardiac arrest due to a ventricular tachyarrhythmia 8 days later. The heart rate time series and frequency spectrum reveal period oscillations. The delay map shows relatively periodic orbits around a central hub. (Adapted with permission from Goldberger et al. [17].)

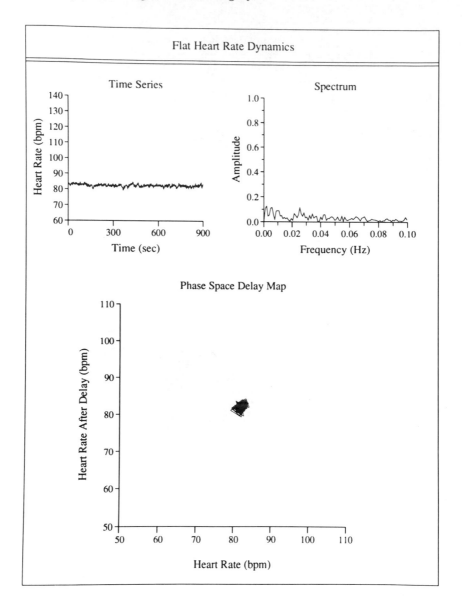

FIGURE 22.12. Two-dimensional phase space representation (delay map) of 900 sec of heart rate data from a patient 13 hr before cardiac arrest. There is a marked overall loss of heart rate variability, indicated by the heart rate time series and frequency spectrum. The delay map is reminiscent of a fixed point attractor. (Adapted with permission from Goldberger et al. [17].)

TABLE 22.4. Four dynamical deceptions in cardiac electrophysiology.

1.	Rhythms that appear regular (periodic) may actually be quite erratic and even chaotic. Example: normal sinus rhythm.
2.	Rhythms that appear completely chaotic may have strongly periodic dynamics. Examples: certain bursts of complex ectopic beats(cardiac ballet), ventricular fibrillation.
3.	Rhythms that are clearly erratic may not be chaotic in the technical sense. Example: ventricular response to atrial fibrillation with normal AV node function.
4.	Period-doublinglike behavior (e.g., alternans phenomena) does not imply that the system will exhibit chaos. Example: ST segment alternans before ventricular fibrillation.

22.2 Dynamical Deceptions

Concepts from nonlinear dynamics are being applied with increasing enthusiasm by dynamically minded cardiologists and physiologically inclined dynamicists. However, caution must be exercised by both camps since cardiac electrophysiology may be quite deceptive (Table 22.4). These problems have led, in our opinion, to the misapplication of chaos theory in the past. In particular, we call attention to the following dynamical caveats.

What appears to be regular may be erratic and even chaotic. Normal sinus rhythm as noted above is usually regarded as highly regular. Yet time series and spectral analysis reveal possibly chaotic dynamics lurking beneath the appearance of periodicity. On the other hand, a second caveat is that *what appears chaotic, such as ventricular fibrillation, may have strongly periodic features.* Another example of this principle is provided by bursts of multiform ectopy, usually seen with severe organic heart disease. High-grade, complex ventricular ectopy of this kind may indicate an increased risk of sudden death. Yet careful analysis of longer rhythm strips sometimes reveals that these apparently chaotic bursts actually recur in a remarkably periodic fashion (Figure 22.13). Another syndrome characterized by repetitions of complex electrocardiographic patterns has been called *cardiac ballet* by Smirk and Ng [42].

A third caveat is that *rhythms that are in fact clearly erratic may not be chaotic in the technical sense.* Probably the best example is the ventricular response to spontaneous atrial fibrillation in humans (Figure 22.14). Clinicians describe this activity as "irregularly irregular." Time series plots confirm a highly erratic dynamic. The spectrum is broad but with nearly equal power across a wide band of frequencies (white noise pattern). Autocorrelation plots indicate that there is no memory from one beat to the

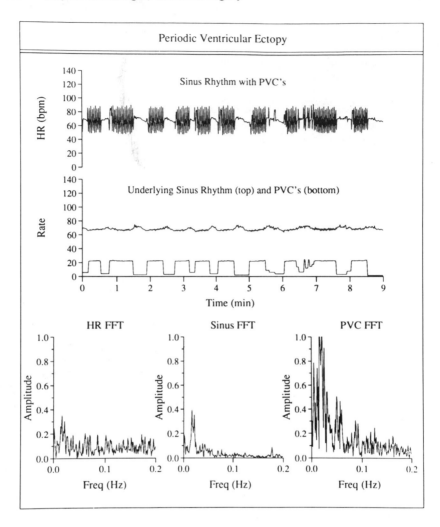

FIGURE 22.13. Time series analysis of complex and apparently "chaotic" bursts of ventricular ectopy may reveal surprisingly periodic dynamics. Top panel shows heart rate (HR) time series from a patient with sinus rhythm and bursts of premature ventricular contractions (PVCs) that give the fine-toothcomb pattern. The middle panel shows the time series of the underlying sinus rate and the PVC rate. The bottom panel shows the frequency spectra of these three time series. Note the frequency peak at about 0.02 Hz in all three spectra. This patient was noted clinically to have the Cheyne-Stokes breathing pattern, suggesting that breathing and heartbeat dynamics were entrained to same frequency.

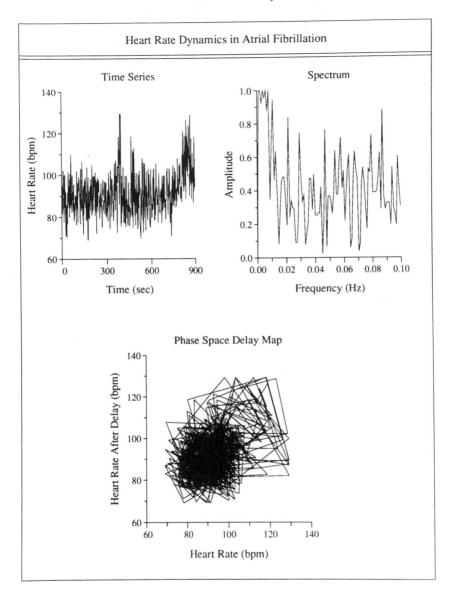

FIGURE 22.14. The venticular response to spontaneous atrial fibrillation in humans seems chaotic. However, the spectrum is more consistent with white noise and the phase representation does not reveal any evident structure (attractor).

next, that is, the autocorrelation drops to 0 after a delay of only one or two beats [7]. These findings indicate that the process leading to atrial fibrillation is more like a random number generator than like nonlinear dynamical systems that produce chaos [21].[3]

A fourth caveat is that *the appearance of period-doublinglike behavior in physiology (e.g., alternans phenomena) does not necessarily imply chaos.* Presumably isolated subharmonic bifurcations are commonly observed in clinical cardiology. Indeed, these kinds of dynamics (Table 22.2) are usually important markers of cardiac electrical or mechanical instability. For example, ST segment alternans may precede the abrupt onset of ventricular fibrillation [43]. However, as noted earlier, there is no convincing evidence that fibrillation represents nonlinear chaos or, for that matter, that an actual period-doubling sequence in any pathologic condition leads to chaotic dynamics.

22.3   The Prognosis For Nonlinear Science in Clinical Cardiology

Nonlinear dynamics promises insights at both the basic and clinical levels. Particularly exciting are the prospects for new approaches to bedside and ambulatory monitoring based on evaluation of beat-to-beat fluctuations in sinus rhythm prior to ventricular arrhythmias. Currently, analysis of Holter monitor data is embarrassingly superficial, focusing only on examination of mean heart rate and range, along with counts and characterization of ectopic beats. Yet, as illustrated in Figure 22.15, two subjects may have nearly identical mean heart rates and heart rate variances, but very different dynamics. Further, these dynamic differences may have important prognostic implications. Whether assessment of Lyapunov exponents, correlation dimensions, and other nonlinear metrics will be clinically useful remains to be determined. Preliminary data from our own and other laboratories indicate that dynamical analyses of heart rate and other physiological variables may be helpful in quantitating physiologic versus chronologic aging [33], prediction of cocaine toxicity [45], analysis of space sickness [19], as well as in the prediction of risk for sudden cardiac death [16].

---

[3] Further caveats must be applied to the analysis of the ventricular response in atrial fibrillation since this is not a dynamically homogeneous entity. While the ventricular response in spontaneous atrial fibrillation in humans appears to be random, there may be interspecies differences [37]. Furthermore, drugs (e.g., digitalis, verapamil) may alter the dynamics, as may underlying AV nodal disease. For example, toxic doses of digitalis may lead to regularization of the ventricular response, as may intrinsic AV node pathology.

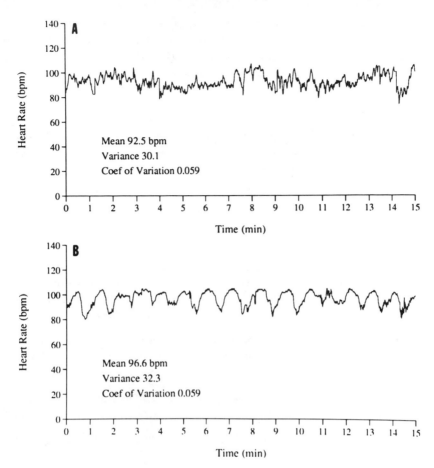

FIGURE 22.15. Two heart rate time series are shown with nearly identical means and variances but very different dynamics. The top panel shows normal, chaotic-appearing heart rate fluctuations. The bottom panel is from a patient with severe heart disease and shows large amplitude, periodic oscillations. New techniques of analyzing and comparing physiological data sets of this type using spectral estimates and dimensional calculations may extend the diagnostic and prognostic ability of patient monitoring systems.

Acknowledgements: This work is supported in part by grants from the National Heart, Lung and Blood Institute (R01HL42172), the National Aeronautics and Space Administration (NAG 2-514), and the G. Harold and Leila Y. Mathers Charitable Foundation.

REFERENCES

[1] D. Adler and Y. Mahler. Modeling mechanical alternans in the beating heart: Advantages of a systems oriented approach. *Am. J. Physiol.*, 253:H690–H698, 1987.

[2] S. Akselrod, D. Gordon, F.A. Ubel, D.C. Shannon, A.C. Barger, and R.J. Cohen. Power spectrum analysis of heart rate fluctuation: A quantitative probe of beat-to-beat cardiovascular control. *Science*, 213:220–222, 1981.

[3] A.M. Albano, A.I. Mees, G.C. de Guzman, and P.E. Rapp. Data requirements for reliable estimation of correlation dimensions. In H. Degn, A.V. Holden, and L.F. Olsen, editors, *Chaos in Biological Systems*, pages 207–220, Plenum Press, New York, 1987.

[4] E.T. Angelakos and G.M. Shepherd. Autocorrelation of electrocardiographic activity during ventricular fibrillation. *Circ. Res.*, 5:657–658, 1957.

[5] E.J. Battersby. Pacemaker periodicity in atrial fibrillation. *Circ. Res.*, 17:296–302, 1965.

[6] D.S. Broomhead and G.P. King. Extracting qualitative dynamics from experimental data. *Physica D.*, 20:217–236, 1986.

[7] R.J. Cohen, R.O. Berger, and T.E. Dushane. A quantitative model for the ventricular response during atrial fibrillation. *IEEE Trans. Biomed. Eng.*, BME-30:769–781, 1983.

[8] G.S. Francis, S.R. Goldsmith, T.B. Levine, M.T. Olivari, and J.N. Cohn. The neurohumoral axis in congestive heart failure. *Ann. Intern. Med.*, 101:370–377, 1984.

[9] A.M. Fraser and H.L. Swinney. Independent coordinates for strange attractors from mutual information. *Phys. Rev. A.*, 33:1134–1140, 1986.

[10] R.K. Freeman and T.J. Garite. *Fetal Heart Rate Monitoring*. Williams & Wilkins, Baltimore, 1981.

[11] L. Glass, A.L. Goldberger, M. Courtemanche, and A. Shrier. Nonlinear dynamics, chaos and complex cardiac arrhythmias. *Phil. Trans. Roy. Soc. A.*, 413:9–26, 1987.

[12] A.L. Goldberger, V. Bhargava, B.J. West, and A.J. Mandell. Non-linear dynamics of the heartbeat II. Subharmonic bifurcations of the cardiac interbeat interval in sinus node disease. *Physica D.*, 17:207–214, 1985.

[13] A.L. Goldberger, V. Bhargava, B.J. West, and A.J. Mandell. Some observations on the question: Is ventricular fibrillation "chaos"? *Physica D.*, 19:282–289, 1986.

[14] A.L. Goldberger and D.R. Rigney. On the nonlinear motions of the heart: Fractals, chaos and cardiac dynamics. In A. Goldbeter, editor, *Cell-to-Cell Signaling: From Experiments to Theoretical Models*, pages 541–550, Academic Press, New York, 1989.

[15] A.L. Goldberger and D.R. Rigney. Sudden death is not chaos. In J.A.S. Kelso, A.J. Mandell, and M.F. Shlesinger, editors, *Dynamic Patterns in Complex Systems*, pages 248–264, World Scientific, Teaneck, NJ, 1988.

[16] A.L. Goldberger, D.R. Rigney, J. Mietus, E.M. Antman, and S. Greenwald. Nonlinear dynamics in sudden cardiac death syndrome. Heart rate oscillations and bifurcations. *Experientia*, 44:983–987, 1988.

[17] A.L. Goldberger, D.R. Rigney, and B.J. West. Fractals and chaos in human physiology. *Sci. Am.*, 262:42, 1990.

[18] A.L. Goldberger, R. Shabetai, V. Bhargava, B.J. West, and A.J. Mandell. Nonlinear dynamics, electrical alternans and pericardial tamponade. *Am. Heart J.*, 107:1297–1299, 1984.

[19] A.L. Goldberger, W. Thornton, W.R. Jarisch, W.J. Manning, D.R. Rigney, and A.J. Mandell. Low frequency heart rate oscillations in shuttle astronauts: A potential new marker of susceptibility to space motion sickness. In *Space Life Sciences Symposium: Three Decades of Life Science Research in Space*, pages 78–80, NASA Life Science, Washington, DC, 1987.

[20] A.L. Goldberger and B.J. West. Fractals in physiology and medicine. *Yale J. Biol. Med.*, 60:421–435, 1987.

[21] A.L. Goldberger, B.J. West, and V. Bhargava. Nonlinear dynamics in physiology and pathophysiology: Toward a dynamical theory of health and disease. In R. Henrickson B. Wahlstrom and N.P. Sundby, editors, *Proceedings of the 11th International Association for Mathematics and Computers in Simulation World Congress, Oslo, Norway*, Vol. 2, pages 239–242, North Holland Publishing Co, Amsterdam, 1985.

[22] P. Grassberger and I. Procaccia. Measuring the strangeness of strange attractors. *Physica D.*, 9:189–208, 1983.

[23] P. Grassberger and I. Proccacia. Estimation of the Kolmogorov entropy from a chaotic signal. *Phys. Rev. A*, 28:2591–2593, 1983.

[24] M.R. Guevara and L. Glass. Phase locking, period doubling bifurcations and chaos in a mathematical model of a periodically driven oscillator: A theory for the entrainment of biological oscillators and the generation of cardiac dysrhythmias. *J. Math. Biol.*, 14:1–23, 1982.

[25] M.R. Guevara, L. Glass, and A. Shrier. Phase locking, period-doubling bifurcations, and irregular dynamics in periodically stimulated cardiac cells. *Science*, 214:1350–1353, 1981.

[26] M.R. Guevara, G. Ward, A. Shrier, and L. Glass. Electrical alternans and period-doubling bifurcations. In *Computers in Cardiology*, pages 167–170, IEEE Computer Society, Long Beach, CA, 1984.

[27] J.M. Herbschleb, R.M. Heethaar, I. van der Tweel, A.N.E. Zimmerman, and F.L. Meijler. Signal analysis of ventricular fibrillation. In *Computers in Cardiology*, pages 49–54, IEEE Computer Society, Long Beach, CA, 1979.

[28] R.E. Ideker, G.J. Klein, L. Harrison, et al. The transition to ventricular fibrillation induced by reperfusion after acute ischemia in the dog: A period of organized epicardial activation. *Circulation*, 63:1371–1379, 1981.

[29] J. Jalife and D.C. Michaels. Modulated parasystolic rhythms as mechanisms of coupled extrasystoles and ventricular tachycardias. In D. Zipes and D.J. Rowlands, editors, *Progress in Cardiology*, pages 47–62, Lea & Febiger, Philadelphia, 1988.

[30] T.N. James, editor. Sudden cardiac death. Fifteenth Bethesda conference report. *J. Am. Coll. Cardiol.*, 5, Number 6, (Suppl), 1985.

[31] D.T. Kaplan. *The Dynamics of Cardiac Electrical Instability*. Ph.D. thesis, Harvard University, 1989.

[32] H. Kusuoka, W.E. Jacobus, and E. Marban. Calcium oscillations in digitalis-induced ventricular fibrillation: Pathogenetic role and metabolic consequences in isolated ferret hearts. *Circ. Res.*, 62:609–619, 1988.

[33] L.A. Lipsitz, J. Mietus, G. Moody, and A.L. Goldberger. Spectral characteristics of heart rate variability before and during postural tilt: Relations to aging and risk of syncope. *Circulation*, 81:1803–1810, 1990.

[34] E.N. Lorenz. Noisy periodicity and reverse bifurcation. *Ann. NY Acad. Sci.*, 357:282–291, 1980.

[35] M.C. Mackey and L. Glass. Oscillation and chaos in physiological control systems. *Science*, 197:287–289, 1977.

[36] A.I. Mees, P.E. Rapp, and L.S. Jennings. Singular-value decomposition and embedding dimension. *Phys. Rev. A.*, 36:340–346, 1987.

[37] F.L. Meijler, J. Kroneman, I. Van der Tweel, J.N. Herbschleb, R.M. Heethaar, and C. Borst. Nonrandom ventricular rhythm in horses with atrial fibrillation and its significance for patients. *J. Am. Coll. Cardiol.*, 4:316–323, 1984.

[38] G.R. Moe, W.D. Reinboldt, and J.A. Abildskov. A computer model of atrial fibrillation. *Am. Heart J.*, 67:200–220, 1964.

[39] D.R. Rigney and A.L. Goldberger. Non-linear mechanics of the heart's swinging during pericardial effusion. *Am. J. Physiol.*, 257:H1292–H1305, 1989.

[40] A.L. Ritzenberg, D.R. Adam, and R.J. Cohen. Period multupling-evidence for nonlinear behavior of the canine heart. *Nature*, 307:159–161, 1984.

[41] D. Ruelle. Strange attractors. *Math. Intelligencer*, 2:126–137, 1980.

[42] F.H. Smirk and J. Ng. Cardiac ballet: Repetitions of complex electrocardiographic patterns. *Br. Heart J.*, 31:426–434, 1969.

[43] J.M. Smith, E.A. Clancy, C.R. Valeri, J.N. Ruskin, and R.J. Cohen. Electrical alternans and cardiac electrical instability. *Circulation*, 77:110–121, 1988.

[44] J.M. Smith and R.J. Cohen. Simple finite-element model accounts for wide range of cardiac dysrhythmias. *Proc. Natl. Acad. Sci. USA*, 81:233–237, 1984.

[45] B.S. Stambler, J.P. Morgan, J. Mietus, G.B. Moody, and A.L. Goldberger. Cocaine alters heart rate dynamics in conscious ferrets. *Yale J. Biol. Med.* (in press).

[46] A. Wolf, J.B. Swift, H.L. Swinney, and J.A. Vastano. Determining Lyapunov exponents from a time series. *Physica D.*, 16:285–317, 1985.

[47] S.J. Worley, J.L. Swain, P.G. Colavita, W.M. Smith, and R.E. Ideker. Development of an endocardial-epicardial gradient of activation rate in electrically induced, sustained ventricular fibrillation in dogs. *Am. J. Cardiol.*, 55:813–820, 1985.

23

A Clinical Perspective on Theory of Heart

Sándor J. Kovács[1]

ABSTRACT The mathematical theory of cardiac mechanics and electrophysiology has not yet had a significant impact in clinical cardiology. Future clinical advances are anticipated in applications to mechanical and electrical disorders of the heart.

23.1 Introduction

One ultimate aim in the application of theoretical methods to problems in cardiology is to minimize the effects caused by heart disease in its myriad of clinical manifestations. An expected beneficial byproduct is the generation of fundamental new insight and understanding in cardiovascular biophysics as it relates to the structure and function of the heart.

We understand the phrase "theoretical methods" to mean the utilization of problem solving, modeling, or computational methods generally recognized to stem from fields *other* than clinical medicine, such as theoretical biology, classical physics, theoretical and applied mechanics, pure and applied mathematics and computer science, among others. The foregoing chapters in these proceedings provide ample evidence of the international scope of the effort as well as the spectrum of problems under consideration.

The "cross talk" between scientists with expertise in mathematical problem solving methods and physiologists is not new. It has a long history, stemming from the days of Helmholtz, with particularly insightful examples evidenced by the work of D'Arcy Thompson [6] at the turn of the century. However, because of the tremendous increase in numerical computational power over the last decade, accompanied by a simultaneous decrease in the cost of computation per unit time and per unit computational step, the effort directed toward mathematical and theoretical problem solving in cardiology has tremendously increased. In addition, certain aspects of cardiology lend themselves in a natural way to physical and mathematical description, modeling, and analysis. As the chapters in this book illustrate,

[1]Departments of Internal Medicine and Physics, Washington University, St. Louis, Missouri 63110

the spectrum is broad. It ranges from the purely mechanical to the purely theoretical, from phase plane techniques for the analysis of nonlinear oscillators as models of the heart's electrical activity to multidimensional coupled nonlinear differential equations used in the analysis of simultaneous ionic currents, to the global topological analysis of the behavior of excitable media. None of these approaches could be pursued without extensive computer simulation and numerical analysis (which was heretofore unavailable) and the active collaboration between cardiologists (clinical as well as basic science oriented) and their mathematically oriented colleagues.

23.2   Present Impact on Theory in the Clinical Arena

In today's clinical practice of cardiology the theoretical methods and techniques mentioned above have not yet had any direct effect. However, the cardiologist's ability to diagnose the presence and severity of an ailment and prescribe appropriate pharmacologic or, in the case of coronary angioplasty or electrophysiological studies, invasive therapy (such as catheter ablation of arrhythmogenic foci), has undergone significant changes. These changes have been primarily driven by technological developments. Selected examples include noninvasive imaging (fast CT scanning, magnetic resonance imaging, Doppler and colorflow echocardiography, positron emission tomography), invasive imaging and therapy [coronary and vascular angiography using digital imaging [5], percutaneous transluminal coronary angiography (PTCA), i.e., mechanical dilation of narrowed coronary arteries, transcatheter ablation], treatment of life threatening arrhythmias refractory to medical therapy [automatic implantable cardioverter defibrillator (AICD) or implantable pacemaker].

A better understanding of the pathophysiology of atherosclerosis and associated thrombosis has resulted in the rapid development of new pharmacological agents such as thrombolytics using techniques stemming from molecular biology and genetic engineering.

We must recall that the practice of clinical cardiology *has been* primarily diagnostic with treatment confined to pharmacologic and to a lesser extent surgical means. The therapeutic areas in cardiology, such as PTCA, ablation of arrhythmogenic foci during electrophysiologic studies, and AICD therapy are relatively recent. All of these innovative modalitites relied, at least in part, on applied modeling and analytical methods especially in the equipment design phase.

Thus, mathematical and physical as well as theoretical analyses have had a major impact on present clinical invasive, noninvasive, and pharmacological modalities in cardiology when translated into easily applied technology.

23.3   Future Considerations

In light of the above comments, there is great promise for the development of sophisticated, noninvasive, and invasive diagnostic modalities as well as techniques to assess and guide follow-up therapy. This includes quantitation of efficacy of pharmacologic or invasive therapies. These developments are likely to stem directly from theoretical and mathematical approaches some of which are exemplified by the contents of this book. Some of the general areas where these types of developments may likely occur are listed below.

Electrical Phenomena.

Mathematical analysis is being carried out on the electrical state of the heart and certain classes of cardiac arrhythmias. This work is based on theories utilizing the description of excitable media and nonlinear dynamics [2]. The intent is the identification of individuals at high risk for sudden death by elucidation of the electromechanical substrate which characterizes these patients. For example, signal averaged electrocardiography and the quantitative measurement of late potentials in the clinical arena is intended to characterize a subset of patients with increased risk of life-threatening arrhythmias [4]. This approach is a direct extension of the use of mathematical signal averaging algorithms that improve signal to noise in the clinical ECG recording. This technique has assisted in the development of clinical indices that correlate with vulnerability to malignant arrhythmias. These methods have been primarily phenomenological so far; in other words, the clinical events motivate the characterization of the arrhythmia. However, once a more complete theory of arrhythmogenesis is available patients may be characterized in a prospective fashion.

Mechanical Phenomena.

Theoretical and numerical analysis of the mechanical factors that influence hemodynamics, pump function, and ventricular geometry may permit characterization of patients at risk of catastrophic mechanical failures such as rupture of chordae tendinae, the ventricular wall or septum, or alteration of ventricular geometry related to breakdown or restructuring of extracellular matrix constituents in acute and chronic disease states. For example, detailed analysis of the passive and active material properties of the heart, particularly in diastole, including ultrastructural assessment using electron microscopy, suggests that the extracellular myocardial connective tissue (elastin and collagen) elements play a significant role in determining diastolic ventricular shape, size, and through the constitutive relations, determine end-diastolic pressure as a function of end-diastolic volume [1]. These structural components are likely to play a role in diastolic suction and determine in part the Doppler-derived clinical indices of diastolic filling (E wave, A wave contours). The alteration of these connective tissue components has been documented in "stunned myocardium" (viable but not actively beating heart muscle segment). Dysfunction of this connective tisue matrix is suspected on hemodynamic grounds in virtually all pathological states that manifest abnormal diastolic function [7].

Pharmacologic Therapy from Theory

The development of novel pharmacologic agents utilizing molecular biologic techniques may ultimately be motivated in part by mathematical models of cardiac function. These models can encompass a range of functions over seven orders of magnitude, from the subcellular (microns) to the global myocardial scale (10 cm). For example, the global kinematic effect of the extracellular (or interstitial) myocardial matrix as a driving mechanism for diastolic suction has been mathematically characterized. Models incorporating this feature of stored elastic stress have clarified the kinematic relation of suction (sometimes referred to as active relaxation or elastic recoil) to the clinically recorded diastolic Doppler velocity profile [3]. In light of the profound influence of the extracellular matrix on clinical diastolic function and its role in determining the Doppler tracing, if the molecular biology and control mechanisms (destruction vs. generation) of the extracellular myocardial connective tissue matrix becomes understood, therapeutic agents may be developed that enhance the interstitial repair process or limit excess matrix deposition and thereby restore normal diastolic mechanical behavior [7].

23.4   Conclusion

The tremendous advances in computational speed and associated memory storage capability over the last 10 years have facilitated substantial growth in the "cross talk" between mathematical and clinical attempts to solve basic clinical problems in cardiology. As of this writing (1990) the effects of these mathematical methods are not yet significantly evident in the day-to-day practice of clinical cardiology; however, applied modeling and theoretical methods are at least in part responsible for the rapid evolution of diagnostic and therapeutic modalities already in use today. The future is bright. It is extremely likely that in the years to come diagnosis, therapy, and follow-up of cardiac patients will critically depend on developments achieved through or motivated by current mathematical and theoretical methods. In addition, these methods are likely to provide fundamental new insights into directions as yet undreamed of, not only in basic science but in clinical medicine as well.

Acknowledgements: Supported in part by Merit Review Grant from the Veterans Administration and BSRG Grant: NIH/DDR S07-RR05491.

REFERENCES

[1] P. Brun, R.S. Chadwick, and B.I. Levy. Cardiovascular dynamics and models. In *Proceedings of NIH-Inserm workshops, Bethesda, September 1985*, Paris, 1988.

[2] L. Glass and M.C. Mackey. *From Clocks to Chaos: The Rhythms of Life*. Princeton University Press, Princeton, 1988.

[3] S.J. Kovács, B. Barzilai, and J.E. Perez. Evaluation of diastolic function with Doppler echocardiography: The PDF formalism. *Am. J. Physiol.*, 252:H178–H187, 1987.

[4] D.L. Kuchar and D.S. Rosenbaum. Noninvasive recording of late potentials: Current state of the art. *PACE*, 12:1538–1549, 1989.

[5] G.B.J. Mancini. *Clinical Applications of Cardiac Digital Angiography*. Raven Press, New York, 1988.

[6] D.W. Thompson. *On Growth and Form*. Cambridge University Press, New York, 1969.

[7] K.T. Weber. Cardiac interstitium in health and disease: The fibrillar collagen network. *J. Am. Coll. Cardiol.*, 13:1637–1652, 1989.